INTRODUCTION TO
MODERN
THERMODYNAMICS

INTRODUCTION TO MODERN THERMODYNAMICS

Dilip Kondepudi
Thurman D Kitchin Professor of Chemistry
Wake Forest University

John Wiley & Sons, Ltd

Copyright © 2008   John Wiley & Sons Ltd, The Atrium, Southern Gate, Chichester,
West Sussex PO19 8SQ, England

Telephone (+44) 1243 779777

Email (for orders and customer service enquiries): cs-books@wiley.co.uk
Visit our Home Page on www.wileyeurope.com or www.wiley.com

Reprinted September 2008

Other Wiley Editorial Offices

John Wiley & Sons Inc., 111 River Street, Hoboken, NJ 07030, USA

Jossey-Bass, 989 Market Street, San Francisco, CA 94103-1741, USA

Wiley-VCH Verlag GmbH, Boschstr. 12, D-69469 Weinheim, Germany

John Wiley & Sons Australia Ltd, 42 McDougall Street, Milton, Queensland 4064, Australia

John Wiley & Sons (Asia) Pte Ltd, 2 Clementi Loop #02-01, Jin Xing Distripark, Singapore 129809

John Wiley & Sons Ltd, 6045 Freemont Blvd, Mississauga, Ontario L5R 4J3, Canada

Wiley also publishes its books in a variety of electronic formats. Some content that appears in print may not be available in electronic books.

Library of Congress Cataloging-in-Publication Data

Kondepudi, D. K. (Dilip K.), 1952–
    Introduction to modern thermodynamics / Dilip Kondepudi.
        p.   cm.
    Includes bibliographical references and index.
    ISBN 978-0-470-01598-8 (cloth) – ISBN 978-0-470-01599-5 (pbk.)
1. Thermodynamics.   I. Title.
    QC311.K658 2007
    536′.7–dc22

                                                2007044834

British Library Cataloguing in Publication Data

A catalogue record for this book is available from the British Library

ISBN 978-0-470-01598-8 (H/B) ISBN 978-9470-01599-5 (Ħ B)

Cover photo of Hurricane Hugo courtesy of NASA

For the memory of Ilya Prigogine,
poet of thermodynamics.

CONTENTS

Preface   xiii

PART I   THE FORMALISM OF MODERN THERMODYNAMICS

1   BASIC CONCEPTS AND THE LAWS OF GASES   3

   Introduction   3
   1.1   Thermodynamic Systems   4
   1.2   Equilibrium and Nonequilibrium Systems   6
   1.3   Biological and Other Open Systems   9
   1.4   Temperature, Heat and Quantitative Laws of Gases   11
   1.5   States of Matter and the van der Waals Equation   19
   1.6   An Introduction to Kinetic Theory of Gases   29
   Appendix 1.1 Partial Derivatives   37
   Appendix 1.2 Elementary Concepts in Probability Theory   39
   Appendix 1.3 Mathematica Codes   41
   References   44
   Examples   44
   Exercises   45

2   THE FIRST LAW OF THERMODYNAMICS   49

   The Idea of Energy Conservation amidst New Discoveries   49
   2.1   The Nature of Heat   50
   2.2   The First Law of Thermodynamics: The Conservation of Energy   55
   2.3   Elementary Applications of the First Law   64
   2.4   Thermochemistry: Conservation of Energy in Chemical Reactions   68
   2.5   Extent of Reaction: A State Variable for Chemical Systems   76
   2.6   Conservation of Energy in Nuclear Reactions and Some
         General Remarks   79
   2.7   Energy Flows and Organized States   81
   Appendix 2.1 Mathematica Codes   87
   References   88
   Examples   88
   Exercises   92

**3   THE SECOND LAW OF THERMODYNAMICS AND THE
      ARROW OF TIME**   97

   3.1   The Birth of the Second Law   97
   3.2   The Absolute Scale of Temperature   106

3.3   The Second Law and the Concept of Entropy   108
3.4   Entropy, Reversible and Irreversible Processes   116
3.5   Examples of Entropy Changes due to Irreversible Processes   125
3.6   Entropy Changes Associated with Phase Transformations   128
3.7   Entropy of an Ideal Gas   129
3.8   Remarks about the Second Law and Irreversible Processes   130
Appendix 3.1 The Hurricane as a Heat Engine   132
Appendix 3.2 Entropy Production in Continuous Systems   135
References   136
Examples   137
Exercises   139

4   ENTROPY IN THE REALM OF CHEMICAL REACTIONS   141

4.1   Chemical Potential and Affinity: The Thermodynamic Force
       for Chemical Reactions   141
4.2   General Properties of Affinity   150
4.3   Entropy Production Due to Diffusion   153
4.4   General Properties of Entropy   155
Appendix 4.1 Thermodynamics Description of Diffusion   158
References   158
Examples   159
Exercises   160

5   EXTREMUM PRINCIPLES AND GENERAL
    THERMODYNAMIC RELATIONS   163

Extremum Principles in Nature   163
5.1   Extremum Principles Associated with the Second Law   163
5.2   General Thermodynamic Relations   173
5.3   Gibbs Energy of Formation and Chemical Potential   176
5.4   Maxwell Relations   179
5.5   Extensivity with Respect to N and Partial Molar Quantities   181
5.6   Surface Tension   183
References   187
Examples   187
Exercises   189

PART II   APPLICATIONS: EQUILIBRIUM AND
NONEQUILIBRIUM SYSTEMS

6   BASIC THERMODYNAMICS OF GASES, LIQUIDS
    AND SOLIDS   195

Introduction   195
6.1   Thermodynamics of Ideal Gases   195
6.2   Thermodynamics of Real Gases   199

6.3   Thermodynamics Quantities for Pure Liquids and Solids   208
Appendix 6.1 Equations of State   211
Reference   211
Examples   212
Exercises   213

7   THERMODYNAMICS OF PHASE CHANGE   215

Introduction   215
7.1   Phase Equilibrium and Phase Diagrams   215
7.2   The Gibbs Phase Rule and Duhem's Theorem   221
7.3   Binary and Ternary Systems   223
7.4   Maxwell's Construction and the Lever Rule   229
7.5   Phase Transitions   231
References   235
Examples   235
Exercises   236

8   THERMODYNAMICS OF SOLUTIONS   239

8.1   Ideal and Nonideal Solutions   239
8.2   Colligative Properties   243
8.3   Solubility Equilibrium   250
8.4   Thermodynamic Mixing and Excess Functions   255
8.5   Azeotropy   259
References   260
Examples   260
Exercises   262

9   THERMODYNAMICS OF CHEMICAL TRANSFORMATIONS   265

9.1   Transformations of Matter   265
9.2   Chemical Reaction Rates   266
9.3   Chemical Equilibrium and the Law of Mass Action   273
9.4   The Principle of Detailed Balance   278
9.5   Entropy Production due to Chemical Reactions   280
9.6   Elementary Theory of Chemical Reaction Rates   285
9.7   Coupled Reactions and Flow Reactors   288
Appendix 9.1 Mathematica Codes   295
References   298
Examples   298
Exercises   300

10   FIELDS AND INTERNAL DEGREES OF FREEDOM   305

The Many Faces of Chemical Potential   305
10.1   Chemical Potential in a Field   305
10.2   Membranes and Electrochemical Cells   311

    10.3   Isothermal Diffusion   319
    Reference   324
    Examples   324
    Exercises   325

11 INTRODUCTION TO NONEQUILIBRIUM SYSTEMS   327

    Introduction   327
    11.1   Local Equilibrium   328
    11.2   Local Entropy Production, Thermodynamic Forces and Flows   331
    11.3   Linear Phenomenological Laws and Onsager Reciprocal
           Relations   333
    11.4   Symmetry-Breaking Transitions and Dissipative Structures   339
    11.5   Chemical Oscillations   345
    Appendix 11.1 Mathematica Codes   352
    References   355
    Further Reading   356
    Exercises   357

PART III   ADDITIONAL TOPICS

12 THERMODYNAMICS OF RADIATION   361

    Introduction   361
    12.1   Energy Density and Intensity of Thermal Radiation   361
    12.2   The Equation of State   365
    12.3   Entropy and Adiabatic Processes   368
    12.4   Wien's Theorem   369
    12.5   Chemical Potential of Thermal Radiation   371
    12.6   Matter-Antimatter in Equilibrium with Thermal Radiation:
           The State of Zero Chemical Potential   373
    References   377
    Examples   377
    Exercises   377

13 BIOLOGICAL SYSTEMS   379

    13.1   The Nonequilibrium Nature of Life   379
    13.2   Gibbs Energy Change in Chemical Transformations   382
    13.3   Gibbs Energy Flow in Biological Systems   385
    13.4   Biochemical Kinetics   399
    References   406
    Further Reading   406
    Examples   406
    Exercises   409

14 THERMODYNAMICS OF SMALL SYSTEMS   411

Introduction   411
14.1   Chemical Potential of Small Systems   411
14.2   Size-Dependent Properties   414
14.3   Nucleation   418
14.4   Fluctuations and Stability   421
References   430
Examples   430
Exercises   430

15 CLASSICAL STABILITY THEORY   433

15.1   Stability of Equilibrium States   433
15.2   Thermal Stability   433
15.3   Mechanical Stability   435
15.4   Stability with Respect to Fluctuations in N 437
References   439
Exercises   439

**16 CRITICAL PHENOMENA AND CONFIGURATIONAL
     HEAT CAPACITY**   441

Introduction   441
16.1   Stability and Critical Phenomena   441
16.2   Stability and Critical Phenomena in Binary Solutions   443
16.3   Configurational Heat Capacity   447
Further Reading   448
Exercises   449

17 ELEMENTS OF STATISTICAL THERMODYNAMICS   451

Introduction   451
17.1   Fundamentals and Overview   452
17.2   Partition Function Factorization   454
17.3   The Boltzmann Probability Distribution and Average Values   455
17.4   Microstates, Entropy and the Canonical Ensemble   457
17.5   Canonical Partition Function and
       Thermodynamic Quantities   462
17.6   Calculating Partition Functions   462
17.7   Equilibrium Constants   469
Appendix 17.1 Approximations and Integrals   471
Reference   472
Examples   472
Exercises   473

LIST OF VARIABLES   475

STANDARD THERMODYNAMIC PROPERTIES   477

PHYSICAL CONSTANTS AND DATA   485

NAME INDEX   487

SUBJECT INDEX   489

PREFACE

What is Modern Thermodynamics?

In almost every aspect of nature, we see irreversible changes. But it is Isaac Newton's mechanical paradigm, the clockwork universe governed by time-reversible laws of mechanics and its grand success in explaining the motion of heavenly bodies, that has dominated our thinking for centuries. During the twentieth century, however, the dominance of the mechanical paradigm slowly began to wane. We now recognize that the paradigm for periodic phenomena in nature is not a 'mechanical clock', such as a pendulum, based on time-reversible laws of mechanics, but a 'thermodynamic clock' based on irreversible processes, such as chemical reactions. The ashing of a fire y, the beating of a heart, and the chirping of a cricket are governed by irreversible processes. Modern thermodynamics is a theory of irreversible processes.

Classical thermodynamics, as it was formulated in the nineteenth century by Carnot, Clausius, Joule, Helmholz, Kelvin, Gibbs and others, was a theory of initial and final *states* of a system, not a theory that included the irreversible *processes* that were responsible for the transformation of one state to another. It was a theory confined to systems in thermodynamic equilibrium. That is the way it is still presented in most introductory texts. Thermodynamics is treated as a subject concerned only with equilibrium states. Computations of changes in entropy and other thermodynamic quantities are done only for idealized reversible processes that take place at an infinitely slow rate. For such processes, the change in entropy $dS = dQ/T$. Time does not explicitly appear in this formalism: there are no expressions for the rate of change of entropy, for instance. For irreversible processes that take place at a nonzero rate, it is only stated that $dS > dQ/T$. The student is left with the impression that thermodynamics only deals with equilibrium states and that irreversible processes are outside its scope. That impression is an inevitable consequence of the way nineteenth-century classical thermodynamics was formulated.

Modern thermodynamics, formulated in the twentieth century by Lars Onsager, Theophile De Donder, Ilya Prigogine and others, is different. It is a theory of irreversible processes that very much includes time: it relates entropy, the central concept of thermodynamics, to irreversible processes. In the modern theory, dS is the change of entropy in a time interval dt. The change in entropy is written as a sum of two terms:

$$dS = d_e S + d_i S$$

in which $d_e S$ is the entropy change due to exchange of energy and matter ($d_e S = dQ/T$ for exchange of heat) and $d_i S$ is the entropy change due to irreversible processes. Both these changes in entropy, $d_e S$ and $d_i S$, are computed using rates at which irreversible processes, such as heat conduction and chemical reactions, occur. Indeed, the rate at which entropy is produced due to irreversible processes, d_iS/dt, is clearly identified – and it is always positive, in accord with the second law. Irreversible processes, such as chemical reactions, diffusion and heat conduction that take place in nonequilibrium systems, are described as thermodynamic flows driven by thermodynamic forces; the rate of entropy production d_iS/dt is, in turn, written in terms of the thermodynamic forces and flows. In contrast, in most physical chemistry texts, since classical thermodynamics does not include processes, students are presented with two separate subjects: thermodynamics and kinetics. Each irreversible process, chemical reactions, diffusion and heat conduction, is treated separately in a phenomenological manner without a unifying framework. In the modern view, all these irreversible processes and the ways in which they interact are under one thermodynamic framework. In addition to all the classical thermodynamic variables, the student is also introduced to the concept of rate of entropy production, a quantity of much current interest in the study of nonequilibrium systems.

Modern thermodynamics also gives us a paradigm for the order and self-organization we see in Nature that is different from the clockwork paradigm of mechanics. The self-organization that we see in the formation of beautiful patterns in convecting fluids and in the onset of oscillations and pattern formation in chemical systems are consequences of irreversible processes. The maintenance of order or structure in such systems comes at the expense of entropy production. While it is true that increase of entropy can be associated with increase in disorder and dissipation of usable energy, entropy-producing irreversible processes can yet generate the ordered structures we see in Nature. Such structures, which are created and maintained by irreversible processes, were termed *dissipative structures* by Ilya Prigogine. It is a topic that fascinates students and excites them with the prospect of making new discoveries in the field of thermodynamics.

The dancing Siva or Nataraja on the jacket of the book by Peter Glansdorff and Ilya Prigogine, *Thermodynamic Theory of Structure, Stability and Fluctuations*, sums up the role of irreversible processes in Nature. In his cosmic dance, Siva carries in one hand a drum, a symbol of creation and order; in another, he holds fire, a symbol of destruction. So it is with irreversible processes: they create order on the one hand and increase disorder (entropy) on the other. I hope this text conveys this enduring view to the student: irreversible processes are creators and destroyers of order.

This text is an offshoot of *Modern Thermodynamics: From Heat Engines to Dissipative Structures*, which I co-authored with Ilya Prigogine in 1998. It is intended for use in a one-semester course in thermodynamics. It is divided into three parts. Part I, Chapters 1–5, contains the basic formalism of modern thermodynamics. Part II, Chapters 6–11, contains basic applications, covering both equilibrium and nonequilibrium systems. Chapter 11 is a concise introduction to linear nonequilibrium thermodynamics, Onsager reciprocal relations and dissipative structures. Part III contains additional topics that the instructor can include in the course, such as thermodynamics of radiation, small systems and biological systems. The text ends with an introductory chapter on statistical thermodynamics, a topic that is often taught along with thermodynamics.

ACKNOWLEDGEMENTS

As student and researcher, I had the privilege of working with Ilya Prigogine, poet of thermodynamics, as his colleagues thought of him. My gratitude for this privilege is more than words can express. I have also had the wonderful opportunity to discuss thermodynamics with Peter Glansdorff, whose knowledge of thermodynamics and its history has greatly enriched me. I profited much by my visits to the Institut Carnot de Bourgogne, University of Burgundy, where I enjoyed the stimulating discussions and warm hospitality of Florence Baras and Malek Mansour. My thanks to Abdou Lachgar, Jed Macosco and Wally Baird for reading some chapters and making suggestions to improve their content and clarity. I am very grateful to Wake Forest University for awarding me Reynolds Leave to complete this book.

PART I

THE FORMALISM OF MODERN THERMODYNAMICS

Introduction to Modern Thermodynamics   Dilip Kondepudi
© 2008 John Wiley & Sons, Ltd

1 BASIC CONCEPTS AND THE LAWS OF GASES

Introduction

Adam Smith's *Wealth of Nations* was published in the year 1776. Seven years earlier James Watt (1736–1819) had obtained a patent for his version of the steam engine. Both men worked at the University of Glasgow. Yet, in Adam Smith's great work the only use for coal was in providing heat for workers [1]. The machines of the eighteenth century were driven by wind, water and animals. Nearly 2000 years had passed since Hero of Alexandria made a sphere spin with the force of steam; but still fire's power to generate motion and drive machines remained hidden. Adam Smith (1723–1790) did not see in coal this hidden wealth of nations.

The steam engine revealed a new possibility. Wind, water and animals converted one form of motion to another. The steam engine was fundamentally different: it converted heat to mechanical motion. Its enormous impact on civilization not only heralded the industrial revolution, it also gave birth to a new science: *thermodynamics*. Unlike the science of Newtonian mechanics, which had its origins in theories of motion of heavenly bodies, thermodynamics was born out of a more practical interest: generating motion from heat.

Initially, thermodynamics was the study of heat and its ability to generate motion; then it merged with the larger subject of energy and its interconversion from one form to another. With time, thermodynamics evolved into a theory that describes transformations of states of matter in general, motion generated by heat being a consequence of particular transformations. It is founded on essentially two fundamental laws, one concerning *energy* and the other *entropy*. A precise definition of energy and entropy, as measurable physical quantities, will be presented in the Chapters 2 and 3 respectively. In the following two sections we will give an overview of thermodynamics and familiarize the reader with the terminology and concepts that will be developed in the rest of the book.

Every system is associated with an energy and an entropy. When matter undergoes transformation from one state to another, the total amount of energy in the system and its exterior is conserved; total entropy, however, can only increase or, in idealized cases, remain unchanged. These two simple-sounding statements have far-reaching consequences. Max Planck (1858–1947) was deeply influenced by the breadth of the conclusions that can be drawn from them and devoted much of himself to the study of thermodynamics. In reading this book, I hope the reader will come to appreciate the significance of the following often-quoted opinion of Albert Einstein (1879–1955):

Introduction to Modern Thermodynamics   Dilip Kondepudi
© 2008 John Wiley & Sons, Ltd

A theory is more impressive the greater the simplicity of its premises is, the more different kinds of things it relates, and the more extended its area of applicability. Therefore the deep impression which classical thermodynamics made upon me. It is the only physical theory of universal content concerning which I am convinced that, within the framework of the applicability of its basic concepts, it will never be over thrown.

The thermodynamics of the nineteenth century, which so impressed Planck and Einstein, described static systems that were in thermodynamic equilibrium. It was formulated to calculate the initial and final entropies when a system evolved from one equilibrium state to another. In this 'Classical Thermodynamics' there was no direct relationship between natural processes, such as chemical reactions and conduction of heat, and the rate at which entropy changed. During the twentieth century, Lars Onsager (1903–1976), Ilya Prigogine (1917–2003) and others extended the formalism of classical thermodynamics to relate the *rate* of entropy change to *rates* of processes, such as chemical reactions and heat conduction. From the outset, we will take the approach of this 'Modern Thermodynamics' in which thermodynamics is a theory of irreversible processes, not merely a theory of equilibrium states.

1.1   Thermodynamic Systems

A thermodynamic description of natural processes usually begins by dividing the world into a 'system' and its 'exterior', which is the rest of the world. This cannot be done, of course, when one is considering the thermodynamic nature of the entire universe – however, although there is no 'exterior', thermodynamics can be applied to the entire universe. The definition of a thermodynamic system depends on the existence of 'boundaries', boundaries that separate the system from its exterior and restrict the way the system interacts with its exterior. In understanding the thermodynamic behavior of a system, the manner in which it exchanges energy and matter with its exterior, is important. Therefore, thermodynamic systems are classified into three types: isolated, closed and open systems (Figure 1.1) according to the way they interact with the exterior.

Isolated systems do not exchange energy or matter with the exterior. Such systems are generally considered for pedagogical reasons, while systems with extremely slow exchange of energy and matter can be realized in a laboratory. Except for the universe as a whole, truly isolated systems do not exist in nature.

Closed systems exchange energy but not matter with their exterior. It is obvious that such systems can easily be realized in a laboratory: A closed flask of reacting chemicals which is maintained at a fixed temperature is a closed system. The Earth, on a time-scale of years, during which it exchanges negligible amounts matter with its exterior, may be considered a closed system; it only absorbs radiation from the sun and emits it back into space.

Open systems exchange both energy and matter with their exterior. All living and ecological systems are open systems. The complex organization in open systems is

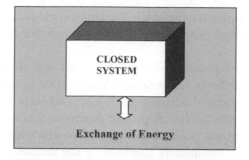

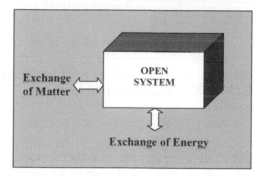

Figure 1.1 Isolated, closed and open systems. Isolated systems exchange neither energy nor matter with the exterior. Closed systems exchange heat and mechanical energy but not matter with the exterior. Open systems exchange both energy and matter with the exterior

a result of exchange of matter and energy and the entropy generating irreversible processes that occur within.

In thermodynamics, the **state** of a system is specified in terms of macroscopic **state variables**, such as volume V, pressure p, temperature T, and moles N_k of chemical constituent k, which are self-evident. These variables are adequate for the description of equilibrium systems. When a system is not in thermodynamic equilibrium, more variables, such as rate of convective flow or of metabolism, may be needed to describe it. The two laws of thermodynamics are founded on the concepts of energy U, and entropy S, which, as we shall see, are **functions of state variables**.

Since the fundamental quantities in thermodynamics are functions of many variables, thermodynamics makes extensive use of calculus of many variables. A brief summary of some basic properties of functions of many variables is given in Appendix A1.1 (at the end of this chapter). Functions of state variables, such as U and S, are often called **state functions**.

It is convenient to classify thermodynamic variables into two categories. Variables such as volume and amount of a substance (moles), which indicate the size of the system, are called **extensive variables**. Variables such as temperature T and pressure p, which specify a local property, which do not indicate the system's size, are called **intensive variables**.

If the temperature is not uniform, then heat will flow until the entire system reaches a state of uniform temperature. Such a state is the state of **thermal equilibrium**. The state of thermal equilibrium is a special state towards which all isolated systems will inexorably evolve. A precise description of this state will be given later in this book. In the state of thermal equilibrium, the values of total internal energy U and entropy S are completely specified by the temperature T, the volume V and the amounts of the system's chemical constituents N_k (moles):

$$U = U(T, V, N_k) \quad \text{or} \quad S = S(T, V, N_k) \tag{1.1.1}$$

The values of an extensive variable, such as total internal energy U or entropy S, can also be specified by other extensive variables:

$$U = U(S, V, N_k) \quad \text{or} \quad S = S(U, V, N_k) \tag{1.1.2}$$

As we shall see in the following chapters, intensive variables can be expressed as derivatives of one extensive variable with respect to another. For example, we shall see that the temperature $T = (\partial U / \partial S)_{V,N_k}$. The laws of thermodynamics and the calculus of many-variable functions give us a rich understanding of many phenomena we observe in nature.

1.2 Equilibrium and Nonequilibrium Systems

It is our experience that if a physical system is isolated, its state – specified by macroscopic variables such as pressure, temperature and chemical composition – evolves *irreversibly* towards a time-invariant state in which we see no further physical or chemical change. This is the state of **thermodynamic equilibrium**. It is a state characterized by a uniform temperature throughout the system. The state of equilibrium is also characterized by several other physical features that we will describe in the following chapters.

The evolution of a system towards the state of equilibrium is due to **irreversible processes**, such as heat conduction and chemical reactions, which act in a specific direction but not its reverse. For example, heat always flows from a higher to a lower temperature, never in the reverse direction; similarly, chemical reactions cause compositional changes in a specific direction not its reverse (which, as we shall see in Chapter 4, is described in terms of 'chemical potential', a quantity similar to temperature, and 'affinity', a thermodynamic force that drives chemical reactions). At equilibrium, these processes vanish. Thus, a nonequilibrium state can be characterized as a state in which irreversible processes are taking place driving the system towards the equilibrium state. In some situations, especially during chemical transformations, the rates at which the state is transforming irreversibly may be extremely small, and an isolated system might appear as if it is time invariant and has reached its state of equilibrium. Nevertheless, with appropriate specification of the chemical reactions, the nonequilibrium nature of the state can be identified.

Two or more systems that interact and exchange energy and/or matter will eventually reach the state of thermal equilibrium in which the temperature within each system is spatially uniform and the temperature of all the systems are the same. If a system A is in thermal equilibrium with system B and if B is in thermal equilibrium with system C, then it follows that A is in thermal equilibrium with C. This 'transitivity' of the state of equilibrium is sometimes called the **zeroth law**. Thus, equilibrium systems have a well-defined, spatially uniform temperature; for such systems, the energy and entropy are functions of state as expressed in (1.1.1).

Uniformity of temperature, however, is not a requirement for the entropy or energy of a system to be well defined. For **nonequilibrium systems**, in which the temperature is not uniform but is well defined locally at every point x, we can define densities of thermodynamic quantities such as energy and entropy. Thus, the energy density at x

$$u[T(x), n_k(x)] = \text{internal energy per unit volume} \qquad (1.2.1)$$

can be defined in terms of the local temperature $T(x)$ and the concentration

$$n_k(x) = \text{moles of constituent } k \text{ per unit volume} \qquad (1.2.2)$$

Similarly, an entropy density $s(T, n_k)$ can be defined. The atmosphere of the Earth, shown in Box 1.1, is an example of a nonequilibrium system in which both n_k and T are functions of position. The total energy U, the total entropy S and the total amount of the substance N_k are

$$S = \int_V s[T(x), n_k(x)] \mathrm{d}V \qquad (1.2.3)$$

$$U = \int_V u[T(x), n_k(x)] \mathrm{d}V \qquad (1.2.4)$$

$$N_k = \int_V n_k(x) \mathrm{d}V \qquad (1.2.5)$$

In nonequilibrium (non-uniform) systems, the total energy U is no longer a function of other extensive variables such as S, V and N_k, as in (1.1.2), and obviously one cannot define a single temperature for the entire system because it may not uniform.

Box 1.1 The Atmosphere of the Earth.

Blaise Pascal (1623–1662) explained the nature of atmospheric pressure. The pressure at any point in the atmosphere is due to the column of air above it. The atmosphere of the Earth is not in thermodynamic equilibrium: its temperature is not uniform and the amounts of its chemical constituents (N_2, O_2, Ar, CO_2, etc.) are maintained at a nonequilibrium value through cycles of production and consumption.

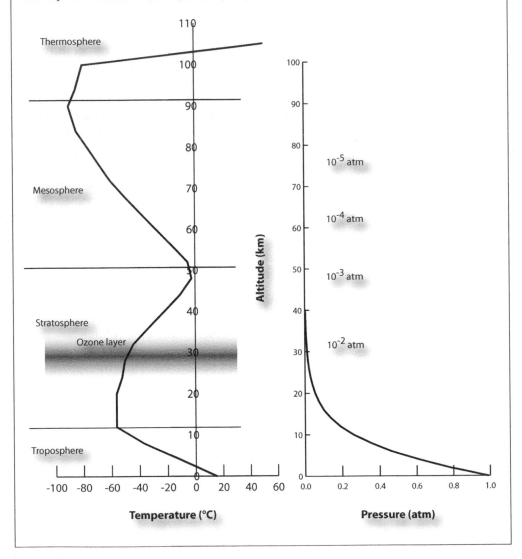

In general, each of the variables, the total energy U, entropy S, the amount of substance N_k and the volume V, is no longer a function of the other three variables, as in (1.1.2). But this does not restrict in any way our ability to assign an entropy to a system that is not in thermodynamic equilibrium, as long as the temperature is locally well defined.

In texts on classical thermodynamics, when it is sometimes stated that entropy of a nonequilibrium system is not defined, it is meant that S is not a function of the variables U, V and N_k. If the temperature of the system is locally well defined, then indeed the entropy of a nonequilibrium system can be defined in terms of an entropy density, as in (1.2.3).

1.3   Biological and Other Open Systems

Open systems are particularly interesting because in them we see spontaneous self-organization. The most spectacular example of self-organization in open systems is life. Every living cell is an open system that exchanges matter and energy with its exterior. The cells of a leaf absorb energy from the sun and exchange matter by absorbing CO_2, H_2O and other nutrients and releasing O_2 into the atmosphere. A biological open system can be defined more generally: it could be a single cell, an organ, an organism or an ecosystem. Other examples of open systems can be found in industry; in chemical reactors, for example, raw material and energy are the inputs and the desired and waste products are the outputs.

As noted in the previous section, when a system is not in equilibrium, processes such as chemical reactions, conduction of heat and transport of matter take place so as to drive the system towards equilibrium. And all of these processes generate entropy in accordance with the Second Law (see Figure 1.2). However, this does not mean that the entropy of the system must always increase: the exchange of energy and matter may also result in the net output of entropy in such a way that the entropy of a system is maintained at a low value. One of the most remarkable aspects of nonequilibrium systems that came to light in the twentieth century is the phenomenon of self-organization. Under certain nonequilibrium conditions, systems can spontaneously undergo transitions to organized states, which, in general, are states with lower entropy. For example, nonequilibrium chemical systems can make a transition to a state in which the concentrations of reacting compounds vary periodically, thus becoming a 'chemical clock'. The reacting chemicals can also spatially organize into patterns with great symmetry. In fact, it can be argued that most of the 'organized' behavior we see in nature is created by irreversible processes that dissipate energy and generate entropy. For these reasons, these structures are called **dissipative structures** [1], and we shall study more about them in Chapter 11. In an open system, these organized states could be maintained indefinitely, but only at the expense of exchange of energy and matter and increase of entropy outside the system.

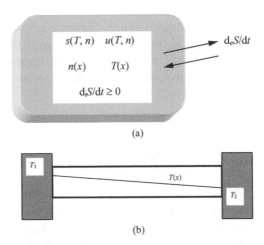

Figure 1.2 (a) In a nonequilibrium system, the temperature and number density $n_k(x)$ may vary with position. The entropy and energy of such a system may be described by an entropy density $s(T, n_k)$ and an energy density $u(T, n_k)$. The total entropy $S = \int_V s[T(x), n_k(x)]dV$, the total energy $U = \int_V u[T(x), n_k(x)]dV$, and the total molenumber $N_k = \int_V n_k(x)dV$. For such a nonequilibrium system, the total entropy S is not a function of U, N and the total volume V. The term d_iS/dt is the rate of change of entropy due to chemical reactions, diffusion, heat conduction and other such irreversible processes; according to the second law, d_iS/dt can only be positive. In an open system, entropy can also change due to exchange of energy and matter; this is indicated by the term d_eS/dt, which can be either positive or negative. (b) A system in contact with thermal reservoirs of unequal temperatures is a simple example of a nonequilibrium system. The temperature is not uniform and there is a flow of heat due to the temperature gradient. The term d_eS/dt is related to the exchange of heat at the boundaries in contact with the heat reservoirs, whereas d_iS/dt is due to the irreversible flow of heat within the system

1.4 Temperature, Heat and Quantitative Laws of Gases

During the seventeenth and eighteenth centuries, a fundamental change occurred in our conception of Nature. Nature slowly but surely ceased to be solely a vehicle of God's will, comprehensible only through theology. The new 'scientific' conception of Nature based on rationalism and experimentation gave us a different world view, a view that liberated the human mind from the confines of religious doctrine. In the new view, Nature obeyed simple and universal laws, laws that humans can know and express in the precise language of mathematics. Right and wrong were decided through experiments and observation. It was a new dialogue with Nature. Our questions became experiments, and Nature's answers were consistent and unambiguous.

It was during this time of great conceptual change that a scientific study of the nature of heat began. This was primarily due to the development of the thermometer, which was constructed and used in scientific investigations since the time of Galileo Galilei (1564–1642) [2, 3]. The impact of this simple instrument was considerable. In the words of Sir Humphry Davy (1778–1829), 'Nothing tends to the advancement of knowledge as the application of a new instrument'.

The most insightful use of the thermometer was made by Joseph Black (1728–1799), a professor of medicine and chemistry at Glasgow. Black drew a clear distinction between temperature, or degree of hotness, and the quantity of heat (in terms of current terminology, temperature is an intensive quantity whereas heat is an extensive quantity). His experiments using the newly developed thermometers established the fundamental fact that *the temperatures of all the substances in contact with each other will eventually reach the same value, i.e. systems that can exchange heat will reach a state of thermal equilibrium.* This idea was not easily accepted by his contemporaries because it seems to contradict the ordinary experience of touch, in which a piece of metal felt colder than a piece of wood even after they had been in contact for a very long time. But the thermometer proved this point beyond doubt. With the thermometer, Black discovered **specific heat**, laying to rest the general belief at his time that the amount of heat required to increase the temperature of substance by a given amount depended solely on its mass, not specific to its makeup. He also discovered latent heats of fusion and evaporation of water – the latter with the enthusiastic help from his pupil James Watt (1736–1819) [4].

Though the work of Joseph Black and others established clearly the distinction between heat and temperature, the nature of heat remained an enigma for a long time. Whether heat was an indestructible substance without mass, called the 'caloric', that moved from substance to substance or whether it was a form of microscopic motion was still under debate as late as the nineteenth century. After considerable debate and experimentation it became clear that heat was a form of energy that could be transformed to other forms, and so the caloric theory was abandoned – though we still measure the amount of heat in 'calories', in addition to using the SI units of joules.

Joseph Black (1728–1799) (Reproduced with permission from the Edgar Fahs Smith Collection, University of Pennsylvania Library)

Temperature can be measured by noting the change of a physical property, such as the volume of a fluid (such as mercury), the pressure of a gas or the electrical resistance of a wire, with degree of hotness. This is an *empirical* definition of temperature. In this case, the uniformity of the unit of temperature depends on the uniformity with which the measured property changes as the substance gets hotter. The familiar Celsius scale, which was introduced in the eighteenth century by Anders Celsius (1701–1744), has largely replaced the Fahrenheit scale, which was also introduced in the eighteenth century by Gabriel Fahrenheit (1686–1736). As we shall see in the following chapters, the development of the Second Law of thermodynamics during the middle of the nineteenth century gave rise to the concept of *an absolute scale of temperature* that is independent of material properties. Thermodynamics is formulated in terms of the absolute temperature. We shall denote this absolute temperature by T.

THE LAWS OF GASES

In the rest of this section we will present an overview of the laws of gases without going into much detail. We assume the reader is familiar with the laws of ideal gases and some basic definitions are given in Box 1.2.

One of the earliest quantitative laws describing the behavior of gases was due to Robert Boyle (1627–1691), an Englishman and a contemporary of Isaac Newton (1642–1727). The same law was also discovered by Edmé Mariotte (1620(?)–1684) in France. In 1660, Boyle published his conclusion in his 'New experiments physico-mechanical, touching the spring of the air and its effects': at a fixed temperature T, the volume V of a gas was inversely proportional to the pressure p, i.e.:

$$V = \frac{f_1(T)}{p} \quad f_1(T) \text{ is some function of the temperature } T \qquad (1.4.1)$$

Robert Boyle (1627–1691) (Reproduced with permission from the Edgar Fahs Smith Collection, University of Pennsylvania Library)

Box 1.2 Basic Definitions.

Pressure is defined as the force per unit area. The pascal is the SI unit of pressure:

$$\textbf{pascal (Pa)} = 1\,\text{N\,m}^{-2}$$

The pressure due to a column of fluid of uniform density ρ and height h equals $h\rho g$, where g is the acceleration due to gravity ($9.806\,\text{m\,s}^{-2}$). The pressure due to the Earth's atmosphere changes with location and time, but it is often close to $10^5\,\text{Pa}$ at the sea level. For this reason, a unit called the **bar** is defined:

$$\textbf{1 bar} = 10^5\,\textbf{Pa} = 100\,\textbf{kPa}$$

The atmospheric pressure at the Earth's surface is also nearly equal to the pressure due to a 760 mm column of mercury. For this reason, the following units are defined:

$$\textbf{torr} = \text{pressure due to } 1.00\,\text{mm column of mercury}$$

$$\textbf{1 atmosphere (atm)} = \textbf{760\,torr} = \textbf{101.325\,kPa}$$

1 atm equals approximately $10\,\text{N\,cm}^{-2}$ (1 kg weight/cm^2 or 15 lb/inch^2). The atmospheric pressure decreases exponentially with altitude (see Box 1.1).

  Temperature is usually measured in kelvin (K), Celsius (°C) or Fahrenheit (°F). The Celsius and Fahrenheit scales are empirical, whereas (as we shall see in Chapter 3) the kelvin scale is an absolute scale based on the Second Law of thermodynamics: 0 K is the absolute zero, the lowest possible temperature. Temperatures measured in these scales are related as follows:

$$T/°\text{C} = (5/9)[(T/°\text{F}) - 32] \qquad T/\text{K} = (T/°\text{C}) + 273.15$$

On the Earth, the highest recorded temperature is 57.8 °C, or 136 °F; it was recorded in El Azizia, Libiya, in 1922. The lowest recorded temperature is −88.3 °C, or −129 °F; it was recorded in Vostok, Antarctica. In the laboratory, sodium gas has been cooled to temperatures as low as $10^{-9}\,\text{K}$, and temperatures as high as $10^8\,\text{K}$ have been reached in nuclear fusion reactors.

  Heat was initially thought to be an indestructible substance called the **caloric**. According to this view, caloric, a fluid without mass, passed from one body to another, causing changes in temperature. However, in the 19th century it was established that heat was not an indestructible caloric but a form of energy that can convert to other forms of energy (see Chapter 2). Hence, heat is measured in the units of energy. In this text we shall mostly use the SI units in which heat is measured in **joules**, though the calorie is an often-used unit of heat. A calorie was originally defined as the amount of heat required to increase the temperature of 1 g of water from 14.5 °C to 15.5 °C. The current practice is to *define* a thermo-chemical calorie as 4.184 J.

  The **gas constant** R appears in the ideal gas law, $pV = NRT$. Its numerical values are:

$$\begin{aligned} R = 8.314\,\text{J\,K}^{-1}\,\text{mol}^{-1}\ (\text{or Pa\,m}^3\,\text{K}^{-1}\,\text{mol}^{-1}) &= 0.08314\,\text{bar\,L\,K}^{-1}\,\text{mol}^{-1} \\ &= 0.0821\,\text{atm\,L\,K}^{-1}\,\text{mol}^{-1} \end{aligned}$$

The **Avogadro number** $N_A = 6.023 \times 10^{23}\,\text{mol}^{-1}$. The **Boltzmann constant** $k_B = R/N_A = 1.3807 \times 10^{-23}\,\text{J\,K}^{-1}$.

(Though the temperature that Boyle knew and used was the empirical temperature, as we shall see in Chapter 3, it is appropriate to use the absolute temperature T in the formulation of the law of ideal gases. To avoid excessive notation we shall use T whenever it is appropriate.) Boyle also advocated the view that heat was not an indestructible substance (caloric) that passed from one object to another but was '. . . intense commotion of the parts . . .' [5].

At constant pressure, the variation of volume with temperature was studied by Jacques Charles (1746–1823) who established that

$$\frac{V}{T} = f_2(p) \quad f_2(p) \text{ is some function of the pressure } p \qquad (1.4.2)$$

In 1811, Amedeo Avogadro (1776–1856) announced his hypothesis that, under conditions of the same temperature and pressure, equal volumes of all gases contained equal numbers of molecules. This hypothesis greatly helped in explaining the changes in pressure due to chemical reactions in which the reactants and products were gases. It implied that, at constant pressure and temperature, the volume of a gas is proportional to the amount of the gas. Hence, in accordance with Boyle's law (1.4.1), for N moles of a gas:

$$V = N \frac{f_1(T)}{p} \qquad (1.4.3)$$

Jacques Charles (1746–1823) (Reproduced with permission from the Edgar Fahs Smith Collection, University of Pennsylvania Library)

A comparison of (1.4.1), (1.4.2) and (1.4.3) leads to the conclusion that $f_1(T)$ is proportional to T and to the well-known **law of ideal gases**:

$$pV = NRT \qquad\qquad (1.4.4)$$

in which R is the gas constant. Note: $R = 8.31441 \, \mathrm{J\,K^{-1}\,mol^{-1}}$ (or $\mathrm{Pa\,m^3\,K^{-1}\,mol^{-1}}$) = $0.08314 \, \mathrm{bar\,L\,K^{-1}\,mol^{-1}} = 0.0821 \, \mathrm{atm\,L\,K^{-1}\,mol^{-1}}$.

As more gases were identified and isolated by the chemists during the eighteenth and nineteenth centuries, their properties were studied. It was found that many obeyed Boyle's law approximately. For most gases, this law describes the experimentally observed behavior fairly well for pressures to about 10 atm. As we shall see in the next section, the behavior of gases under a wider range of pressures can be described by modifications of the ideal gas law that take into consideration the molecular size and intermolecular forces.

For a *mixture* of ideal gases, we have the **Dalton's law of partial pressures**, according to which the pressure exerted by each component of the mixture is independent of the other components of the mixture, and each component obeys the ideal gas equation. Thus, if p_k is the partial pressure due to component k, we have

$$p_k V = N_k RT \qquad\qquad (1.4.5)$$

Joseph-Louis Gay-Lussac (1778–1850), who made important contributions to the laws of gases, discovered that a dilute gas expanding into vacuum did so without change in temperature. James Prescott Joule (1818–1889) also verified this fact in his series of experiments that established the equivalence between mechanical energy and heat. In Chapter 2 we will discuss Joule's work and the law of conservation of energy in detail. When the concept of energy and its conservation was established, the implication of this observation became clear. Since a gas expanding into vacuum does not do any work during the processes of expansion, its energy does not change. The fact that the temperature does not change during expansion into vacuum while the volume and pressure do change implies that the energy of a given amount of ideal gas depends only on its temperature T, not on its volume or pressure. Also, a change in the ideal gas temperature occurs only when its energy is changed through exchange of heat or mechanical work. These observations lead to the conclusion that the energy of a given amount of ideal gas is a function only of its temperature T. Since the amount of energy (heat) needed to increase the temperature of an ideal gas is proportional to the amount of the gas, the energy is proportional to N, the amount of gas in moles. Thus, the energy of the ideal gas, $U(T, N)$, is a function only of the temperature T and the amount of gas N. It can be written as

$$U(T, N) = N U_{\mathrm{m}}(T) \qquad\qquad (1.4.6)$$

in which U_{m} is the total internal energy per mole, or **molar energy**. For a mixture of gases the total energy is the sum of the energies of the components:

Joseph-Louis Gay-Lussac (1778–1850) (Reproduced with permission from the Edgar Fahs Smith Collection, University of Pennsylvania Library)

$$U(T, N) = \sum_k U_k(T, N_k) = \sum_k N_k U_{mk}(T) \qquad (1.4.7)$$

in which the components are indexed by k. Later developments established that

$$U_m = cRT + U_0 \qquad (1.4.8)$$

to a good approximation, in which U_0 is a constant. For monatomic gases, such as He and Ar, $c = 3/2$; for diatomic gases, such as N_2 and O_2, $c = 5/2$. The factor c can be deduced from the kinetic theory of gases, which relates the energy U to the motion of a gas molecules.

The experiments of Gay-Lussac also showed that, at constant pressure, the relative change in volume $\delta V/V$ due to increase in temperature had nearly the same value for all dilute gases; it was equal to $1/273\,°C^{-1}$. Thus, a gas thermometer in which the volume of a gas at constant pressure was the indicator of temperature t had the quantitative relation

$$V = V_0(1 + \alpha t) \tag{1.4.9}$$

in which $\alpha = 1/273$ is the coefficient of expansion at constant pressure. In Chapter 3 we will establish the relation between the temperature t, measured by the gas thermometer, and the absolute temperature T.

The above empirical laws of gases played an important part in the development of thermodynamics. They are the testing ground for any general principle and are often used to illustrate these principles. They were also important for developments in the atomic theory of matter and chemistry.

For most gases, such as CO_2, N_2, and O_2, the ideal gas law was found to be an excellent description of the experimentally observed relation between p, V and T only for pressures up to about 20 atm. Significant improvements in the laws of gases did not come until the molecular nature of gases was understood. In 1873, more than 200 years after Boyle published his famous results, Johannes Diderik van der Waals (1837–1923) proposed an equation in which he incorporated the effects of attractive forces between molecules and molecular size on the pressure and volume of a gas. We shall study van der Waals' equation in detail in the next section, but here we would like to familiarize the reader with its basic form so that it can be compared with the ideal gas equation. According to van der Waals, p, V, N and T are related by

$$\left(p + a\frac{N^2}{V^2}\right)(V - Nb) = NRT \tag{1.4.10}$$

In this equation, the constant a is a measure of the attractive forces between molecules and b is proportional to the size of the molecules. For example, the values of a and b for helium are smaller than the corresponding values for a gas such as CO_2. The values of the constants a and b for some of the common gases are given in Table 1.1. Unlike the ideal gas equation, this equation explicitly contains molecular parameters and it tells us how the ideal gas pressure and volume are to be 'corrected' because of the molecular size and intermolecular forces. We shall see how van der Waals arrived at this equation in the next section. At this point, students are encouraged to pause and try deriving this equation on their own before proceeding to the next section.

As one might expect, the energy of the gas is also altered due to forces between molecules. In Chapter 6 we will see that the energy U_{vw} of a van der Waals gas can be written as

$$U_{vw} = U_{ideal} - a\left(\frac{N}{V}\right)^2 V \tag{1.4.11}$$

Table 1.1 Van der Waals constants a and b and critical constants T_c, p_c and V_{mc} for selected gases

Gas	a/bar L^2 mol^{-2}	b/L mol^{-1}	T_c/K	p_c/bar	V_{mc}/L mol^{-1}
Acetylene (C_2H_2)	4.516	0.0522	308.3	61.39	0.113
Ammonia (NH_3)	4.225	0.0371	405.5	113.5	0.072
Argon (Ar)	1.355	0.0320	150.9	49.55	0.075
Carbon dioxide (CO_2)	3.658	0.0429	304.1	73.75	0.094
Carbon monoxide (CO)	1.472	0.0395	132.9	34.99	0.093
Chlorine (Cl_2)	6.343	0.0542	416.9	79.91	0.123
Ethanol (C_2H_5OH)	12.56	0.0871	513.9	61.32	0.167
Helium (He)	0.0346	0.0238	5.19	2.22	0.057
Hydrogen (H_2)	0.245	0.0265	32.97	12.93	0.065
Hydrogen chloride (HCl)	3.700	0.0406	324.7	83.1	0.081
Methane (CH_4)	2.300	0.0430	190.5	46.04	0.099
Nitric oxide (NO)	1.46	0.0289	180	64.8	0.058
Nitrogen (N_2)	1.370	0.0387	126.2	33.9	0.090
Oxygen (O_2)	1.382	0.0319	154.59	50.43	0.073
Propane (C_3H_8)	9.385	0.0904	369.82	42.50	0.203
Sulfur dioxide (SO_2)	6.865	0.0568	430.8	78.84	0.122
Sulfur hexafluoride (SF_6)	7.857	0.0879	318.69	37.7	0.199
Water (H_2O)	5.537	0.0305	647.14	220.6	0.056

Source: An extensive listing of van der Waals constants can be found in D.R. Lide (ed.), *CRC Handbook of Chemistry and Physics*, 75th edition. 1994, CRC Press: Ann Arbor, MI.

The van der Waals equation was a great improvement over the ideal gas law, in that it described the observed liquefaction of gases and the fact that, above a certain temperature, called the critical temperature, gases could not be liquefied regardless of the pressure, as we will see in the following section. But still, it was found that the van der Waals equation failed at very high pressures (Exercise 1.13). The various improvements suggested by Clausius, Berthelot and others are discussed in Chapter 6.

1.5 States of Matter and the van der Waals Equation

The simplest transformations of matter caused by heat is the melting of solids and the vaporization of liquids. In thermodynamics, the various states of matter (solid, liquid, gas) are often referred to as **phases**. Every compound has a definite temperature T_m at which it melts and a definite temperature T_b at which it boils. In fact, this property can be used to identify a compound or separate the constituents of a mixture. With the development of the thermometer, these properties could be studied with precision. As noted earlier, Joseph Black and James Watt discovered another interesting phenomenon associated with the changes of phase: at the melting or the boiling temperature, the heat supplied to a system does not result in an increase in temperature; it only has the effect of converting the substance from one phase to another. This heat that lays 'latent' or hidden without increasing the temperature was called the **latent heat**. When a liquid solidifies, for example, this heat is given out to the surroundings. This phenomenon is summarized in Figure 1.3.

Clearly, the ideal gas equation, good as it is in describing many properties of gases, does not help us to understand why gases convert to liquids when compressed. An ideal gas remains a gas at all temperatures and its volume can be compressed without limit. In 1822, Gay-Lussac's friend Cagniard de la Tour (1777–1859) discovered that a gas does not liquefy when compressed unless its temperature is below a critical value, called the **critical temperature**. This behavior of gases was studied in detail by Thomas Andrews (1813–1885), who published his work in 1869. During this time, atomic theory was gaining more and more ground, while Maxwell, Clausius and others advanced the idea that heat was related to molecular motion and began to find an explanation of the properties of gases, such as pressure and viscosity, in the random motion of molecules. It was in this context that Johannes Diderik van der Waals (1837–1923) sought a single equation of state for the liquid and gas phases of a substance. In 1873, van der Waals presented his doctoral thesis titled 'On the continuity of the gas and liquid state', in which he brilliantly explained the conver-

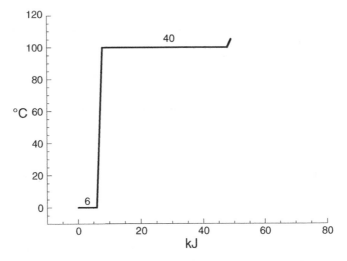

Figure 1.3 The change in temperature of 1 mol of H_2O versus amount of heat, at a pressure of 1 atm. At the melting point, absorption of heat does not increase the temperature until all the ice melts. It takes about 6 kJ to melt 1 mol of ice, the 'latent heat' discovered by Joseph Black. Then the temperature increases until the boiling point is reached, at which point it remains constant until all the water turns to steam. It takes about 40 kJ to convert 1 mol of water to steam

sion of a gas to a liquid and the existence of critical temperature as the consequence of forces between molecules and molecular volume.

Van der Waals realized that two main factors modify the ideal gas equation: the effect of molecular volume and the effect of intermolecular forces. Since molecules have a nonzero volume, the volume of a gas cannot be reduced to an arbitrarily small value by increasing p. The corresponding modification of the ideal gas equation would be $(V - bN) = NRT/p$, in which the constant b is the limiting volume of 1 mol of the gas, as $p \to \infty$. The constant b is sometimes called the 'excluded volume'. The effect of intermolecular forces, van der Waals noted, is to decrease the pressure, as illustrated in Figure 1.4. Hence, the above 'volume-corrected' equation is further modified to

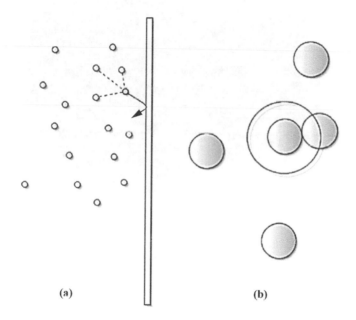

(a)   (b)

Figure 1.4 Van der Waals considered molecular interaction and molecular size to improve the ideal gas equation. (a) The pressure of a real gas is less than the ideal gas pressure because intermolecular attraction decreases the speed of the molecules approaching the wall. Therefore, $p = p_{ideal} - \delta p$. (b) Since the molecules of a gas have a nonzero size, the volume available to molecules is less than the volume of the container. Each molecule has a volume around it that is not accessible to other molecules because the distance between the centers of the molecules cannot be less than the sum of the molecular radii. As a result, the volume of the gas cannot decrease below this 'excluded volume'. Thus, V in the ideal gas equation is replaced with $(V - bN)$ so that as $p \to \infty$, $V \to bN$

Johannes van der Waals (1837–1923) (Reproduced with permission from the Edgar Fahs Smith Collection, University of Pennsylvania Library)

$$p = \frac{NRT}{V - bN} - \delta p$$

Next, van der Waals related the factor δp to the number density N/V using the kinetic theory of gases, which showed how molecular collisions with container walls cause pressure. Pressure depends on the number of molecules that collide with the walls per unit area, per unit time; therefore, it is proportional to the number density N/V (as can be seen from the ideal gas equation). In addition, each molecule that is close to a container wall and moving towards it experiences the retarding attractive forces of molecules behind it (see Figure 1.4); this force would also be

STATES OF MATTER AND THE VAN DER WAALS EQUATION

proportional to number density N/V; hence, δp should be proportional to two factors of N/V, so that one may write $\delta p = a(N/V)^2$, in which the constant a is a measure of the intermolecular forces. The expression for pressure that van der Waals proposed is

$$p = \frac{NRT}{V - bN} - a\frac{N^2}{V^2}$$

or, as it is usually written:

$$\left(p + a\frac{N^2}{V^2}\right)(V - Nb) = NRT \tag{1.5.1}$$

This turns out to be an equation of state for both the liquid and the gas phase. Van der Waals' insight revealed that the two phases, which were considered distinct, can, in fact, be described by a single equation. Let us see how.

For a given T, a p–V curve, called the **p–V isotherm**, can be plotted. Such isotherms for the van der Waals equation (1.5.1) are shown in Figure 1.5. They show an important feature: the **critical temperature** T_c studied by Thomas Andrews. If the temperature T is greater than T_c then the p–V curve is always single valued, much like the ideal gas isotherm, indicating that there is no transition to the liquid state. But for lower temperatures, $T < T_c$, the isotherm has a maximum and a minimum. There are two extrema because the van der Waals equation is cubic in V. This region represents a state in which the liquid and the gas phases coexist in thermal equilibrium. On the p–V curve shown in Figure 1.5, the gas begins to condense into a liquid at point A; the conversion of gas to liquid continues until point C, at which all the gas has been converted to liquid. Between A and C, the actual state of the gas does not follow the path AA'BB'C along the p–V curve because this curve represents an unstable supersaturated state in which the gas condenses to a liquid. The actual state of the gas follows the straight line ABC. As T increases, the two extrema move closer and finally coalesce at $T = T_c$. For a mole of a gas, the point (p, V) at which the two extrema coincide is defined as the **critical pressure** p_c and **critical molar volume** V_{mc}. For T higher than T_c, there is no **phase transition** from a gas to a liquid; the distinction between gas and liquid disappears. (This does not happen for a transition between a solid and a liquid because a solid is more ordered than a liquid; the two states are always distinct.) Experimentally, the **critical constants** p_c, V_{mc} and T_c can be measured and they are tabulated (Table 1.1 lists some examples). We can relate the critical parameters to the van der Waals parameters a and b by the following means. We note that if we regard $p(V, T)$ as a function of V, then, for $T < T_c$, the derivative $(\partial p/\partial V)_T = 0$ at the two extrema. As T increases, at the point where the two extrema coincide, i.e. at the critical point $T = T_c$, $p = p_c$ and $V = V_{mc}$, we have an *inflection point*. At an inflection point, the first and second derivatives of a function vanish. Thus, at the critical point:

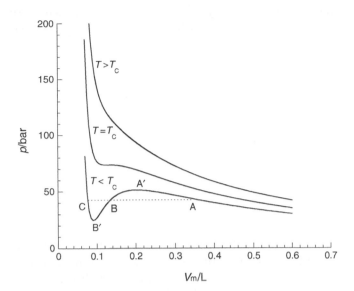

Figure 1.5 The van der Waals isotherms for CO_2 ($T_c =$ 304.14 K). When $T < T_c$, there is a region AA'BB'C in which, for a given value of p, the van der Waals equation does not specify a unique volume V; in this region, the gas transforms to a liquid. The segment A'BB' is an unstable region; states corresponding to points on this segment are not experimentally realizable. Experimentally realizable states are on the dotted line ABC. The observed state follows the path ABC. A detailed description of this region is discussed in Chapter 7

$$\left(\frac{\partial p}{\partial V}\right)_T = 0 \quad \left(\frac{\partial^2 p}{\partial V^2}\right)_T = 0 \tag{1.5.2}$$

Using these equations one can obtain the following relations between the critical constants and the constants a and b (Exercise 1.17):

$$a = \frac{9}{8} RT_c V_{mc} \quad b = \frac{V_{mc}}{3} \tag{1.5.3}$$

in which V_{mc} is the molar critical volume. Conversely, we can write the critical constants in terms of the van der Waals constants a and b (Exercise 1.17):

$$T_c = \frac{8a}{27Rb} \quad p_c = \frac{a}{27b^2} \quad V_{mc} = 3b \tag{1.5.4}$$

Table 1.1 contains the values of a and b and critical constants for some gases.

THE LAW OF CORRESPONDING STATES

Every gas has a characteristic temperature T_c, pressure p_c, and volume V_{mc} which depend on the molecular size and intermolecular forces. In view of this, one can introduce dimensionless **reduced variables** defined by

$$T_r = \frac{T}{T_c} \quad V_{mr} = \frac{V}{V_{mc}} \quad p_r = \frac{p}{p_c} \qquad (1.5.5)$$

Van der Waals showed that, if his equation is rewritten in terms of these reduced variables, one obtains the following 'universal equation' (Exercise 1.18), which is independent of the constants a and b:

$$p_r = \frac{8T_r}{3V_r - 1} - \frac{3}{V_r^2} \qquad (1.5.6)$$

This is a remarkable equation because it implies that gases have corresponding states: *at a given value of reduced volume and reduced temperature, all gases have the same reduced pressure*. This statement is called the **law of corresponding states** or **principle of corresponding states**, which van der Waals enunciated in an 1880 publication. Noting that the reduced variables are defined wholly in terms of the experimentally measured critical constants, p_c, V_{mc} and T_c, he conjectured that the principle has a general validity, independent of his equation of state. According to the principle of corresponding states, then, at a given T_r and V_{mr} the reduced pressures p_r of all gases should be the same (which is not necessarily the value given by (1.5.6)).

The deviation from ideal gas behavior is usually expressed by defining a **compressibility factor**:

$$Z = \frac{V_m}{V_{m,ideal}} = \frac{pV_m}{RT}$$

which is the ratio between the actual volume of a gas and that of the ideal gas at a given T and p. Ideal gas behavior corresponds to $Z = 1$. For real gases, at low pressures and temperatures, it is found that $Z < 1$; but for higher pressures and temperatures, $Z > 1$. It is also found that there is a particular temperature, called the Boyle temperature, at which Z is nearly 1 and the relationship between p and V is close to that of an ideal gas (Exercise 1.11). One way to verify the law of corresponding states experimentally is to plot Z as a function of reduced pressure p_r at a given reduced temperature T_r. The compressibility factor Z can be written in terms of the reduced variables: $Z = (p_c V_{mc}/RT_c)(p_r V_{mr}/T_r)$; if the value of $(p_c V_{mc}/RT_c) = Z_c$ is the same for all gases (for the van der Waals gas $Z_c = (p_c V_{mc}/RT_c) = 3/8$ (Exercise 1.18)) then Z is a function of the reduced variables. Experimental values of Z for different gases could be plotted as a functions of p_r for a fixed T_r. If the law of corresponding states is valid, then at a given value of T_r and p_r the value of Z must be the same for all gases. The plot shown in Figure 1.6 indicates that the validity of the law of corresponding states is fairly general.

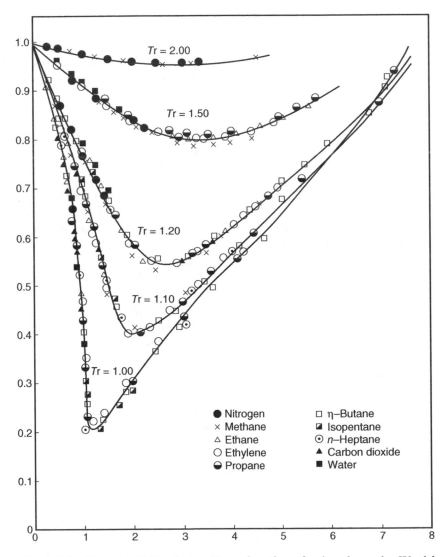

Figure 1.6 Compressibility factor Z as a function of reduced van der Waals' variables. (Reprinted with permission from Goug-Jen Su, *Industrial and Engineering Chemistry*, **38** (1946) 803. Copyright 1946, American Chemical Society)

The van der Waals equation and the law of corresponding states, however, have their limitations, which van der Waals himself noted in his 1910 Nobel Lecture:

But closer examination showed me that matters were not so simple. To my surprise I realized that the amount by which the volume must be reduced is variable, that in extremely dilute state this amount, which I notated *b*, is fourfold the molecular volume* – but that this amount decreases with decreasing external volume and gradually falls to about half. But the law governing this decrease has still not been found.

Van der Waals also noted that the experimental value of $Z_c = (p_c V_{mc}/RT_c)$ for most gases was not $3/8 = 0.375$, as predicted by his equation, but was around 0.25 (0.23 for water and 0.29 for Ar). Furthermore, it became evident that the van der Waals constant *a* depended on the temperature – Rudolf Clausius even suggested that *a* was inversely proportional to *T*. Thus, the parameters *a* and *b* might themselves be functions of gas density and temperature. As a result, a number of alternative equations have been proposed for the description of real gases. For example, engineers and geologists often use the following equation, known as the **Redlich–Kwong equation**:

$$p = \frac{NRT}{V-b} - \frac{a}{\sqrt{T}}\frac{N^2}{V(V-Nb)} = \frac{RT}{V_m-b} - \frac{a}{\sqrt{T}}\frac{1}{V_m(V_m-b)} \tag{1.5.7}$$

The constants *a* and *b* in this equation differ from those in the van der Waals equation; they can be related to the critical constants and they are tabulated just as the van der Waals *a* and *b* are. We will discuss other similar equations used to describe real gases in Chapter 6.

The limitation of van der Waals-type equations and the principle of corresponding states lies in the fact that molecular forces and volume are quantified with just two parameters, *a* and *b*. As explained below, two parameters can characterize the forces between small molecules fairly well, but larger molecules require more parameters.

MOLECULAR FORCES AND THE LAW OF CORRESPONDING STATES

From a molecular point of view, the van der Waals equation has two parameters, *a* and *b*, that describe molecular forces, often called **van der Waals' forces**. These forces are attractive when the molecules are far apart but are repulsive when they come into contact, thus making the condensed state (liquid or solid) hard to compress. It is the repulsive core that gives the molecule a nonzero volume. The typical potential energy between two molecules is expressed by the so-called Lennard–Jones energy:

---

*Molecular volume is the actual volume of the molecules ($N_A 4\pi r^3/3$ for a mole of spherical molecules of radius *r*).

$$U_{LJ}(r) = 4\varepsilon \left[\left(\frac{\sigma}{r} \right)^{12} - \left(\frac{\sigma}{r} \right)^{6} \right] \qquad\qquad (1.5.8)$$

Figure 1.7 shows a plot of this potential energy as a function of the distance r between the centers of the molecules. As the distance between the molecules decreases, so U_{LJ} decreases, reaches a minimum, and sharply increases. The decreasing part of U_{LJ} is due to the term $-(\sigma/r)^6$, which represents an attractive force, and the sharply increasing part is due to the term $(\sigma/r)^{12}$, which represents a repulsive core. The Lennard–Jones energy reaches a minimum value of $-\varepsilon$ when $r = 2^{1/6}\sigma$ (Exercise 1.20). The two van der Waals parameters, a and b, are related to ε and σ respectively, the former being a measure of the molecular attractive force and the latter a measure of the molecular size. In fact, using the principles of statistical thermodynamics, for a given ε and σ the values of a and b can be calculated. Such a relationship between the molecular interaction potential and the parameters in the van der Waals equation of state gives us an insight into the limitations of the law of corresponding states, which depends on just two parameters, a and b. If more than two parameters are needed to describe the forces between two molecules adequately, then we can also expect the equation of state to depend on more than two parameters. Lennard–Jones-type potentials that use two parameters are good approximations for small

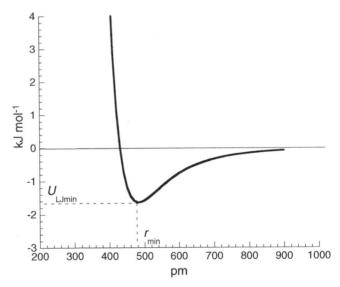

Figure 1.7   Lennard–Jones '6–12' potential energy between two molecules as a function of the distance between their centers. It is common to specify ε in units of kelvin using the ratio ε/k_B, in which k_B is the Boltzmann constant. The assumed Lennard–Jones parameter values for the above curve are $\varepsilon/k_B = 197\,\text{K}$ (which corresponds to $\varepsilon N_A = 1.638\,\text{kJ}\,\text{mol}^{-1}$) and $\sigma = 430\,\text{pm}$. These values represent the approximate interaction energy between CO_2 molecules

molecules; for larger molecules the interaction energy depends not only on the distance between the molecules, but also on their relative orientation and other factors that require more parameters. Thus, significant deviation from the law of corresponding states can be observed for larger molecules.

1.6   An Introduction to Kinetic Theory of Gases

When Robert Boyle published his study on the nature of the 'spring of the air' (what we call pressure today) and argued that heat was an 'intense commotion of the parts', he did not know how pressure actually arose. During the seventeenth century, a gas was thought to be a continuous substance. A century later, Daniel Bernoulli (1700–1782) published the idea that the mechanism that caused pressure is the rapid collisions of molecules with the walls of the container [5]. In his 1738 publication, *Hydrodynamica*, Bernoulli presented his calculation of the average force on the container walls due to molecular collisions and obtained a simple expression for the pressure: $p = (mnv_{avg}^2/3)$, in which m is the molecular mass, n is the number of molecules per unit volume and v_{avg} is the average speed of molecules. At that time, no one had any idea how small gas molecules were or how fast they moved, but Bernoulli's work was an important step in explaining the properties of a gas in terms of molecular motion. It was the beginnings of a subject that came to be known as the kinetic theory of gases.

The kinetic theory of gases was largely developed in the late nineteenth century. Its goal was to explain the observed properties of gases by analyzing the random motion of molecules. Many quantities, such as pressure, diffusion constant and the coefficient of viscosity, could be related to the average speed of molecules, their mass, size, and the average distance they traversed between collisions (called the mean free path). As we shall see in this section, the names of James Clerk Maxwell (1831–1879) and Ludwig Boltzmann (1844–1906) are associated with some of the basic concepts in this field, while, as is often the case in science, several others contributed to its development [4, 5]. In this introductory section we shall deal with some elementary aspects of kinetic theory, such as the mechanism that causes pressure and the relation between average kinetic energy and temperature.

KINETIC THEORY OF PRESSURE

As Daniel Bernoulli showed, using the basic concepts of force and randomness, it is possible to relate the pressure of a gas to molecular motion: pressure is the average force per unit area exerted on the walls by colliding molecules.

We begin by noting some aspects of the random motion of molecules. First, if all directions have the same physical properties, then we must conclude that motion along all directions is equally probable: the properties of molecules moving in one direction will be the same as the properties of molecules moving in any other direction. Let us assume the average speed of the gas molecules is v_{avg}. We denote its x, y and z components of the by v_{xavg}, v_{yavg}, and v_{zavg}. Thus:

$$v_{\text{avg}}^2 = v_{x\,\text{avg}}^2 + v_{y\,\text{avg}}^2 + v_{z\,\text{avg}}^2 \tag{1.6.1}$$

Because all directions are equivalent, we must have

$$v_{x\,\text{avg}}^2 = v_{y\,\text{avg}}^2 = v_{z\,\text{avg}}^2 = \frac{v_{\text{avg}}^2}{3} \tag{1.6.2}$$

The following quantities are necessary for obtaining the expression for pressure:

$$
\begin{aligned}
N &= \text{amount of gas in moles}\\
V &= \text{gas volume}\\
M &= \text{Molar mass of the gas}\\
m &= \text{mass of a single molecule} = M/N_A\\
n &= \text{number of molecules per unit volume} = NN_A/V\\
N_A &= \text{Avogadro number}
\end{aligned}
\tag{1.6.3}
$$

Now we calculate the pressure by considering the molecular collisions with the wall. In doing so, we will approximate the random motion of molecules with molecules moving with an average speed v_{avg}. (A rigorous derivation gives the same result.) Consider a layer of a gas, of thickness Δx, close to the wall of the container (see Figure 1.8). When a molecule collides with the wall, which we assume is perpendicular to the x-axis, the change in momentum of the molecule in the x direction equals $2mv_{x\,\text{avg}}$. In the layer of thickness Δx and area A, because of the randomness of molecular motion, about half the molecules will be moving towards the wall; the rest will be moving away from the wall. Hence, in a time $\Delta t = \Delta x/v_{\text{avg}}$ about half the molecules in the layer will collide with the wall. The number of molecules in the

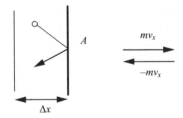

Figure 1.8 Rapid collisions of gas molecules with the walls of the container give rise to pressure. By computing the average momentum transferred to the wall by colliding molecules, pressure can be related to the average of the square of molecular velocity

layer is $(\Delta x A)n$ and the number of molecules colliding with the walls is $(\Delta x A)n/2$. Now, since each collision imparts a momentum $2mv_{x\,\mathrm{avg}}$, in a time Δt, the total momentum imparted to the wall is $2mv_{x\,\mathrm{avg}}(\Delta x A)n/2$. Thus, the average force F on the wall of area A is

$$F = \frac{\text{Momentum imparted}}{\Delta t} = \frac{2mv_{x\,\mathrm{avg}}\Delta x A}{\Delta t}\frac{n}{2} = \frac{mv_{x\,\mathrm{avg}}\Delta x A n}{(\Delta x/v_{x\,\mathrm{avg}})} = mv_{x\,\mathrm{avg}}^2 nA \quad (1.6.4)$$

Pressure p, which is the force per unit area, is thus

$$p = \frac{F}{A} = mv_{x\,\mathrm{avg}}^2 n \quad (1.6.5)$$

Since the direction x is arbitrary, it is better to write this expression in terms of the average speed of the molecule rather than its x component. By using (1.6.2) and the definitions (1.6.3), we can write the pressure in terms the macroscopic variables M, V and N:

$$p = \frac{1}{3}mnv_{\mathrm{avg}}^2 = \frac{1}{3}M\frac{N}{V}v_{\mathrm{avg}}^2 \quad (1.6.6)$$

This expression relates the pressure to the square of the average speed. A rigorous description of the random motion of molecules leads to the same expression for the pressure with the understanding that v_{avg}^2 is to be interpreted as the average of the square of the molecular velocity, a distinction that will become clear when we discuss the Maxwell velocity distribution. When Daniel Bernoulli published the above result in 1738, he did not know how to relate the molecular velocity to temperature; that connection had to wait until Avogadro stated his hypothesis in 1811 and the formulation of the ideal gas law based on an empirical temperature that coincides with the absolute temperature that we use today (see Equation (1.4.9)). On comparing expression (1.6.6) with the ideal gas equation, $pV = NRT$, we see that

$$RT = \frac{1}{3}Mv_{\mathrm{avg}}^2. \quad (1.6.7)$$

Using the Boltzmann constant $k_B = R/N_A = 1.3807 \times 10^{-23}\,\mathrm{J\,K^{-1}}$ and noting $M = mN_A$, we can express (1.6.7) as a relation between the kinetic energy and temperature:

$$\frac{1}{2}mv_{\mathrm{avg}}^2 = \frac{3}{2}k_B T \quad (1.6.8)$$

This is a wonderful result because it relates temperature to molecular motion, in agreement with Robert Boyle's intuition. It shows us that the average kinetic energy of a molecule equals $3k_BT/2$. It is an important step in our understanding of the meaning of temperature at the molecular level.

From (1.6.8) we see that the total kinetic energy of 1 mol of a gas equals $3RT/2$. Thus, for monatomic gases, whose atoms could be thought of as point particles that have neither internal structure nor potential energy associated with intermolecular forces (He and Ar are examples), the total molar energy of the gas is entirely kinetic; this implies $U_m = 3RT/2$. The molar energy of a gas of polyatomic molecules is larger. A polyatomic molecule has additional energy in its rotational and vibrational motion. In the nineteenth century, as kinetic theory progressed, it was realized that random molecular collisions result in equal distribution of energy among each of the independent modes of motion. According to this **equipartition theorem**, the energy associated with each independent mode of motion equals $k_BT/2$. For a point particle, for example, there are three independent modes of motion, corresponding to motion along each of the three independent spatial directions x, y and z. According to the equipartition theorem, the average kinetic energy for motion along the x direction is $mv_{x\,avg}^2 = k_BT/2$, and similarly for the y and z directions, making the total kinetic energy $3(k_BT/2)$ in agreement with (1.6.8). For a diatomic molecule, which we may picture as two spheres connected by a rigid rod, there are two independent modes of rotational motion in addition to the three modes of kinetic energy of the entire molecule. Hence, for a diatomic gas the molar energy $U_m = 5RT/2$, as we noted in the context of Equation (1.4.8). The independent modes of motion are often called **degrees of freedom**.

MAXWELL–BOLTZMANN VELOCITY DISTRIBUTION

A century after Bernoulli's *Hydrodynamica* was published, the kinetic theory of gases began to make great inroads into the nature of the randomness of molecular motion. Surely molecules in a gas move with different velocities. According to (1.6.8), the measurement of pressure only tells us the average of the square of the velocities. It does not tell us what fraction of molecules have velocities with a particular magnitude and direction. In the later half of the nineteenth century, James Clerk Maxwell (1831–1879) directed his investigations to the **probability distribution of molecular velocity** that specifies such details. We shall denote the probability distribution of the molecular velocity **v** by $P(\mathbf{v})$. The meaning of $P(\mathbf{v})$ is as follows:

$P(\mathbf{v})\,dv_x dv_y dv_z$ is the fraction of the total number of molecules whose velocity vectors have their components in the range $(v_x, v_x + dv_x)$, $(v_y, v_y + dv_y)$ and $(v_z, v_z + dv_z)$.

As shown in the Figure 1.9, each point in the velocity space corresponds to a velocity vector; $P(\mathbf{v})\,dv_x dv_y dv_z$ is the probability that the velocity of a molecule lies within an elemental volume dv_x, dv_y and dv_z at the point (v_x, v_y, v_z). $P(\mathbf{v})$ is called the **probability density** in the velocity space.

The mathematical form of $P(\mathbf{v})$ was obtained by James Clerk Maxwell; the concept was later generalized by Ludwig Boltzmann (1844–1906) to the probability

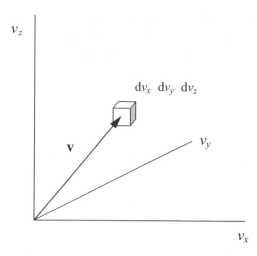

Figure 1.9 The probability distribution for the velocity is defined in the velocity space. $P(\mathbf{v})dv_x dv_y dv_z$ is the probability that the velocity of a molecule is within the shown cube

distribution of the total energy E of the molecule. According to the principle discovered by Boltzmann, when a system reaches thermodynamic equilibrium, the probability that a molecule is in a state with energy E is proportional to $\exp(-E/k_B T)$. If $\rho(E)$ is the number of different states in which the molecule has energy E, then

$$P(E) \propto \rho(E) e^{-E/k_B T} \tag{1.6.9}$$

The quantity $\rho(E)$ is called the **density of states**. Relation (1.6.9), called the **Boltzmann principle**, is one of the fundamental principles of physics. Using this principle, equilibrium thermodynamic properties of a substance can be derived from molecular energies E – a subject called statistical thermodynamics, presented in Chapter 17. In this introductory section, however, we will only study some elementary consequences of this principle.

The energy of a molecule $E = E_{trans} + E_{rot} + E_{vib} + E_{int} + \ldots$, in which E_{trans} is the kinetic energy of translational motion of the whole molecule, E_{rot} is the energy of rotational motion, E_{vib} is the energy of vibrational motion, E_{int} is the energy of the molecule's interaction with other molecules and fields such as electric, magnetic or gravitational fields, and so on. According to the Boltzmann principle, the probability that a molecule will have a translational kinetic energy E_{trans} is proportional to $\exp(-E_{trans}/k_B T)$ (the probabilities associated with other forms of energy are factors that multiply this term). Since the kinetic energy due to translational motion of the molecule is $mv^2/2$, we can write the probability as a function of the velocity $\mathbf{v}$ by which we mean probability that a molecule's velocity is in an elemental cube in velocity space, as shown in the Figure 1.9. For a continuous variable, such as velocity, we must define a probability density $P(\mathbf{v})$ so that the probability that a molecule's velocity is in an elemental cube of volume $dv_x dv_y dv_z$ located at the tip of the velocity vector $\mathbf{v}$ is $P(\mathbf{v}) dv_x dv_y dv_z$. According to the Boltzmann principle, this probability is

$$P(\mathbf{v})dv_x\, dv_y\, dv_z = \frac{1}{z}e^{-mv^2/2k_BT}\, dv_x\, dv_y\, dv_z \qquad (1.6.10)$$

in which $v^2 = v_x^2 + v_y^2 + v_z^2$.

Here, z is the **normalization factor**, defined by

$$\int_{-\infty}^{\infty}\int_{-\infty}^{\infty}\int_{-\infty}^{\infty} e^{-mv^2/2k_BT}\, dv_x\, dv_y\, dv_z = z \qquad (1.6.11)$$

so that a requirement of the very definition of a probability, $\int_{-\infty}^{\infty}\int_{-\infty}^{\infty}\int_{-\infty}^{\infty} P(\mathbf{v})dv_x dv_y dv_z$

$= 1$, is met. The normalization factor z, as defined in (1.6.11), can be calculated using the definite integral:

$$\int_{-\infty}^{\infty} e^{-ax^2}dx = \left(\frac{\pi}{a}\right)^{1/2}$$

which gives

$$\frac{1}{z} = \left(\frac{m}{2\pi k_BT}\right)^{3/2} \qquad (1.6.12)$$

(Some integrals that are often used in kinetic theory are listed at the end of this chapter in Appendix 1.2.) With the normalization factor thus determined, the probability distribution for the velocity can be written explicitly as

$$P(\mathbf{v})dv_x dv_y dv_z = \left(\frac{m}{2\pi k_BT}\right)^{3/2} e^{-mv^2/2k_BT}\, dv_x dv_y dv_z \qquad (1.6.13)$$

This is the **Maxwell velocity distribution**. Plots of this function show the well-known Gaussian or 'bell-shaped' curves shown in Figure 1.10a. It must be noted that this velocity distribution is that of a gas at thermodynamic equilibrium. The width of the distribution is proportional to the temperature. A gas not in thermodynamic equilibrium has a different velocity distribution and the very notion of a temperature may not be well defined; but such cases are very rare. In most situations, even if the temperature changes with location, the velocity distribution locally is very well approximated by (1.6.13). Indeed, in computer simulations of gas dynamics it is found that any initial velocity distribution evolves into the Maxwell distribution very quickly, in the time it takes a molecule to undergo few collisions, which in most cases is less than 10^{-8} s.

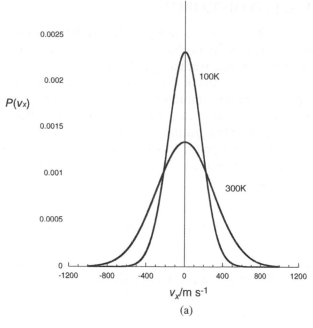

$P(v_x)$

v_x/m s^{-1}

(a)

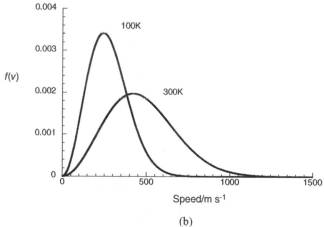

$f(v)$

Speed/m s^{-1}

(b)

Figure 1.10 Probability distributions of nitrogen. (a) Maxwell velocity distribution of the x-component of the velocity v_x at $T = 100\,K$ and $300\,K$. The width of the distribution is proportional to T. (b) $f(v)$ versus speed at $T = 100\,K$ and $300\,K$

THE MAXWELL SPEED DISTRIBUTION

The average velocity of a molecule is clearly zero because every direction of velocity and its opposite are equally probable (but the average of the square of the velocity is not zero). However, the average speed, which depends only on the magnitude of the velocity, is not zero. From the Maxwell velocity distribution (1.6.13) we can obtain the probability distribution for molecular speed, i.e. the probability that a molecule will have a speed in the range $(v, v + dv)$ regardless of direction. This can be done by summing or integrating $P(\mathbf{v})$ over all the directions in which the velocity of a fixed magnitude can point. In spherical coordinates, since the volume element is $v^2 \sin\theta\, d\theta\, d\varphi\, dv$, the probability is written as $P(\mathbf{v})v^2 \sin\theta\, d\theta\, d\varphi\, dv$. The integral over all possible directions is

$$\int_{\theta=0}^{\pi}\int_{\varphi=0}^{2\pi} P(\mathbf{v})v^2 \sin\theta\, d\theta\, d\varphi\, dv = 4\pi P(\mathbf{v})v^2 dv \tag{1.6.14}$$

The quantity $4\pi P(\mathbf{v})v^2$ is the probability density for the molecular speed. We shall denote it by $f(v)$. With this notation, the probability distribution for molecular speeds can be written explicitly as

$$f(v)dv = 4\pi\left(\frac{m}{2\pi k_{B}T}\right)^{3/2} e^{-\beta v^2} v^2\, dv$$
$$\beta = \frac{m}{2k_{B}T} \tag{1.6.15}$$

Because the molar mass $M = mN_A$ and $R = k_B N_A$, the above expressions can also be written as

$$f(v)dv = 4\pi\left(\frac{M}{2\pi RT}\right)^{3/2} e^{-\beta v^2} v^2\, dv$$
$$\beta = \frac{M}{2RT} \tag{1.6.16}$$

The shape of the function $f(v)$ is shown in the Figure 1.10b. This graphs show that, at a given temperature, there are a few molecules with very low speeds and a few with large speeds; we can also see that $f(v)$ becomes broader as T increases. The speed v at which $f(v)$ reaches its maximum is the most probable speed.

With the above probability distributions we can calculate several average values. We shall use the notation in which the average value of a quantity X is denoted by $\langle X\rangle$. The average speed is given by the integral

$$\langle v\rangle = \int_{0}^{\infty} v f(v)dv \tag{1.6.17}$$

For the probability distribution (1.6.15), such integrals can be calculated using integral tables or *Mathematica* or *Maple*. While doing such calculations, it is convenient to write the probability $f(v)$ as

$$f(v)\mathrm{d}v = \frac{4\pi}{z}\mathrm{e}^{-\beta v^2}v^2\,\mathrm{d}v$$

$$\beta = \frac{M}{2RT} \qquad \frac{1}{z} = \left(\frac{M}{2\pi RT}\right)^{3/2}$$

(1.6.18)

Using the appropriate integral in Appendix 1.2 at the end of this chapter, the average speed can be obtained in terms of T and the molar mass M (Exercise 1.23):

$$\langle v \rangle = \frac{4\pi}{z}\int_0^\infty v^3\,\mathrm{e}^{-\beta v^2}\,\mathrm{d}v = \frac{4\pi}{z}\frac{1}{2\beta^2} = \sqrt{\frac{8RT}{\pi M}}$$

(1.6.19)

Similarly, one can calculate the average energy of a single molecule using m and k_B instead of M and R (Exercise 1.23):

$$\left\langle \frac{1}{2}mv^2 \right\rangle = \frac{m4\pi}{2z}\int_0^\infty v^4\,\mathrm{e}^{-\beta v^2}\,\mathrm{d}v = \frac{m2\pi}{z}\frac{3\sqrt{\pi}}{8\beta^{5/2}} = \frac{3}{2}k_B T$$

(1.6.20)

A rigorous calculation of the pressure using the Maxwell–Boltzmann velocity distribution leads to the expression (1.6.6) in which $v_{avg}^2 = \langle v^2 \rangle$. Also, the value of v at which $f(v)$ has a maximum is the most probable speed. This can easily be determined by setting $\mathrm{d}f/\mathrm{d}v = 0$, a calculation left as an exercise.

What do the above calculations tell us? First, we see that the average speed of a molecule is directly proportional to the square root of the absolute temperature and inversely proportional to its molar mass. This is one of the most important results of the kinetic theory of gases. Another point to note is the simple dependence of the average kinetic energy of a molecule on the absolute temperature (1.6.20). It shows that the average kinetic energy of a gas molecule depends only on the temperature and is independent of its mass.

Appendix 1.1 Partial Derivatives

DERIVATIVES OF MANY VARIABLES

When a variable such as energy $U(T, V, N_k)$ is a function of many variables V, T and N_k, its *partial derivative* with respect to each variables is defined by holding all other variables constant. Thus, for example, if $U(T, V, N) = (5/2)NRT - a(N^2/V)$ then the partial derivatives are

$$\left(\frac{\partial U}{\partial T}\right)_{V,N} = \frac{5}{2}NR \qquad\qquad (A1.1.1)$$

$$\left(\frac{\partial U}{\partial N}\right)_{V,T} = \frac{5}{2}RT - a\frac{2N}{V} \qquad\qquad (A1.1.2)$$

$$\left(\frac{\partial U}{\partial V}\right)_{N,T} = a\frac{N^2}{V^2} \qquad\qquad (A1.1.3)$$

The subscripts indicate the variables that are held constant during the differentiation. In cases where the variables being held constant are understood, the subscripts are often dropped. The change in U, i.e. the differential dU, due to changes in N, V and T is given by

$$dU = \left(\frac{\partial U}{\partial T}\right)_{V,N} dT + \left(\frac{\partial U}{\partial V}\right)_{T,N} dV + \left(\frac{\partial U}{\partial N}\right)_{V,T} dN \qquad\qquad (A1.1.4)$$

For functions of many variables, there is a second derivative corresponding to every pair of variables: $\partial^2 U/\partial T\partial V$, $\partial^2 U/\partial N\partial V$, $\partial^2 U/\partial T^2$, etc. For the 'cross-derivatives' such as $\partial^2 U/\partial T\partial V$, which are derivatives with respect to two different variables, the order of differentiation does not matter. That is:

$$\frac{\partial^2 U}{\partial T\partial V} = \frac{\partial^2 U}{\partial V\partial T} \qquad\qquad (A1.1.5)$$

The same is valid for all higher derivatives, such as $\partial^3 U/\partial T^2\partial V$; i.e. the order of differentiation does not matter.

BASIC IDENTITIES

Consider three variables x, y and z, each of which can be expressed as a function of the other two variables, $x = x(y, z)$, $y = y(z, x)$ and $z = z(x, y)$. (p, V and T in the ideal gas equation $pV = NRT$ is an example.) Then the following identities are valid:

$$\left(\frac{\partial x}{\partial y}\right)_z = \frac{1}{(\partial y/\partial x)_z} \qquad\qquad (A1.1.6)$$

$$\left(\frac{\partial x}{\partial y}\right)_z \left(\frac{\partial y}{\partial z}\right)_x \left(\frac{\partial z}{\partial x}\right)_y = -1 \qquad\qquad (A1.1.7)$$

Consider a functions of x and y, $f = f(x, y)$, other than z. Then:

$$\left(\frac{\partial f}{\partial x}\right)_z = \left(\frac{\partial f}{\partial x}\right)_y + \left(\frac{\partial f}{\partial y}\right)_x \left(\frac{\partial y}{\partial x}\right)_z \qquad \text{(A1.1.8)}$$

Appendix 1.2 Elementary Concepts in Probability Theory

In the absence of a deterministic theory that enables us to calculate the quantities of interest to us, one uses probability theory. Let x_k, in which $k = 1, 2, 3, \ldots, n$, represent all possible n values of a random variable x. For example, x could be the number of molecules at any instant in a small volume of $1\,\text{nm}^3$ within a gas or the number of visitors at a website at any instant of time. Let the corresponding probabilities for these n values of x be $P(x_k)$. Since x_k, $k = 1, 2, \ldots, n$, represents all possible states:

$$\sum_{k=1}^{n} P(x_k) = 1 \qquad \text{(A1.2.1)}$$

AVERAGE VALUES

We shall denote the average value of a quantity A by $\langle A \rangle$. Thus, the average value of x would be

$$\langle x \rangle = \sum_{k=1}^{n} x_k P(x_k) \qquad \text{(A1.2.2)}$$

Similarly, the average value of x^2 would be

$$\langle x^2 \rangle = \sum_{k=1}^{n} x_k^2 P(x_k) \qquad \text{(A1.2.3)}$$

More generally, if $f(x_k)$ is a function of x, its average value would be

$$\langle f \rangle = \sum_{k=1}^{n} f(x_k) P(x_k)$$

If the variable x takes continuous values in the range (a, b), then the average values are written as integrals:

$$\langle x \rangle = \int_a^b x P(x)\,\mathrm{d}x \quad \langle f \rangle = \int_a^b f(x) P(x)\,\mathrm{d}x \qquad \text{(A1.2.4)}$$

For a given probability distribution, the **standard deviation** s is defined as

$$S = \sqrt{\langle (x - \langle x \rangle)^2 \rangle} \qquad \text{(A1.2.5)}$$

SOME COMMON PROBABILITY DISTRIBUTIONS

Binomial distribution. This is the probability distribution associated with two outcomes H and T (such as a coin toss) with probabilities p and $(1 - p)$. The probability that, in N trials, m are H and $(N - m)$ are T is given by

$$P(N, m) = \frac{N!}{m!(N - m)!} p^m (1 - p)^{N-m} \qquad (A1.2.6)$$

Poisson distribution. In many random processes the random variable is a number n. For example, the number of gas molecules in a small volume within a gas will vary randomly around an average value. Similarly, so is the number of molecules undergoing chemical reaction in a given volume per unit time. The probability of n in such processes is given by the Poisson distribution:

$$P(n) = e^{-\alpha} \frac{\alpha^n}{n!} \qquad (A1.2.7)$$

The Poisson distribution has one parameter, α; it is equal to the average value of n, i.e. $\langle n \rangle = \alpha$.

Gaussian distribution. When a random variable x is a sum of many variables, its probability distribution is generally a Gaussian distribution:

$$P(x)dx = \left(\frac{1}{2\pi\sigma^2} \right)^{1/2} \exp\left(-\frac{(x - x_0)^2}{2\sigma^2} \right) dx \qquad (A1.2.8)$$

The Gaussian distribution has two parameters, x_0 and σ. The average value of x is equal to x_0 and the standard deviation equals σ.

SOME USEFUL INTEGRALS

(a)
$$\int_0^\infty e^{-ax^2} \, dx = \frac{1}{2} \left(\frac{\pi}{a} \right)^{1/2}$$

(b)
$$\int_0^\infty x e^{-ax^2} \, dx = \frac{1}{a}$$

(c)
$$\int_0^\infty x^2 e^{-ax^2} \, dx = \frac{1}{4a} \left(\frac{\pi}{a} \right)^{1/2}$$

(d)
$$\int_0^\infty x^3 e^{-ax^2} \, dx = \frac{1}{2a^2}$$

More generally:

(e)
$$\int_0^\infty x^{2n}\, e^{-ax^2}\, dx = \frac{1\times 3\times 5\times \ldots \times (2n-1)}{2^{n+1}a^n}\left(\frac{\pi}{a}\right)^{1/2}$$

(f)
$$\int_0^\infty x^{2n+1}e^{-ax^2}dx = \frac{n!}{2}\left(\frac{1}{a^{n+1}}\right)$$

Appendix 1.3 Mathematica Codes

The following Mathematica codes show how to define functions, plot them using the 'Plot' command, create numerical text output files using the 'Export' command and do algebraic calculations and evaluate derivatives.

CODE A: EVALUATING AND PLOTTING PRESSURE USING THE EQUATION OF STATE

```
(* Values of a and b for CO2; We set N=1 *)
a=3.658;      (* L^2.bar.mol^-2*)
b=0.0429;     (* L.mol^-1*)
R=0.0831;     (* L.bar.K^-1.mol^-1 *)

PVW[V_,T_]:= (R*T/(V-b)) - (a/(V^2));
PID[V_,T_]:= R*T/V;

PID[1.5,300]
PVW[1.5,300]
TC=(8/27)*(a/(R*b))
16.62
15.4835
304.027
```

Using the functions defined above, p–V curves could be plotted using the following command:

```
Plot[{PVW[V,270],PVW[V,304],PVW[V,330]},{V,0.06,0.6}]
```

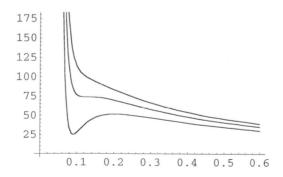

```
- Graphics -
```

To write output files for spreadsheets use the 'Export' command and the file format 'CSV'. For more detail, see the Mathematica help file for the 'Export' command. In the command below, the output filename is data.txt. This file can be read by most spreadsheets and graphing software.

```
Export["data.txt", Table[{x, PVW[x, 270], PVW[x, 304], PVW[x, 350]
     {x, 0.07, 0.6, 0.005}], "CSV"]
```

data.txt

```
Table[{x, PVW[x, 300]}, {x, 0.06, 0.1, 0.01}]//TableForm
0.06    441.784
0.07    173.396
0.08    100.405
0.09    77.6944
0.1     70.8025
```

CODE B: MATHEMATICA CODE FOR OBTAINING CRITICAL CONSTANTS FOR THE VAN DER WAALS EQUATION

```
Clear[a,b,R,T,V];
p[V_,T_]:=(R*T/(V-b)) -(a/V^2);

(* At the critical point the first and second derivatives of p
with respect to V are zero*)

(* First derivative *)
D[p[V,T],V]
```

$$\frac{2a}{V^3} - \frac{RT}{(-b+V)^2}$$

```
(* Second derivative *)
D[p[V,T],V,V]
```

$$-\frac{6a}{V^4} + \frac{2RT}{(-b+V)^3}$$

```
Solve[{(-6*a)/V^4 + (2*R*T)/(-b + V)^3==0,
(2*a)/V^3 - (R*T)/(-b + V)^2==0},{T,V}]
```

$$\left\{\left\{T \to \frac{8a}{27bR}, V \to 3b\right\}\right\}$$

Now we can substitute these values in the equation for p and obtain p_c.

```
T = (8*a)/(27*b*R); V = 3*b;
p[V,T]
```

$$\frac{a}{27b^2}$$

Thus, we have all the critical variables: $p_c = a/27b^2$, $T_c = 8a/27bR$, $V_c = 3b$.

CODE C: MATHEMATICA CODE FOR THE LAW OF CORRESPONDING STATES

```
Clear[a,b,R,T,V];
 T = Tr*(8*a)/(27*b*R);  V = Vr*3*b;  pc = a/(27*b^2);

 (* In terms of these variables the reduced pressure pr = p/pc. This can
now be calculated*)

 p[V,T]/pc
```

$$\frac{27b^2\left(-\dfrac{a}{9b^2Vr^2}+\dfrac{8aTr}{27b(-b+3bVr)}\right)}{a}$$

$$\text{FullSimplify}\left[\frac{27b^2\left(-\dfrac{a}{9b^2Vr^2}+\dfrac{8aTr}{27b(-b+3bVr)}\right)}{a}\right]$$

$$-\frac{3}{Vr^2}+\frac{8Tr}{-1+3Vr}$$

Thus, we have the following relation for the reduced variables, which is the law of corresponding states: $p_r = [8T_r/(3V_r - 1)] - 3/V_r^2$.

CODE D: THE MAXWELL–BOLTZMANN SPEED DISTRIBUTION FOR A GAS OF MASS M AT TEMPERATURE T CAN BE PLOTTED USING THE FOLLOWING CODE

```
Clear[a,b,R,T,V];
M = 28.0*10⁻³;  (*molar mass in kg*)  R = 8.314(*J/K.mol*)
b = M/(2*R);
```

$$p[v\_, T\_] := (4.0*\text{Pi})\left(\frac{M}{2*\text{Pi}*R*T}\right)^{3/2} v^2 * \text{Exp}\left[\frac{-b*v^2}{T}\right];$$

```
Plot[{p[v,300], p[v,100]}, {v,0,1500}]
```

8.314

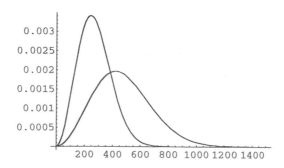

```
- Graphics -
```

References

1. Prigogine, I., Stengers, I., *Order Out of Chaos*. 1984, Bantam: New York.
2. Mach, E., *Principles of the Theory of Heat*. 1986, D. Reidel: Boston, MA.
3. Conant, J.B. (ed.), *Harvard Case Histories in Experimental Science*, Vol. 1. 1957, Harvard University Press: Cambridge, MA.
4. Mason, S.F., *A History of the Sciences*. 1962, Collier Books: New York.
5. Segrè, E., *From Falling Bodies to Radio Waves*. 1984, W.H. Freeman: New York; 188.

Examples

Example 1.1   The atmosphere consists of 78.08% by volume of N_2 and 20.95% of O_2. Calculate the partial pressures due to the two gases.

Solution   The specification 'percentage by volume' may be interpreted as follows. If the components of the atmosphere were to be separated, at the pressure of 1 atm, the volume occupied by each component is specified by the volume percent. Thus, if we isolate the N_2 in 1.000 L of dry air, at a pressure of 1 atm its volume will be 0.781 L. According to the ideal gas law, at a fixed pressure and temperature, the amount of gas $N = V(p/RT)$, i.e. the molar amount is proportional to the volume. Hence, percentage by volume is the same as percentage in N, i.e. 1.000 mol of air consists of 0.781 mol of N_2. According to the Dalton's law (see (1.4.5)), the partial pressure is proportional to N; so, the partial pressure of N_2 is 0.781 atm and that of O_2 is 0.209 atm.

Example 1.2   Using the ideal gas approximation, estimate the change in the total internal energy of 1.00 L of N_2 at $p = 2.00$ atm and $T = 298.15$ K if its temperature is increased by 10.0 K. What is the energy required to heat 1.00 mol of N_2 from 0.0 K to 298 K?

Solution   The energy of an ideal gas depends only on the amount of gas N and the temperature T. For a diatomic gas such as N_2 the energy per mole equals $(5/2)RT + U_0$. Hence, for N moles of N_2 the change in energy ΔU for a change in temperature from T_1 to T_2 is

$$\Delta U = N(5/2)R(T_2 - T_1)$$

In the above case

$$N = \frac{pV}{RT} = \frac{2.00 \text{ atm} \times 1.00 \text{ L}}{0.0821 \text{ L atm mol}^{-1} \text{ K}^{-1}(298.15 \text{ K})} = 8.17 \times 10^{-2} \text{ mol}$$

Hence:

$$\Delta U = (8.17 \times 10^{-2} \text{ mol})\frac{5}{2}(8.314 \text{ J mol}^{-1} \text{ K}^{-1})(10.0 \text{ K})$$
$$= 17.0 \text{ J}$$

(Note the different units of R used in this calculation.)

The energy required to heat 1.00 mol of N_2 from 0 K to 298 K is

$$(5/2)RT = (5/2)(8.314\,J\,K^{-1}\,mol^{-1})(298\,K) = 6.10\,kJ\,mol^{-1}$$

Example 1.3   At $T = 300\,K$, 1.00 mol of CO_2 occupies a volume of 1.50 L. Calculate the pressures given by the ideal gas equation and the van der Waals equation. (The van der Waals constants a and b can be obtained from Table 1.1.)
Solution   The ideal gas pressure is

$$p = \frac{1.00\,mol \times 0.0821\,atm\,L\,mol^{-1}\,K^{-1} \times 300\,K}{1.50\,L} = 16.4\,atm$$

The pressure according to the van der Waals equation is

$$p = \frac{NRT}{V - Nb} - a\frac{N^2}{V^2}$$

Since the van der Waals constants a and b given in Table 1.1 are in units of $L^2\,atm\,mol^{-2}$ and $L\,mol^{-2}$ respectively, we will use the value or $R = 0.0821\,atm\,L\,mol^{-1}\,K^{-1}$. This will give the pressure in atmospheres:

$$p = \frac{1.00(0.0821)300}{1.50 - 1.00(0.0421)} - 3.59\frac{1.00}{1.50^2} = 15.3\,atm$$

Exercises

1.1   Describe an experimental method, based on the ideal gas law, to obtain the molecular mass of a gas.

1.2   The density of dry air at $p = 1.0$ bar and $T = 300\,K$ is $1.161\,kg\,m^{-3}$. Assuming that it consists entirely of N_2 and O_2 and using the ideal gas law, determine the amount of each gas in moles in a volume of $1\,m^3$ and their mole fractions.

1.3   The molecule density of interstellar gas clouds is about 10^4 molecules/mL. The temperature is approximately 10 K. Calculate the pressure. (The lowest vacuum obtainable in the laboratory is about three orders of magnitude larger.)

1.4   A sperm whale dives to a depth of more than 1.5 km into the ocean to feed. Estimate the pressure the sperm whale must withstand at this depth. (Express your answer in atmospheres.)

46

1.5 (a) Calculate the amount of gas in moles per cubic meter of atmosphere at $p = 1$ tm and $T = 298$ K using the ideal gas equation.
(b) The atmospheric content of CO_2 is about 360 ppmv (parts per million by volume). Assuming a pressure of 1.00 atm, estimate the amount of CO_2 in a 10.0 km layer of the atmosphere at the surface of the Earth. The radius of the Earth is 6370 km. (The actual amount of CO_2 in the atmosphere is about 6.0×10^{16} mol.)
(c) The atmospheric content of O_2 is 20.946% by volume. Using the result in part (b), estimate the total amount of O_2 in the atmosphere.
(d) Life on Earth consumes about 0.47×10^{16} mol of O_2 per year. What percentage of the O_2 in the atmosphere does life consume in a year?

1.6 The production of fertilizers begins with the Haber processes, which is the reaction $3H_2 + N_2 \rightarrow 2NH_3$ conducted at about 500 K and a pressure of about 300 atm. Assume that this reaction occurs in a container of fixed volume and temperature. If the initial pressure due to 300.0 mol H_2 and 100.0 mol N_2 is 300.0 atm, what will the final pressure be? What will the final pressure be if initially the system contained 240.0 mol H_2 and 160.0 mol N_2? (Use the ideal gas equation even though the pressure is high.)

1.7 The van der Waals constants for N_2 are $a = 1.370\,L^2\,atm\,mol^{-2}$ and $b = 0.0387\,L\,mol^{-1}$. Consider 0.5 mol of $N_2(g)$ is in a vessel of volume 10.0 L. Assuming that the temperature is 300 K, compare the pressures predicted by the ideal gas equation and the van der Waals equation.
(a) What is the percentage error in using the ideal gas equation instead of the van der Waals equation?
(b) Keeping $V = 10.0$ L, use Maple/Mathematica to plot p versus N for $N = 1$ to 100, using the ideal gas and the van der Waals equations. What do you notice regarding the difference between the pressure predicted by the two equations?

1.8 For 1.00 mol of Cl_2 in a volume of 2.50 L, calculate the difference in the energy between U_{ideal} and U_{vw}. What is the percentage difference when compared with U_{ideal}?

1.9 (a) Using the ideal gas equation, calculate the volume of 1 mol of a gas at a temperature of 25 °C and a pressure of 1 atm. This volume is called the *Avogadro volume*.
(b) The atmosphere of Venus is 96.5% $CO_2(g)$. The surface temperature is about 730 K and the pressure is about 90 atm. Using the ideal gas equation, calculate the volume of 1 mol of $CO_2(g)$ under these conditions (Avogadro volume on Venus).
(c) Use Maple/Mathematica and the van der Waals equation to obtain the Avogadro volume on Venus and compare it (find the percentage difference) with the result obtained using the ideal gas equation.

1.10 The van der Waals parameter b is a measure of the volume excluded due to the finite size of the molecules. Estimate the size of a single molecule from the data in Table 1.1.

1.11 For the van der Waals equation, express the pressure as a power series in $1/V_m$. Using this expression, determine the Boyle temperature T_B at which $p \approx RT_B/V_m$.

1.12 For the Redlich–Kwong equation

$$p = \frac{RT}{V_m - b} - \frac{a}{\sqrt{T}} \frac{1}{V_m(V_m - b)}$$

show that there is a critical temperature above which there is no transition to a liquid state.

1.13 Though the van der Waals equation was a big improvement over the ideal gas equation, its validity is also limited. Compare the following experimental data with the predictions of the van der Waals equation for $1\,mol$ of CO_2 at $T = 40°C$. (*Source*: I. Prigogine and R. Defay, *Chemical Thermodynamics*. 1967, London: Longmans.)

p/atm	V_m/L mol^{-1}
1	25.574
10	2.4490
25	0.9000
50	0.3800
80	0.1187
100	0.0693
200	0.0525
500	0.0440
1000	0.0400

1.14 (a) Use Mathematica/Maple to plot the van der Waals p–V curves for Ar, N_2 and C_3H_8 using the data listed in Table 1.1 (see Appendix 1.3 for sample programs). In particular, compare the van der Waals curves for CO_2 and He and the ideal gas equation.

1.15 For CO_2, plot the compressibility factor $Z = pV_m/RT$ as function of the reduced pressure p_r for fixed reduced temperatures $T_r = 1.2$ and $T_r = 1.7$. Verify that the Z–p_r curves are the same for all van der Waals' gases. (This can be plotted using Parametric Plots.)

1.16 Using Table 1.1 and the relations (1.5.4) obtain the critical temperature T_c, critical pressure p_c and critical molar volume V_{mc} for CO_2, H_2 and CH_4. Write a *Maple/Mathematica* code to calculate the van der Waals constants a and b given T_c, p_c and V_{mc} for any gas.

1.17 (a) From the van der Waals equation, using (1.5.2) obtain (1.5.3) and (1.5.4). (These calculations may also be done using Mathematica/Maple). (b) Show that $Z_c = (p_c V_{mc}/RT_c) = 3/8$ a constant for all van der Waals gases.

1.18 Using Mathematica/Maple, obtain Equation (1.5.6) from (1.5.5).

1.19 For CO_2, plot p–V isotherms for the van der Waals and Redlich–Kwong equations on the same graph for $T = 200\,K$, $300\,K$ and $400\,K$. The table below lists some constants a and b for the Redlich–Kwong equation (*Source*: J.H. Noggle, *Physical Chemistry*. 1996, Harper Collins):

	a/bar L^2 mol^{-1} K$^{-1/2}$	b/L mol^{-1}
Ar	16.71	0.0219
CO_2	64.48	0.0296
O_2	17.36	0.0221

1.20 Show that the Lennard–Jones energy

$$U_{LJ}(r) = 4\varepsilon \left[\left(\frac{\sigma}{r} \right)^{12} - \left(\frac{\sigma}{r} \right)^{6} \right]$$

has a minimum value equal to $-\varepsilon$ at $r = 2^{1/6}\sigma$.

1.21 Estimate the average distance between molecules at $T = 300\,K$ and $p = 1.0\,atm$. (Hint: consider a cube of side $10\,cm$ in which the molecules occupy points on a three-dimensional cubic lattice.)

1.22 According to the Graham's law of diffusion, the rate of diffusion of gas molecules is inversely proportional to the square root of its mass. Explain why this is so using the kinetic theory of gases. How would you expect the diffusion coefficient to depend on the temperature?

1.23 (a) Using the integrals in Appendix 1.2, obtain the average speed (1.6.19) and kinetic energy (1.6.20) of a gas molecule.
(b) Using the Maxwell probability distribution $f(v)$, obtain the most probable speed of a molecule of molar mass M at a temperature T.

2 THE FIRST LAW OF THERMODYNAMICS

The Idea of Energy Conservation amidst New Discoveries

The concepts of kinetic energy, associated with motion, and potential energy, associated with conservative forces such as gravitation, were well known at the beginning of nineteenth century. For a body in motion, the conservation of the sum of kinetic and potential energy is a direct consequence of Newton's laws (Exercise 2.1). But this concept had no bearing on the multitude of thermal, chemical and electrical phenomena that were being investigated at that time. And, during the final decades of the eighteenth and initial decades of the nineteenth century, new phenomena were being discovered at a rapid pace.

The Italian physician Luigi Galvani (1737–1798) discovered that a piece of charged metal could make the leg of a dead frog twitch! The amazed public was captivated by the idea that electricity can generate life as dramatized by Mary Shelley (1797–1851) in her *Frankenstein*. Summarizing the results of his investigations in a paper published in 1791, Galvani attributed the source of electricity to animal tissue. But it was the physicist Alessandro Volta (1745–1827) who recognized that the 'galvanic effect' is due to the passage of electric current. In 1800, Volta went on to construct the so-called Volta's pile, the first 'chemical battery'; electricity could now be generated from chemical reactions. The inverse effect, the driving of a chemical reaction by electricity, was demonstrated by Michael Faraday (1791–1867) in the 1830s. The newly discovered electric current could also produce heat and light. To this growing list of interrelated phenomena, the Danish physicist Hans Christian Oersted (1777–1851) added the generation of magnetic field by an electrical current in 1819. In Germany, in 1822, Thomas Seebeck (1770–1831) (who helped Goethe in his scientific investigations) demonstrated that 'thermoelectric effect', the generation of electricity by heat. The well-known Faraday's law of induction, the generation of an electrical current by a changing magnetic field, came in 1831. All these discoveries presented a great web of interrelated phenomena in heat, electricity, magnetism and chemistry to the nineteenth-century scientists (Figure 2.1).

Soon, within the scientific community that faced this multitude of new phenomena, the idea that all these effects really represented the transformation of one indestructible quantity, 'the energy', began to take shape [1]. This law of conservation of energy is the First Law of thermodynamics. We will see details of its formulation in the following sections.

The mechanical view of nature holds that all energy is ultimately reducible to kinetic and potential energy of interacting particles. Thus, the law of conservation of energy may be thought of as essentially the law of conservation of the sum of

Introduction to Modern Thermodynamics   Dilip Kondepudi
© 2008 John Wiley & Sons, Ltd

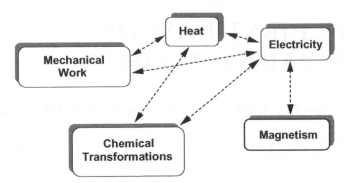

Figure 2.1   Interrelations between various phenomena discovered in the nineteenth century

kinetic and potential energies of all the constituent particles. A cornerstone for the formulation of the First Law is the decisive experiments of James Prescott Joule (1818–1889) of Manchester, a brewer and an amateur scientist. Here is how Joule expressed his view of conservation of energy [2, 3]:

Indeed the phenomena of nature, whether mechanical, chemical or vital, consist almost entirely in a continual conversion of attraction through space,* living force† and heat into one another. Thus it is that order is maintained in the universe – nothing is deranged, nothing ever lost, but the entire machinery, complicated as it is, works smoothly and harmoniously. And though, as in the awful vision of Ezekiel, '. . . wheel may be in the middle of wheel . . .', and everything may appear complicated and involved in the apparent confusion and intricacy of an almost endless variety of causes, effects, conversion, and arrangements, yet is the most perfect regularity preserved – the whole being governed by the sovereign will of God.

In practice, however, we measure energy in terms of heat and changes in macroscopic variables, such as chemical composition, electrical voltage and current, not the kinetic and potential energies of molecules. Energy can take many forms, e.g. mechanical work, heat, chemical energy, and it can reside in electric, magnetic and gravitational fields. For each of these forms we can specify the energy in terms of macroscopic variables, and the changes of energy in each form have a mechanical equivalent.

2.1   The Nature of Heat

Though the distinction between temperature and heat was recognized in the eighteenth century as a result of the work of Joseph Black and others, the nature of heat was not clearly understood until the middle of the nineteenth century. Robert Boyle,

---

* Potential energy.
† Kinetic energy.

James Prescott Joule (1818–1889) (Reproduced courtesy of the AIP Emilio Segre Visual Archive, Physics Today Collection)

Isaac Newton and others held the view that heat was the microscopic chaotic motion of particles. An opposing view, which prevailed in France, was that heat was an indestructible fluid-like substance without mass that was exchanged between material bodies. This indestructible substance was called **caloric** and it was measured in 'calories' (see Box 2.1). In fact, such figures as Antoine-Laurent Lavoisier (1743–1794), Jean Baptiste Joseph Fourier (1768–1830), Pierre-Simon de Leplace (1749–1827) and Siméon-Denis Poisson (1781–1840) all supported the caloric theory of heat. Even Sadi Carnot (1796–1832), in whose insights the Second Law originated, initially used the concept of caloric, though he later rejected it.

Box 2.1 Basic Definitions.

Heat can be measured by the change in temperature it causes in a body. In this text we shall mostly use the SI units in which heat is measured in joules, though the calorie is an often-used unit of heat.

The calorie. The calorie, a word derived from the caloric theory of heat, was origi-nally defined as the amount of heat required to increase the temperature of 1 g of water by 1 °C. When it was realized that this amount depended on the initial temperature of the water, the following definition was adopted: a calorie is the amount of heat required to increase the temperature of 1 g of water from 14.5 °C to 15.5 °C at a pressure of 1 bar. **The current practice is to define 1 cal as 4.184 J.** In fact, the International Union of Pure and Applied Chemistry (IUPAC) defines three types of calorie: the thermo-chemical calorie, $cal_{th} = 4.184$ J; the international calorie, $cal_{IT} = 4.1868$ J; the 15 °C calorie, $cal_{15} \approx 4.1855$ J.

Work and heat. In classical mechanics, when a body undergoes a displacement $d\mathbf{s}$ by a force $\mathbf{F}$, the mechanical work done $dW = \mathbf{F} \cdot d\mathbf{s}$. Work is measured in joules. Dis-sipative forces, such as friction between solids in contact, or viscous forces in liquids, convert mechanical energy to heat. Joule's experiments demonstrated that a certain amount of mechanical work, regardless of the manner in which it is performed, always produces the same amount of heat. Thus, an equivalence between work and heat was established.

Heat capacity. The heat capacity C of a body is the ratio of the heat absorbed dQ to the resulting increase in temperature dT:

$$C = \frac{dQ}{dT}$$

For a given dQ, the change in temperature dT depends on whether the substance is maintained at constant volume or at constant pressure. The corresponding heat capaci-ties are denoted by C_V and C_p respectively. Heat capacities are generally functions of temperature.

Molar heat capacity is the heat capacity of 1 mol of the substance. We shall denote it by C_{mV} or C_{mp}.

Specific heat of a substance is the heat required to change the temperature of a unit mass (usually 1.0 g or 1.0 kg) of the substance by 1 °C.

The true nature of heat as a form of energy that can interconvert to other forms of energy was established after much debate. One of the most dramatic demonstra-tions of the conversion of mechanical energy to heat was performed by Benjamin Thompson, an American born in Woburn, Massachusetts, whose adventurous life took him to Bavaria where he became Count Rumford (1753–1814) [4]. Rumford immersed metal cylinders in water and drilled holes in them. The heat produced due to mechanical friction could bring the water to a boil! He even estimated that the production of 1 cal of heat requires about 5.5 J of mechanical work [5].

It was the results of the careful experiments of James Prescott Joule, reported in 1847, that established beyond doubt that heat was not an indestructible substance, that, in fact, it can be transformed to mechanical energy and vice versa [5, 6]. Furthermore, Joule showed that there is an equivalence between heat and mechanical energy in the following sense: a certain amount of mechanical energy, regardless of the particular means of conversion, always produces the same amount of heat (4.184 J produce 1 cal of heat). This meant heat and mechanical energy can be thought of as different manifestations of the same physical quantity, the 'energy'.

But still, what is heat? One could say that physical and chemical processes have a natural tendency to convert all other forms of energy to heat. In the classical picture of particle motion, it is the kinetic energy associated with chaotic motion, as we saw in Chapter 1. Molecules in incessant motion collide and randomize their kinetic energy and the Maxwell–Boltzmann velocity distribution is quickly established; the average kinetic energy, which equals $3k_BT/2$, generally increases with absorption of heat. However, heat does not change the temperature of the body during phase transformations, but transforms the phase.

But that is not all we can say about heat. In additions to particles, we also have fields. The interaction between the particles is described by fields, such as electromagnetic fields. Classical physics had established that electromagnetic radiation was a physical quantity that can carry energy and momentum. So when particles gain or lose energy, some of it can transform into the energy of the field. The energy associated with electromagnetic radiation is an example. The interaction between matter and radiation also leads to a state of thermal equilibrium in which a temperature can be associated with radiation. Radiation in thermal equilibrium with matter is called 'heat radiation' or 'thermal radiation'. So heat can also be in the form of radiation. We shall study the thermodynamics of thermal radiation in some detail in Chapter 12.

During the twentieth century, our view of particles and fields has been unified by modern quantum field theory. According to quantum field theory, all particles are excitations of quantum fields. We now know, for example, that the electromagnetic field is associated with particles we call photons, though it also has a wave nature. Similarly, other fields, such as those associated with nuclear forces, have corresponding particles. Just as photons are emitted or absorbed by molecules undergoing a transition from one state to another (see Figure 2.2) – which in the classical picture corresponded to emission or absorption of radiation – other particles, such as mesons, can be absorbed and emitted by nuclear particles in high-energy processes. The energy density of thermal radiation depends only on the temperature.

One of the most remarkable discoveries of modern physics is that every particle has an antiparticle. When a particle encounters its antiparticle they may annihilate each other, converting their energy into other forms, such as photons. All this has expanded our knowledge of the possible states of matter. As mentioned above, the average kinetic energy of particles is proportional to temperature. At the temperatures we normally experience, collisions between molecules result in the emission

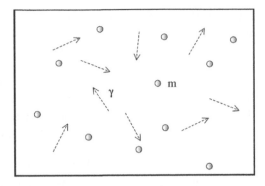

Figure 2.2 Classical picture of a gas of molecules (m) at low temperatures in equilibrium with radiation (γ)

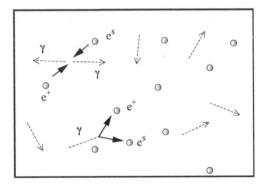

Figure 2.3 A gas of electrons (e⁻) and positrons (e⁺) in equilibrium with radiation (γ) at very high temperatures. At temperatures over 10^{10} K, particle–antiparticle pair creation and annihilation begins to occur and the total number of particles is no longer a constant. At these temperatures, electrons, positrons and photons are in a state called thermal radiation. The energy density of thermal radiation depends only on temperature

of photons, but not other particles. At sufficiently high temperatures (greater than 10^{10} K), other particles can also be similarly created as a result of collisions. Particle creation is often in the form of particle–antiparticle pairs (see Figure 2.3). Thus, there are states of matter in which there is incessant creation and annihilation of particle–antiparticle pairs, a state in which the number of particles does not remain constant. This state of matter is a highly excited state of a field. The notion of thermodynamic equilibrium and a temperature should apply to such a state as well.

Fields in thermal equilibrium can be more generally referred to as **thermal radiation**. One of the characteristic properties of thermal radiation is that its energy density is only a function of temperature; unlike the ideal gas, the number of particle of each kind itself depends on the temperature. 'Blackbody radiation', the study of which led Max Planck (1858–1947) to the quantum hypothesis, is thermal radiation associated with the electromagnetic field. At high enough temperatures, all particles (electrons and positrons, protons and anti-protons) can exist in the form of thermal radiation. Immediately after the big bang, when the temperature of the universe was unimaginably high, the state of matter in the universe was in the form of thermal radiation. As the universe expanded and cooled, the photons remained in the state of thermal radiation which can be associated with a temperature, but the protons, electrons and neutrons are no longer in that state. In its present state, the radiation that fills the universe is in an equilibrium state of temperature about 2.7 K, but the observed abundance of elements in the universe is not that expected in a state of thermodynamic equilibrium [7].

2.2   The First Law of Thermodynamics: The Conservation of Energy

As mentioned at the beginning of this chapter, though mechanical energy (kinetic energy plus potential energy) and its conservation was known from the time of Newton and Leibnitz, energy was not thought of as a general and universal quantity until the nineteenth century [5, 8].

With the establishment of the mechanical equivalence of heat by Joule, it became accepted that heat is form of energy that could be converted to work and vice versa. It was in the second half of the nineteenth century that the concept of *conservation* of energy was clearly formulated. Many contributed to this idea, which was very much 'in the air' at that time. For example, the law of 'constant summation of heats of reaction' formulated in 1840 by the Russian chemist Germain Henri Hess (1802–1850). This was essentially the law of energy conservation in chemical reactions. This law, now called Hess's law, is routinely used to calculate heats of chemical reactions.

It can be said that the most important contributions to the idea of conservation of energy as a universal law of nature came from Julius Robert von Mayer (1814–1878), James Prescott Joule (1818–1889) and Hermann von Helmholtz (1821–1894). Two of the landmarks in the formulation of the law of conservation of energy are a paper by Robert von Mayer titled 'Bermerkungen über die Kräfte der unbelebten Natur' ('Remarks on the forces of inanimate nature'), published in 1842, and a paper by Helmholtz titled 'Uber die Erhaltung der Kraft' ('On the conservation of force') that appeared in 1847 [5, 6].

The law of conservation of energy can be stated and utilized entirely in terms of macroscopic variables. A transformation of state may occur due to exchange of heat,

Hermann von Helmholtz (1821–1894) (Reproduced with permission from the Edgar Fahs Smith Collection, University of Pennsylvania Library)

performance of work and change in chemical composition and other such macroscopic processes. Each of these processes is associated with a change in energy, and the First Law of thermodynamics could be stated as:

When a system undergoes a transformation of state, the algebraic sum of the different energy changes, heat exchanged, work done, etc., is independent of the manner of the transformation. It therefore depends only on the initial and final states of the transformation.

For example, as shown in Figure 2.4, when N_k is constant, a transformation of volume and temperature of a gas mixture from the state O to the state X may occur via different paths, each following different intermediate volumes and temperatures. For each path, the total amount of heat exchanged and the mechanical work done

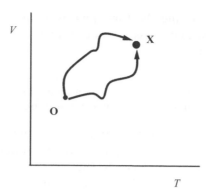

Figure 2.4 The change of energy U_x during a transformation from normal or reference state 'O' to the state 'X' is independent of the manner of transformation. In the figure, the state of a system is specified by its volume V and temperature T.

will be different. But, as the First Law states, the sum of the two will be the same, independent of the path. Since the total change in energy is independent of the path, the infinitesimal change dU associated with any transformation must depend solely on the initial and final states. An alternative way of stating this assertion is that in every cyclic process (closed path) that restores the system to its initial state, the integral of the energy change is zero:

$$\oint dU = 0 \qquad (2.2.1)$$

Equation (2.2.1) may also be considered a statement of the First Law. Since changes in U are independent of the transformation path, its change from a fixed state O to any final state X is entirely specified by X. The state X of many systems is specified by the state variables T, V and N_k. For such systems, if the value of U at the state O is arbitrarily defined as U_0, then U is a function of the state X:

$$U = U(T, V, N_k) + U_0 \qquad (2.2.2)$$

If more variables (such as electric or magnetic fields) are needed to specify the state of a system, then U will be a function of those variables as well. In this formulation, the energy U can only be defined up to an arbitrary additive constant. Its absolute value cannot be specified.

Yet another way of stating the First Law is as an 'impossibility', a restriction nature imposes on physical processes. For example, in Max Planck's treatise [9], the First Law is stated thus:

. . . it is in no way possible, either by mechanical, thermal, chemical, or other devices, to obtain perpetual motion, i.e. it is impossible to construct an engine which will work in a cycle and produce continuous work, or kinetic energy, from nothing (author's italics).

It is easy to see that this statement is equivalent to the above formulation summarized in Equation (2.2.1). We note again that this statement is entirely in macroscopic, operational terms and has no reference whatsoever to the microscopic structure of matter. The process described above is called *perpetual motion of the first kind*.

For a closed system, the energy exchanged with its exterior in a time dt may be divided into two parts: dQ, the amount of heat, and, dW, the amount of mechanical energy. Unlike the change in total energy dU, the quantities dQ and dW are *not independent* of the manner of transformation; we cannot specify dQ or dW simply by knowing the initial and final states because their values depend on the path or the process that causes the energy exchange. Hence, it is not possible to define a function Q that depends only on the initial and final states, i.e. heat is not a state function. While every system can be said to possess a certain amount of energy U, the same cannot be said of heat Q or work W. But there is no difficulty in specifying the amount of heat exchanged in a particular transformation. The process that causes the heat exchange enables us to specify dQ as the heat exchanged in a time interval dt.

Most introductory texts on thermodynamics do not include irreversible processes; they describe all transformations of state as idealized, infinitely slow changes. In that formulation, dQ cannot be defined in terms of a time interval dt because the transformation does not occur in finite time; in fact, classical thermodynamics does not contain time at all. This point is clearly stated in the well-known physical chemistry text by Alberty and Silbey [10]: 'Thermodynamics is concerned with equilibrium states of matter and has nothing to do with time'. It is a theory based solely on *states* with no explicit inclusion of irreversible *processes*, such as heat conduction. This poses a problem: because Q is not a state function, the heat exchanged dQ cannot be uniquely specified by initial and final states. To overcome this difficulty, an 'imperfect differential' dQ is defined to represent the heat exchanged in a transformation, a quantity that depends on the initial and final states *and* the path of transformation. In our approach we will avoid the use of imperfect differentials. The heat flow is described by processes that occur at a finite time and, with the assumption that the rate of heat flow is known, the heat exchanged dQ in a time dt is well defined. The same is true for the work dW. Idealized, infinitely slow reversible processes still remain useful for some conceptual reasons and we will use them occasionally, but we will not restrict our presentation to reversible processes as many texts do.

The total change in energy dU of a closed system in a time dt is

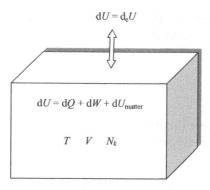

$$dU = d_e U$$

$$dU = dQ + dW + dU_{\text{matter}}$$

$$T \quad V \quad N_k$$

Figure 2.5 The law of conservation of energy: the total energy of an isolated system U remains a constant. The change in the energy dU of a system, in a time dt, can only be due to exchange of energy $d_e U$ with the exterior in the form of heat, mechanical work dW, and through the exchange of matter dU_{matter}. The energy change of the system is equal and opposite to that of the exterior

$$\boxed{dU = dQ + dW} \tag{2.2.3}$$

The quantities dQ and dW can be specified in terms of the rate laws for heat transfer and the forces that do the work. For example, the heat supplied in a time dt by a heating coil of resistance R carrying a current I is given by $dQ = VI\,dt = (I^2R)\,dt$, in which V is the voltage drop across the coil.

For open systems there is an additional contribution due to the flow of matter dU_{matter} (Figure 2.5):

$$dU = dQ + dW + dU_{\text{matter}} \tag{2.2.4}$$

Also, for open systems we define the volume not as the volume occupied by a fixed amount of substance, but by the boundary of the system, e.g. as a membrane. Since the flow of matter into and out of the system can be associated with mechanical work (as, for instance, the flow of molecules into the system through a semi-permeable membrane due to excess external pressure), dW is not necessarily associated with changes in the system volume. The calculation of changes in energy dU in open systems does not pose any fundamental difficulty. In any process, if changes in T, V and N_k can be computed, then the change in energy can be calculated. The total change in the energy can then be obtained by integrating $U(T, V, N_k)$ from the initial state A to the final state B:

$$\int_A^B dU = U_B - U_A \tag{2.2.5}$$

Because U is a state function, this integral is independent of the path.

Let us now consider some specific examples of exchange of energy in forms other than heat.

• For closed systems, if dW is the mechanical work due to a volume change, then we may write

$$dW_{mech} = -p\,dV \tag{2.2.6}$$

in which p is the pressure at the moving surface and dV is the change in volume (see Box 2.2).

• For transfer of charge dq across a potential difference ϕ

$$dU_q = \phi\,dq \tag{2.2.7}$$

• For dielectric systems, the change of electric dipole moment dP in the presence of an electric field E is associated with a change of energy

$$dU_{elect} = -E\,dP \tag{2.2.8}$$

• For magnetic systems, the change of magnetic dipole moment dM in the presence of a magnetic field B is associated with a change of energy

$$dU_{mag} = -B\,dM \tag{2.2.9}$$

• For a change of surface area $d\Sigma$ with an associated interfacial energy γ (interfacial energy per unit area)

$$dU_{surf} = \gamma\,d\Sigma \tag{2.2.10}$$

In general, the quantity dW is a sum of all the various forms of 'work', each term being a product of an intensive variable and a differential of an extensive variable.

Thus, in general, the change in the total internal energy may be written as

$$dU = dQ - p\,dV + \phi\,dq + E\,dP + \ldots \tag{2.2.11}$$

Box 2.2 Mechanical Work due to Change in Volume.

Mechanical work: $dW = \mathbf{F} \cdot d\mathbf{s}$

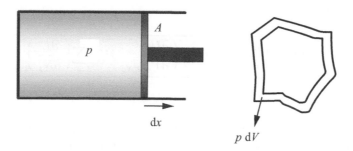

The force on the piston of area A due to a pressure p is pA. An expanding gas does work; hence, its energy decreases. The decrease in the energy when the gas pressure moves the piston by an amount dx is

$$dW = -pA\ dx = -p\,dV$$

in which dV is the change in volume of the gas. The negative sign is used to ensure that the energy of the gas decreases when V increases. By considering small displacements of the surface of a body at pressure p, the above expression for the work done by a gas can be shown to be generally valid.

ISOTHERMAL VOLUME CHANGE

By keeping a gas in contact with a reservoir at temperature T and slowly changing its volume, a constant-temperature or *isothermal* process can be realized. For such a process, the change in the energy of the gas equals the isothermal work given by the expression:

$$\text{Work} = \int_{V_i}^{V_f} -p\,dV = \int_{V_i}^{V_f} -\frac{NRT}{V}\,dV = -NRT \ln\left(\frac{V_f}{V_i}\right)$$

The negative sign indicates that an expanding gas transfers its energy to the exterior. During an isothermal expansion, flow of heat from the reservoir to the gas keeps T constant.

This change of energy of a system is a function of state variables such as T, V and N_k.

For systems undergoing chemical transformations, the total energy $U = U(T, V, N_k)$ may be expressed as a function of T, V and the molar amounts of the constituents N_k. As a function of T, V and N_k, total differential of U can be written as

$$\boxed{dU = \left(\frac{\partial U}{\partial T}\right)_{V,N_k} dT + \left(\frac{\partial U}{\partial V}\right)_{T,N_k} dV + \sum_k \left(\frac{\partial U}{\partial N_k}\right)_{V,T,N_{i\neq k}} dN_k}$$
$$= dQ + dW + dU_{matter} \tag{2.2.12}$$

The exact form of the function $U(T, V, N_k)$ for a particular system is obtained empirically. One way of obtaining the temperature dependence of U is the measurement of the *molar heat capacity* C_{mV} *at constant volume* (see Box 2.1 for basic definitions of heat capacity and specific heat). At constant volume, since no work is performed, $dU = dQ$. Hence:

$$C_{mV}(T,V) \equiv \left(\frac{dQ}{dT}\right)_{V=const} = \left(\frac{\partial U}{\partial T}\right)_{V,N=1} \tag{2.2.13}$$

If C_{mV} is determined experimentally (Box 2.3), then the internal energy $U(T, V)$ is obtained through integration of C_{mV}:

$$U(T,V,N) - U(T_0,V,N) = N \int_{T_0}^{T} C_{mV}(T,V)\, dT \tag{2.2.14}$$

in which T_0 is a the temperature of a reference state. If, for example, C_{mV} is independent of temperature and volume, as is the case for an ideal gas, then we have

$$U_{Ideal} = C_{mV} NT + U_0 \tag{2.2.15}$$

in which U_0 is an arbitrary additive constant. As noted earlier, U can only be defined up to an additive constant. For ideal monatomic gases $C_{mV} = (3/2)R$ and for diatomic gases $C_{mV} = (5/2)R$.

The notion of total internal energy is not restricted to homogeneous systems in which quantities such as temperature are uniform. For many systems, temperature is locally well defined but may vary with the position x and time t. In addition, the equations of state may remain valid in every elemental volume δV (i.e. in a small volume element defined appropriately at every point x) in which all the state variables are specified as densities. For example, corresponding to the energy $U(T, V, N_k)$ we may define the energy destiny $u(x, T)$, energy per unit volume, at the point x at time t, which can be expressed as a function the local temperature $T(x, t)$ and the

---

Box 2.3 Calorimetry.

Calorimeter. Heat evolved or absorbed during a transformation, such as a chemical reaction, is measured using a calorimeter. The transformation of interest is made to occur inside a chamber which is well insulated from the environment to keep heat loss to a minimum. To measure the heat generated by a process, first the heat capacity of the calorimeter should be determined. This is done by noting the increase in the temperature of the calorimeter due to a process for which the heat evolved is known. The heat produced by a current-carrying resistor, for example, is known to be I^2R joules per second, in which I is the current in amps and R is the resistance in ohms. (Using Ohm's law, $V = IR$, in which V is the voltage across the resistor in volts, the heat generated per second may also be written as VI.) If the heat capacity C_{cal} of the calorimeter is known, then one only needs to note the change in the temperature of the calorimeter to determine the heat generated by a process.

Calorimetry is widely used in present-day laboratories.

Bomb calorimeter. The heat of combustion of a compound is determined in a bomb calorimeter. In a bomb calorimeter, the combustion takes place in a chamber pressurized to about 20 atm with pure oxygen to ensure that the combustion is complete.

Isothermal calorimeter. In this type of calorimeter, the sample that absorbs or generates heat due to a physico-chemical process is maintained at a constant temperature using a sensitive heat exchanger that can record the amount of heat exchanged. This technique is highly developed and sensitive enough to measure enthalpy changes as low as a few nanojoules. It is a method widely used to study the thermodynamics of biological systems.

---

molar density $n_k(x, t)$ (moles of k per unit volume, also called number density) which in general are functions of both position x and time t:

$$u(x,t) = u(T(x,t), n_k(x,t)) \qquad (2.2.16)$$

The law of conservation of energy is a *local conservation law*: the change in energy in a small volume can only be due to a flow of energy into or out of the volume.

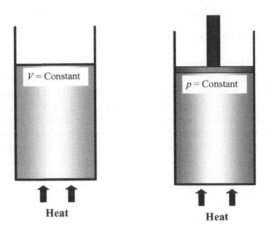

Figure 2.6 Molar heat capacity at constant pressure is larger than that at constant volume

Two spatially separated regions cannot exchange energy unless the energy passes through the region connecting the two parts.*

2.3   Elementary Applications of the First Law

RELATION BETWEEN C_{mP} AND C_{mV}

The First Law of thermodynamics leads to many simple and useful conclusions. It leads to a relation between the *molar heat capacities* at constant pressure C_{mp} and at constant volume C_{mV} (Figure 2.6 and Table 2.1). Consider a one-component substance. Then, using (2.2.3) and (2.2.6), and the fact that U is a function of the volume and temperature, the change in the energy dU can be written as

$$dU = dQ - p\,dV = \left(\frac{\partial U}{\partial T}\right)_V dT + \left(\frac{\partial U}{\partial V}\right)_T dV \qquad (2.3.1)$$

From this it follows that the heat exchanged by the gas can be written as

$$dQ = \left(\frac{\partial U}{\partial T}\right)_V dT + \left[p + \left(\frac{\partial U}{\partial V}\right)_T\right]dV \qquad (2.3.2)$$

---

*One might wonder why energy conservation does not take place non-locally, disappearing at one location and simultaneous appearing at another. Such conservation, it turns out, is not compatible with the theory of relativity. According to relativity, events that are simultaneous but occurring at different locations to one observer may not be simultaneous to another. Hence, the simultaneous disappearance and appearance of energy as seen by one observer will not be simultaneous for all. For some observers, energy would have disappeared at one location first and only some time later would it reappear at the other location, thus violating the conservation law during the time interval separating the two events.

Table 2.1 Molar heat capacities C_{mV} and C_{mp} for some substances at $T = 298.15$ K and $p = 1$ bar

Substance	C_{mp}/J mol^{-1} K^{-1}	C_{mV}/J mol^{-1} K^{-1}
Ideal monatomic	$(5/2)R$	$(3/2)R$
Ideal diatomic	$(7/2)R$	$(5/2)R$
Noble gases (He, Ne, Ar, Kr, Xe)	20.8	12.5
$N_2(g)$	29.17	20.82
$O_2(g)$	29.43	21.07
$CO_2(g)$	37.44	28.93
$H_2(g)$	28.83	20.52
$H_2O(l)$	75.33	
$CH_3OH(l)$	81.21	
$C_6H_6(l)$	132.9	
$Cu(s)$	24.47	
$Fe(s)$	25.09	

Source: P.J. Linstrom and W.G. Mallard (eds), *NIST Chemistry WebBook, NIST Standard Reference Database Number 69*, June 2005, National Institute of Standards and Technology, Gaithersburg, MD (http://webbook.nist.gov).

If the gas is heated at a constant volume, then, since no work is done, the change in the energy of the gas is entirely due to the heat supplied. Therefore:

$$C_V \equiv \left(\frac{dQ}{dT}\right)_V = \left(\frac{\partial U}{\partial T}\right)_V \tag{2.3.3}$$

On the other hand, if the gas is heated at constant pressure, then from (2.3.2) we have:

$$C_p \equiv \left(\frac{dQ}{dT}\right)_p = \left(\frac{\partial U}{\partial T}\right)_V + \left[p + \left(\frac{\partial U}{\partial V}\right)_T\right]\left(\frac{\partial V}{\partial T}\right)_p \tag{2.3.4}$$

Comparing (2.3.3) and (2.3.4), we see that C_V and C_p are related by

$$\boxed{C_p - C_V = \left[p + \left(\frac{\partial U}{\partial V}\right)_T\right]\left(\frac{\partial V}{\partial T}\right)_p} \tag{2.3.5}$$

The right-hand side of (2.3.5) is equal to the additional amount of heat required to raise the temperature in a constant-pressure, or 'isobaric', process to compensate for the energy expended due to expansion of volume.

Relation (2.3.5) is generally valid for all substances. For an ideal gas, it reduces to a simple form. As mentioned in Chapter 1 (see Equations (1.4.6) and (1.4.8)), the energy U is only a function of the temperature and is independent of the volume. Hence, in (2.3.5), $(\partial U/\partial V)_T = 0$; for *1 mol* of an ideal gas, since $pV = RT$, the remaining term $p(\partial V/\partial T)_p = R$. Therefore, for the molar heat capacities, (2.3.5) reduces to the simple relation

$$\boxed{C_{mp} - C_{mV} = R}$$

(2.3.6)

ADIABATIC PROCESSES IN AN IDEAL GAS

In an **adiabatic process**, the state of a system changes without any exchange of heat. Using the equation $dU = dQ - p\,dV$, we can write

$$dQ = dU + p\,dV = \left(\frac{\partial U}{\partial T}\right)_V dT + \left(\frac{\partial U}{\partial V}\right)_T dV + p\,dV = 0$$

(2.3.7)

For an ideal gas, since U is a function of temperature but not of volume, and because $(\partial U/\partial T)_V = NC_{mV}$, (2.3.7) reduces to

$$C_{mV} N\,dT + p\,dV = 0$$

(2.3.8)

If the change in volume occurs so that the ideal gas equation is valid during the process of change, then we have

$$C_{mV}dT + \frac{RT}{V}dV = 0$$

(2.3.9)

(For very rapid changes in volume, the relation between p, V and T may deviate from the ideal gas law.) But since $R = C_{mp} - C_{mV}$ for an ideal gas, we can write (2.3.9) as

$$\frac{dT}{T} + \frac{C_{mp} - C_{mV}}{C_{mV}V}dV = 0$$

(2.3.10)

Integration of (2.3.10) gives

$$\boxed{TV^{(\gamma-1)} = \text{const.}} \quad \text{where} \quad \boxed{\gamma = \frac{C_{mp}}{C_{mV}}}$$

(2.3.11)

Using $pV = NRT$, the above relation can be transformed into

$$\boxed{pV^{\gamma} = \text{const.}} \quad \text{or} \quad \boxed{T^{\gamma}p^{1-\gamma} = \text{const.}}$$

(2.3.12)

Thus, the First Law gives us (2.3.11) and (2.3.12), which characterize adiabatic processes in an ideal gas. Table 2.2 lists the ratio of heat capacities γ for some gases. We shall discuss adiabatic processes in real gases in Chapter 6.

Table 2.2 Ratios of molar heat capacities and speed of sound C_{sound} at $T = 298.15$ K and $p = 1$ bar

Gas	C_{mp}/J mol^{-1} K^{-1}	C_{mV}/J mol^{-1} K^{-1}	$\gamma = C_{mp}/C_{mV}$	C_{sound}/m s^{-1}
Ar(g)	20.83	12.48	1.669	321.7
CO_2(g)	37.44	28.93	1.294	268.6
H_2(g)	28.83	20.52	1.405	1315
He(g)	20.78	12.47	1.666	1016
N_2(g)	29.17	20.82	1.401	352.1
O_2(g)	29.43	21.07	1.397	328.7

Source: E.W. Lemmon, M.O. McLinden and D.G. Friend, Thermophysical properties of fluid systems. In *NIST Chemistry WebBook, NIST Standard Reference Database Number 69*, P.J. Linstrom and W.G. Mallard (eds), June 2005, National Institute of Standards and Technology, Gaithersburg MD, (http://webbook.nist.gov).

SOUND PROPAGATION

An example of an adiabatic process in nature is the rapid variations of pressure during the propagation of sound. These pressure variations, which are a measure of the sound intensity, are small. A measure of these pressure variations is p_{rms}, the root-mean-square value of the sound pressure with respect to the atmospheric pressure, i.e. p_{rms} is the square root of the average value of $(p - p_{atms})^2$. The unit for measuring sound intensity is the *bel* (B; named in honor of Alexander Graham Bell). The usual practice is to express the intensity in units of *decibels* (dB). The decibel can be expressed as a logarithmic measure of the pressure variations defined by

$$I = 10 \log_{10} \left(\frac{p_{rms}^2}{p_0^2} \right) \tag{2.3.13}$$

in which the reference pressure $p_0 = 2 \times 10^{-8}$ kPa ($= 2 \times 10^{-10}$ bar) roughly corresponds to audibility threshold in humans – an astoundingly low threshold, which in units of energy intensity equals a mere 2×10^{-12} W m^{-2}. The logarithmic scale is used because it corresponds roughly to the sensitivity of the ear. We normally encounter sound whose intensity is in the range 10–100 dB, corresponding to a pressure variations in the range 6×10^{-10} to 2×10^{-5} bar. These small variations of pressure for audible sound occur in the frequency range 20 Hz–20 kHz (music being in the range 40–4000 Hz).

Owing to the rapidity of pressure variations, hardly any heat is exchanged by the volume of air that is undergoing the pressure variations and its surroundings. It is essentially an adiabatic process. As a first approximation, we may assume that the ideal gas law is valid for these rapid changes. In introductory physics texts it is shown that the speed of sound C_{sound} in a medium depends on the bulk modulus B and the density ρ according to the relation

$$C_{sound} = \sqrt{\frac{B}{\rho}} \tag{2.3.14}$$

The bulk modulus $B = -\delta p/(\delta V/V)$, relates the relative change in the volume of a medium $\delta V/V$ due to a change in pressure δp; the negative sign indicates that for positive δp the change δV is negative. If the propagation of sound is an adiabatic process, in the ideal gas approximation, then the changes in volume and pressure are such that $pV^\gamma = $ constant. By differentiating this relation, one can easily see that the bulk modulus B for an adiabatic process is

$$B = -V\frac{\mathrm{d}p}{\mathrm{d}V} = \gamma p \qquad (2.3.15)$$

For an ideal gas of density ρ and molar mass M, we have

$$p = \frac{NRT}{V} = \frac{NM}{V}\frac{RT}{M} = \frac{\rho RT}{M}$$

Hence:

$$B = \frac{\gamma \rho RT}{M} \qquad (2.3.16)$$

Using this expression in (2.3.14) we arrive at the conclusion that, if the propagation of sound is an adiabatic process, the velocity C is given by

$$\boxed{C_{\text{sound}} = \sqrt{\frac{\gamma RT}{M}}} \qquad (2.3.17)$$

Experimental measurements of sound confirm this conclusion to a good approximation. The velocities of sound in some gases are listed in Table 2.2.

2.4 Thermochemistry: Conservation of Energy in Chemical Reactions

During the first half of the nineteenth century, chemists mostly concerned themselves with the analysis of compounds and chemical reactions and paid little attention to the heat evolved or absorbed in a chemical reaction. Though the early work of Lavoisier (1743–1794) and Laplace (1749–1827) established that heat absorbed in a chemical reaction was equal to the heat released in the reverse reaction, the relation between heat and chemical reactions was not investigated very much. The Russian chemist Germain Henri Hess (1802–1850) was rather an exception among the chemists of his time in regard to his interest in the heat released or absorbed by chemical reactions [11]. Hess conducted a series of studies in neutralizing acids and

Germain Henri Hess (1802–1850) (Reproduced with permission from the Edgar Fahs Smith Collection, University of Pennsylvania Library)

measuring the heats released (see Box 2.4). This, and several other experiments on the heats of chemical reactions, led Hess to his 'law of constant summation', which he published in 1840, 2 years before the appearance of Robert von Mayer's paper on the conservation of energy:

The amount of heat evolved during the formation of a given compound is constant, independent of whether the compound is formed directly or indirectly in one or in a series of steps [12].

Hess's work was not very well known for many decades after its publication. The fundamental contribution of Hess to thermochemistry was made known to chemists largely through Wilhelm Ostwald's (1853–1932) *Textbook of General Chemistry*,

---

Box 2.4 The Experiments of Germain Henry Hess.

Hess conducted a series of studies in which he first diluted sulfuric acid with different amounts of water and then neutralized the acid by adding a solution of ammonia. Heat was released in both steps. Hess found that, depending on the amount of water added during the dilution, different amounts of heat were released during the dilution and the subsequent neutralization with ammonia. However, the sum of the heats released in the two processes was found to be the same [11]. The following example, in which ΔH are the heats released, illustrates Hess's experiments:

$$1\text{L of } 2\text{M } H_2SO_4 \xrightarrow[\Delta H_1]{\text{Dilution}} 1.5\text{M } H_2SO_4 \xrightarrow[\Delta H_2]{NH_3 \text{ solution}} 3\text{L Neutral solution}$$

$$1\text{L of } 2\text{M } H_2SO_4 \xrightarrow[\Delta H_1']{\text{Dilution}} 1.0\text{M } H_2SO_4 \xrightarrow[\Delta H_2']{NH_3 \text{ solution}} 3\text{L Neutral solution}$$

Hess found that $\Delta H_1 + \Delta H_2 = \Delta H_1' + \Delta H_2'$ to a good approximation.

---

published in 1887. The above statement, known as **Hess's law**, was amply confirmed in the detailed work of Mercellin Berthelot (1827–1907) and Julius Thompsen (1826–1909) [13]. As we shall see below, Hess's law is a consequence of the law of conservation of energy and is most conveniently formulated in terms of a state function called *enthalpy*.

Hess's law refers to the heat evolved in a chemical reaction under constant (atmospheric) pressure. Under such conditions, a part of the energy released during the reaction may be converted to work $W = \int_{V_1}^{V_2} - p\,dV$, if there is a change in volume from V_1 to V_2. Using the basic equation of the First Law, $dU = dQ - Vdp$, the heat evolved ΔQ_p during a chemical transformation at a constant pressure can be written as

$$\Delta Q_p = \int_{U_1}^{U_2} dU + \int_{V_1}^{V_2} p\,dV = (U_2 - U_1) + p(V_2 - V_1) \qquad (2.4.1)$$

From this we see that the heat released can be written as a difference between two terms, one referring to the initial state (U_1, V_1) the other to the final state (U_2, V_2):

$$\Delta Q_p = (U_2 + pV_2) - (U_1 + pV_1) \qquad (2.4.2)$$

Since U, p and V are specified by the state of the system and are independent of the manner in which that state was reached, the quantity $U + pV$ is a state function, fully specified by the state variables. According to (2.4.2), the heat evolved ΔQ_p is the difference between the values of the function $(U + pV)$ at the initial and the final states. The state function $(U + pV)$ is called **enthalpy** H:

$$\boxed{H \equiv U + pV} \tag{2.4.3}$$

The heat released by a chemical reaction at constant pressure $\Delta Q_p = H_2 - H_1$. Since ΔQ_p depends only on the values of enthalpy at the initial and final states, it is independent of the 'path' of the chemical transformation, in particular if the transformation occurs 'directly or indirectly in one or in a series of steps', as Hess concluded.

As a simple example, consider the reaction

$$2P(s) + 5Cl_2(g) \rightarrow 2PCl_5(s) \tag{2.A}$$

in which 2 mol of P reacts with 5 mol of Cl_2 to produce 2 mol of PCl_5. In this reaction, 886 kJ of heat is released. The reaction can occur directly when an adequate amount of Cl_2 is present or it could be made to occur in two steps:

$$2P(s) + 3Cl_2(g) \rightarrow 2PCl_3(l) \tag{2.B}$$

$$2PCl_3(g) + 2Cl_2(g) \rightarrow 2PCl_5(s) \tag{2.C}$$

In reactions (2.B) and (2.C), for the molar quantities shown in the reaction, the heats evolved are 640 kJ and 246 kJ respectively. If the change in the enthalpies between the initial and final states of reactions (2.A), (2.B) and (2.C) are denoted by ΔH_{rA}, ΔH_{rB} and ΔH_{rC} respectively, then we have:

$$\Delta H_{rA} = \Delta H_{rB} + \Delta H_{rC} \tag{2.4.4}$$

The heat evolved or enthalpy change in a chemical reaction under constant pressure is usually denoted by ΔH_r and is called the **enthalpy of reaction**. The enthalpy of reaction is *negative for exothermic reactions* and it is *positive for endothermic reactions.*

The First Law of thermodynamics applied to chemical reactions in the form of Hess's law gives us a very useful way of predicting the heat evolved or absorbed in a chemical reaction if we can formally express it as a sum of chemical reactions for which the enthalpies of reaction are known. In fact, if we can assign a value for the enthalpy of 1 mol of each compound, then the heats of reactions can be expressed as the difference between sums of enthalpies of the initial reactants and the final products. In reaction (2.C), for example, if we can assign enthalpies for a mole of $PCl_3(g)$, $Cl_2(g)$ and $PCl_5(s)$, then the enthalpy of this reaction will be the difference between the enthalpy of the product $PCl_5(s)$ and the sum of the enthalpies of the reactants $PCl_3(g)$ and $Cl_2(g)$. But from the definition of enthalpy in (2.4.3) it is clear that it could only be specified with respect to a reference or normal state because U can only be defined in this way.

A convenient way to calculate enthalpies of chemical reactions at a specified temperature has been developed by defining a **standard molar enthalpy of formation** $\Delta H_f^0[X]$ for each compound X as described in Boxes 2.5 and 2.6.

Box 2.5 Basic Definitions of Standard States

Like the energy U, the quantitative specification of enthalpy and other thermodynamic quantities that we will discuss in later chapters can be done with reference to a **standard state** at a specified temperature T, standard pressure p^0, standard molality m^0 and standard concentration c^0. Though the choice of p^0, m^0 and c^0 depends on the system under consideration, the most common choice for tabulating data is

$$T = 298.15, \ p^0 = 1 \text{ bar} = 10^5 \text{ Pa}; \ m^0 = 1 \text{ mol kg}^{-1} \text{ and } c^0 = 1 \text{ mol dm}^{-3}$$

The standard state of a pure substance at particular temperature is its most stable state (gas, liquid or solid) at a pressure of 1 bar (10^5 Pa).

Notation used to indicate the standard state: g = gas; l = liquid; s = pure crystalline solid.

In a **gas phase**, the standard state of a substance, as a pure substance or as a component in a gas mixture, is the hypothetical state exhibiting ideal gas behavior at $p = 1$ bar. (Note that this definition implies that real gases at $p = 1$ bar are not in their standard state.*)

In a **condensed phase** (solid or liquid), the standard state of a substance, as a pure substance or as component of a mixture, is the state of the pure substance in the liquid or solid phase at the standard pressure p^0.

For a **solute** in a solution, the standard state is a hypothetical ideal solution of standard concentration c^0 at the standard pressure p^0. Notation used to indicate the standard state: ai = completely dissociated electrolyte in water; ao = undissociated compound in water.

---

*Since the energy of an ideal gas depends only on the temperature, the standard state energy and enthalpy of a gas depend only on the temperature. This implies that real gases at a pressure of 1 bar are not in their standard state; their energies and enthalpies differ from that of the standard state of a gas at that temperature. For a real gas, at a temperature T and pressure of 1 bar, the energy $U_{real}(T) = U^0_{ideal}(T) + \Delta U(T)$, in which $\Delta U(T)$ is the correction due to the nonideality of the gas; it is sometimes called 'internal energy imperfection'. Similarly, the enthalpy of a real gas at a temperature T and pressure of 1 bar is $H_{real}(T) = H^0_{ideal}(T) + \Delta H(T)$. The corrections, $\Delta U(T)$ and $\Delta H(T)$, are small, however, and they can be calculated using the equation of state such as the van der Waals equation.

Standard enthalpies of formation of compounds can be found in tables of thermodynamic data [14]. Using these tables and Hess's law, the standard enthalpies of reactions can be computed by viewing the reaction as 'dismantling' of the reactants to their constituent elements and recombining them to form the products. Since the enthalpy for the dismantling step is the negative of the enthalpy of formation, the enthalpy of the reaction

$$aX + bY \rightarrow cW + dZ \tag{2.4.5a}$$

for example, can be written as

$$\Delta H_r^0 = -a\Delta H_f^0[X] - b\Delta H_f^0[Y] + c\Delta H_f^0[W] + d\Delta H_f^0[Z] \tag{2.4.5b}$$

Box 2.6 Basic Definitions used in Thermochemistry

Standard reaction enthalpies at a specified temperature are reaction enthalpies in which the reactants and products are in their standard states.

 Standard molar enthalpy of formation $\Delta H_f^0[X]$ of a compound X, at a specified temperature T, is the enthalpy of formation of the compound X from its constituent elements in their standard state. Consider the example where X = $CO_2(g)$:

$$C(s) + O_2(g) \xrightarrow{\Delta H_f^0[CO_2(g)]} CO_2(g)$$

The enthalpies of formation of elements in their standard state are defined to be zero at any temperature.

 Thus, the enthalpies of formation $\Delta H_f^0[H_2]$, $\Delta H_f^0[O_2]$ and $\Delta H_f^0[Fe]$ are defined to be zero at all temperatures.

 The consistency of the above definition is based on the fact that in 'chemical reactions' elements do not transform among themselves, i.e. reactions between elements do not result in the formation of other elements (though energy is conserved in such a reaction).

Enthalpies of various chemical transformations are discussed in detail in later chapters and in the exercises at the end of this chapter.

 Though it is most useful in thermochemistry, the enthalpy H, as defined in (2.4.3), is a function of state that has a wider applicability. For example, we can see that the constant-pressure heat capacity C_p can be expressed in terms of H as follows. Since the heat exchanged in a process that takes place at constant pressure is equal to the change in the system's enthalpy:

$$dQ_p = dU + p\,dV = dH_p \tag{2.4.6}$$

in which the subscripts denote a constant-pressure process. If the system consists of 1 mol of a substance, and if the change in temperature due to the exchange of heat is dT, then it follows that

$$\left(\frac{dQ}{dT}\right)_p = C_{mp} = \left(\frac{\partial H_m}{\partial T}\right)_p \tag{2.4.7}$$

Also, in general, the change in enthalpy in a chemical reaction (not necessarily occurring at constant pressure) can be written as

$$\Delta H_r = H_f - H_i = (U_f - U_i) + (p_f V_f - p_i V_i)$$
$$= \Delta U_r + (p_f V_f - p_i V_i) \tag{2.4.8}$$

in which the subscripts 'i' and 'f' denote the initial and final states. In an isothermal processes occurring at temperature T, if all the gaseous components in the reaction can be approximated to be ideal gases and if the change in volume of the non-gaseous components can be neglected, then the changes of enthalpy and energy are related by

$$\Delta H_r = \Delta U_r + \Delta N_r RT \qquad (2.4.9)$$

in which ΔN_r is the change in the total molar amount of the gaseous reactants, a relation used in obtaining enthalpies of combustion using a bomb colorimeter.

VARIATION OF ENTHALPY WITH TEMPERATURE

Being a state function, enthalpy is a function of T. Using the relation (2.4.7), the dependence of enthalpy on T can be expressed in terms of the molar heat capacity C_{mp}:

$$H(T, p, N) - H(T_0, p, N) = N \int_{T_0}^{T} C_{mp}(T) dT \qquad (2.4.10)$$

Though the variation of C_{mp} with temperature is generally small, the following equation, called the Shomate equation, is often used:

$$C_{mp} = A + BT + CT^2 + DT^3 + \frac{E}{T^2} \qquad (2.4.11)$$

Values of the coefficients A, B, C, D and E for some gases are shown in Table 2.3.
From (2.2.14) and (2.4.10) it is clear that the temperature dependence of the total internal energy U and enthalpy H of any particular gas can be obtained if its heat capacity is known as a function of temperature. Sensitive calorimetric methods are available to measure heat capacities experimentally.
Using the relation (2.4.10), if the reaction enthalpy at a pressure p_0 (which could be the standard pressure $p^0 = 1$ bar) is known at one temperature T_0, then the reaction enthalpy at any other temperature T can be obtained if the molar heat capacities

Table 2.3 Values of constants A, B, C, D and E in (2.4.11) for some gases. The range of validity is 300 to 1200 K ($p = 1$ bar)

Gas	A/J mol^{-1} K^{-1}	B/10^{-3} J mol^{-1} K^{-2}	C/10^{-6} J mol^{-1} K^{-3}	D/10^{-9} J mol^{-1} K^{-4}	E/10^6 J mol^{-1} K
$O_2(g)$	29.66	6.137	−1.186	0.0958	−0.2197
$N_2(g)$	29.09	8.218	−1.976	0.1592	0.0444
$CO_2(g)$	24.99	55.19	−33.69	7.948	−0.1366

Source: P.J. Linstrom and W.G. Mallard (eds), *NIST Chemistry WebBook, NIST Standard Reference Database Number 69*, June 2005, National Institute of Standards and Technology, Gaithersburg, MD (http://webbook.nist.gov).

C_{mp} for the reactants and the products are known. The enthalpies of reactants or products X at the temperatures T and T_0 are related according to (2.4.10):

$$H_X(T, p_0, N_X) - H_X(T_0, p_0, N_X) = N_X \int_{T_0}^{T} C_{mp}(T)\,dT \qquad (2.4.12)$$

in which the subscript X identifies the reactants or products. Then, by subtracting the sum of the enthalpies of reactants from the sum of the enthalpies of the products (as shown in (2.4.5b)) we arrive at the following relation between the reaction enthalpies $\Delta H_r(T, p_0)$ and $\Delta H_r(T_0, p_0)$:

$$\Delta H_r(T, p_0) - \Delta H_r(T_0, p_0) = \int_{T_0}^{T} \Delta C_p(T)\,dT \qquad (2.4.13)$$

in which ΔC_p is the difference in the heat capacities of the products and the reactants. Thus, the $\Delta H_r(T, p_0)$ at any arbitrary temperature T can be obtained knowing $\Delta H_r(T_0, p_0)$ at a reference temperature T_0. Relation (2.4.13) was first noted by Gustav Kirchhoff (1824–1887), and is sometimes called **Kirchhoff's law**. The change in reaction enthalpy with temperature is generally small.

VARIATION OF ENTHALPY WITH PRESSURE

The variation of H with pressure, at a fixed temperature, can be obtained from the definition $H = U + pV$. Generally, H and U can be expressed as functions of p, T and N. For changes in H we have

$$\Delta H = \Delta U + \Delta(pV) \qquad (2.4.14)$$

At constant T_0, and N, in the ideal gas approximation, $\Delta H = 0$ for gases. This is because U and the product pV are functions only of temperature (see Chapter 1); hence $H = U + pV$ is a function only of T and is independent of pressure. The change in H due to a change in p is mainly due to intermolecular forces, and it becomes significant only for large densities. These changes in H can be calculated, for example, using the van der Waals equation.

For most solids and liquids, at a constant temperature, the total energy U does not change much with pressure. Since the change in volume is rather small unless the changes in pressure are very large, the change in enthalpy ΔH due a change in pressure Δp can be approximated by

$$\Delta H \approx V \Delta p \qquad (2.4.15)$$

A more accurate estimate can be made from a knowledge of the compressibility of the compound.

The First Law thus provides a powerful means of understanding the heats of chemical reactions. It enables us to compute the heats of reactions of an enormous number of reactions using the heats of formation of compounds at a standard temperature and pressure. The table entitled 'Standard Thermodynamic Properties' at the end of the book lists the standard heats of formation of some compounds. In addition, with a knowledge of heat capacities and compressibilities, the heats of reactions at any temperature and pressure can be calculated given those at a reference temperature and pressure.

COMPUTATION OF ΔH_r USING BOND ENTHALPIES

The concept of a chemical bond gives us a better understanding of the nature of a chemical reaction: it is essentially the breaking and making of bonds between atoms. The heat evolved or absorbed in a chemical reaction can be obtained by adding the heat absorbed in the breaking of bonds and the heat evolved in the making of bonds. The heat or enthalpy needed to break a bond is called the **bond enthalpy**.

For a particular bond, such as a C—H bond, the bond enthalpy varies from compound to compound, but one can meaningfully use an average bond enthalpy to estimate the enthalpy of a reaction. For example, the reaction $2H_2(g) + O_2(g) \rightarrow 2H_2O(g)$ can be written explicitly indicating the bonds as:

$$2(H—H) + O=O \rightarrow 2(H—O—H)$$

This shows that the reaction involves the breaking of two H—H bonds and one O=O bond and the making of four O—H bonds. If the bond enthalpy of the H—H bond is denoted by $\Delta H[H—H]$ etc., the reaction enthalpy ΔH_r may be written as

$$\Delta H_r = 2\Delta H[H—H] + \Delta H[O=O] - 4\Delta H[O—H]$$

This is a good way of estimating the reaction enthalpy of a large number of reactions using a relatively small table of average bond enthalpies. Table 2.4 lists some average bond enthalpies which can be used to estimate the enthalpies of a large number of reactions.

2.5 Extent of Reaction: A State Variable for Chemical Systems

In each chemical reaction, the changes in the mole numbers N_k are related through the stoichiometry of the reaction. In fact, *only one parameter is required to specify the changes in N_k resulting from a particular chemical reaction.* This can be seen as follows. Consider the elementary chemical reaction:

$$H_2(g) + I_2(g) \rightleftharpoons 2HI(g) \tag{2.5.1}$$

which is of the form

$$A + B \rightleftharpoons 2C \tag{2.5.2}$$

Table 2.4 Average bond enthalpies for some common bonds

	H	C	N	O	F	Cl	Br	I	S	P	Si
				Bond enthalpy/kJ mol^{-1}							
H	436										
C (single)	412	348									
C (double)		612									
C (triple)		811									
C (aromatic)		518									
N (single)	388	305	163								
N (double)		613	409								
N (triple)		890	945								
O (single)	463	360	157	146							
O (double)		743		497							
F	565	484	270	185	155						
Cl	431	338	200	203	254	252					
Br	366	276				219	193				
I	299	238				210	178	151			
S	338	259				250	212		264		
P	322									172	
Si	318		374								176

Source: L. Pauling, *The Nature of the Chemical Bond*. 1960, Cornell University Press: Ithaca.

In this case the changes in the molar amounts dN_A, dN_B and dN_C of the components A, B and C are related by the stoichiometry. We can express this relation as

$$\frac{dN_A}{-1} = \frac{dN_B}{-1} = \frac{dN_C}{2} \equiv d\xi \tag{2.5.3}$$

in which we have introduced a single variable $d\xi$ that expresses all the changes in the mole numbers due to the chemical reaction. This variable ξ introduced by Theophile De Donder [15, 16] is basic for the thermodynamic description of chemical reactions and is called the **extent of reaction** or **degree of advancement**. The **rate of conversion** (or reaction velocity) is the rate at which the extent of reaction changes with time:

$$\text{Rate of conversion (or reaction velocity)} = \frac{d\xi}{dt} \tag{2.5.4}$$

If the initial values of N_k are written as N_{k0}, then the values of all N_k during the reactions can be specified by the extent of reaction ξ:

$$N_k = N_{k0} + v_k\xi \tag{2.5.5}$$

in which v_k is the stoichiometric coefficient of the reacting component N_k. v_k is negative for reactants and positive for products. In this definition $\xi = 0$ for the initial state.

If the changes in the N_k in a system are due to chemical reactions, then the total internal energy U of such a system can be expressed in terms of the initial N_{k0}, which are constants, and the extents of reaction ξ_i defined for each of the reactions. For example, consider a system consisting of three substances A, B and C undergoing a single reaction (2.5.2). Then the molar amounts can be expressed as: $N_A = N_{A0} - \xi$, $N_B = N_{B0} - \xi$, and $N_C = N_{C0} + 2\xi$. The value of ξ completely specifies all the molar amounts N_A, N_B and N_C. Hence, the total energy U may be regarded as a function $U(T, V, \xi)$ with the understanding that the initial molar amounts N_{A0}, N_{B0} and N_{C0} are constants in the function U. If more than one chemical reaction is involved, then one extent of reaction ξ_i is defined for each independent reaction i and each mole number is specified in terms of the extents of reaction of all the chemical reactions in which it takes part. Clearly, the ξ_i are state variables, and internal energy can be expressed as a function of T, V and ξ_i: $U(T, V, \xi_i)$.

In terms of the state variables T, V, and ξ_i, the total differential of U becomes

$$dU = \left(\frac{\partial U}{\partial T}\right)_{V,\xi_k} dT + \left(\frac{\partial U}{\partial V}\right)_{T,\xi_k} dV + \sum_k \left(\frac{\partial U}{\partial \xi_k}\right)_{V,T,\xi_{i \neq k}} d\xi_k \qquad (2.5.6)$$

Using the First Law, the partial derivatives of U can be related to 'thermal coefficients' which characterize the system's response to heat under various conditions. Consider a system with one chemical reaction. We have one extent of reaction ξ. Then, by using the First Law:

$$dQ - p\,dV = dU = \left(\frac{\partial U}{\partial T}\right)_{V,\xi} dT + \left(\frac{\partial U}{\partial V}\right)_{T,\xi} dV + \left(\frac{\partial U}{\partial \xi}\right)_{T,V} d\xi \qquad (2.5.7)$$

which can be written as

$$dQ = \left(\frac{\partial U}{\partial T}\right)_{V,\xi} dT + \left[p + \left(\frac{\partial U}{\partial V}\right)_{T,\xi}\right] dV + \left(\frac{\partial U}{\partial \xi}\right)_{T,V} d\xi \qquad (2.5.8)$$

Just as the partial derivative $(\partial U/\partial T)_V$ has the physical meaning of being the heat capacity at constant volume C_V, the other derivatives, called thermal coefficients, can be related to experimentally measurable quantities. The derivative $r_{T,V} \equiv (\partial U/\partial \xi)_{V,T}$, for example, is the amount of *heat evolved* per unit change in the extent of reaction (one equivalent of reaction) at constant V and T. If it is negative, then the reaction is exothermic; if it is positive, then the reaction is endothermic. Just as we derived the relation (2.3.6) between the thermal coefficients C_p and C_V, one can derive several interesting relations between these thermal coefficients as a consequence of the First Law [17].

Also, since the extent of reaction is a state variable, the enthalpy of a reacting system can be expressed as function of the extent of reaction:

$$H = H(p,T,\xi) \tag{2.5.9}$$

The heat of reaction per unit change of ξ, which we shall denote as $h_{p,T}$, is the derivative of H with respect to ξ:

$$\boxed{h_{p,T} = \left(\frac{\partial H}{\partial \xi}\right)_{p,T}} \tag{2.5.10}$$

2.6 Conservation of Energy in Nuclear Reactions and Some General Remarks

At terrestrial temperatures, transformations of states of matter are mostly chemical, radioactivity being an exception. Just as molecules collide and react at terrestrial temperatures, at very high temperatures that exceed 10^6 K, typical of temperatures attained in the stars, nuclei collide and undergo nuclear reactions. At these temperatures, the electrons and nuclei of atoms are completely torn apart. Matter turns into a state that is unfamiliar to us and the transformations that occur are between nuclei, which is why it is called 'nuclear chemistry'.

All the elements heavier than hydrogen on our and other planets are a result of nuclear reactions, generally referred to as nucleosynthesis, that occurred in the stars [18]. Just as we have unstable molecules that dissociate into other more stable molecules, some of the nuclei that were synthesized in the stars are unstable and they disintegrate: these are the 'radioactive' elements. The energy released by radioactive elements turns into heat, and this is a source of heat for the Earth's interior. For example, the natural radioactivity in granite due to ^{238}U, ^{235}U, ^{232}Th and ^{40}K produces a small amount of heat equal to about 5μcal per gram of granite per year; however, accumulation of such heat over billions of years in the interior of the Earth is significant.

Under special circumstances, nuclear reactions can occur on the Earth, as in the case of nuclear fission of uranium and in the nuclear fusion of hydrogen in special reactors. Nuclear reactions release vastly greater amounts of energy than chemical reactions. The energy released in a nuclear reaction can be calculated from the difference in the rest mass of the reactants and the products using the famous relation $E^2 = p^2c^2 + m_0^2c^4$, derived by Einstein, in which E is the total energy of a particle, p is its momentum, m_0 its rest mass and c is the velocity of light *in vacuo*. If the total rest mass of the products is lower than total rest mass of the reactants, then the difference in energy due to change in the rest mass turns into the kinetic energy of the products. This excess kinetic energy turns into heat due to collisions. If the difference in the kinetic energy of the reactants and products is negligible, then the heat released $\Delta Q = \Delta m_0 c^2$, in which Δm_0 is the difference in the rest mass between the reactants and the products. In nuclear fusion, two deuterium nuclei ^{2}H can combine to form a helium nucleus and a neutron is released:

$$^2H + {^2}H \rightarrow {^3}He + n$$

$$\begin{aligned}\Delta m_0 &= 2(\text{mass of } {^2}H) - (\text{mass of } {^3}He + \text{mass of n}) \\ &= 2(2.0141)\,\text{amu} - (3.0160 + 1.0087)\,\text{amu} \\ &= 0.0035\,\text{amu}\end{aligned}$$

where amu stands for atomic mass unit. Since $1\,\text{amu} = 1.6605 \times 10^{-27}\,\text{kg}$, when $2\,\text{mol}$ of ^{2}H react to produce $1\,\text{mol}$ of ^{3}He and $1\,\text{mol}$ of n, the difference in mass $\Delta m_0 = 3.5 \times 10^{-6}\,\text{kg}$. The corresponding heat released is:

$$\Delta E = \Delta m_0 c^2 = 3.14 \times 10^8\,\text{kJ mol}^{-1}$$

If a nuclear process occurs at constant pressure, then the heat released is equal to the enthalpy and all the thermodynamic formalism that applies to the chemical reactions also applies to nuclear reactions. Needless to say, in accordance with the First Law, Hess's law of additivity of reaction enthalpies is also valid for nuclear reactions.

GENERAL REMARKS

Thermodynamically, energy is only defined up to an additive constant. In physical processes, it is only the change in energy (2.2.11) that can be measured, which leaves the absolute value of energy undetermined. With the advent of the theory of relativity, which has given us the relation between rest mass, momentum and energy, $E^2 = p^2 c^2 + m_0^2 c^4$, the definition of energy has become as absolute as the definition of mass and momentum. The absolute value of the energy of elementary particles can be used to describe matter in the state of thermal radiation that we discussed in Section 2.1.

The conservation of energy has become the founding principle of physics. During the early days of nuclear physics, studies of β radiation, or 'β decay' as it is often called, showed initially that the energy of the products was not equal to the energy of the initial nucleus. This resulted in some reexamination of the law of conservation of energy, with some physicists wondering if it could be violated in some processes. Asserting the validity of the conservation of energy, Wolfgang Pauli (1900–1958) suggested in 1930 that the missing energy was carried by a new particle which interacted extremely weakly with other particles and, hence, was difficult to detect. This particle later acquired the name neutrino. Pauli was proven right 25 years later. Experimental confirmation of the existence of the neutrino came in 1956 from the careful experiments conducted by Frederick Reines and (the now late) Clyde Cowen. Since then our faith in the law of conservation of energy has become stronger than ever. Frederick Reines received the Physics Noble Prize in 1995 for the discovery of the elusive neutrino; for the interesting history behind the discovery of the neutrino, see Ref. [19].

2.7   Energy Flows and Organized States

In Nature, the role of energy is much more than just a conserved quantity: *energy flows* are crucial to life, ecosystems and human activity that we call 'economy'. It could be said that energy flows have a creative role, in that out of these flows emerge complex processes that range from global biogeochemical cycles to the photosynthetic bacteria. In this section, we present a brief introduction to energy flow and some of its consequences. We will discuss more about nonequilibrium systems which become organized spontaneously in Chapter 11.

SELF-ORGANIZATION

At the outset we must note that what is of interest to us in a thermodynamic system is not only its *state*, but also the *processes* that take place in it and the way the system interacts with its exterior. The state of thermodynamic equilibrium is static, devoid of processes: in this state there is no flow of energy or matter from one point to another and no chemical change takes place. When a system is driven out of equilibrium by energy and matter flows, however, *irreversible* processes begin to emerge within the system. These processes are 'irreversible' in that the transformations they cause have a definite direction. Heat conduction is an example of an irreversible process: heat always flows towards a region at a lower temperature, never in the opposite direction. The concept of entropy, which will be introduced in the following chapters, makes the notion of irreversibility more precise; but even without the concept of entropy, one can see through simple examples how irreversible processes can create structure and organization in a system. One such example involving heat flow is illustrated in Figure 2.7. It consists of a fluid placed between two metal plates.

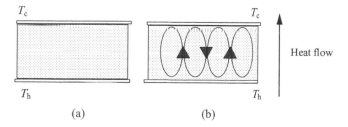

(a)   (b)

Figure 2.7   Energy flows can cause self-organized patterns to emerge. A fluid is placed between two plates and heated from below. The temperature difference $\Delta T = T_h - T_c$ between the two plates drives a heat flow. (a) When ΔT is small, the heat flow is due to conduction and the fluid is static. (b) When ΔT exceeds a critical value, organized convection patterns emerge spontaneously

The lower plate is maintained at a temperature T_h, which is higher than that of the upper plate temperature T_c. The temperature difference will cause a flow of heat through the fluid. If the difference in the temperature $\Delta T = (T_h - T_c)$ is increased, there is a point at which a well-organized pattern of convection cells emerges. The threshold value of ΔT depends on the fluid properties, such as thermal expansion coefficient and viscosity. What is remarkable about this familiar convection pattern is that it emerges entirely out of chaotic motion associated with heat. Furthermore, the fluid's organized convection pattern now serves a 'function': it increases the rate of heat flow. This is an example in which the energy flow drives a system to an organized state which in turn increases the energy flow.

The convection pattern exists as long as there is heat flow; if the heat flow is stopped, then the system evolves to equilibrium and the pattern disappears. Such patterns in nonequilibrium systems should be distinguished from patterns that we might see in a system in equilibrium, such as layers of immiscible fluids separated by differences in density. Organized states in nonequilibrium systems are maintained by flow of energy and matter and, as we shall see in later chapters, production of entropy.

In the formulation of modern thermodynamics, flows of matter and energy are thermodynamic flows. The laws that govern them can be formulated in thermodynamic terms, as we shall see in later chapters. Empirical laws governing heat flow and radiative cooling have been known for centuries. Some commonly used laws are summarized in Box 2.7. These laws can be used to analyze heat flows in various systems.

PROCESS FLOWS

An important application of the First Law is to the analysis of energy flows associated with fluids in industrial processes and engines. Energy flows in this case include the kinetic and potential energies of the fluid in addition to the thermodynamic internal energy U. The First Law applies to the total energy

$$E = U + \frac{1}{2}Mv^2 + \Psi \tag{2.7.1}$$

in which M is the mass of the system and Ψ is its potential energy. In describing energy flows, it is convenient to use energy and mass densities:

Internal energy density u $(\mathrm{J\,m^{-3}})$
Mass density ρ $(\mathrm{kg\,m^{-3}})$

When the change in potential energy is insignificant, the energy flowing in and out of a system is in the form of heat, mechanical work and kinetic and internal energy of matter (Figure 2.8). Let us assume that matter with energy density u_i is flowing into the system under a pressure p_i and velocity v_i. Consider a displacement $dx_i = v_i\,dt$ of the matter flowing in through an inlet with area of cross-section A_i, in time dt

Box 2.7 Laws of Heat Flow

Heat flow or heat current $\mathbf{J_q}$ is defined as the amount of heat flowing per unit surface area per unit time.

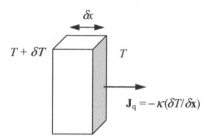

CONDUCTION

Jean Baptiste Joseph Fourier (1768–1830) proposed a general law of heat conduction in his 1807 memoir which states that the heat current is proportional to the gradient of temperature:

$$\mathbf{J_q} = -k\nabla T = -k\left(\hat{\mathbf{i}}\frac{\partial T}{\partial x} + \hat{\mathbf{j}}\frac{\partial T}{\partial y} + \hat{\mathbf{k}}\frac{\partial T}{\partial z}\right)$$

in which k $(\mathrm{W\,m^{-1}\,K^{-1}})$ is the thermal conductivity and $\hat{\mathbf{i}}$, $\hat{\mathbf{j}}$ and $\hat{\mathbf{k}}$ are unit vectors. The SI units of $\mathbf{J_q}$ are $\mathrm{W\,m^{-2}}$.

CONVECTION

A law of cooling due to convection, attributed to Newton, states that the rate of heat loss $\mathrm{d}Q/\mathrm{d}t$ of a body at temperature T surrounded by a fluid at temperature T_0 is proportional to the difference $(T - T_0)$ and the body's surface area A:

$$\frac{\mathrm{d}Q}{\mathrm{d}t} = -hA(T - T_0)$$

in which h $(\mathrm{W\,m^{-2}\,K^{-1}})$ is the heat transfer coefficient. This law is a good approximation when heat loss is mainly due to convection and that due to radiation can be ignored.

RADIATION

When thermal equilibrium between matter and radiation (Chapter 12) is analyzed, it is found that heat radiated by a body at temperature T is proportional to T^4 and the heat radiation absorbed from the surroundings at temperature T_0 is proportional to T_0^4. The net radiative heat loss is equal to

$$\frac{\mathrm{d}Q}{\mathrm{d}t} = -\sigma e A(T^4 - T_0^4)$$

in which $\sigma = 5.67\ 10^{-8}\,\mathrm{W\,m^{-2}\,K^{-4}}$ is the Stefan–Boltzmann constant, A is the surface area of the body and e is the body's emissivity (the maximum $e = 1$ for a black body). At high temperatures, the cooling of bodies is due to convective and radiative heat losses.

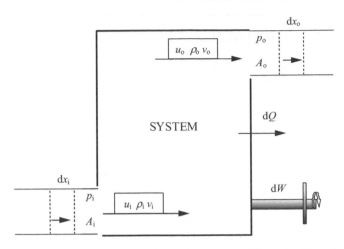

Figure 2.8 Energy flow through a system. The subscripts i and o identify the variables for the inflow and outflow respectively. p is the pressure, A is the area of cross-section, u is the energy density, ρ is the mass density and v is the flow velocity. The change in the energy of the system in a time dt is given by (2.7.3), in which h is the enthalpy density

(see Figure 2.8). The amounts of the various forms of energy entering the system through the inlet due to this displacement are:

Internal energy $u_i A_i \, dx_i$
Kinetic energy $(1/2)\rho v_i^2 A_i \, dx_i$
Mechanical work $p_i A_i \, dx_i$

Similar considerations apply for the energy flowing through the outlet, for which we use the subscript 'o' instead of 'i'. In addition, we assume, in time dt, there is a net heat output dQ and mechanical energy output dW that are not associated with matter flow (which are positive for a net output and negative for a net input). Taking all these into consideration, we see the total change in energy dU in a time dt is

$$dU = -dW - dQ + u_i A_i \, dx_i + \frac{1}{2}\rho_i v_i^2 A_i dx_i + p_i A_i dx_i - u_o A_o dx_o - \frac{1}{2}\rho_o v_o^2 A_o dx_o$$
$$- p_o A_o \, dx_o \qquad (2.7.2)$$

By defining enthalpy per unit volume $h = u + p$, and by noting that $dx_i = v_i dt$ and $dx_o = v_o dt$, the above expression can be written as

$$\boxed{\frac{dU}{dt} = -\frac{dW}{dt} - \frac{dQ}{dt} + \left(h_i + \frac{1}{2}\rho_i v_i^2\right) A_i v_i - \left(h_o + \frac{1}{2}\rho_o v_o^2\right) A_o v_o} \qquad (2.7.3)$$

In many situations, the system may reach a *steady state* in which all its thermodynamic quantities are constant, i.e. $dU/dt = 0$. Also, in such a state, the mass of matter flowing into the system is equal to the mass flowing out. Since the mass of matter flowing into and out of the system in time dt is $(\rho_i v_i A_i)\, dt$ and $(\rho_o v_o A_o)\, dt$ respectively, we have $(\rho_i v_i A_i)\, dt = (\rho_o v_o A_o)\, dt$ for a steady state. Hence, we can rewrite (2.7.3) as

$$\boxed{\frac{dW}{dt} + \frac{dQ}{dt} = \left[\left(\frac{h_i}{\rho_i} + \frac{v_i^2}{2}\right) - \left(\frac{h_o}{\rho_o} + \frac{v_o^2}{2}\right)\right]\frac{dm}{dt}} \qquad (2.7.4)$$

in which $dm/dt = \rho_i v_i A_i = \rho_o v_o A_o$ is the mass flow rate. In this expression, dW/dt is the rate of work output and dQ/dt is the rate of heat output. Thus, we obtain a relation between the work and heat output and the change of enthalpy densities and the kinetic energy of the matter flowing through the system.

This general equation can be applied to various situations. In a steam turbine, for example, steam enters the system at a high pressure and temperature and leaves the system at a lower pressure and temperature, delivering its energy to the turbine, which converts it to mechanical energy. In this case the heat output dQ/dt is negligible. We then have

$$\frac{dW}{dt} = \left[\left(\frac{h_i}{\rho_i} + \frac{v_i^2}{2}\right) - \left(\frac{h_o}{\rho_o} + \frac{v_o^2}{2}\right)\right]\frac{dm}{dt} \qquad (2.7.5)$$

The ratio h/ρ is the **specific enthalpy** (enthalpy per unit mass) and its values at a given pressure and temperature are tabulated in 'steam tables'. The term dm/dt $(\mathrm{kg\,s^{-1}})$ is the rate of mass flow through the system in the form of steam and, in many practical situations, the term $(v_i^2 - v_o^2)$ is small compared with the other terms in (2.7.5). Thus, the rate of work output in a turbine is related to the rate of steam flow and the specific enthalpy through the simple relation

$$\frac{dW}{dt} \approx \left[\left(\frac{h_i}{\rho_i}\right) - \left(\frac{h_o}{\rho_o}\right)\right]\frac{dm}{dt} \qquad (2.7.6)$$

Using steam tables, the expected work output can be calculated. Note that if there are heat losses, i.e. if $dQ/dt > 0$, then the work output dW/dt is correspondingly less. Explicit examples of the application of (2.7.4) are given at the end of this chapter.

SOLAR ENERGY FLOW

In discussing energy flows, it is useful to have a quantitative idea of the energy flows on a global scale. Figure 2.9 summarizes the flows that result from the solar energy incident on the Earth. The energy from the sun traverses $150 \times 10^6\,\mathrm{km}$ (93 million miles) before it reaches the Earth's atmosphere. The amount of solar energy reaching the Earth (called the 'total solar radiance') is about $1300\,\mathrm{W\,m^{-2}}$ which amounts to a total of about $54.4 \times 10^{20}\,\mathrm{kJ\,year^{-1}}$. About 30% of this energy is reflected back into space by clouds and other reflecting surfaces, such as snow. A significant fraction of the solar energy entering the Earth's surface goes to drive the water cycle, the

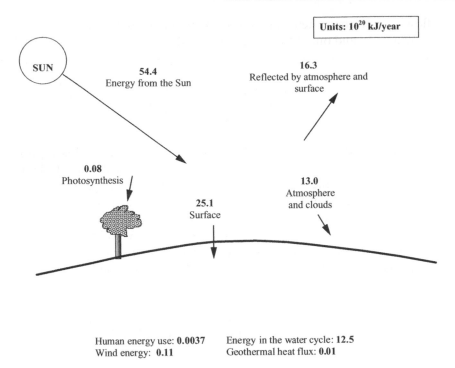

Figure 2.9   Annual solar energy flow through the Earth's atmosphere and the surface. (*Source*: T.G. Spiro and W.M. Stigliani, *Chemistry of the Environment*, second edition. 2003, Prentice Hall: Upper Saddle River, NJ)

evaporation and condensation as rain (Exercise 2.20). Of the solar energy not lost due to reflection, it is estimated that only a small fraction, about 0.08×10^{20} kJ year^{-1}, or 0.21%, goes into the biosphere through photosynthesis. The energy consumed by human economies is estimated to be about 0.0037×10^{20} kJ year^{-1}, which is about 5% of the energy that flows into photosynthesis. The interior of the Earth also has a vast amount of geothermal energy that flows to the surface at a rate of about 0.01×10^{20} kJ year^{-1}. The solar energy entering the Earth system is ultimately radiated back into space, and the total energy contained in the crust and the atmosphere is essentially in a steady state which is not a state of thermodynamic equilibrium. This flow of 38.1×10^{20} kJ year^{-1} drives the winds, creates rains and drives the cycle of life.

ENERGY FLOWS IN BIOLOGICAL SYSTEMS

The process of life is impossible without the flow of energy. Energy enters the biosphere through photosynthesis, which results in the production of biochemical compounds, such as carbohydrates, from CO_2, H_2O and other simple compounds containing nitrogen and other elements. Photosynthesis releases O_2 into the atmosphere while removing CO_2 (Figure 2.10). The solar energy is captured in the

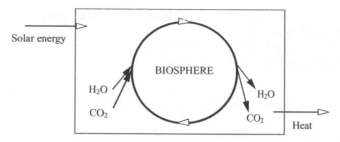

Figure 2.10   Energy flow through the biosphere

biomolecules which contain more energy than the compounds from which they are synthesized. The 'high-energy' products of photosynthesis are in turn the energy source for organisms that feed on them. Photosynthesis, the primary source of 'food', drives a complex food chain that sustains 'higher organisms' and ultimately a complex ecosystem. The energy flow in higher organisms is through the conversion of carbohydrates back to CO_2 and H_2O; this flow drives life processes: feeding, reproducing, flying, running, etc. While living cells do not exist in a steady state but go through a cycle of life and death, the ecosystems as a whole could be in a self-sustaining 'steady state' on a long time-scale. As energy flows through the biosphere, it is converted to heat and is returned to the atmosphere. The metabolic processes in a human, for example, generate about $100\,\mathrm{J\,s^{-1}}$ of heat. The heat generated in the biosphere is ultimately radiated back into space. We shall discuss the thermodynamic aspects of life processes in more detail in Chapter 13.

Appendix 2.1 *Mathematica* Codes

CODE A: MATHEMATICA CODE FOR EVALUATING WORK DONE IN THE ISOTHERMAL EXPANSION OF A GAS

```
Clear[p,V,T,a,b]
p[V_,T_]:=(R*T/(V-b))-(a/V^2);
```

$$\int_{V1}^{V2} p[V,T]\,dlv$$

$$\frac{aV1-aV2-R\ T\ V1\ V2\ Log[-b+V1]+R\ T\ V1\ V2\ Log[-b+V2]}{V1V2}$$

```
Simplify[%]
```

$$a\left(-\frac{1}{V1}+\frac{1}{V2}\right)-R\ T\ Log[-b+V1]+R\ T\ Log[-b+V2]$$

The same integral can also be evaluated using the 'Integrate' command:

```
Clear[p,V,T,a,b];
p[V_,T_]:=(R*T/(V-b))-(a/V^2);
Integrate[p[V,T],{V,V1,V2}]
```

$$\frac{aV1-aV2-R\ T\ V1\ V2\ Log[-b+V1]+R\ T\ V1\ V2\ Log[-b+V2]}{V1V2}$$

```
Simplify[%]
```

$$a\left(-\frac{1}{V1}+\frac{1}{V2}\right)-R\ T\ Log[-b+V1]+R\ T\ Log[-b+V2]$$

References

1. Kuhn, T. Energy conservation as an example of simultaneous discovery. In *The Essential Tension*. 1977, Chicago: University of Chicago Press.
2. Steffens, H.J. *James Prescott Joule and the Concept of Energy*. 1979, Science History Publications: New York; 134.
3. Prigogine, I. and Stengers, I. *Order Out of Chaos*. 1984, Bantam: New York; 108.
4. Dornberg, J. Count Rumford: the most successful Yank abroad, ever. *Smithsonian*, **25** (1994) 102–115.
5. Segre 2, E. *From Falling Bodies to Radio Waves*. 1984, W.H. Freeman: New York.
6. Mach, E. *Principles of the Theory of Heat*. 1986, D. Reidel: Boston, MA.
7. Weinberg, S. *The First Three Minutes*. 1980, Bantam: New York.
8. Mason, S.F. *A History of the Sciences*. 1962, Collier Books: New York.
9. Planck, M. *Treatise on Thermodynamics*, third edition. 1945, Dover: New York.
10. Alberty, A.A. and Silbey, R.J. *Physical Chemistry*, first edition. 1992, Wiley: New York; 1.
11. Leicester, H.M. Germain Henri Hess and the foundations of thermochemistry. *J. Chem. Ed.*, **28** (1951) 581–583.
12. Davis, T.W. A common misunderstanding of Hess's law. *J. Chem. Ed.*, **28** (1951) 584–585.
13. Leicester, H.M. *The Historical Background of Chemistry*. 1971, Dover: New York.
14. Lide D.R. (ed.). *CRC Handbook of Chemistry and Physics*, 75th edition. 1994, CRC Press: Ann Arbor, MI. Also see website of National Institute for Standards and Technology: http://webbook.nist.gov.
15. De Donder, T. *Lecons de Thermodynamique et de Chimie-Physique*. 1920, Gauthiers-Villars: Paris.
16. De Donder, T. and Van Rysselberghe, P. *Affinity*. 1936, Stanford University Press: Menlo Park, CA.
17. Prigogine, I. and Defay, R. *Chemical Thermodynamics*, fourth edition. 1967, Longmans: London; 542.
18. Mason, S.F. *Chemical Evolution*. 1991, Clarendon Press: Oxford.
19. *Physics Today*, Dec. (1995) 17–19.

Examples

Example 2.1 A bullet of mass $20.0\,g$ traveling at a speed of $350.0\,m\,s^{-1}$ is lodged into a block of wood. How many calories of heat are generated in this process?

Solution In this process, the kinetic energy (KE) of the bullet is converted to heat.

$$\text{KE}_{\text{bullet}} = mv^2/2 = (1/2)20.0\times10^{-3}\,\text{kg} \times (350\,\text{m s}^{-1})^2 = 1225\,\text{J}$$

$$1225\,\text{J} = 1225\text{J}/4.184\,\text{J cal}^{-1} = 292.6\,\text{cal}$$

Example 2.2 Calculate the energy ΔU required to increase the temperature of 2.50 mol of an ideal monatomic gas from 15.0 °C to 65.0 °C.
Solution Since the specific heat $C_V = (\partial U/\partial T)_V$, we see that

$$\Delta U = \int_{T_i}^{T_f} C_V \mathrm{d}T = C_V (T_f - T_i)$$

Since C_V for a monatomic ideal gas is $(3/2)R$:

$$U = (3/2)(8.314\,\text{J mol}^{-1}\,\text{K}^{-1})(2.5\,\text{mol})(65.0 - 15.0)\ \text{K} = 1559\,\text{J}$$

Example 2.3 The velocity of sound in CH_4 at 41.0 °C was found to be 466.0 m s^{-1}. Calculate the value of γ, the ratio of specific heats, at this temperature.
Solution Equation (2.3.17) gives the relation between γ and the velocity of sound:

$$\gamma = \frac{MC^2_{\text{sound}}}{RT} = \frac{16.04\times10^{-3}\,\text{kg}\times(466\,\text{m s}^{-1})^2}{8.314\times314.15\,\text{K}} = 1.33$$

Example 2.4 1 mol of $N_2(g)$ at 25.0 °C and a pressure of 1.0 bar undergoes an isothermal expansion to a pressure of 0.132 bar. Calculate the work done.
Solution For an isothermal expansion:

$$\text{Work} = -NRT\ln\left(\frac{V_f}{V_i}\right)$$

For an ideal gas, at constant T, $P_i V_i = P_f V_f$. Hence:

$$\begin{aligned}
\text{Work} &= -NRT\ln\left(\frac{V_f}{V_i}\right) \\
&= -NRT\ln\left(\frac{P_i}{P_f}\right) = -1.0(8.314\,\text{J K}^{-1})\ln\left(\frac{1.0\,\text{bar}}{0.132\,\text{bar}}\right) = -5.03\,\text{kJ}
\end{aligned}$$

Example 2.5 Calculate the heat of combustion of propane in the reaction at 25 °C:

$$C_3H_8(g) + 5O_2(g) \rightarrow 3CO_2(g) + 4H_2O(l)$$

Solution   From the table of heats of formation at 298.15 K we obtain

$$\Delta H_r^0 = -\Delta H_f^0[C_3H_8] - 5\Delta H_f^0[O_2] + 3\Delta H_f^0[CO_2] + 4\Delta H_f^0[H_2O]$$

$$= -(-103.85\,\text{kJ}) - (0) + 3(-393.51\,\text{kJ}) + 4(-285.83\,\text{kJ}) = -2220\,\text{kJ}$$

Example 2.6   For the reaction $N_2(g) + 3H_2(g) \rightarrow 2NH_3(g)$, at $T = 298.15\,\text{K}$ the standard enthalpy of reaction is $-46.11\,\text{kJ}\,\text{mol}^{-1}$. At constant volume, if 1.0 mol of $N_2(g)$ reacts with 3.0 mol of $H_2(g)$, what is the energy released?

Solution   The standard enthalpy of reaction is the heat released at constant pressure of 1.0 bar. At constant volume, since no mechanical work is done, the heat released equals the change in internal energy ΔU. From Equation (2.4.9) we see that

$$\Delta H_r = \Delta U_r + \Delta N_r RT$$

In the above reaction, $\Delta N_r = -2$. Hence:

$$\Delta U_r = \Delta H_r - (-2)RT = -46.11\,\text{kJ} + 2(8.314\,\text{J}\,\text{K}^{-1})298.15 = -41.15\,\text{kJ}$$

Example 2.7   Apply the energy flow equation to a thermal power station for which the energy flow is as shown in the figure below. The power station takes in heat to run an electrical power generator.

Solution   A thermal power plant may be considered as a system with the following properties: heat flows into the system, a part of it is converted to mechanical energy that runs an electrical power generator, and the unused heat is expelled. There is no flow of matter. Applying Equation (2.7.4) to this system, we see that $dm/dt = 0$, and we are left with $dW/dt = -dQ/dt$. Since dQ/dt is the *net* outflow of heat, a negative value of dQ/dt means that the inflow of heat is larger than the outflow, i.e. part of the heat flowing into the system is converted to mechanical energy that runs the power generator. What fraction of the heat energy flowing into the system is converted to mechanical energy depends on the efficiency of the power plant. In Chapter 3 we will discuss Sadi Carnot's discovery that conversion of heat to mechanical energy has limitations; it is impossible to convert 100% of the heat flowing into the system in to mechanical energy.

Example 2.8 N_2 is flowing into a nozzle with a velocity $v_i = 35.0\,\text{m}\,\text{s}^{-1}$ at $T = 300.0\,\text{K}$. The temperature of the gas flowing out of the nozzle is 280.0 K. Calculate the velocity of the gas flowing out of the nozzle. (Assume the ideal gas law for the flowing gas.)

Solution   For flow through a nozzle, there is no net output of heat or work. Applying Equation (2.7.4) to this system, we see that $dW/dt = 0$ and $dQ/dt = 0$. Hence:

$$\frac{h_i}{\rho_i} + \frac{v_i^2}{2} = \frac{h_o}{\rho_o} + \frac{v_o^2}{2}$$

in which the subscripts 'i' and 'o' denote the quantities for inflow and outflow respectively. Using the given values of T for the flows, the specific enthalpies h/ρ of the gas flowing in and out of the nozzle can be calculated as follows. For an ideal gas, enthalpy $H = U + pV = cNRT + NRT = (c + 1)RTN$. ($c = 5/2$ for a diatomic gas such as N_2.) If the molar mass of the gas is M, then

$$h = \frac{H}{V} = \frac{(c+1)RT}{M}\frac{NM}{V} = \frac{(c+1)RT}{M}\rho$$

i.e.

$$\frac{h}{\rho} = \frac{(c+1)RT}{M}$$

Now we can write the specific enthalpies in terms of temperature in the above expression and obtain

$$\frac{v_o^2}{2} = \frac{v_i^2}{2} + \frac{(c+1)R}{M}(T_i - T_o)$$

For a diatomic gas N_2, $c = 5/2$. We have $v_i = 35.0\,\mathrm{m\,s^{-1}}$, $M = 28 \times 10^{-3}\,\mathrm{kg\,mol^{-1}}$, $T_i = 300.0\,\mathrm{K}$ and $T_0 = 280.0\,\mathrm{K}$. Using these values, v_o can be calculated: $v_o = 206\,\mathrm{m\,s^{-1}}$.

Example 2.10   A steam turbine operates under the following conditions: steam flows into a turbine through an inlet pipe of radius 2.50 cm, at a velocity 80.0 m s^{-1}, at $p = 6.0\,\mathrm{MPa}$ and $T = 450.0\,°C$. The spent steam flows out at $p = 0.08\,\mathrm{MPa}$, $T = 93.0\,°C$ through an outlet pipe of radius 15.0 cm. Assuming steady-state conditions, calculate the power output using the following data from steam tables:

at $p = 6.0\,\text{MPa}$, $T = 450.0\,°\text{C}$, the specific volume $1/\rho = 0.052\,\text{m}^3\text{kg}^{-1}$ and $h/\rho = 3301.4\,\text{kJ}\,\text{kg}^{-1}$
at $p = 0.08\,\text{MPa}$, $T = 93.0\,°\text{C}$, the specific volume $1/\rho = 2.087\,\text{m}^3\text{kg}^{-1}$ and $h/\rho = 2665.4\,\text{kJ}\,\text{kg}^{-1}$

Solution At steady state, the mass flowing in must equal the mass flowing out (mass balance). Hence, $A_i v_i \rho_i = A_o v_o \rho_o$. Using this equation and the given data, we can calculate the velocity of the steam in the outlet:

$$v_o = \frac{\pi(0.025\,\text{m})^2(19.2\,\text{kg}\,\text{m}^{-3})}{\pi(0.15\,\text{m})^2(0.479\,\text{kg}\,\text{m}^{-3})}80.0\,\text{m}\,\text{s}^{-1} = 89.0\,\text{m}\,\text{s}^{-1}$$

The rate of mass flow is

$$\frac{\text{d}m}{\text{d}t} = A_i v_i \rho_i = \pi(0.025\,\text{m})^2(19.2\,\text{kg}\,\text{m}^{-3})80\,\text{m}\,\text{s}^{-1} = 3.01\,\text{kg}\,\text{s}^{-1}$$

Now we can apply Equation (2.7.4) to calculate the power output. In this case, mechanical energy is the output and we may assume negligible heat losses, i.e. $\text{d}Q/\text{d}t = 0$. We then have

$$\frac{\text{d}W}{\text{d}t} = \left[\left(\frac{h_i}{\rho_i} + \frac{v_i^2}{2}\right) - \left(\frac{h_o}{\rho_o} + \frac{v_o^2}{2}\right)\right]\frac{\text{d}m}{\text{d}t}$$

Using the steam-table data, we see that $h_i/\rho_i = 3301.4\,\text{kJ}\,\text{kg}^{-1}$ and $h_o/\rho_o = 2665.4\,\text{kJ}\,\text{kg}^{-1}$. Thus, the power output is

$$\frac{\text{d}W}{\text{d}t} = [(3301.4 - 2665.4)10^3 + 0.5(80.0^2 - 89.0^2)]3.01 = 1915\,\text{kJ}\,\text{s}^{-1} = 1.9\,\text{MW}$$

Exercises

2.1 For a conservative force $F = -\partial V(x)/\partial x$, in which $V(x)$ is the potential, using Newton's laws of motion, show that the sum of kinetic and potential energy is a constant.

2.2 How many joules of heat are generated by the brakes of a 1000 kg car when it is brought to rest from a speed of $50\,\text{km}\,\text{h}^{-1}$? If we use this amount of heat to heat 1.0 L of water from an initial temperature of $30\,°\text{C}$, estimate the final temperature assuming that the heat capacity of water is about $1\,\text{cal}\,\text{mL}^{-1}$ (1 cal $= 4.184\,\text{J}$).

2.3   The manufacturer of a heater coil specifies that it is a 500 W device.
(a) At a voltage of 110 V, what is the current through the coil?
(b) Given that the latent heat of fusion of ice is about 6.0 kJ mol^{-1}, how long will it take for this heater to melt 1.0 kg of ice at 0 °C.

2.4   Use the relation $dW = -p\,dV$ to show that:
(a) The work done in an *isothermal* expansion of N moles of an ideal gas from initial volume V_i to the final volume V_f is: Work $= -NRT \ln(V_f/V_i)$
(b) For 1 mol of an ideal gas, calculate the work done in an isothermal expansion of 1 mol from $V_i = 10.0$ L to $V_f = 20.0$ L at temperature $T = 350$ K.
(c) Repeat the calculation of part (a) using the van der Waals equation in place of the ideal gas equation and show that

$$\text{Work} = -NRT \ln\left(\frac{V_f - Nb}{V_i - Nb}\right) + aN^2\left(\frac{1}{V_i} - \frac{1}{V_f}\right)$$

2.5   Given that for the gas Ar the heat capacity $C_V = (3R/2) = 12.47$ J K^{-1} mol^{-1}, calculate the velocity of sound in Ar at $T = 298$ K using the ideal-gas relation between C_p and C_V. Do the same for N_2, for which $C_V = 20.74$ J K^{-1} mol^{-1}.

2.6   Calculate the sound velocities of He, N_2 and CO_2 using (2.3.17) and the values of γ in Table 2.2 and compare them with the experimentally measured velocities shown in the same table.

2.7   The human ear can detect an energy intensity of about 2×10^{-12} W m^{-2}. Consider a light source whose output is 100 W. At what distance is its intensity equal to 2×10^{-12} W m^{-2}?

2.8   A **monatomic** ideal gas is initially at $T = 300$ K, $V = 2.0$ L and $p = 1.0$ bar. If it is expanded adiabatically to $V = 4.0$ L, what will its final T be?

2.9   We have seen (Equation (2.3.5)) that, for any system

$$C_p - C_V = \left[p + \left(\frac{\partial U}{\partial V}\right)_T\right]\left(\frac{\partial V}{\partial T}\right)_p$$

For the van der Waals gas the energy $U_{vw} = U_{ideal} - a(N/V)^2 V$, in which $U_{ideal} = C_V NT + U_0$ (Equation (2.2.15)). Use these two expressions and the van der Waals equation to obtain an explicit expression for the difference between C_p and C_V.

2.10   For nitrogen at $p = 1$ atm, and $T = 298$ K, calculate the change in temperature when it undergoes an adiabatic compression to a pressure of 1.5 atm. $\gamma = 1.404$ for nitrogen.

2.11 Using Equation (2.4.11) and Table 2.3, calculate the change in enthalpy of 1.0 mol of $CO_2(g)$ when it is heated from 350.0 K to 450.0 K at $p = 1$ bar.

2.12 Using the Standard Thermodynamic Properties table at the back of the book, which contains heats of formation of compounds at $T = 298.15$ K, calculate the standard heats of reaction for the following reactions:
(a) $H_2(g) + F_2(g) \rightarrow 2HF(g)$
(b) $C_7H_{16}(l) + 11O_2(g) \rightarrow 7CO_2(g) + 8H_2O(l)$
(c) $2NH_3(g) + 6NO(g) \rightarrow 3H_2O_2(l) + 4N_2(g)$

2.13 Gasoline used as motor fuel consists of a mixture of the hydrocarbons heptane (C_7H_{16}), octane (C_8H_{18}) and nonane (C_9H_{20}). Using the bond energies in Table 2.4, estimate the enthalpy of combustion of 1 g of each of these fluids. (In a combustion reaction, an organic compound reacts with $O_2(g)$ to produce $CO_2(g)$ and $H_2O(g)$.)

2.14 Calculate the amount of energy released in the combustion of 1 g of glucose and compare it with the mechanical energy needed to lift 100 kg through 1 m. (Combustion of glucose: $C_{12}H_{22}O_{11} + 12O_2 \rightarrow 11H_2O + 12CO_2$.)

2.15 Consider the reaction $CH_4(g) + 2O_2(g) \rightarrow CO_2(g) + 2H_2O(l)$. Assume that initially there are 3.0 mol CH_4 and 2.0 mol O_2 and that the extent of reaction $\xi = 0$. When the extent of reaction $\xi = 0.25$ mol, what are the amounts of the reactants and the products? How much heat is released at this point? What is the value of ξ when all the O_2 has reacted?

2.16 The sun radiates energy approximately at a rate of 3.9×10^{26} J s^{-1}. What will be the change in its mass in 1 million years if it radiates at this rate?

2.17 Calculate the energy released in the reaction

$$2\,^1H + 2n \rightarrow\ ^4He$$

given the following masses: mass of $^1H = 1.0078$ amu, mass of $n = 1.0087$ amu, mass of $^4He = 4.0026$ amu. (1 amu = 1.6605×10^{-27} kg.)

2.18 O_2 is flowing into a nozzle with a velocity $v_i = 50.0$ m s^{-1} at $T = 300.0$ K. The temperature of the gas flowing out of the nozzle is 270.0 K. (a) Assume the ideal gas law for the flowing gas and calculate the velocity of the gas flowing out of the nozzle. (b) If the inlet diameter is 5.0 cm, what is the outlet's diameter?

2.19 A steam turbine has the following specifications: inlet diameter 5.0 cm; steam inflow is at $p = 4.0$ MPa, at $T = 450.0$ °C at a velocity of $v_i = 150$ m s^{-1}. The outlet pipe has a diameter of 25.0 cm and the steam flows out at $p = 0.08$ MPa, $T = 93.0$ °C. (a) Assuming steady-state conditions, calculate the output power using the data given below from steam tables. (b) Show that the change in kinetic energy between the inflow and the outflow is negligible compared with the change in the specific enthalpy.

Data from steam tables:

at $p = 4.0\,\text{MPa}$, $T = 450.0\,°\text{C}$, the specific volume $1/\rho = 0.080\,\text{m}^3\,\text{kg}^{-1}$ and $h/\rho = 3330.1\,\text{kJ}\,\text{kg}^{-1}$.

at $p = 0.08\,\text{MPa}$, $T = 93.0\,°\text{C}$, the specific volume $1/\rho = 2.087\,\text{m}^3\,\text{kg}^{-1}$ and $h/\rho = 2665.4\,\text{kJ}\,\text{kg}^{-1}$.

2.20 The amount of solar energy driving the water cycle is approximately $12.5 \times 10^{20}\,\text{kJ}\,\text{year}^{-1}$. Estimate the amount of water, in moles and liters, evaporated per day in the water cycle.

3 THE SECOND LAW OF THERMODYNAMICS AND THE ARROW OF TIME

3.1 The Birth of the Second Law

James Watt (1736–1819), the most famous of Joseph Black's pupils, obtained a patent for his modifications of Thomas Newcomen's steam engine in the year 1769.

James Watt (1736–1819) (Reproduced with permission from the Edgar Fahs Smith Collection, University of Pennsylvania Library)

Introduction to Modern Thermodynamics Dilip Kondepudi
© 2008 John Wiley & Sons, Ltd

Soon, this invention brought unimagined power and speed to everything: mining of coal, transportation, agriculture and industry. This revolutionary generation of motion from heat that began in the British Isles quickly crossed the English Channel and spread throughout Europe.

Nicolas Léonard Sadi Carnot (1796–1832), a brilliant French military engineer, lived in this rapidly industrializing Europe. 'Every one knows', he wrote in his memoirs, 'that heat can produce motion. That it possesses vast motive-power no one can doubt, in these days when the steam-engine is everywhere so well known' [1, p. 3]. The name Carnot is well known in France. Sadi Carnot's father, Lazare Carnot (1753–1823), held many high positions during and after the French Revolution and was known for his contributions to mechanics and mathematics. Lazare Carnot had a strong influence on his son Sadi. Both had their scientific roots in engineering, and both had a deep interest in general principles in the tradition of

Sadi Carnot (1796–1832) (Reproduced with permission from the Edgar Fahs Smith Collection, University of Pennsylvania Library)

the French Encyclopedists. It was his interest in general principles that led Sadi Carnot to his abstract analysis of heat engines. Carnot pondered over the principles that governed the working of the steam engine and identified the *flow of heat* as the fundamental process required for the generation of 'motive power' – 'work' in today's terminology. He analyzed the fundamental processes that underlie **heat engines**, engines that performed mechanical work through the flow of heat, and realized that there was a fundamental limit to the amount of work generated from the flow of a given amount of heat. Carnot's great insight was that this limit was *independent* of the machine and the manner in which work was obtained: it depended only on the temperatures that caused the flow of heat. As explained in the following sections, further development of this principle resulted in the formulation of the second law of thermodynamics.

Carnot described his general analysis of heat engines in his only scientific publication '*Réflexions sur la puissance motrice du feu, et sur les machines propres a développer cette puissance*' ('Reflections on the motive force of fire and on the machines fitted to develop that power') [1]. Six hundred copies of this work were published in 1824, at Carnot's own expense. At that time, Carnot was a well-known name in the French scientific community due to the fame of Sadi's father, Lazare Carnot. Still, Sadi Carnot's book did not attract much attention at the time of its publication. Eight years after the publication of his 'Reflexions', Sadi Carnot died of cholera. A year later, Émile Clapeyron (1799–1864) was to come across Carnot's book and realize its fundamental importance and make it known to the scientific community.

Carnot's analysis proceeded as follows. First, Carnot observed 'Wherever there exists a difference of temperature, motive force can be produced' [1, p. 9]. Every heat engine that produced work from the flow of heat operated between two heat reservoirs of unequal temperatures. In the processes of transferring heat from a hot to a cold reservoir, the engine performed mechanical work (see Figure 3.1). Carnot then specified the following condition for the production of maximum work [1, p. 13]:

The necessary condition of the maximum (work) is, *that in the bodies employed to realize the motive power of heat there should not occur any change of temperature which may not be due to a change of volume*. Reciprocally, every time that this condition is fulfilled the maximum will be attained. This principle should never be lost sight of in the construction of a heat engine; it is its fundamental basis. If it cannot be strictly observed, it should at least be departed from as little as possible.

Thus, for maximum work generation, all changes in volume – such as the expansion of a gas (steam) that pushes a piston – should occur with minimal temperature gradients so that changes in temperature are almost all due to volume expansion and not due to flow of heat caused by temperature gradients. This is achieved in heat engines that absorb and discard heat during very slow changes in volume, keeping their internal temperature as uniform as possible.

Furthermore, in the limit of infinitely slow transfer of heat during changes of volume, with infinitesimal temperature difference between the source of heat (the

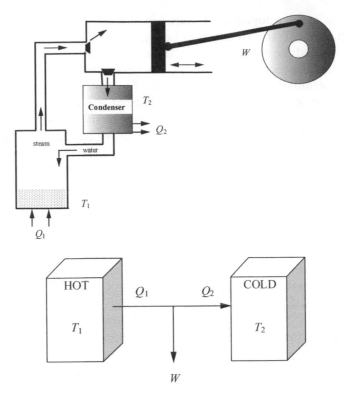

Figure 3.1   The upper figure shows a schematic of a steam engine. The lower figure shows the essential process that governs heat engines, engines that convert heat to work. It illustrates the fundamental observation made by Sadi Carnot: 'Wherever there exists a difference of temperature, motive force can be produced' [1, p. 9]. The heat engine absorbs heat Q_1 from a hot reservoir (heat source), converts part of it to work W, and discards heat Q_2 to a cold reservoir (heat sink). The efficiency η is given by $W = \eta Q_1$. (According to the caloric theory of heat that Carnot used, $Q_1 = Q_2$, but an analysis consistent with the First Law gives $W = Q_1 - Q_2$)

'heat reservoir') and the engine, the operation of such an engine is a **reversible processes**, which means that the series of states the engine goes through can be retraced in the exact opposite order. A reversible engine can perform mechanical work W by transferring heat form a hot to a cold reservoir or it can do the exact reverse of transferring the *same* amount of heat from a cold reservoir to a hot reservoir by using the *same* amount of work W.

   The next idea Carnot introduced is that of a **cycle**: during its operation, the heat engine went through a cycle of states so that, after producing work from the flow of heat, it returned to its initial state, ready to go through the cycle once again. A modern version of Carnot's reversible cycle will be discussed later in this section.

Carnot argued that the reversible cyclic heat engine must produce the maximum work ('motive force'), but he did so using the caloric theory of heat, according to which heat was an indestructible substance. If any engine could produce a greater amount of work than that produced by a reversible cyclic heat engine, then it was possible to produce work endlessly by the following means. Begin by moving heat from the hot reservoir to a cold reservoir using the more efficient engine. Then move the same amount of heat back to the hot reservoir using the reversible engine. Because the forward process does more work than is needed to perform the reverse process, there is a net gain in work. In this cycle of operations, a certain amount of heat was simply moved from the hot to the cold reservoir and back to the hot reservoir, with a net gain of work. By repeating this cycle, an unlimited amount of work can be obtained simply by moving a certain amount of heat back and forth between a hot and a cold reservoir. This, Carnot asserted, was impossible:

This would be not only perpetual motion, but an unlimited creation of motive power without consumption either of caloric or of any other agent whatever. Such a creation is entirely contrary to ideas now accepted, to laws of mechanics and of sound physics. It is inadmissible [1, p. 12].

Hence, *reversible cyclic engines must produce the maximum amount of work*. A corollary of this conclusion is that *all* reversible cyclic engines must produce the same amount of work regardless of their construction. Furthermore, and most importantly, since all reversible engines produce the same amount of work from a given amount of heat, the amount of work generated by a reversible heat engine is independent of the material properties of the engine: it can depend only on the temperatures of the hot and cold reservoirs. This brings us to the most important of Sadi Carnot's conclusions [1, p. 20]:

The motive power of heat is independent of the agents employed to realize it; its quantity is fixed solely by the temperatures of the bodies between which is effected, finally, the transfer of caloric.

Carnot did not derive a mathematical expression for the maximum efficiency attained by a reversible heat engine in terms of the temperatures between which it operated. This was done later by others who realized the importance of his conclusion. Carnot did, however, find a way of calculating the maximum work that can be generated. For example, he concluded that '1000 units of heat passing from a body maintained at the temperature of 1 degree to another body maintained at zero would produce, in acting upon the air, 1.395 units of motive power' [1, p. 42].

Though Sadi Carnot used the caloric theory of heat to reach his conclusions, his later scientific notes reveal his realization that the caloric theory was not supported by experiments. In fact, Carnot understood that heat is converted to mechanical work and even estimated the conversion factor to be approximately $3.7 \, \text{J cal}^{-1}$ (the more accurate value being $4.18 \, \text{J cal}^{-1}$) [1–3]. Unfortunately, Sadi Carnot's brother, Hippolyte Carnot, who was in possession of Sadi's scientific notes after his death in 1832, did not make them known to the scientific community until 1878 [3]. That was the year in which Joule published his last paper. By then, the equivalence between heat and work and the law of conservation of energy were well known through the work of Joule, Helmholtz, von Mayer and others. (1878 was also the year in which Gibbs published his famous work *On the Equilibrium of Heterogeneous Substances.*)

Sadi Carnot's brilliant insight went unnoticed until Émile Clapeyron (1799–1864) came across Carnot's book in 1833. Realizing its importance, he reproduced the main ideas in an article that was published in the *Journal de l'Ecole Polytechnique* in 1834. Clapeyron represented Carnot's example of a reversible engine in terms of a *p–V* diagram (which is used today) and described it with mathematical detail. Clapeyron's article was later read by Lord Kelvin and others who realized the fundamental nature of Carnot's conclusions and investigated its consequences. These developments led to the formulation of the Second Law of thermodynamics as we know it today.

To obtain the efficiency of a reversible heat engine, we shall not follow Carnot's original reasoning because it considered heat as an indestructible substance. Instead, we shall modify it by incorporating the First Law. For the heat engine represented in Figure 3.1, the law of conservation of energy gives $W = Q_1 - Q_2$. This means, a fraction η of the heat Q_1 absorbed from the hot reservoir is converted into work W, i.e. $\eta = W/Q_1$. The fraction η is called *the efficiency of the heat engine*. Since $W = (Q_1 - Q_2)$ in accordance with the first law, $\eta = (Q_1 - Q_2)/Q_1 = (1 - Q_2/Q_1)$. Carnot's discovery that the reversible engine produces maximum work amounts to the statement that its efficiency is maximum. This efficiency is independent of the properties of the engine and is a function only of the temperatures of the hot and the cold reservoirs:

$$\eta = 1 - \frac{Q_2}{Q_1} = 1 - f(t_1, t_2) \qquad (3.1.1)$$

in which $f(t_1, t_2)$ is a function only of the temperatures t_1 and t_2 of the hot and cold reservoirs. The scale of the temperatures t_1 and t_2 (Celsius or other) is not specified here. Equation (3.1.1) is **Carnot's theorem**. In fact, as described below, Carnot's observation enables us to define an absolute scale of temperature that is independent of the material property used to measure it.

EFFICEIENCY OF A REVERSIBLE ENGINE

Now we turn to the task of obtaining the efficiency of reversible heat engines. Since the efficiency of a reversible heat engine is the maximum, all of them must have the same efficiency. Hence, obtaining the efficiency of one reversible engine will suffice. The following derivation also makes it explicit that the efficiency of Carnot's engine is only a function of temperature.

Carnot's reversible engine consists of an ideal gas that operates in a cycle between hot and cold reservoirs, at temperatures θ_1 and θ_2 respectively. Until their identity is established below, we shall use θ for the temperature that appears in the ideal gas equation and T for the absolute temperature (which, as we shall see in the next section, is defined by the efficiency of a reversible cycle). Thus, the ideal gas equation is written as $pV = NR\theta$, in which θ is the temperature measured by noting the change of some quantity such as volume or pressure. (Note that measuring temperature by volume expansion is purely empirical; each unit of temperature is simply correlated with a certain change in volume.) In the following, the work done *by the gas* will be a positive quantity and the work done *on the gas* will be a negative quantity, so that the net work obtained in a cycle is positive for a net heat transfer from the hot to the cold reservoir. The reversible cycle we consider consists of the following four steps (Figure 3.2).

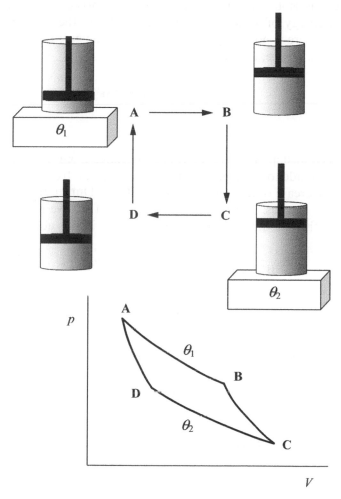

Figure 3.2   The Carnot cycle. The upper part shows the four steps of the Carnot cycle, during which the engine absorbs heat from the hot reservoir, produces work and returns heat to the cold reservoir. The lower part shows the representation of this process in a *p–V* diagram used by Clapeyron in his exposition of Carnot's work

Step 1

The gas has an initial volume of V_A and it is in contact with the hot reservoir at temperature θ_1. At a constant temperature θ_1 due to its contact with the reservoir, the gas undergoes an infinitely slow *reversible expansion* (as Carnot specified it) to the state B, of volume V_B. The work done by the gas during this expansion is

$$W_{AB} = \int_{V_A}^{V_B} p \, dV = \int_{V_A}^{V_B} \frac{NR\theta_1}{V} \, dV = NR\theta_1 \ln\left(\frac{V_B}{V_A}\right) \tag{3.1.2}$$

During this isothermal processes, heat is absorbed from the reservoir. Since the internal energy of an ideal gas depends only on the temperature (see (1.3.8) and (2.2.15)), there is no change in the energy of the gas; the heat absorbed equals the work done. Hence, the heat absorbed:

$$Q_{AB} = W_{AB} \tag{3.1.3}$$

Step 2

In the second step, the gas is thermally insulated from the reservoir and the environment and made to undergo an adiabatic expansion from state B to a state C, resulting in a decrease of temperature from θ_1 to θ_2. During this adiabatic process, work is done by the gas. Noting that on the adiabat BC we have $pV^\gamma = p_B V_B^\gamma = p_C V_C^\gamma$, we calculate the amount of work done by the gas in this adiabatic expansion:

$$\begin{aligned} W_{BC} &= \int_{V_B}^{V_C} p \, dV = \int_{V_B}^{V_C} \frac{p_B V_B^\gamma}{V^\gamma} \, dV = \frac{p_C V_C^\gamma V_C^{1-\gamma} - p_B V_B^\gamma V_B^{1-\gamma}}{1-\gamma} \\ &= \frac{p_C V_C - p_B V_B}{1-\gamma} \end{aligned}$$

Using $pV = NR\theta$, the above equation can be further simplified to

$$W_{BC} = \frac{NR(\theta_1 - \theta_2)}{\gamma - 1} \tag{3.1.4}$$

in which θ_1 and θ_2 are the initial and final temperatures during the adiabatic expansion.

Step 3

In the third step, the gas is in contact with the reservoir of temperature θ_2 and it undergoes an isothermal compression to the point D, at which the volume V_D is such that an adiabatic compression can return it to the initial state A. (V_D can be specified by finding the point of intersection of the adiabat through the point A and the isotherm at temperature θ_2.) During this isothermal process, the work done *on the gas* is transferred as heat Q_{CD} to the reservoir (since the energy of the ideal gas depends only on its temperature):

$$W_{CD} = \int_{V_C}^{V_D} p\, dV = \int_{V_C}^{V_D} \frac{NR\theta_2}{V}\, dV = NR\theta_2 \ln\left(\frac{V_D}{V_C}\right) = -Q_{CD} \tag{3.1.5}$$

Step 4

In the final step, an adiabatic compression takes the gas from the state D to its initial state A. Since this process is similar to step 2, we can write

$$W_{DA} = \frac{NR(\theta_2 - \theta_1)}{\gamma - 1} \tag{3.1.6}$$

The total work done in this reversible Carnot cycle in which heat Q_{AB} was absorbed from the reservoir at a temperature of θ_1 and heat Q_{CD} was transferred to the reservoir at temperature θ_2 is

$$\begin{aligned} W &= W_{AB} + W_{BC} + W_{CD} + W_{DA} = Q_{AB} - Q_{CD} \\ &= NR\theta_1 \ln\left(\frac{V_B}{V_A}\right) - NR\theta_2 \ln\left(\frac{V_C}{V_D}\right) \end{aligned} \tag{3.1.7}$$

The efficiency $\eta = W/Q_{AB}$ can now be written using (3.1.2), (3.1.3) and (3.1.7):

$$\eta = \frac{W}{Q_{AB}} = 1 - \frac{NR\theta_2 \ln(V_C/V_D)}{NR\theta_1 \ln(V_B/V_A)} \tag{3.1.8}$$

For the two isothermal processes, we have $p_A V_A = p_B V_B$ and $p_C V_C = p_D V_D$; and for the two adiabatic processes, we have $p_B V_B^{\gamma} = p_C V_C^{\gamma}$ and $p_A V_A^{\gamma} = p_D V_D^{\gamma}$. Using these relations, it can easily be seen that $(V_C/V_D) = (V_B/V_A)$. Using this relation in (3.1.8), we arrive at a simple expression for the efficiency:

$$\eta = \frac{W}{Q_{AB}} = 1 - \frac{\theta_2}{\theta_1} \tag{3.1.9}$$

In this expression for the efficiency, θ is the temperature defined by one particular property (such as volume at a constant pressure) and we assume that it is the temperature in the ideal gas equation. The temperature t measured by any other empirical means, such as measuring the volume of mercury, is related to θ; that is, θ can be expressed as a function of t, i.e. $\theta(t)$. Thus, the temperature t measured by one means corresponds to $\theta = \theta(t)$, measured by another means. In terms of any other temperature t, the efficiency may take a more complex form. In terms of the temperature θ that obeys the ideal gas equation, however, the efficiency of the reversible heat engine takes a particularly simple form (3.1.9).

3.2   The Absolute Scale of Temperature

The fact that the efficiency of a reversible heat engine is independent of the physical and chemical nature of the engine has an important consequence which was noted by Lord Kelvin (William Thomson (1824–1907)). Following Carnot's work, Lord Kelvin introduced the *absolute scale of temperature*. The efficiency of a reversible heat engine is a function only of the temperatures of the hot and cold reservoirs, independent of the material properties of the engine. Furthermore, the efficiency

William Thomson/Lord Kelvin (1824–1907) (Reproduced with permission from the Edgar Fahs Smith Collection, University of Pennsylvania Library)

cannot exceed unity, in accordance with the First Law. These two facts can be used to define an *absolute scale of temperature that is independent of any material properties*.

First, by considering two successive Carnot engines, one operating between t_1 and t' and the other operating between t' and t_2, we can see that the function $f(t_2, t_1)$ in Equation (3.1.1) is a ratio of a functions of t_1 and t_2: if Q' is the heat exchanged at temperature t', then we can write

$$f(t_2, t_1) = \frac{Q_2}{Q_1} = \frac{Q_2}{Q'}\frac{Q'}{Q_1} = f(t_2, t')f(t', t_1) \tag{3.2.1}$$

This relation, along with $f(t, t) = 1$, implies that we can write the function $f(t_2, t_1)$ as the ratio $f(t_2)/f(t_1)$. Hence, the efficiency of a reversible Carnot engine can be written as

$$\eta = 1 - \frac{Q_2}{Q_1} = 1 - \frac{f(t_2)}{f(t_1)} \tag{3.2.2}$$

One can now define a temperature $T \equiv f(t)$ based solely on the efficiencies of reversible heat engines. In terms of this temperature scale, the efficiency of a reversible engine is given by

$$\boxed{\eta = 1 - \frac{Q_2}{Q_1} = 1 - \frac{T_2}{T_1}} \tag{3.2.3}$$

in which T_1 and T_2 are the absolute temperatures of the cold and hot reservoirs respectively. An efficiency of unity defines the absolute zero of this scale. **Carnot's theorem** is the statement that reversible engines have the maximum efficiency given by (3.2.3).

Comparing expression (3.2.3) with (3.1.9), we see that the ideal gas temperature coincides with the absolute temperature and, hence, we can use the same symbol, i.e. T, for both.*

In summary, for an idealized, *reversible heat engine* that absorbs heat Q_1 from a hot reservoir at absolute temperature T_1 and discards heat Q_2 to a cold reservoir at absolute temperature T_2, we have from (3.2.3)

$$\boxed{\frac{Q_1}{T_1} = \frac{Q_2}{T_2}} \tag{3.2.4}$$

---

* The empirical temperature t of a gas thermometer is defined through the increase in volume at constant pressure (see (1.3.9)): $V = V_0(1 + \alpha t)$. Gay-Lussac found that $\alpha \approx (1/273)\,°C^{-1}$. From this equation it follows that $dV/V = \alpha dt/(1 + \alpha t)$. On the other hand, from the ideal gas equation $pV = NRT$, we have, at constant p, $dV/V = dT/T$. This enables us to relate the absolute temperature T to the empirical temperature t by $T = (1 + \alpha t)/\alpha$.

All *real* heat engines that go through a cycle in finite time must involve irreversible processes such as flow of heat due to a temperature gradient. They are less efficient. Their efficiency η' is less than the efficiency of a reversible heat engine, i.e. $\eta' = 1 - (Q_2/Q_1) < 1 - (T_2/T_1)$. This implies $T_2/T_1 < Q_2/Q_1$ whenever irreversible processes are involved. Therefore, while the equality (3.2.4) is valid for a reversible cycle, for the operation of an *irreversible cycle* that we encounter in reality we have the *inequality*

$$\frac{Q_1}{T_1} < \frac{Q_2}{T_2} \qquad (3.2.5)$$

As we shall see below, irreversibility in Nature, or a sense of an 'arrow of time', is manifest in this inequality.

A spectacular example of a 'heat engine' in Nature is the hurricane. In a hurricane, heat is converted to kinetic energy of the hurricane wind. As summarized in Box 3.1, and described in more detail in Appendix 3.1, by using Carnot's theorem one can obtain an upper bound to the velocity of the hurricane wind.

3.3   The Second Law and the Concept of Entropy

The far-reaching import of the concepts originating in Carnot's '*Reflexions on the motive force of fire*' was realized in the generalizations made by Rudolf Clausius (1822–1888), who introduced the concept of *entropy*, a new physical quantity as fundamental and universal as energy.

Clausius began by generalizing expression (3.2.4) that follows from Carnot's theorem to an arbitrary cycle. This was done by considering composites of Carnot cycles in which the corresponding isotherms differ by infinitesimal amount ΔT, as shown in Figure 3.3 (a). Let Q_1 be the heat absorbed during the transformation from A to A', at temperature T_1 and let Q_1' be the heat absorbed during the transformation A'B at temperature $(T_1 + \Delta T)$. Similarly we define Q_2' and Q_2 for the transformations CC' and C'D occurring at temperatures $T_2 + \Delta T$ and T_2 respectively. Then the reversible cycle AA'BCC'DA can be thought of as a sum of the two reversible cycles AA'C'DA and A'BCC'A' because the adiabatic work A'C' in one cycle cancels that of the second cycle, C'A. For the reversible cycle AA'BCC'D, we can therefore write

$$\frac{Q_1}{T_1} + \frac{Q_1'}{T_1 + \Delta T} - \frac{Q_2}{T_2} - \frac{Q_2'}{T_2 + \Delta T} = 0 \qquad (3.3.1)$$

The above composition of cycles can be extended to an arbitrary closed path (as shown in Figure 3.3b) by considering it as a combination of a large number of infinitesimal Carnot cycles. With the notation $dQ > 0$ if heat is absorbed by the

Box 3.1 The Hurricane as a Heat Engine

The mechanism of a hurricane is essentially that of a heat engine as shown in the figure below in the cycle ABCD. The maximum intensity of a hurricane, i.e. the maximum hurricane wind speed, can be predicted using Carnot's theorem for the efficiency of a heat engine.

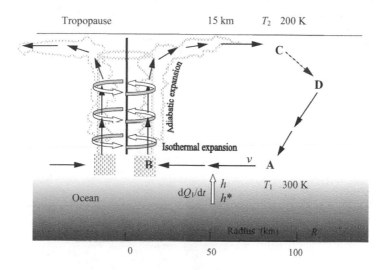

In a hurricane, as the wind spirals inwards towards the eye at low pressure, enthalpy (heat) is absorbed at the warm ocean–air interface in an essentially isothermal processes: water vaporizes and mixes with the air, carrying with it the enthalpy of vaporization (segment AB). When this moist air reaches the hurricane's eyewall, it rises rapidly to about 15 km along the eyewall. Since the pressure decreases with altitude, it expands adiabatically and cools (segment BC). As the rising moist air's temperature drops, the water vapor in it condenses as rain, releasing the enthalpy of vaporization (latent heat), part of which is radiated into outer space. In a real hurricane, the air at the higher altitude flows out into the weather system. Theoretically, in order to close the Carnot cycle, it could be assumed that the enthalpy of vaporization is lost in an isothermal process (segment CD). The last step (segment DA) of the cycle is an adiabatic compression of dry air. During the cycle, part of the enthalpy absorbed from the ocean is converted into mechanical energy of the hurricane wind.

The 'hurricane heat engine' operates between the ocean surface temperature T_1 (about 300 K) and the lower temperature T_2 (about 200 K) at the higher altitude, close to the upper boundary of the troposphere (tropopause). The conversion of the heat of vaporization to mechanical energy of the hurricane wind can now be analyzed. In a time dt, if dQ_1 is the heat absorbed at the ocean surface, dQ_2 is the heat radiated at the higher altitude and dW is the amount of heat converted to mechanical energy of the hurricane wind, then, according to Carnot's theorem:

$$\frac{dW}{dt} \leq \left(1 - \frac{T_2}{T_1}\right)\frac{dQ_1}{dt}$$

Appendix 3.1 shows that the use of this expression in an analysis of the energetics of a hurricane leads to the following estimate for the maximum hurricane wind speed $|v_{max}|$:

$$|v_{max}|^2 \approx \left(\frac{T_1 - T_2}{T_2}\right)\frac{C_h}{C_D}(h^* - h)$$

Here, C_h and C_D are constants, h^* is the specific enthalpy (enthalpy per unit mass) of the air saturated with moisture close to the ocean surface and h is the specific enthalpy of dry wind above the ocean surface (see figure above). All the terms on the right-hand side are experimentally measured or theoretically estimated. The ratio $C_h/C_D \approx 1$. Kerry Emanual, the originator of the above theory, has demonstrated that the above expression for v_{max} leads to remarkably good estimates of the hurricane wind speeds [4]. More details can be found in Appendix 3.1 and in the cited articles.

Rudolf Clausius (1822–1888) (Reproduced with permission from the Edgar Fahs Smith Collection, University of Pennsylvania Library)

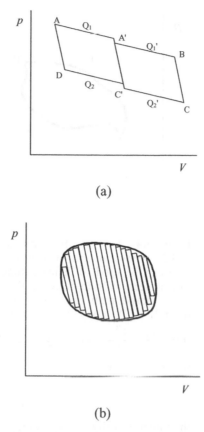

(a)

(b)

Figure 3.3 Clausius's generalization of Carnot cycle. (a) Two Carnot cycles can be combined to obtain a larger cycle. (b) Any closed path can be thought of as a combination of a large number of infinitesimal Carnot cycles

system and $dQ < 0$ if it is discarded, the generalization of (3.3.1) of an arbitrary closed path gives

$$\oint \frac{dQ}{T} = 0 \tag{3.3.2}$$

This equation has an important consequence: it means that the integral of the quantity dQ/T along a path representing a reversible process from a state A to a state B depends only on the states A and B and is independent of the path, as described in Figure 3.4. Thus, Clausius saw that one can define a function S that *depends only*

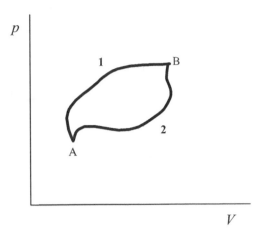

Figure 3.4 That any function, such as Q/T in (3.3.2), whose integral over any closed path is zero can be used to define a function of state can be seen as follows. Since the total integral for a closed path from A to B along 1 and from B to A along 2 is zero, it follows that $\oint \mathrm{d}Q/\mathrm{d}T = \int_{A,\,\text{path}\,1}^{B} \mathrm{d}Q/T + \int_{B,\,\text{path}\,2}^{A} \mathrm{d}Q/T = 0.$ Now we note that $\int_{A}^{B} \mathrm{d}Q/T = -\int_{B}^{A} \mathrm{d}Q/T$ along paths 1 or 2. Hence, $\int_{A,\,\text{path}\,1}^{B} \mathrm{d}Q/T = \int_{A,\,\text{path}\,2}^{B} \mathrm{d}Q/T$, i.e.

the integral of $\mathrm{d}Q/T$ from point A to point B is independent of the path; it depends only on the points A and B. Hence, if the initial reference state is fixed, the integral, which is a function only of the final state, is a state function

on the initial and final states of a reversible process. If S_A and S_B are the values of this function in the states A and B, then we can write

$$S_B - S_A = \int_A^B \frac{\mathrm{d}Q}{T} \quad \text{or} \quad \mathrm{d}S = \frac{\mathrm{d}Q}{T} \tag{3.3.3}$$

By defining a reference state 'O', the new function of state S could be defined for any state X as the integral of $\mathrm{d}Q/T$ for a *reversible process* transforming the state O to the state X.

Clausius introduced this new quantity S in 1865 stating 'I propose to call the magnitude S the *entropy* of the body, from the Greek word τποπη, *transformation*.' [5, p. 357]. The usefulness of the above definition depends on the assumption that any two states can be connected by a reversible transformation.

If the temperature remains fixed, then it follows from (3.3.3) that, for a reversible flow of heat Q, the change in entropy is Q/T. In terms of entropy, Carnot's theorem (3.2.3) is equivalent to the statement that the sum of the *entropy changes in a reversible cycle is zero*:

$$\boxed{\frac{Q_1}{T_1} - \frac{Q_2}{T_2} = 0}$$
(3.3.4)

In a reversible process, since the temperatures of the system and the reservoirs are the same when heat is exchanged, the change of entropy of the reservoir in any part of the cyclic process is the negative of the entropy change of the system.

In a less efficient *irreversible cycle*, a smaller fraction of Q_1 (the heat absorbed form the hot reservoir) is converted into work. This means that the amount of heat delivered to the cold reservoir by an irreversible cycle Q_2^{irr} is greater than Q_2. Therefore, We have

$$\frac{Q_1}{T_1} - \frac{Q_2^{irr}}{T_2} < 0$$
(3.3.5)

Since the cyclic engine returns to its initial state whether it is reversible or irreversible, there is no change in its entropy. On the other hand, since the heats transferred to the reservoirs and to the irreversible engine have opposite sign, the *total change of entropy of the reservoirs is*

$$\frac{-Q_1}{T_1} - \frac{-Q_2^{irr}}{T_2} > 0$$
(3.3.6)

if the reservoir temperatures can be assumed to be the same as the temperatures at which the engine operates. In fact, for heat to flow at a nonzero rate, the reservoir temperatures T_1' and T_2' must be such that $T_1' > T_1$ and $T_2' < T_2$. In this case, the increase in entropy is even larger than (3.3.6).

Generalizing the above result, for a system that goes through an arbitrary cycle, with the equalities holding for a reversible processes, we have

$$\oint \frac{dQ}{T} \leq 0 \quad \text{(system)}$$
(3.3.7)

For the 'exterior' with which the system exchanges heat, since dQ has the opposite sign, we then have

$$\oint \frac{dQ}{T} \geq 0 \quad \text{(exterior)}$$
(3.3.8)

At the end of the cycle, be it reversible or irreversible, there is no change in the system's entropy because it has returned to its original state. For irreversible cycles this means that the system expels more heat to the exterior. This is generally a conversion of mechanical energy into heat through irreversible processes. Consequently, the entropy of the exterior increases. Thus, the entropy changes may be summarized as follows:

reversible cycle
$$\mathrm{d}S = \frac{\mathrm{d}Q}{T} \quad \oint \mathrm{d}S = \oint \frac{\mathrm{d}Q}{T} = 0 \tag{3.3.9}$$

irreversible cycle
$$\mathrm{d}S > \frac{\mathrm{d}Q}{T} \quad \oint \mathrm{d}S = 0, \quad \oint \frac{\mathrm{d}Q}{T} < 0 \tag{3.3.10}$$

As we shall see in the following section, this statement can be made more precise by expressing the entropy change $\mathrm{d}S$ as a sum of two parts:

$$\boxed{\mathrm{d}S = \mathrm{d_e}S + \mathrm{d_i}S} \tag{3.3.11}$$

Here, $\mathrm{d_e}S$ is the change of the system's entropy due to exchange of energy and matter and $\mathrm{d_i}S$ is the change in entropy due to irreversible processes within the system. For a closed system that does not exchange matter, $\mathrm{d_e}S = \mathrm{d}Q/T$. The quantity $\mathrm{d_e}S$ could be positive or negative, but $\mathrm{d_i}S$ can only be equal to or greater than zero. In a cyclic process that returns the system to its initial state, since the net change in entropy must be zero, we have

$$\oint \mathrm{d}S = \oint \mathrm{d_e}S + \oint \mathrm{d_i}S = 0 \tag{3.3.12}$$

Since $\mathrm{d_i}S \geq 0$, we must have $\oint \mathrm{d_i}S \geq 0$. For a closed system, from (3.3.12) we immediately obtain the previous result (3.3.10):

$$\oint \mathrm{d_e}S = \oint \frac{\mathrm{d}Q}{T} \leq 0$$

This means that, for the system to return to its initial state, the entropy $\oint \mathrm{d_i}S$ generated by the irreversible processes within the system has to be discarded through the expulsion of heat to the exterior. There is no real system in nature that can go through a cycle of operations and return to its initial state without increasing the entropy of the exterior, or more generally the 'universe'. Every process in Nature increases the entropy, thus establishing a distinction between the past and future. *The Second Law establishes an arrow of time: the increase of entropy distinguishes the future from the past.*

STATEMENTS OF THE SECOND LAW

The limitation to the convertibility of heat to work that Carnot discovered is one manifestation of a fundamental limitation in all natural processes: it is the Second Law of thermodynamics. The second law can be formulated in many equivalent ways. For example, as a statement about a macroscopic impossibility, without any reference to the microscopic nature of matter:

It is impossible to construct an engine which will work in a complete cycle, and convert *all* the heat it absorbs from a reservoir into mechanical work.

A statement perfectly comprehensible in macroscopic, operational terms. A cyclic engine that converts all heat to work is shown in Figure 3.5. Since the reservoir or the 'exterior' only loses heat, inequality (3.3.8) is clearly violated. Such an engine is sometimes called a *perpetual motion machine of the second kind* and the Second Law is the statement that such a machine is impossible. The equivalence between this statement and Carnot's theorem can be seen easily and it is left as an exercise for the reader.

Another way of stating the Second Law is due to Rudolf Clausius (1822–1888):

Heat cannot by itself pass from a colder to a hotter body.

If heat could pass spontaneously from a colder body to a hotter body, then a perpetual motion machine of the second kind can be realized by simply making the heat Q_2 expelled by a cyclic heat engine to the colder reservoir pass by itself to the hotter reservoir. The result would be the complete conversion of the heat $(Q_1 - Q_2)$ to work.

As we have seen above, any real system that goes through a cycle of operations and returns to its initial state does so only by *increasing* the entropy of its exterior with which it is interacting. This also means that in no part of the cycle, the *sum* of entropy changes of the system and its exterior can be negative because if it were so, we could complete the rest of the cycle through a reversible transformation, which does not contribute to the change of entropy. The net result is a *decrease* of entropy in a cyclic process. Thus, the Second Law may also be stated as

The sum of the entropy changes of a system and its exterior can never decrease.

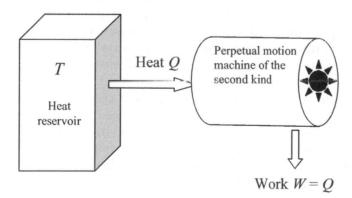

Figure 3.5 A perpetual motion machine of the second kind absorbs heat Q and converts all of it to work W. Such a machine, though consistent with the First Law, is impossible according to the Second Law. The existence of such a machine would violate the inequalities (3.3.7) and (3.3.8)

Thus, the universe as a whole can never return to its initial state. Remarkably, Carnot's analysis of heat engines has led to the formulation of a cosmological principle. The two laws of thermodynamics are best summarized by Rudolf Clausius thus:

The energy of the universe is a constant.
The entropy of the universe approaches a maximum.

3.4   Entropy, Reversible and Irreversible Processes

The usefulness of the concept of entropy and the second law depends on our ability to define entropy for a physical system in a calculable way. Using (3.3.3), if the entropy S_0 of a *reference or standard state is defined*, then the entropy of an arbitrary state S_X can be obtained through a *reversible process* that transforms the state O to the state X (see Figure 3.6):

$$S_X = S_0 + \int_0^X \frac{dQ}{T}$$

(3.4.1)

(In practice dQ is measured with the knowledge of the heat capacity using $dQ = C\, dT$.) In a real system, the transformation from the state O to the state X occurs in a finite time and it involves irreversible processes along the path I. In this process, the entropy of the system, and hence the universe, increases. *In classical thermodynamics it is assumed that every irreversible transformation that a system undergoes can also be achieved through a reversible transformation for which (3.4.1) is valid.* In other words, it is assumed that every irreversible transformation that results in a certain change in the entropy of the system can be exactly reproduced through a reversible process in which the entropy change is solely due to the exchange of heat. Since the change in entropy of the system depends only on the initial and final states, the change in entropy calculated using a reversible path will be equal to the entropy change produced by the irreversible processes. However, it must be noted that a reversible transformation from an initial state O to the final state X (Figure 3.6) may give the right value for the change in entropy of the system, but it leaves the entropy of the universe unchanged; in a reversible process, the change in entropy of the system is compensated by the opposite change in the entropy of the exterior leaving the entropy of the universe unchanged. On the other hand, the naturally occurring irreversible transformation from O to X increases the entropy of the universe. (Some authors restrict the above assumption to transformations between equilibrium states; this restriction excludes chemical reactions, in which the transformations are often from nonequilibrium state to an equilibrium state (see Chapters 4 and 7).)

A process is reversible only in the limit of infinite slowness: as perfect reversibility is approached, the speed of the process approaches zero. As Max Planck notes in his treatise [6, p. 86], 'Whether reversible processes exist in nature or not, is not *a priori* evident or demonstrable'. But, irreversibility, if it exists, has to be universal

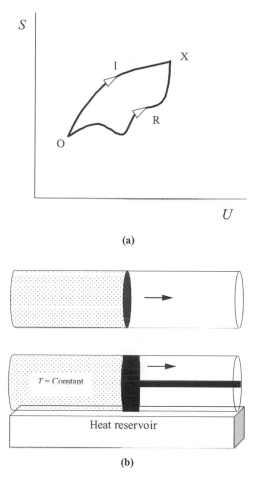

(a)

(b)

Figure 3.6 Reversible and irreversible processes: (a) The system reaches the state X from the standard state O through a path I involving irreversible processes. It is assumed that the same transformation can be achieved through a reversible transformation R. (b) An example of an irreversible process is the spontaneous expansion of a gas into a vacuum, as shown in the upper part. The same change can be achieved reversibly through an isothermal expansion of a gas that occurs infinitely slowly so that the heat absorbed from the reservoir equals the work done on the piston. In the latter case, the change in entropy can be calculated using $dS = dQ/T$

because a spontaneous decrease of entropy in one system could be utilized to decrease in entropy of any other system through appropriate interaction; a spontaneous decrease of entropy of one system implies a spontaneous decrease of entropy of all systems. *Hence, either all systems are irreversible, or none are* as Max Planck emphasized [6].

The notion of an idealized reversible path provides a convenient way for calculating entropy changes. But it is also lacking in providing the real connection between physical processes and entropy. Addressing this issue in his 1943 monograph *The Nature of Thermodynamics*, P.W. Bridgman wrote [7, p. 133]:

It is almost always emphasized that thermodynamics is concerned with reversible processes and equilibrium states and that it can have nothing to do with irreversible processes or systems out of equilibrium in which changes are progressing at a finite rate. The reason for the importance of equilibrium states is obvious enough when one reflects that temperature itself is defined in terms of equilibrium states. But the admission of general impotence in the presence of irreversible processes appears on reflection to be a surprising thing. Physics does not usually adopt such an attitude of defeatism.

Today, in most texts on thermodynamics, an irreversible transformation is usually identified by the **Clausius inequality**:

$$\boxed{dS \geq \frac{dQ}{T}} \qquad (3.4.2)$$

which we saw in the last section. But the fact that Clausius considered irreversible processes as an integral part of formulating the Second Law is generally not mentioned. In his ninth memoir, Clausius included irreversible processes explicitly into the formalism of entropy and replaced the inequality (3.4.2) by an equality [8, p. 363, eq. (71)]:

$$N = S - S_0 - \int \frac{dQ}{T} \qquad (3.4.3)$$

in which S is the entropy of the final state and S_0 is the entropy of the initial state. He identified the change in entropy due to exchange of heat with the exterior by the term dQ/T (which is compensated by equal gain or loss of heat by the exterior). Clausius wrote: 'The magnitude N thus determines the uncompensated *transformation*' ('uncompensirte Verwandlung') [8, p. 363]. It is the entropy produced by irreversible processes within the system. While dQ can be positive or negative, the Clausius inequality (3.4.2) implies that the change in entropy due to irreversible processes can only be positive:

$$N = S - S_0 - \int \frac{dQ}{T} > 0 \qquad (3.4.4)$$

Clausius also stated the Second Law as: 'Uncompensated transformations can only be positive' [8, p. 247].

Perhaps Clausius hoped to, but did not, provide a means of computing the entropy N associated with irreversible processes. Nineteenth-century thermodynamics remained in the restricted domain of idealized reversible transformation and without a theory that related entropy explicitly to irreversible processes. Some expressed the view that entropy is a physical quantity that is spatially distributed and transported, e.g. [9], but still no theory relating irreversible processes to entropy was formulated in the nineteenth century.

Noticing the importance of relating entropy to irreversible processes, Pierre Duhem (1861–1916) began to develop a formalism. In his extensive and difficult two-volume work titled *Energétique* [10], Duhem explicitly obtained expressions for the entropy produced in processes involving heat conductivity and viscosity [11]. Some of these ideas of calculating the 'uncompensated heat' also appeared in the work of the Polish researcher L. Natanson [12] and the Viennese school led by G. Jaumann [13–15], where the notions of entropy flow and entropy production were developed.

Formulation of a theory of entropy along these lines continued during the twentieth century, and today we do have a theory in which the entropy change can be calculated in terms of the variables that characterize the irreversible processes. It is a theory applicable to all systems in which the temperature is well defined at every location. For example, the modern theory relates the rate of change of entropy to the rate of heat conduction or the rates of chemical reaction. *To obtain the change in entropy, it is not necessary to use infinitely slow reversible processes.*

With reference to Figure 3.6, in the classical formulation of entropy it is often stated that, along the irreversible path I, the entropy may not be a function of the total energy and the total volume and hence it is not defined. However, for a large class of systems, the notion of *local equilibrium* makes entropy a well-defined quantity, even if it is not a function of the total energy and volume. We shall discuss the foundations of this and other approaches in Chapter 11. For such systems, entropy, which represents irreversibility in nature, can be related directly to irreversible processes.

In his pioneering work on the thermodynamics of chemical processes, Théophile De Donder (1872–1957) [16–18] incorporated the 'uncompensated transformation' or 'uncompensated heat' of Clausius into the formalism of the Second Law through the concept of affinity (which is presented in Chapter 4). Unifying all these developments, Ilya Prigogine (1917–2003) formulated the 'modern approach' incorporating irreversibility into the formalism of the second law by providing general expressions for the computation of entropy produced by irreversible processes [19–21], thus giving the 'uncompensated heat' of Clausius a sound theoretical basis. Thus, thermodynamics evolved into a theory of irreversible processes in contrast to classical thermodynamics, which is a theory of equilibrium states. We shall follow this more general approach in which, along with thermodynamic *states*, irreversible *processes* appear explicitly in the formalism.

The basis of the modern approach is the notion of *local equilibrium*. For a very large class of systems that are not in thermodynamic equilibrium, thermodynamic quantities such as temperature, concentration, pressure, internal energy remain well-defined concepts locally, i.e. one could meaningfully formulate a thermodynamic description of a system in which intensive variables such as temperature and pressure are well defined in each elemental volume, and extensive variables such as entropy and internal energy are replaced by their corresponding *densities*. Thermodynamic variables can thus be functions of position and time. This is the assumption of *local equilibrium*. There are, of course, systems in which this assumption is not a good approximation, but such systems are exceptional. In most hydrodynamic and chemical systems, local equilibrium is an excellent approximation. Modern computer simulations of molecular dynamics have shown that if initially the system is in such a state that temperature is not well defined, then in a very short time (few molecular collisions) the system relaxes to a state in which temperature is a well-defined quantity [22].

Ilya Prigogine began the modern formalism by expressing the changes in entropy as a sum of two parts [19]:

$$\boxed{dS = d_e S + d_i S}$$
(3.4.5)

in which $d_e S$ is the entropy change due exchange of matter and energy with the exterior and $d_i S$ is the entropy change due to 'uncompensated transformation', the entropy produced by the irreversible processes in the interior of the system (Figure 3.7).

Our task now is to obtain explicit expressions for $d_e S$ and $d_i S$ in terms of experimentally measurable quantities. Irreversible processes can in general be thought of as **thermodynamic forces** and **thermodynamic flows**. The thermodynamic flows are a

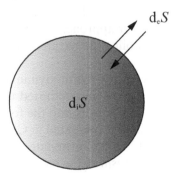

Figure 3.7 Entropy changes in a system consist of two parts: $d_i S$ due to irreversible processes and $d_e S$ due to exchange of energy and matter. According to the second law, the change $d_i S$ can only be positive. The entropy change $d_e S$ can be positive or negative

consequence of the thermodynamic forces. For example, temperature gradient is the thermodynamic *force* that causes an irreversible *flow* of heat; similarly, a concentration gradient is the thermodynamic force that causes the flow of matter (Figure 3.8). In general, the irreversible change $d_i S$, is associated with a flow dX of a quantity, such as heat or matter, that has occurred in a time dt. For the flow of heat, $dX = dQ$, the amount of heat that flowed in time dt; for the case of matter flow, $dX = dN$, moles of the substance that flowed in time dt. In each case, the change in entropy can be written in the form

$$d_i S = F \, dX \qquad (3.4.6)$$

in which F is the thermodynamic force. The thermodynamic forces are expressed as functions of thermodynamic variables such as temperature and concentrations. In the following section we shall see that, for the flow of heat shown in Figure 3.8, the thermodynamic force takes the form $F = (1/T_{cold} - 1/T_{hot})$. For the flow of matter, the corresponding thermodynamic force is expressed in terms of **affinity**, which, as noted above, is a concept developed in Chapter 4. All irreversible processes can be described in terms of thermodynamic forces and thermodynamic flows. In general, the irreversible increase in entropy $d_i S$ is the sum of all the increases due to irreversible flows dX_k. We then have the general expression

$$\boxed{d_i S = \sum_k F_k \, dX_k \geq 0} \quad \text{or} \quad \boxed{\frac{d_i S}{dt} = \sum_k F_k \frac{dX_k}{dt} \geq 0} \qquad (3.4.7)$$

Equation (3.4.7) is modern statement the Second Law of thermodynamics. The rate of entropy production due to each irreversible process is a product of the corresponding thermodynamic force F_k and the flow $J_k = dX_k/dt$ and it can only be positive.

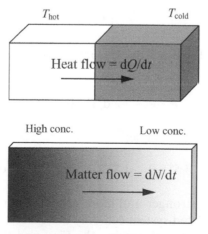

Figure 3.8 Flow of heat and diffusion of matter are examples of irreversible processes

The entropy exchange with the exterior d_eS is expressed in terms of the flow of heat and matter: For **isolated systems**, since there is no exchange of energy or matter:

$$d_eS = 0 \quad \text{and} \quad d_iS \geq 0 \tag{3.4.8}$$

For **closed systems**, which exchange energy but not matter:

$$d_eS = \frac{dQ}{T} = \frac{dU + p\,dV}{T} \quad \text{and} \quad d_iS \geq 0 \tag{3.4.9}$$

In this expression, dQ is the amount of heat exchanged by the system in a time dt. (By defining dQ in this way, we avoid the 'imperfect differentials' used in classical thermodynamics.)

For **open systems**, which exchange both matter and energy:

$$d_eS = \frac{dU + p\,dV}{T} + (d_eS)_{\text{matter}} \quad \text{and} \quad d_iS \geq 0 \tag{3.4.10}$$

d_eS_{matter} is the exchange of entropy due to matter flow. This term can be written in terms of *chemical potential*, a concept that will be developed Chapter 4. (When there is a flow of matter, as discussed in Section 2.7, $dU + p\,dV \neq dQ$, because the internal and kinetic energies of the matter flowing through the system must be included.)

Whether we consider isolated, closed or open systems, $d_iS \geq 0$. It is the statement of the Second Law in its most general form. There is another important aspect to this statement: it is valid for all subsystems, not just for the entire system. For example, if we assume that the entire system is divided into two subsystems, we not only have

$$d_iS = d_iS^1 + d_iS^2 \geq 0 \tag{3.4.11}$$

in which d_iS^1 and d_iS^2 are the entropy productions in each of the subsystems, but we also have

$$d_iS^1 \geq 0 \quad d_iS^2 \geq 0 \tag{3.4.12}$$

We cannot have, for example:

$$d_iS^1 > 0, \quad d_iS^2 < 0 \quad \text{but} \quad d_iS = d_iS^1 + d_iS^2 \geq 0 \tag{3.4.13}$$

This statement is stronger and more general than the classical statement that the entropy of an isolated system can only increase.

In summary, for closed systems, the First and the Second Laws can be stated as

$$dU = dQ + dW \tag{3.4.14}$$

$$dS = d_iS + d_eS \quad \text{in which} \quad d_iS \geq 0, \quad d_eS = dQ/T \tag{3.4.15}$$

If a transformation of the state is assumed to take place through a reversible process, $d_iS = 0$ and the entropy change is solely due to flow of heat. We then obtain the equation

$$\boxed{dU = T\,dS + dW = T\,dS - p\,dV} \tag{3.4.16}$$

which is found in texts that confine the formulation of thermodynamics to idealized reversible processes. For open systems, the changes in energy and entropy have additional contributions due to the flow of matter. In this case, though the definition of heat and work need careful consideration, there is no fundamental difficulty in obtaining dU and d_eS.

Finally, we must note that the above formulation enables us to calculate only the *changes* of entropy. It does not give us a way to obtain the absolute value of entropy. In this formalism, entropy can be known only up to an additive constant. However, in 1906, Walther Nernst (1864–1941) formulated a law which stated that *the entropy of all systems approaches zero as the temperature approaches zero* [23]:

$$S \to 0 \quad \text{as} \quad T \to 0 \tag{3.4.17}$$

Walther Nernst (1864–1941) (Reproduced with permission from the Edgar Fahs Smith Collection, University of Pennsylvania Library)

This law is often referred to as the **Third Law** of thermodynamics or the **Nernst heat theorem**. Its validity has been well verified by experiment.

The Third Law enables us to give the absolute value for the entropy. The physical basis of this law lies in the behavior of matter at low temperature that can only be explained by quantum theory. It is remarkable that the theory of relativity gave us means to define absolute values of energy and quantum theory enables us to define absolute values of entropy.

The concept of entropy has its foundation in macroscopic processes. No mention has been made about its meaning at a molecular level. In order to explain what entropy is at a molecular level, Ludwig Boltzmann (1844–1906) introduced the statistical interpretation of entropy. Box 3.2 gives an introduction to this topic; a more detailed discussion of this topic is in Chapter 17.

Box 3.2 Statistical Interpretation of Entropy

As we have seen in this chapter, the foundation of the concept of entropy as a state function is entirely macroscopic. The validity of the Second Law is rooted in the reality of irreversible processes. In stark contrast to the irreversibility of processes we see all around us, the laws of both classical and quantum mechanics possess no such irreversibility. Classical and quantum laws of motion are time symmetric: if a system can evolve from a state A to a state B then its time reversed evolution, from B to A, is also admissible. The laws of mechanics make no distinction between evolution into the future and evolution into the past. For example, the spontaneous flow of gas molecules from a location at higher concentration to a location at lower concentration and its reverse (which violates the Second Law) are both in accord with the laws of mechanics. Processes that are ruled impossible by the Second Law of thermodynamics do not violate the laws of mechanics. Yet all irreversible macroscopic processes, such as the flow of heat, are the consequence of motion of atoms and molecules that are governed by the laws of mechanics; the flow of heat is a consequence of molecular collisions that transfer energy. How can irreversible macroscopic processes emerge from reversible motion of molecules? What is the relation between entropy and the microscopic constituents of matter? The energy of a macroscopic system is the sum of the energies of its microscopic constituents. What about entropy? Addressing these questions, Ludwig Boltzmann (1844–1906) proposed an extraordinary relation; entropy is a logarithmic measure of the number of microscopic states that correspond to the macroscopic state:

$$S = k_B \ln W$$

in which W is the number of microstates corresponding to the macrostate whose entropy is S. (We shall discuss this relation in detail in Chapter 17.) The constant k_B is now called the Boltzmann constant;[*] $k_B = 1.381 \times 10^{-23} \, \text{J} \, \text{K}^{-1}$. The gas constant $R = k_B N_A$, in which N_A is the Avogadro number. The following example will illustrate the meaning of W. Consider the macrostate of a box containing a gas with N_1 molecules in one half and N_2 in the other (see figure below). Each molecule can be in one half or the other. The total number of ways in which the $(N_1 + N_2)$ molecules can be distributed between the two halves such that N_1 molecules are in one and N_2 molecules in the other is equal to W. The number of distinct 'microstates' with N_1 molecules in one half and N_2 in the other is

$$W = \frac{(N_1 + N_2)!}{N_1! N_2!}$$

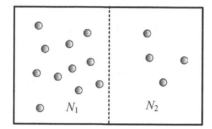

According to Boltzmann, macrostates with larger W are more probable. The irreversible increase of entropy then corresponds to evolution to states of higher probability in the future. Equilibrium states are those for which W is a maximum. In the above example, it can be shown that W reaches a maximum when $N_1 = N_2$.

*Ter Harr notes that it was Max Planck who introduced k_B in the above form; Planck also determined its numerical value (D. ter Haar, *The Old Quantum Theory*. 1967, Pergamon Press: London; 12.)

3.5   Examples of Entropy Changes due to Irreversible Processes

To illustrate how entropy changes are related to irreversible processes, we shall consider some simple examples. The examples we consider are 'discrete systems' in which the system consists of two parts that are not mutually in equilibrium. An example of a continuous system (whose description generally requires vector calculus) is presented in Appendix 3.2.

Heat Conduction

Consider an *isolated* system which we assume (for simplicity) consists of two parts, each part having a well-defined temperature, i.e. each part is locally in equilibrium. Let the temperatures of the two parts be T_1 and T_2 (as shown in Figure 3.9), with T_1 being greater than T_2. Let dQ be the amount of heat flow from the hotter part to the colder part in a time dt. Since this isolated system does not exchange entropy with the exterior, $d_eS = 0$. Also, since the volume of each part is a constant, $dW = 0$. The energy change in each part is due solely to the flow of heat: $dU_i = dQ_i$, $i = 1, 2$. In accordance with the First Law, the heat gained by one part is equal to the heat lost by the other. Therefore, $-dQ_1 = dQ_2 = dQ$. Both parts are locally in equilibrium with a well-defined temperature and entropy. The total change in entropy d_iS of the system is the sum of the changes of entropy in each part due to the flow of heat:

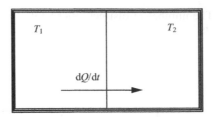

Fig. 3.9 Entropy production due to heat flow. The irreversible flow of heat between parts of unequal temperature results in an increase in entropy. The rate at which entropy is produced, d_iS/dt, is given by (3.5.3)

$$d_iS = -\frac{dQ}{T_1} + \frac{dQ}{T_2} = \left(\frac{1}{T_2} - \frac{1}{T_1}\right)dQ \qquad (3.5.1)$$

Since the heat flows irreversibly from the hotter to the colder part, dQ is positive if $T_1 > T_2$. Hence, $d_iS > 0$. In expression (3.5.1), dQ and $(1/T_1 - 1/T_2)$ respectively correspond to dX and F in (3.4.6). In terms of the *rate* of flow of heat dQ/dt, the rate of entropy production can be written as

$$\frac{d_iS}{dt} = \left(\frac{1}{T_2} - \frac{1}{T_1}\right)\frac{dQ}{dt} \qquad (3.5.2)$$

Now the rate of flow of heat or the heat current $J_Q \equiv dQ/dt$ is given by the laws of heat conduction. For example, according to the Fourier law of heat conduction, $J_Q = \alpha(T_1 - T_2)$, in which α is the coefficient of heat flow (it can be expressed in terms of the coefficient of heat conductivity and the area of cross section). Note that the 'thermodynamic flow' J_Q is driven by the 'thermodynamic force' $F = (1/T_2 - 1/T_1)$. For the rate of entropy production we have from (3.5.2):

$$\frac{d_iS}{dt} = \left(\frac{1}{T_2} - \frac{1}{T_1}\right)\alpha(T_1 - T_2) = \frac{\alpha(T_1 - T_2)^2}{T_1T_2} \geq 0 \qquad (3.5.3)$$

Owing to the flow of heat, the two temperatures eventually become equal and the entropy production ceases. This is the state of equilibrium. *Entropy production must vanish in the state of equilibrium,* which implies that the force F and the corresponding flux J_Q both vanish. In fact, we can deduce the properties of the equilibrium state by stipulating that all entropy production must vanish in that state.

From (3.5.3) we see that the entropy production rate d_iS/dt is a quadratic function of the deviation $\Delta \equiv (T_1 - T_2)$. In the state of equilibrium, the entropy production rate takes its minimum value equal to zero. This is indicated graphically in Figure 3.10a.

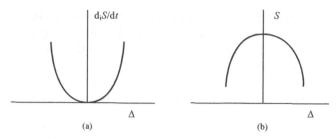

Figure 3.10 Two equivalent properties that characterize the state of equilibrium. (a) The entropy production rate d_iS/dt as a function of the difference in temperatures $\Delta \equiv (T_1 - T_2)$ of the two parts of the system shown in Figure 3.6. At equilibrium, the entropy production rate is zero. (b) At equilibrium the entropy reaches its maximum value. Both properties can be used to identify a system in equilibrium

A nonequilibrium state in which $T_1 \neq T_2$ evolves to the equilibrium state in which $T_1 = T_2 = T$ through continuous increase of entropy. Therefore, the entropy of the equilibrium state must be larger than the entropy of any nonequilibrium state. In Chapter 15, we will see explicitly that for a small deviation $\Delta = (T_1 - T_2)$ from the state of equilibrium the corresponding change ΔS is a quadratic function of Δ attaining a maximum at $\Delta = 0$ (see Fig. 3.10b).

This example illustrates the general assertion that the state of equilibrium can be characterized either by the principle of minimum (equal to zero) rate of entropy production, or the principle of maximum entropy.

IRREVERSIBLE EXPANSION OF A GAS

In a reversible expansion of a gas, the pressure of the gas and that on the piston are assumed to be the same. If we consider an isothermal expansion of a gas which is constant temperature T by virtue of its contact with a heat reservoir, the change in entropy of the gas $d_eS = dQ/T$, in which dQ is the heat flow from the reservoir to the gas that is necessary to maintain the temperature constant. This is an ideal situation. In any real expansion of a gas that takes place in a finite time, the pressure of the gas is greater than that on the piston. If p_{gas} is the pressure of the gas and p_{piston} that the pressure on the piston, the difference $(p_{gas} - p_{piston})$ is the force per unit area that moves the piston. The irreversible increase in entropy in this case is given by

$$d_iS = \frac{p_{gas} - p_{piston}}{T} dV > 0 \qquad (3.5.4)$$

In this case, the term $(p_{gas} - p_{piston})/T$ corresponds to the 'thermodynamic force' and dV/dt the corresponding "flow"'. The term $(p_{gas} - p_{piston}) dV$ may be identified as the 'uncompensated heat' of Clausius. Since the change in the volume and $(p_{gas} - p_{piston})$ have the same sign, d_iS is always positive. In this case, $dS = d_eS + d_iS = dQ/T + (p_{gas} - p_{piston})\, dV/T$. In the case of an ideal gas, since the energy is only a function of T, the initial and final energies of the gas remain the same; the heat absorbed is equal to the work done in moving the piston $p_{piston}\, dV$. For a given change in volume, the maximum work is obtained for a reversible process in which $p_{gas} = p_{piston}$.

3.6   Entropy Changes Associated with Phase Transformations

In this section we will consider a simple example of entropy exchange d_eS. Changes in the phase of a system, from a solid to a liquid or a liquid to vapor (as shown in Figure 1.3), provide a convenient situation because, at the melting or boiling point, the temperature remains constant even when heat is being exchanged. Hence, in the expression for the entropy change associated with the heat exchange, $d_eS = dQ/T$, the temperature T remains constant. The total entropy change ΔS due to the exchange of heat ΔQ is now easy to determine. In a solid-to-liquid transition, for example, if the melting temperature is T_m, we have

$$\Delta S = \int_0^{\Delta Q} \frac{dQ}{T_m} = \frac{\Delta Q}{T_m} \qquad (3.6.1)$$

As was discovered by Joseph Black (see Section 1.5), the heat absorbed, 'the latent heat', converts the solid to a liquid at the fixed temperature. Generally, this change happens at a fixed pressure and, hence, we may equate ΔQ to ΔH, the enthalpy change associated with melting. The enthalpy associated with the conversion of 1 mol of the solid to liquid is called the **molar enthalpy of fusion** ΔH_{fus}. The corresponding change in entropy, the **molar entropy of fusion** ΔS_{fus}, can now be written as

$$\boxed{\Delta S_{fus} = \frac{\Delta H_{fus}}{T_m}} \qquad (3.6.2)$$

Water, for example, has a heat of fusion of $6.008\,kJ\,mol^{-1}$ and a melting temperature of $273.15\,K$ at a pressure of $1.0\,atm$. When 1 mol of ice turns to water, the entropy change $\Delta S_{fus} = 21.99\,J\,K^{-1}\,mol^{-1}$.

Similarly, if the conversion of a liquid to vapor occurs at a constant pressure at its boiling point T_b, then the **molar entropy of vaporization** ΔS_{vap} and the **molar enthalpy of vaporization** ΔH_{vap} are related by

$$\boxed{\Delta S_{vap} = \frac{\Delta H_{vap}}{T_b}} \qquad (3.6.3)$$

The heat of vaporization of water is $40.65\,kJ\,mol^{-1}$. Since the boiling point is $373.15\,K$ at a pressure of $1.0\,atm$, from the above equation it follows that the molar entropy

Table 3.1 Enthalpies of fusion of and vaporization at $p = 101.325$ kPa $= 1.0$ atm and the corresponding transition temperatures

Substance	T_m/K	ΔH_{fus}/kJ mol^{-1}	T_b/K	ΔH_{vap}/kJ mol^{-1}
H_2O	273.15	6.01	373.15	40.65
CH_3OH	175.5	3.18	337.7	35.21
C_2H_5OH	159.0	5.02	351.4	38.56
CH_4	90.69	0.94	111.7	8.19
CCl_4	250.15	3.28	349.9	29.82
NH_3	195.4	5.66	239.8	23.33
CO_2 (sublimes)	$T_{sub} = 194.65$			$\Delta H_{sub} = 25.13$
CS_2	161.6	4.40	319.1	26.74
N_2	63.15	0.71	77.35	5.57
O_2	54.36	0.44	90.19	6.82

Source: D. R. Lide (ed.) *CRC Handbook of Chemistry and Physics*, 75th edition. 1994, CRC Press: Ann Arbor.

change $\Delta S_{vap} = 108.96$ J K^{-1} mol^{-1}, about five times the entropy change associated with the melting of ice. Since entropy increases with volume, the large increase in volume from about 18 mL (volume of 1 mol of water) to about 30 L (volume of 1 mol of steam at $p = 1$ atm) is partly responsible for this larger change. The molar enthalpies of fusion and vaporization of some compounds are given in Table 3.1.

3.7 Entropy of an Ideal Gas

In this section we shall obtain the entropy of an ideal gas. Being a state function, entropy of an ideal gas can be expressed as a function of its volume, temperature and the amount in moles. For a closed system in which the changes of entropy are only due to flow of heat, if we assume that the changes in volume V and temperature T take place so as to make $d_iS = 0$, then we have seen that (see eqn. (3.4.16)) dU = TdS + dW. If dW = −pdV, and if we express dU as function of V and T we obtain:

$$T dS = \left(\frac{\partial U}{\partial V}\right)_T dV + \left(\frac{\partial U}{\partial T}\right)_V dT + p \, dV \qquad (3.7.1)$$

For an ideal gas, $(\partial U/\partial V)_T = 0$, because the energy U is only a function of T – as was demonstrated in the experiments of Joule and Gay-Lussac and others (see Section 1.3, Equation (1.3.6)). Also, by definition $(\partial U/\partial T)_V = NC_{mV}$ in which C_{mV} is the molar heat capacity at constant volume, which is found to be a constant. Hence (3.7.1) may be written as

$$dS = \frac{p}{T} dV + NC_{mV} \frac{dT}{T} \qquad (3.7.2)$$

Using the ideal gas law, $pV = NRT$, (3.7.2) can be integrated to obtain

$$S(V,T,N) = S_0(V_0,T_0,N) + NR\ln(V/V_0) + NC_{mV}\ln(T/T_0) \qquad (3.7.3)$$

in which S_0 in the entropy of the initial state (V_0, T_0). Since $U = C_{mV}NT + U_0$ for an ideal gas, entropy can be written as a function of V, N and U. As described in Box 3.3, entropy is an extensive function. In expression (3.7.3), the extensivity of S as a function of V and N is not explicit because $S_0(V_0, T_0, N)$ contains terms that make S extensive. The requirement that entropy is extensive, i.e. $\lambda S(V, T, N) = S(\lambda V, T, \lambda N)$, can be used to show (Exercise 3.10) that the entropy of an ideal gas has the form

$$\boxed{S(V,T,N) = N[s_0 + R\ln(V/N) + C_{mV}\ln(T)]} \qquad (3.7.4)$$

in which s_0 is a constant. In this form, the extensivity of S is explicit and it is easy to verify that $\lambda S(U, T, N) = S(\lambda U, T, \lambda N)$.

3.8   Remarks about the Second Law and Irreversible Processes

As was emphasized by Planck [24], the statement of the Second Law and the concept of entropy can be made entirely macroscopic. It is perhaps why Einstein was convinced that thermodynamics, 'within the framework of applicability of its basic concepts, it will never be overthrown'. Many modern expositions present the Second Law and entropy starting with their microscopic definitions based on probability that belie their independence from microscopic theories of matter.

The Second Law is universal. In fact, its universality gives us a powerful means to understand the thermodynamic aspects of real systems through the usage of ideal systems. A classic example is Planck's analysis of radiation in thermodynamic equilibrium with matter (the 'black-body radiation') in which Planck considered idealized simple harmonic oscillators interacting with radiation. Planck considered simple harmonic oscillators not because they are good approximations of molecules, but because the properties of radiation in thermal equilibrium with matter are universal, regardless of the particular nature of matter that it is interacting with. The conclusions one arrives at using idealized oscillators and the laws of thermodynamics must also be valid for all other forms of matter, however complex.

In the modern context, the formulation summarized in Figure 3.7 is fundamental for understanding thermodynamic aspects of self-organization, evolution of order and life that we see in Nature. When a system is isolated, $d_eS = 0$. In this case, the entropy of the system will continue to increase due to irreversible processes and reach the maximum possible value, the state of thermodynamic equilibrium. In the state of equilibrium, all irreversible processes cease. When a system begins to exchange entropy with the exterior, then, in general, it is driven away from equilibrium and the entropy-producing irreversible processes begin to operate. The exchange of entropy is due to exchange of heat and matter. The entropy flowing out of the

Box 3.3 Extensivity of Energy and Entropy

At a fixed pressure and temperature, if the amount of substance N is changed by a factor λ, the volume V also changes by the same factor. In many cases, the system's entropy S and energy U also change by the same factor λ. This property is called *extensivity*. Entropy is an extensive function of U, V and N: $S = S(U, V, N)$. That entropy is an *extensive function* can be expressed mathematically as

$$\lambda S(U, V, N) = S(\lambda U, \lambda V, \lambda N)$$

Similarly, energy is a function of S, V and N: $U = U(S, V, N)$ and

$$\lambda U(S, V, N) = U(\lambda S, \lambda V, \lambda N)$$

Physically, extensivity implies that the combining of λ identical systems results in a larger system whose entropy is λ times the entropy of each of the systems. It means the processes of combining λ identical systems is reversible with no entropy or energy change. Here is an example. Initially, two identical compartmentalized subsystems contain an ideal gas, both at the same p and T (see figure below). The process of removing the wall between the two subsystems and creating a system that is twice as large requires neither work nor heat. Hence, the energy of the larger system is the sum of the energies of the subsystems.

Also, since the wall does not contribute to entropy, the process is reversible with no entropy change: $d_e S = d_i S = 0$. Therefore, we deduce that the initial entropy, which is the sum of the entropies of the two identical systems, equals the entropy of the final larger system. In this sense the entropy and energy of most systems can be assumed to be extensive functions.

On the other hand, entropy and energy are *not* extensive functions as expressed in the equations above when the process of combining identical systems to create a larger system involves a change in energy and entropy. Such is the case for very small systems, whose *surface* energy and entropy cannot be ignored as they can be for larger systems. When two small drops of liquid are brought into contact, for example, they spontaneously coalesce to form a larger drop (see figure above). Because the surface of the larger drop is not equal to the sum of the surfaces of the two initial drops, the energy of the larger drop does not equal the sum of energies of the two smaller drops. As we shall see in later chapters, $d_i S > 0$ in this process. Note also that it requires work to break the bigger drop into two smaller drops. Hence, neither entropy nor energy obeys the above equations. However, there is no fundamental difficulty in taking the energy and entropy of the surface into account and formulating the thermodynamics of small systems.

system is always larger than the entropy flowing into the system, the difference arising due to entropy produced by irreversible processes within the system. As we shall see in the following chapters, systems that exchange entropy with their exterior do not simply increase the entropy of the exterior, but may undergo dramatic spontaneous transformations to 'self-organization'. *The irreversible processes that produce entropy create these organized states.* Such self-organized states range from convection patterns in fluids to life. Irreversible processes are the driving force that creates this order.

Appendix 3.1 The Hurricane as a Heat Engine

The mechanism of a hurricane is essentially that of a heat engine, as shown in Figure A3.1 in the cycle ABCD. The maximum intensity of a hurricane, i.e. the maximum hurricane wind speed (Table A3.1), can be predicted using Carnot's theorem for the efficiency of a heat engine.

In a hurricane, as the wind spirals inwards towards the eye at low pressure, enthalpy (heat) is absorbed at the warm ocean–air interface in an essentially isothermal processes: water vaporizes and mixes with the air, carrying with it the enthalpy of vaporization (segment AB). When this moist air reaches the hurricane's eyewall, it rises rapidly about 15 km along the eyewall. Since the pressure decreases with altitude, it expands adiabatically and cools (segment BC). As the rising moist air's temperature drops, the water vapor in it condenses as rain, releasing the enthalpy of vaporization (latent heat) a part of which is radiated into outer space. In a real hurricane, the air at the higher altitude flows out into the weather system. Theoretically, in order to close the Carnot cycle, it could be assumed that the enthalpy of

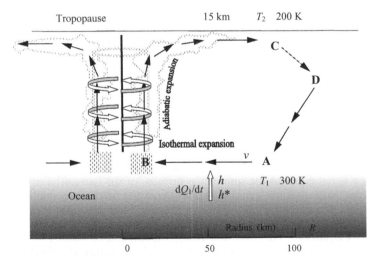

Figure A3.1 The hurricane operates as a heat engine, converting part of the heat absorbed at the ocean surface to mechanical energy of the hurricane wind

Table A3.1   The Saffir–Simpson hurricane intensity scale

Category	Min. central pressure (kPa)	Max. sustained wind speed	
		m s^{-1}	mph
1	>98.0	33–42	74–95
2	97.9–96.5	43–49	96–110
3	96.4–94.5	50–58	111–130
4	94.4–92.0	59–69	131–155
5	<92.0	>70	>156

vaporization is lost in an isothermal process (segment CD). The last step (segment DA) of the cycle is an adiabatic compression of dry air. During the cycle, a part of the enthalpy absorbed from the ocean is converted into mechanical energy of the hurricane wind.

The 'hurricane heat engine' operates between the ocean surface temperature T_1 (about 300 K) and the lower temperature T_2 (about 200 K) at the higher altitude, close to the upper boundary of the troposphere (tropopause). Let us look at the relationship between the heat absorbed at the ocean surface and the mechanical energy of the hurricane wind. In a time dt, if dQ_1 is the heat absorbed at the ocean surface, dQ_2 is the heat radiated at the higher altitude and dW is the amount of heat converted to mechanical energy of the hurricane wind, then, according to the First Law:

$$\frac{dQ_1}{dt} = \frac{dW}{dt} + \frac{dQ_2}{dt} \qquad (A3.1.1)$$

Furthermore, according to Carnot's theorem:

$$\frac{dW}{dt} \leq \left(1 - \frac{T_2}{T_1}\right)\frac{dQ_1}{dt} \qquad (A3.1.2)$$

In a hurricane, the mechanical energy in the wind is converted to heat due to wind friction, almost all of it at the ocean surface. This heat in turn contributes to dQ_1/dt, the rate at which heat is absorbed at the ocean surface. When the hurricane is in a steady state, i.e. when all the flows are constant, all the mechanical energy entering the system as wind motion is converted to heat at the ocean surface: the rate of heat generation due to wind friction is equal to dW/dt. Thus, the rate at which heat enters the Carnot cycle, dQ_1/dt, consists of two parts:

$$\frac{dQ_1}{dt} = \frac{dQ_{10}}{dt} + \frac{dW}{dt} \qquad (A3.1.3)$$

dQ_{10}/dt is the rate at which heat enters the system in the absence of heating due to wind friction. Using (A.3.1.3) in Equation (A3.1.2), it is easy to see that

$$\frac{\mathrm{d}W}{\mathrm{d}t} \leq \left(\frac{T_1 - T_2}{T_2}\right)\frac{\mathrm{d}Q_{10}}{\mathrm{d}t} \qquad (A3.1.4)$$

A detailed study of the physics of the hurricane wind shows that the rate of heat generation per unit area of the ocean surface (i.e. vertically integrated heating) is equal to $C_D\rho|v|^3$, in which C_D is a constant, ρ is the air density and v is the wind velocity. The total amount of heat generated is obtained by integrating over the circular surface of radius R (from the center of the eye to the outer edge of the hurricane), which is the area of contact between the hurricane wind and the ocean. At steady state, since this integral equals $\mathrm{d}W/\mathrm{d}t$, we have

$$\frac{\mathrm{d}W}{\mathrm{d}t} = 2\pi\int_0^R C_D\rho|v|^3\, r\, \mathrm{d}r \qquad (A3.1.5)$$

The term $\mathrm{d}Q_{10}/\mathrm{d}t$ is the rate at which enthalpy enters the inflowing dry air (segment AB). This energy is essentially the enthalpy of vaporization. It is proportional to the difference between specific enthalpies (enthalpies per unit mass) of the air saturated with moisture very close to the ocean surface h^* and the enthalpy of the inflowing dry air h (see Figure A3.1); it is also proportional to the wind velocity at the ocean surface. Thus, the enthalpy entering the system per unit area is $C_h\rho(h^* - h)$ $|v|$. The total amount of enthalpy $\mathrm{d}Q_{10}/\mathrm{d}t$ entering the hurricane system in this process equals the integral of this expression over the circular surface of radius R:

$$\frac{\mathrm{d}Q_{10}}{\mathrm{d}t} = 2\pi\int_0^R C_h\rho(h^* - h)|v|r\, \mathrm{d}r \qquad (A3.1.6)$$

in which C_h is constant. Combining (A3.1.4), (A3.1.5) and (A3.1.6) we obtain

$$\int_0^R C_D\rho|r|^3\, r\mathrm{d}r \leq \left(\frac{T_1 - T_2}{T_2}\right)\int_0^R C_h\rho(h^* - h)|v|r\, \mathrm{d}r$$

If we assume that the dominant contribution to this integral comes from the region where the velocity is maximum, we can write

$$C_D\rho|v_{\max}|^3 \leq \left(\frac{T_1 - T_2}{T_2}\right)C_h\rho(h^* - h)|v_{\max}|$$

Thus, we arrive at the result

$$|v_{\max}|^2 \approx \left(\frac{T_1 - T_2}{T_2}\right)\frac{C_h}{C_D}(h^* - h) \qquad (A3.1.7)$$

Bister and Emanuel [25] have shown that the above result can be obtained through a more rigorous calculation. All the terms on the right-hand side are experimentally measured or theoretically estimated. Comparison of theory and experimental data suggests that the ratio C_k/C_D is in the range 0.75–1.5 [26]. Kerry Emanuel, the originator of the above theory, has demonstrated that (A3.1.7) leads to remarkably good estimates of the hurricane wind speeds [4, 27].

When the system is in a steady state, the heat converted into mechanical energy of the hurricane wind balances the conversion of the wind energy back to heat. Under these conditions, if expression (A3.1.3) is used in (A3.1.1) we obtain $dQ_{10}/dt = dQ_2/dt$, which implies heat of vaporization absorbed by the hurricane wind at the ocean surface is released at higher altitude where the water condenses. This heat is ultimately radiated out of Earth's atmosphere. Thus, the vaporization and condensation of water vapor is a mechanism that transports heat from the oceans to higher altitudes where it is radiated into outer space. If this mechanism did not exist, the heat would be transported entirely through air currents, currents that would be very intense.

Appendix 3.2 Entropy Production in Continuous Systems

We consider a nonequilibrium situation in which a heat-conducting material is in contact with a hot reservoir on one side and a cold reservoir on the other (see Figure A3.2). We further assume that the conductor is insulated in such a way that it exchanges heat only with the heat reservoirs. After going through an initial transient change in temperature, such a system will settle into a steady state in which there is a uniform temperature gradient and a steady flow of heat. We will calculate the rate of entropy production at this steady state.

As each elemental quantity of heat dQ flows through the system the entropy increases. At steady state, there is a steady flow of heat J_Q which is the amount of heat flowing per unit area per second ($\mathrm{J\,m^{-2}\,s^{-1}}$). Since only one space direction is involved in this problem, we shall ignore the vectorial aspect of J_Q. For simplicity, we shall assume that the conductor has an area of cross-section equal to unity. In this case the rate of flow of heat $dQ/dt = J_Q$. For continuous systems, the entropy production due to flow of heat given by (3.5.2) should be replaced by the entropy production due to flow of heat through each infinitesimal segment of the heat conductor of width dx. The corresponding *entropy production per unit volume* at the

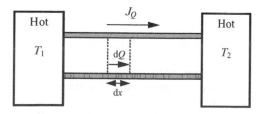

Figure A3.2 The continuous flow of heat is associated with entropy production

point x is denoted by $\sigma(x)$. The quantity $(1/T_1 - 1/T_2)$ is now replaced by the change of the quantity $1/T$ over the length dx, namely $(\partial/\partial x)(1/T)dx$. Combining all these terms, we can now write the entropy production for flow of heat across a segment dx:

$$\sigma(x)dx = J_Q\left(\frac{\partial}{\partial x}\frac{1}{T}\right)dx \qquad (A3.2.1)$$

According to the Fourier law of heat conduction, $J_Q = -\kappa(\partial T/\partial x)$ in which κ is the heat conductivity.

Substituting this expression into (A3.2.1) we can obtain

$$\sigma(x)dx = \alpha\frac{1}{T^2}\left(\frac{\partial T}{\partial x}\right)^2 dx \geq 0 \qquad (A3.2.2)$$

The above expression gives the entropy production at each location x, i.e. the local entropy production. It is the entropy produced per unit time due to flow of heat through the segment of width dx at the location x. As required by the Second Law, it is positive. At steady state, the temperature of the segment is constant. Hence, the entropy of the segment itself is not increasing; the entropy increase is due to the flow of heat down a temperature difference dT across the segment.

To obtain the total rate of entropy production due to the flow of heat from one end of the conductor to the other, we integrate the expression (A3.2.1) over the length l of the conductor. It is easy to see that the result can be written as

$$\frac{d_iS}{dt} = \int_0^l \sigma(x)dx = \int_0^l J_Q\left(\frac{\partial}{\partial x}\frac{1}{T}\right)dx \qquad (A3.2.3)$$

When the system has reached steady state, since J_Q is constant, we can integrate this expression to obtain

$$\frac{d_iS}{dt} = J_Q\left(\frac{1}{T_2} - \frac{1}{T_1}\right) \qquad (A3.2.4)$$

References

1. Mendoza, E. (ed.), *Reflections on the Motive Force of Fire by Sadi Carnot and Other Papers on the Second Law of Thermodynamics by É. Clapeyron and R. Clausius*. 1977, Peter Smith: Gloucester, MA.
2. Kastler, A., L'Oeuvre posthume de Sadi Carnot. In *Sadi Carnot et l'Essor de la Thermodynamique*. 1976, Editions du CNRS: Paris.
3. Segrè, E., *From Falling Bodies to Radio Waves*. 1984, W.H. Freeman: New York.
4. Emanuel, K.A., Thermodynamic control of hurricane intensity. *Nature*, **401** (1999) 665–669.

5. Clausius, R., *Mechanical Theory of Heat*. 1867, John van Voorst: London; 357.
6. Planck, M., *Treatise on Thermodynamics*, third edition. 1945, Dover: New York; 86.
7. Bridgman, P.W., *The Nature of Thermodynamics*. 1943, Harvard University Press: Cambridge, MA; 133.
8. Clausius, R., *Mechanical Theory of Heat*. 1867, John van Voorst: London.
9. Bertrand, J.L.F., *Thermodynamique*. 1887, Gauthier-Villars: Paris.
10. Duhem, P., *Energétique*. 1911, Gauthiers-Villars: Paris.
11. Brouzeng, P., Duhem's Contribution to the development of modern thermodynamics. In *Thermodynamics: History and Philosophy*, K. Martinás, L. Ropolyi and P. Szegedi (eds). 1991, World Scientific: London; 72–80.
12. Natanson, L., *Zsch. Physik. Chem.*, **21** (1896) 193.
13. Lohr, E., *Math. Naturw. Kl.*, **339** (1916) 93.
14. Jaumann, G., *Sitz. Ber. Akad. Wiss. Wien, Math. Naturw. Kl., Abt. IIA*, **120**(2) (1911) 385.
15. Jaumann, G., *Denkschr. Akad. Wiss. Wien, Math. Naturw. Kl.*, **95** (1918) 461.
16. De Donder, T., *Lecons de Thermodynamique et de Chimie-Physique*. 1920, Gauthiers-Villars: Paris.
17. De Donder, T., *L'Affinité*. 1927, Gauthiers-Villars: Paris.
18. De Donder, T. and Van Rysselberghe, P., *Affinity*. 1936, Stanford University Press: Menlo Park, CA.
19. Prigogine, I., *Etude Thermodynamique des Processus Irreversibles*. 1947, Desoer: Liege.
20. Prigogine, I., *Introduction to Thermodynamics of Irreversible Processes*. 1967, Wiley: New York.
21. Prigogine, I. and Defay, R., *Chemical Thermodynamics*. 1954, Longmans: London.
22. Alder, B.J. and Wainright, T., Molecular dynamics by electronic computers. In *Proceedings of the International Symposium on Transport Processes in Statistical Mechanics*, I. Prigogine (ed.). 1958. Interscience: New York.
23. Nernst, W., *A New Heat Theorem*. Dover: New York; 85.
24. Planck, M., Treatise on Thermodynamics. Third edition ed. 1945, New York: Dover.
25. Bister, M. and Emanuel, K.A., Dissipative heating and hurricane intensity. *Meteorol. Atmos. Phys.*, **50** (1998) 233–240.
26. Emanuel, K.A., Sensitivity of tropical cyclones to surface exchange coefficients and a revised steady-state model incorporating eye dynamics. *J. Atmos. Sci.*, **52** (1995) 3969–3976.
27. Emanuel, K.A., Tropical cyclones. *Annu. Rev. Earth. Planet. Sci.*, **31** (2003) 75–104.

Examples

Example 3.1   Draw the S versus T diagram for the Carnot cycle.
Solution   During the reversible adiabatic changes the change in entropy is zero. Hence, the S–T graph is as shown:

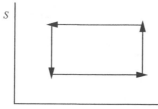

Example 3.2   A heat pump is used to maintain the inside temperature of a house at 20.0°C when the outside temperature is 3.0°C. What is the minimum amount of work necessary to transfer 100.0 J of heat to the inside of the house.

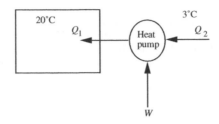

Solution   The ideal heat pump is the Carnot's engine running in reverse, i.e. it uses work to pump heat from a lower temperature to a higher temperature. For an ideal pump, $Q_1/T_1 = Q_2/T_2$. Thus, if $Q_1 = 100.0$ J and $T_2 = 293.0$ K, we have $T_1 = 276.0$ K:

$$Q_2 = 276.0\,\mathrm{K}(100.0\,\mathrm{J}/293.0\,\mathrm{K}) = 94.0\,\mathrm{J}$$

Thus, the heat pump absorbs 94.0 J from the outside and delivers 100.0 J to the inside. Form the first law it follows that the work $W = Q_1 - Q_2$ necessary is 100.0 J − 94.0 J = 6.0 J.

Example 3.3   The heat capacity of a solid is $C_p = 125.48\,\mathrm{J\,K^{-1}}$. What is the change in its entropy if it is heated from 273.0 K to 373.0 K?
Solution   This is simple case of heat transfer. $\mathrm{d}_e S = \mathrm{d}Q/T$. Hence:

$$S_{\mathrm{final}} - S_{\mathrm{initial}} = \int_{T_i}^{T_f} \frac{\mathrm{d}Q}{T} = \int_{T_i}^{T_f} \frac{C_p\,\mathrm{d}T}{T} = C_p \ln\left(\frac{T_f}{T_i}\right)$$
$$= 125.48\,\mathrm{J\,K^{-1}} \ln(373/273) = 39.2\,\mathrm{J\,K^{-1}}$$

Example 3.4   A container with N moles of ideal gas with an initial volume V_i is in contact with a heat reservoir at T_0 K. The gas expands isothermally to a volume V_f. Calculate: (a) the amount of heat absorbed by the gas in this expansion; (b) the increase in the entropy of the gas.

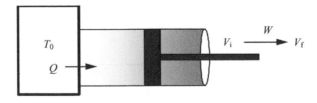

Solution   The energy of an ideal gas depends only on its temperature. Hence, the heat absorbed Q must equal the work done W by the gas. The work done by the gas is

$$W = \int_{V_i}^{V_f} p \, dV = \int_{V_i}^{V_f} \frac{NRT_0}{V} dV = NRT_0 \ln\left(\frac{V_f}{V_i}\right) = Q$$

Since the process occurs isothermally, the change in entropy is

$$S_f - S_i = \int_{S_i}^{S_f} \frac{dQ}{T_0} = \frac{Q}{T_0} = NR \ln\left(\frac{V_f}{V_i}\right)$$

Note that the change in entropy can also be calculated using (3.7.4).

Exercises

3.1   Show the equivalence between a perpetual motion machine of the second kind and Carnot's theorem.

3.2   A refrigerator operating reversibly extracts 45.0 kJ of heat from a thermal reservoir and delivers 67.0 kJ as heat to a reservoir at 300 K. Calculate the temperature of the reservoir from which heat was removed.

3.3   What is the maximum work that can be obtained from 1000.0 J of heat supplied to a steam engine with a high-temperature reservoir at 120.0 °C if the condenser is at 25.0 °C?

3.4   Using the data shown in Figure 2.9, estimate the amount of entropy radiated by the earth per hour.

3.5   The heat of combustion of gasoline is approximately $47 \, kJ \, g^{-1}$. If a gasoline engine operated between 1500 K and 750 K, what is the maximum height that 5.0 g of gasoline can lift an aircraft that weighs 400 kg?

3.6   The heat capacity C_p of a substance is given by

$$C_p = a + bT$$

where $a = 20.35 \, J \, K^{-1}$ and $b = 0.2 \, J \, K^{-2}$. Calculate the change in entropy in increasing the temperature of this substance from 298.15 K to 304.0 K.

3.7   When 0.5 J of heat passes between two large bodies in contact at temperatures 70 °C and 25 °C, what is the change of entropy? If this occurs in 0.23 s, what is the rate of change of entropy d_iS/dt?

3.8 What is the entropy of $1.00\,L$ of $N_2(g)$ at $T = 350.0\,K$ and $p = 20.25\,atm$ given that the standard ($p = 1.00$ bar, $T = 298.15\,K$) molar entropy $S^0_m = 191.6\,J\,K^{-1}\,mol^{-1}$? (Calculate the molar amount of N_2 using the ideal gas equation.)

3.9 Which of the following are *not* extensive functions:

$$S_1 = (N/V)[S_0 + C_V \ln T + R \ln V]$$

$$S_2 = N[S_0 + C_V \ln T + R \ln(V/N)]$$

$$S_3 = N^2[S_0 + C_V \ln T + R \ln(V/N)]$$

3.10 Apply the condition $S(\lambda V, T, \lambda N) = \lambda S(V, T, N)$ to

$$S(V, T, N) = S_0(V_0, T_0, N) + NR \ln(V/V_0) + NC_{mV} \ln(T/T_0),$$

differentiate it with respect to λ, set $\lambda = 1$, solve the resulting differential equation for S_0 and show that

$$S(V, T, N) = N[s_0 + R \ln(V/N) + C_{mV} \ln(T)]$$

3.11 (i) Find out how much solar energy reaches the surface of the Earth per square meter per second. (This is called the 'solar constant'.)
(ii) The present cost of electricity in the USA is in the range $0.12–0.18/kWh ($1\,kWhour = 10^3 \times 3600\,J$). Assume that the efficiency of commercial solar cells is only about 10%, that they can last 30 years and that they can produce power for 5 h/day on average. How much should $1\,m^2$ of solar cells cost so that the total energy it can produce amounts to about $0.15/kWh. (Make reasonable assumptions for any other quantity that is not specified.)

4 ENTROPY IN THE REALM OF CHEMICAL REACTIONS

4.1 Chemical Potential and Affinity: The Thermodynamic Force for Chemical Reactions

Nineteenth-century chemists did not pay much attention to the developments in thermodynamics, while experiments done by chemists – such as Gay-Lussac's on the expansion of a gas into vacuum – were taken up and discussed by the physicists for their thermodynamic implications. The interconversion of heat into other forms of energy was a matter of great interest mostly to the physicists. Among the chemists, the concept of heat as an indestructible caloric, a view supported by Lavoisier, largely prevailed [1]. As we noted in Chapter 2, the work of the Russian chemist Germain Hess on heats of reaction was an exception.

Motion is explained by the Newtonian concept of force, but what is the 'driving force' that was responsible for chemical change? Why do chemical reactions occur at all, and why do they stop at certain points? Chemists called the 'force' that caused chemical reactions *affinity*, but it lacked a clear physical meaning and definition. For the chemists who sought quantitative laws, defining of affinity, as precisely as Newton's defined mechanical force, was a fundamental problem. In fact, this centuries-old concept had different interpretations at different times. 'It was through the work of the thermochemists and the application of the principles of thermodynamics as developed by the physicists', notes the chemistry historian Henry M. Leicester 'that a quantitative evaluation of affinity forces was finally obtained' [1 (p. 203)]. The thermodynamic formulation of affinity as we know it today is due to Théophile De Donder (1872–1957), the founder of the Belgian school of thermodynamics.

De Donder's formulation of chemical affinity [2, 3] was founded on the concept *chemical potential*, one of the most fundamental and far-reaching concepts in thermodynamics that was introduced by Josiah Willard Gibbs (1839–1903). There were earlier attempts: in the nineteenth century, the French chemist Marcellin Berthelot (1827–1907) and the Danish Chemist Julius Thompsen (1826–1909) attempted to quantify affinity using heats of reaction. After determining the heats of reactions for a large number of compounds, in 1875 Berthelot proposed a 'principle of maximum work' according to which 'all chemical changes occurring without intervention of outside energy tend toward the production of bodies or of a system of bodies which liberate more heat' [1 (p. 205)]. But this suggestion met with criticism from Hermann Helmholtz and Walther Nernst (1864–1941), who noted that the principle could not apply to spontaneous endothermic chemical change that absorbed heat. The controversy continued until the concept of a chemical potential

Introduction to Modern Thermodynamics   Dilip Kondepudi
© 2008 John Wiley & Sons, Ltd

formulated by Gibbs (who was a professor at Yale University) became known in Europe. Later, it became clear that it was not the heat of reaction that characterized the evolution to the state of equilibrium, but another thermodynamic quantity called 'free energy'. As we shall describe in detail, De Donder gave a precise definition of affinity using the concept of chemical potential and, through his definition of affinity, obtained a relation between the rate of entropy change and chemical reaction rate. In De Donder's formulation, the Second Law implies that chemical reactions drive the system to a state of thermodynamic equilibrium in which the affinities of the reactions equal zero.

J Willard Gibbs (1839–1903) (Reproduced with permission from the Edgar Fahs Smith Collection, University of Pennsylvania Library)

CHEMICAL POTENTIAL

Josiah Willard Gibbs introduced the idea of chemical potential in his famous work titled *On the Equilibrium of Heterogeneous Substances*, published in 1875 and 1878 [4–6]. Gibbs published his work in the *Transactions of the Connecticut Academy of Sciences*, a journal that was not widely read. This fundamental work of Gibbs remained in relative obscurity until it was translated into German by Wilhelm Ostwald (1853–1932) in 1892 and into French by Henri Le Châtelier (1850–1936) in 1899 [1]. Much of the present-day presentation of classical equilibrium thermodynamics can be traced back to this important work of Gibbs.

Gibbs considered a heterogeneous system (Figure 4.1) that consisted of several homogeneous parts, each part containing various substances $s_1, s_2, \ldots, s_n$ of masses $m_1, m_2, \ldots, m_n$. His initial consideration did not include chemical reactions between these substances, but was restricted to their exchange between different homogeneous parts of a system. Arguing that the change in energy dU of a homogeneous part must be proportional to changes in the masses of the substances, $dm_1, dm_2, \ldots, dm_n$, Gibbs introduced the equation

$$dU = T\,dS - p\,dV + \mu_1 dm_1 + \mu_2 dm_2 + \ldots + \mu_n dm_n \qquad (4.1.1)$$

for each homogeneous part. The coefficients μ_k are called the **chemical potentials**. The heterogeneous systems considered included different phases of a substance that exchanged matter. The considerations of Gibbs, however, were restricted to transformations between states in equilibrium. This restriction is understandable from the viewpoint of the classical definition of entropy, which required the system to be in equilibrium and the transformations between equilibrium states to be reversible so that $dQ = T\,dS$. In the original formulation of Gibbs, the changes in the masses dm_k in Equation (4.1.1) were due to exchange of the substances between the homo-

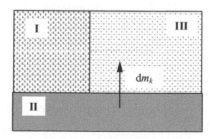

Figure 4.1 A heterogeneous system considered by Gibbs in which substances were exchanged between the parts I, II and III. The change in energy dU of any part when matter was exchanged reversibly is given by (4.1.1)

geneous parts, a situation encountered when various phases of a substance exchange matter and reach equilibrium.

It is more convenient to describe chemical reactions by the change in the molar amounts of the reactants rather than the change in their masses, because chemical reaction rates and the laws of diffusion are most easily formulated in terms of molar amounts. Therefore, we shall rewrite Equation (4.1.1) in terms of the molar amounts N_k of the constituent substances:

$$dU = TdS - pdV + \mu_1 dN_1 + \mu_2 dN_2 + \ldots + \mu_n dN_n$$

i.e.

$$\boxed{dU = T\,dS - p\,dV + \sum_1^n \mu_k\,dN_k} \tag{4.1.2}$$

The above equation implies that U is a function of S, V and N_k, and that coefficients of dS, dV and dN_k are the corresponding derivatives:

$$\boxed{\left(\frac{\partial U}{\partial S}\right)_{V,N_k} = T} \quad \boxed{\left(\frac{\partial U}{\partial V}\right)_{S,N_k} = -p} \quad \boxed{\left(\frac{\partial U}{\partial N_k}\right)_{S,V,N_{j\neq k}} = \mu_k} \tag{4.1.3}$$

CHEMICAL REACTIONS

Though Gibbs did not consider irreversible chemical reactions, Equation (4.1.1) he introduced contained all that was needed, which included all that was needed for a theory of irreversibility and entropy production in chemical processes. By making the important distinction between the entropy change $d_e S$ due to reversible exchange of matter and energy with the exterior, and irreversible increase of entropy $d_i S$ due to chemical reactions, De Donder formulated the thermodynamics of irreversible chemical transformations [2, 3]. Using the concept of chemical potential, De Donder took the 'uncompensated heat' of Clausius in the context of chemical reactions and gave it a clear expression.

Let us look at Equation (4.1.2) from the point of view of reversible entropy flow $d_e S$ and irreversible entropy production $d_i S$ that was introduced in the previous chapter. To make a distinction between irreversible chemical reactions and reversible exchange with the exterior, we express the change in the molar amounts dN_k as a sum of two parts:

$$dN_k = d_i N_k + d_e N_k \tag{4.1.4}$$

in which $d_i N_k$ is the change due to irreversible chemical reactions and $d_e N_k$ is the change due to exchange of matter with the exterior. In Equation (4.1.2), Gibbs

1927 Solvay Conference (Reproduced courtesy of the Solvay Institute, Brussels, Belgium)

considered *reversible* exchange of heat and matter. Because this corresponds to d_eS, we may write (see Equation (3.4.10))

$$d_eS = \frac{dU + p\,dV}{T} - \frac{\sum_1^n \mu_k d_e N_k}{T} \tag{4.1.5}$$

De Donder recognized that, in a closed system, if the change of molar amounts dN_k were due to irreversible chemical reactions, then the resulting entropy production d_iS can be written as

$$d_iS = -\frac{\sum_1^n \mu_k d_i N_k}{T} \tag{4.1.6}$$

This is the 'uncompensated heat' of Clausius in the realm of chemical reactions. The validity of this equation lies in the fact that chemical reactions occur in such a way that d_iS is always positive in accordance with the Second Law. For the total change in entropy dS we have

$$\boxed{dS = d_e S + d_i S}$$ (4.1.7)

in which

$$\boxed{d_e S = \frac{dU + p\,dV}{T} - \frac{1}{T}\sum_1^n \mu_k d_e N_k}$$ (4.1.8)

and

$$\boxed{d_i S = -\frac{1}{T}\sum_1^n \mu_k d_i N_k > 0}$$ (4.1.9)

For a closed system, which by definition does not exchange matter, $d_e N_k = 0$. Since the rates of chemical reaction specify dN_k/dt, the rate of entropy production can be written as

$$\boxed{\frac{d_i S}{dt} = -\frac{1}{T}\sum_1^n \mu_k \frac{dN_k}{dt} > 0}$$ (4.1.10)

If we sum (4.1.8) and (4.1.9) we recover (4.1.2):

$$\boxed{dU = T\,dS - p\,dV + \sum_1^n \mu_k dN_k}$$ (4.1.11)

Further development of this theory relates chemical potential to measurable system variables such as p, T and N_k. The pioneering work of De Donder established a clear connection between entropy production and irreversible chemical reactions: the rate of entropy production $d_i S/dt$ is related directly to the rates of chemical reactions that specify dN_k/dt. In a closed system, if initially the system is not in chemical equilibrium, then chemical reactions will take place that will irreversibly drive the system towards equilibrium. And, according to the Second Law of thermodynamics, this will happen in such a way that (4.1.10) is satisfied.

AFFINITY

De Donder also defined the *affinity* of a chemical reaction, which enables us to write expression (4.1.10) in an elegant form, as the product of a thermodynamic force and a thermodynamic flow. The concept of affinity can be understood through the following simple example.

In a closed system, consider a chemical reaction of the form

$$X + Y \rightleftharpoons 2Z$$ (4.1.12)

In this case the changes in the molar amounts dN_X, dN_Y and dN_Z of the components X, Y and Z are related by the reaction stoichiometry. We can express this relation as

$$\frac{dN_X}{-1} = \frac{dN_Y}{-1} = \frac{dN_Z}{2} \equiv d\xi \tag{4.1.13}$$

in which $d\xi$ is the change in the extent of reaction ξ, which was introduced in Section 2.5. Using (4.1.11), the total entropy change and the entropy change due to irreversible chemical reactions can now be written as

$$dS = \frac{dU + p\,dV}{T} + \frac{1}{T}(\mu_X + \mu_Y - 2\mu_Z)\,d\xi \tag{4.1.14}$$

$$d_iS = \frac{\mu_X + \mu_Y - 2\mu_Z}{T}\,d\xi > 0 \tag{4.1.15}$$

For a chemical reaction $X + Y \rightleftharpoons 2Z$, De Donder defined a new state variable called **affinity** as [1 (p. 203), 2]

$$A \equiv \mu_X + \mu_Y - 2\mu_Z \tag{4.1.16}$$

This affinity is the driving force for chemical reactions. A nonzero affinity implies that the system is not in thermodynamic equilibrium and that chemical reactions will continue to take place driving the system towards equilibrium. In terms of affinity A, the rate of increase of entropy is written as

$$\boxed{\frac{d_iS}{dt} = \left(\frac{A}{T}\right)\frac{d\xi}{dt} > 0} \tag{4.1.17}$$

As in the case of entropy production due to heat conduction, the entropy production due to a chemical reaction is a product of a thermodynamic force A/T and a thermodynamic flow $d\xi/dt$. The flow in this case is the conversion of reactants to products (or vice versa) which is caused by the force A/T. We shall refer to the thermodynamic flow $d\xi/dt$ as the **velocity of reaction** or **rate of conversion**.

Though a nonzero affinity means that there is a driving force for chemical reactions, the velocity $d\xi/dt$ at which these chemical reactions will occur is not specified by the value of affinity A. The velocities of chemical reactions are usually known through empirical means; there is no general relationship between the affinity and the velocity of a reaction.

At equilibrium, the thermodynamic flows and, hence, the entropy production must vanish. This implies that *in the state of equilibrium the affinity of a chemical reaction $A = 0$*. Thus, we arrive at the conclusion that, at thermodynamic equilibrium, the chemical potentials of the compounds X, Y and Z will reach values such that

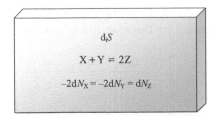

Figure 4.2 The changes in entropy d_iS due to irreversible chemical reactions is formulated using the concept of affinity. For the above reaction, the affinity $A \equiv \mu_X + \mu_Y - 2\mu_Z$, in which μ is the chemical potential

$$A \equiv \mu_X + \mu_Y - 2\mu_Z = 0 \qquad (4.1.18)$$

In Chapter 9, which is devoted to the thermodynamics of chemical processes, we will see how chemical potentials can be expressed in terms of experimentally measurable quantities such as concentrations and temperature. Equations such as (4.1.18) are specific predictions regarding the states of chemical equilibrium. These predictions have been amply verified by experiment, and today they are routinely used in chemistry.

For a general chemical reaction of the form

$$a_1A_1 + a_2A_2 + a_3A_3 + \ldots + a_nA_n \rightleftharpoons b_1B_1 + b_2B_2 + b_3B_3 + \ldots + b_mB_m \quad (4.1.19)$$

the changes in the molar amounts of the reactants A_k and the products B_k are related in such a way that a change dX in one of the species (reactants or products) completely determines the corresponding changes in all the other species. Consequently, there is only one independent variable, which can be defined as

$$\frac{dN_{A_1}}{-a_1} = \frac{dN_{A_2}}{-a_2} = \ldots \frac{dN_{A_n}}{-a_n} = \frac{dN_{B_1}}{b_1} = \frac{dN_{B_2}}{b_2} \ldots = \frac{dN_{B_m}}{b_m} = d\xi \qquad (4.1.20)$$

The affinity A of the reaction (4.1.19) is defined as

$$A \equiv \sum_{k=1}^{n} a_k \mu_{A_k} - \sum_{k=1}^{m} b_k \mu_{B_k} \qquad (4.1.21)$$

in which μ_{A_k} is the chemical potential of the reacting species A_k, etc. If several simultaneous reactions occur in a closed system, then an affinity A_k and a degree of advancement ξ_k can be defined for each reaction and the change of entropy is written as

$$dS = \frac{dU + p\,dV}{T} + \sum_k \frac{A_k}{T} d\xi_k \qquad (4.1.22)$$

$$d_iS = \sum_k \frac{A_k}{T} d\xi_k \geq 0 \qquad (4.1.23)$$

For the rate of entropy production we have the expression

$$\boxed{\frac{d_iS}{dt} = \sum_k \frac{A_k}{T} \frac{d\xi_k}{dt} \geq 0} \qquad (4.1.24)$$

At thermodynamic equilibrium, the affinity A and the velocity $d\xi/dt$ of each reaction are zero. We will consider explicit examples of entropy production due to chemical reactions in Chapter 9.

In summary, when chemical reactions are included, the entropy is a function of the energy U, volume V, and the molar amounts N_k, $S = S(U, V, N_k)$. For a closed system, following equation (4.1.22), it can be written as a function of U, V, and the extent of reaction ξ_k: $S = S(U, V, \xi_k)$.

We conclude this section with a historical remark. In Chapter 5 we will introduce a quantity called the Gibbs free energy. The Gibbs free energy of 1 mol of X can also be interpreted as the chemical potential of X. The conversion of a compound X to a compound Z causes a decrease in the Gibbs free energy of X and an increase in the Gibbs free energy of Z. Thus, the affinity of a reaction, $X + Y \rightleftharpoons 2Z$, defined as $A \equiv (\mu_X + \mu_Y - 2\mu_Z)$, can be interpreted as the negative of the change in Gibbs free energy when 1 mol of X and 1 mol of Y react to produce 2 mol of Z. This change in the Gibbs free energy, called the 'Gibbs free energy of reaction', is related to affinity A by a simple negative sign, but there is a fundamental conceptual difference between the two: *affinity is a concept that relates irreversible chemical reactions to entropy*, whereas *Gibbs free energy is primarily used in connection with equilibrium states and reversible processes*. Nevertheless, in many texts the Gibbs free energy is used in the place of affinity and no mention is made about the relation between entropy and reaction rates (for comments on this point, see Gerhartl [7]). Leicester, in his well-known book *The Historical Background of Chemistry* [1 (p. 206)], traces the origin of this usage to the textbook [8] by Gilbert Newton Lewis (1875–1946) and Merle Randall (1888–1950):

The influential textbook of G.N. Lewis (1875–1946) and Merle Randall (1888–1950) which presents these ideas has led to the replacement of the term 'affinity' by the term 'free energy' in much of the English-speaking world. The older term has never been entirely replaced in thermodynamics literature, since after 1922 the Belgian school under Theéophile De Donder (1872–1957) has made the concept of affinity still more precise.

De Donder's affinity has an entirely different conceptual basis: it relates entropy to irreversible chemical processes that occur in Nature. It is clearly a more general view of entropy, one that does not restrict the idea of entropy to infinitely slow ('quasi-static') reversible processes and equilibrium states.

4.2   General Properties of Affinity

The affinity of a reaction is a state function, completely defined by the chemical potentials. In the following chapters we will see how the chemical potential of a substance can be expressed in terms of state variables such as pressure, temperature and concentration. Thus, affinity can be expressed as a function of p, T and N_k or it can also be expressed as a function of V, T and N_k. For a closed system, since all the changes in N_k can only be due to chemical reactions, it can be expressed in terms of V, T, ξ_k and the initial values of the molar amounts N_{k0}. There are some general properties of affinities that follow from the fact that chemical reactions can be interdependent when a substance is a reactant in more than one reaction.

AFFINITY AND DIRECTION OF REACTION

The sign of affinity can be used to predict the direction of reaction. Consider the reaction $X + Y \rightleftharpoons 2Z$. The affinity is given by $A = \mu_X + \mu_Y - 2\mu_Z$. The sign of the velocity of reaction $d\xi/dt$ indicates the direction of reaction, i.e. whether the net conversion is from $X + Y$ to $2Z$ or from $2Z$ to $X + Y$. From the definition of ξ it follows that if $d\xi/dt > 0$ then the reaction 'proceeds to the right': $X + Y \rightarrow 2Z$; if $d\xi/dt < 0$ then the reaction 'proceeds to the left': $X + Y \leftarrow 2Z$. The Second Law requires that $A(d\xi/dt) \geq 0$. Thus, we arrive at the following relation between the sign of A and the direction of the reaction:

- if $A > 0$, the reaction proceeds to the right;
- if $A < 0$, the reaction proceeds to the left.

ADDITIVITY OF AFFINITIES

A chemical reaction can be the net result of two or more successive chemical reactions. For instance:

$$2C(s) + O_2(g) \rightleftharpoons 2CO(g) \qquad A_1 \tag{4.2.1}$$

$$\underline{2CO(g) + O_2(g) \rightleftharpoons 2CO_2(g) \quad A_2} \tag{4.2.2}$$

$$2[C(s) + O_2(g) \rightleftharpoons CO_2(g)] \quad 2A_3 \tag{4.2.3}$$

which shows that reaction (4.2.3) is the net result or 'sum' of the other two. By definition the affinities of the above three reactions are:

$$A_1 = 2\mu_C + \mu_{O_2} - 2\mu_{CO} \tag{4.2.4}$$

$$A_2 = 2\mu_{CO} + \mu_{O_2} - 2\mu_{CO_2} \tag{4.2.5}$$

$$A_3 = \mu_C + \mu_{O_2} - \mu_{CO_2} \tag{4.2.6}$$

From these definitions it is easy to see that

$$A_1 + A_2 = 2A_3 \tag{4.2.7}$$

Clearly this result can be generalized to many reactions. We thus have the general result: *the sum of affinities of a sequence of reactions equals the affinity of the net reaction.*

The rate of entropy production for the above reactions (4.2.1) and (4.2.2) is the sum of the rates at which entropy is produced in the two reactions:

$$\frac{d_i S}{dt} = \frac{A_1}{T}\frac{d\xi_1}{dt} + \frac{A_2}{T}\frac{d\xi_2}{dt} > 0 \tag{4.2.8}$$

in which ξ_1 and ξ_2 are the corresponding extents of reactions. Note that for the net reaction (4.2.3), because the net conversion from $(C + O_2)$ to CO_2 goes through the intermediate CO, $-dN_C \neq dN_{CO_2}$; the loss of carbon is due to its conversion to CO and CO_2, not just CO_2. As a consequence, the corresponding extent of reaction '$d\xi_3$' is not well defined and we cannot write $-dN_C = dN_{CO_2} \neq d\xi_3$. Therefore, the rate of total entropy production cannot be written as '$(A_3/T)(d\xi_3/dt)$' in general. However, if the reaction velocities $d\xi_1/dt$ and $d\xi_2/dt$ are equal, then the total rate of entropy production (4.2.8) may be written as

$$\frac{d_i S}{dt} = \frac{A_1 + A_2}{T}\frac{d\xi_1}{dt} = \frac{2A_3}{T}\frac{d\xi_1}{dt} = \frac{A_3}{T}\frac{d\xi_3}{dt} > 0 \tag{4.2.9}$$

in which $d\xi_3/dt = 2(d\xi_1/dt)$, the reaction velocity of (4.2.3). The condition $d\xi_1/dt = d\xi_2/dt$ means the rate of production of the intermediate CO in reaction (4.2.1) is balanced by the consumption of CO in reaction (4.2.2), i.e. N_{CO}, the amount of CO, remains constant. When the production of a substance X is exactly balanced by its consumption, it is said to be in a **steady state** (which can be expressed mathematically as $dN_X/dt = 0$). In many chemical reactions, the intermediate reactants are often in a steady state or nearly so. In a series of reactions in which intermediate compounds are produced and consumed, if all the intermediates are in a steady state, then it is possible to define an extent of reaction for the net reaction and write the rate of entropy production in terms of the affinity and the velocity of the net reaction.

COUPLING BETWEEN AFFINITIES

In reactions coupled to each other through common reactants, it may appear as if one reaction with positive entropy production is compensating the negative entropy production of the other in such a way that the total entropy production is positive, in accord with the Second Law. Consider the following example:

$$X + Y \rightleftharpoons Z + W \qquad A_4 > 0 \tag{4.2.10}$$

for which, as indicated, the corresponding affinity A_4 is assumed to be positive. We then expect the reaction to proceed to the right so that $d\xi_4/dt > 0$. It is possible to drive the reaction (4.2.10) effectively to the left, making $d\xi_4/dt < 0$, by 'coupling' it to another reaction:

$$T \rightleftharpoons D \quad A_5 > 0, \quad A_5(d\xi_5/dt) > 0 \tag{4.2.11}$$

The two reactions (4.2.10) and (4.2.11) could be coupled so that their total entropy production $A_4(d\xi_4/dt) + A_5(d\xi_5/dt) > 0$ but $A_4(d\xi_4/dt) < 0$. An example of a mechanism that makes such reaction reversal possible is (Figure 4.3)

$$Z + T \rightleftharpoons Z^* + D \qquad A_6 > 0, \quad A_6(d\xi_6/dt) > 0 \tag{4.2.12}$$

$$Z^* + W \rightleftharpoons X + Y \qquad A_7 > 0, \quad A_7(d\xi_7/dt) > 0 \tag{4.2.13}$$

$$Z + W + T \rightleftharpoons X + Y + D \quad A > 0, \quad A(d\xi/dt) > 0 \tag{4.2.14}$$

Once again, as indicated, the affinities and velocities of reactions (4.2.11)–(4.2.13) are assumed positive. The net reaction $Z + W + T \rightleftharpoons X + Y + D$ is an effective

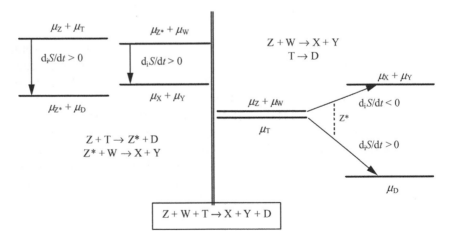

Figure 4.3 Entropy production in coupled reactions. The left and right panels show different ways of representing the same net reaction $Z + W + T \rightarrow X + Y + D$ resulting from two reaction steps. The left panel shows a reaction scheme and the corresponding chemical potentials in which entropy production of both reaction steps are positive. The right panel shows a reinterpretation of the same net reaction when the intermediate Z that couples the two reactions is in a steady state. In this case, the entropy production of one reaction is positive and the other is negative, but their sum, the total entropy production, remains positive.

reversal of $X + Y \rightleftharpoons Z + W$ accompanied by $T \rightleftharpoons D$. This way of interpreting the net reaction can be expressed in terms of the affinities by noting that

$$A = A_6 + A_7 = -A_4 + A_5 \qquad (4.2.15)$$

For the net reaction $Z + W + T \rightleftharpoons X + Y + D$, as discussed above, the corresponding velocity of reaction $d\xi/dt$ can be defined only when the intermediate Z^* is in a steady state, i.e. $d\xi_6/dt = d\xi_7/dt = d\xi/dt$. Under these steady-state conditions, we will now show that the rate of entropy production can be written as if it is due to two coupled reactions $Z + W \rightleftharpoons X + Y$ and $T \rightleftharpoons D$, each proceeding with velocity $d\xi/dt$.

The total rate of entropy production due to the two coupled reactions (4.2.12) and (4.2.13) is

$$\frac{d_i S}{dt} = \frac{A_6}{T}\frac{d\xi_6}{dt} + \frac{A_7}{T}\frac{d\xi_7}{dt} \geq 0 \qquad (4.2.16)$$

Now, if $d\xi_6/dt = d\xi_7/dt = d\xi/dt$, expression (4.2.16) can be rewritten in terms of the affinities A_4 and A_5 of reactions (4.2.10) and (4.2.11) using the equality (4.2.15):

$$\frac{d_i S}{dt} = \frac{A_6 + A_7}{T}\frac{d\xi}{dt} = -\frac{A_4}{T}\frac{d\xi}{dt} + \frac{A_5}{T}\frac{d\xi}{dt} \geq 0 \qquad (4.2.17)$$

In this expression, the affinities A_4 and A_5 are positive and, since we have assumed the net reaction (4.2.14) proceeds to the right, $d\xi/dt > 0$. Thus, the first term on the right-hand side of (4.2.17) is negative but the second term is positive. It can easily be seen that the steady-state condition $d\xi_6/dt = d\xi_7/dt = d\xi/dt$ also implies that $-d\xi_4/dt = d\xi_5/dt = d\xi/dt$, which enables us to rewrite (4.2.17) as

$$\frac{d_i S}{dt} = \frac{A_4}{T}\frac{d\xi_4}{dt} + \frac{A_5}{T}\frac{d\xi_5}{dt} \geq 0 \quad \text{in which} \quad \frac{A_4}{T}\frac{d\xi_4}{dt} < 0 \quad \text{and} \quad \frac{A_5}{T}\frac{d\xi_5}{dt} > 0 \qquad (4.2.18)$$

Such coupled reactions are common in biological systems.

4.3 Entropy Production Due to Diffusion

The concepts of chemical potential and affinity not only describe chemical reactions, but also flow of matter from one region of space to another. With the concept of chemical potential, we are now in a position to obtain an expression for the entropy change due to diffusion, an example of an irreversible process we saw in Chapter 3 (see Figure 3.8). The concept of chemical potential turns out to have wide reach. Other irreversible processes that can be described using a chemical potential will be discussed in Chapter 10. Here, we shall see how it can be used to describe diffusion.

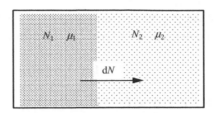

Figure 4.4 The irreversible process of diffusion can be described thermodynamically using chemical potential. The variation of chemical potential with location corresponds to an affinity that drives a flow of matter. The corresponding entropy production is given by (4.3.4)

When chemical potentials of a substance in adjacent parts of a system are unequal, diffusion of that substance takes place until the chemical potentials in the two parts equalize. The process is similar to flow of heat due to a difference in temperature. Diffusion is another irreversible process for which we can obtain the rate of increase in entropy in terms of chemical potentials.

DISCRETE SYSTEMS

For simplicity, let us consider a system consisting of two parts of equal temperature T, one with chemical potential μ_1 and molar amount N_1 and the other with chemical potential μ_2 and molar amount N_2 as shown in Figure 4.4. The flow of particles from one part to another can also be associated with an 'extent of reaction', though no real chemical reaction is taking place here:

$$-dN_1 = dN_2 = d\xi \tag{4.3.1}$$

Following Equation (4.1.14), the entropy change for this process can be written as

$$d_iS = \frac{dU + p\,dV}{T} - \frac{\mu_2 - \mu_1}{T}d\xi \tag{4.3.2}$$

$$= \frac{dU + p\,dV}{T} + \frac{A}{T}d\xi \tag{4.3.3}$$

in which the corresponding affinity $A = \mu_1 - \mu_2$. If $dU = dV = 0$, then the transport of particles results in the change of entropy given by

$$d_iS = \frac{\mu_2 - \mu_1}{T}d\xi > 0 \tag{4.3.4}$$

The positivity of this quantity required by the Second Law implies that particle transport is from a region of high chemical potential to a region of low chemical potential. This is, of course, the process of diffusion of particles from a region of higher concentration to a region of lower concentration in many cases, but it must be emphasized that the *driving force for diffusion is the gradient of chemical potential*, not the gradient of concentration as is often stated (see Appendix 4.1).

4.4 General Properties of Entropy

Entropy, as formulated in this and the previous chapter, encompasses all aspects of transformations of matter: changes in energy, volume and composition. Thus, every system in Nature, be it a gas, an aqueous solution, a living cell or a neutron star, is associated with certain entropy. We shall obtain explicit expressions for entropies of various systems in the following chapters and study how entropy production is related to irreversible processes. At this stage, however, we shall note some general properties of entropy as a function of state.

The entropy of a system is a function of its total energy U, volume V, and molar amounts N_k of its constituents:

$$S = S(U, V, N_1, N_2, \ldots N_s) \tag{4.4.1}$$

As a function of variables U, V and N_k, the differential dS can be written as

$$dS = \left(\frac{\partial S}{\partial U}\right)_{V,N_k} dU + \left(\frac{\partial S}{\partial V}\right)_{U,N_k} dV + \left(\frac{\partial S}{\partial N_k}\right)_{U,V,N_{j\neq k}} dN_k \tag{4.4.2}$$

Furthermore, from the general relation $dU = T\,dS - p\,dV + \sum_1^n \mu_k dN_k$ (cf. (4.1.2)), it follows that

$$dS = \frac{1}{T}dU + \frac{p}{T}dV - \sum_k \frac{\mu_k}{T}dN_k \tag{4.4.3}$$

(Here we have combined the change in N_k due to chemical reactions and the change due to exchange with the exterior). Comparing (4.4.2) and (4.4.3) we immediately see that

$$\left(\frac{\partial S}{\partial U}\right)_{V,N_k} = \frac{1}{T} \quad \left(\frac{\partial S}{\partial V}\right)_{U,N_k} = \frac{p}{T} \quad \left(\frac{\partial S}{\partial N_k}\right)_{U,V,N_{j\neq k}} = -\frac{\mu_k}{T} \tag{4.4.4}$$

If the change in molar amounts N_k is only due to a chemical reaction, then the entropy can also be expressed as a function of U, V and ξ (see Example 4.1). Then one can show that

$$\left(\frac{\partial S}{\partial \xi}\right)_{U,V} = \frac{A}{T} \tag{4.4.5}$$

In addition, for any function of many variables, the 'cross-derivatives' must be equal, i.e. we must have equalities of the type

$$\frac{\partial^2 S}{\partial V \partial U} = \frac{\partial^2 S}{\partial U \partial V} \tag{4.4.6}$$

Relations (4.4.4) then imply that

$$\frac{\partial}{\partial V}\frac{1}{T} = \frac{\partial}{\partial U}\frac{p}{T} \tag{4.4.7}$$

Many such relations can be similarly derived because entropy is function of state.

For homogeneous systems, we have seen in Chapter 3 (Box 3.3) that *entropy is an extensive variable*. Mathematically, this means that entropy S is a homogeneous function of the variables U, V and N_k, i.e. it has the following property:

$$S(\lambda U, \lambda V, \lambda N_{x1}, \lambda N_{x2}, \ldots, \lambda N_{xs}) = \lambda S(U, V, N_{x1}, N_{x2}, \ldots, N_{xs}) \tag{4.4.8}$$

Differentiating (4.4.8) with respect to λ and setting $\lambda = 1$, we obtain the well-known **Euler's theorem** for homogeneous functions:

$$S = \left(\frac{\partial S}{\partial U}\right)_{V,N_k} U + \left(\frac{\partial S}{\partial V}\right)_{U,N_k} V + \sum_k \left(\frac{\partial S}{\partial N_k}\right)_{U,V,N_{i\neq k}} N_k \tag{4.4.9}$$

Using relations (4.4.4) we can write this relation as

$$S = \frac{U}{T} + \frac{pV}{T} - \sum_k \frac{\mu_k N_k}{T} \tag{4.4.10}$$

In (4.4.9) and (4.4.10), we have expressed entropy as a function of U, V and N_k. Since U can be expressed as function of T, V and N_k, entropy can also be expressed as function of T, V and N_k: $S = S(T, V, N_k)$. (The temperature and volume dependence of the energy U and enthalpy H of each component is obtained by using the empirical values of the heat capacity as described in Chapter 2.) Since T, V and N_k are directly measurable state variables, it is often more convenient to express thermodynamic quantities such as entropy and energy as functions of these state variables.

As a function of T, V and N_k, the derivatives of entropy can be obtained by expressing dU in (4.4.3) as a function of V, T and N_k:

$$TdS = dU + p\,dV - \sum_k \mu_k dN_k$$

$$= \left(\frac{\partial U}{\partial T}\right)_{V,N_k} dT + \left(\frac{\partial U}{\partial V}\right)_{T,N_k} dV + \sum_k \left(\frac{\partial U}{\partial N_k}\right)_{V,T,N_{j\neq k}} dN_k + p\,dV - \sum_k \mu_k dN_k$$

i.e.

$$dS = \frac{1}{T}\left[\left(\frac{\partial U}{\partial V}\right)_{T,N_k} + p\right]dV + \frac{1}{T}\left(\frac{\partial U}{\partial T}\right)_{V,N_k} dT - \sum_k \left(\frac{\mu_k}{T}\right)dN_k + \frac{1}{T}\sum_k \left(\frac{\partial U}{\partial N_k}\right)_{V,T,N_{j\neq k}} dN_k$$

(4.4.11)

In Equation (4.4.11), since the coefficient of dV must equal $(\partial S/\partial V)_{T,N_k}$, etc., we can make the following identification:

$$\left(\frac{\partial S}{\partial V}\right)_{T,N_k} = \frac{1}{T}\left(\frac{\partial U}{\partial V}\right)_{T,N_k} + \frac{p}{T}$$

(4.4.12)

$$\left(\frac{\partial S}{\partial T}\right)_{V,N_k} = \frac{1}{T}\left(\frac{\partial U}{\partial T}\right)_{V,N_k} = \frac{C_V}{T}$$

(4.4.13)

$$\left(\frac{\partial S}{\partial N_k}\right)_{V,T,N_{j\neq k}} = -\frac{\mu_k}{T} + \frac{1}{T}\left(\frac{\partial U}{\partial N_k}\right)_{V,T,N_{j\neq k}}$$

(4.4.14)

Similar relations can be derived for U as a function of T, V and N_k.

The above relations are valid for homogeneous systems with uniform temperature and pressure. These relations can be extended to inhomogeneous systems as long as one can associate a well-defined temperature to every location. The thermodynamics of an inhomogeneous system can be formulated in terms of entropy density $s(T(\mathbf{x}), n_k(\mathbf{x}))$, which is a function of the temperature and the molar densities $n_k(\mathbf{x})$ $(\mathrm{mol\,m^{-3}})$ at the point $\mathbf{x}$. If $u(\mathbf{x})$ is the energy density, then following (4.4.4) we have the relations

$$\left(\frac{\partial s}{\partial u}\right)_{n_k} = \frac{1}{T(\mathbf{x})} \quad \left(\frac{\partial s}{\partial n_k}\right)_u = -\frac{\mu(\mathbf{x})}{T(\mathbf{x})}$$

(4.4.15)

in which the positional dependence of the variables is explicitly shown.

An empirically more convenient way is to express both entropy and energy densities as functions of the local temperature $T(\mathbf{x})$ and molar density $n_k(\mathbf{x})$, both of which can be directly measured:

$$u = u(T(\mathbf{x}), n_k(\mathbf{x})) \quad s = s(T(\mathbf{x}), n_k(\mathbf{x}))$$

(4.4.16)

The total entropy and energy of the system are obtained by integrating the corresponding densities over the volume of the system:

$$S = \int_V s(T(\mathbf{x}), n_k(\mathbf{x})) dV \quad U = \int_V u(T(\mathbf{x}), n_k(\mathbf{x})) dV \tag{4.4.17}$$

Since the system as a whole is not in thermodynamic equilibrium, the total entropy S in general is not a function of the total energy U and the total volume V. Nevertheless, a thermodynamic description is still possible as long as the temperature is well defined at each location $\mathbf{x}$.

Appendix 4.1 Thermodynamics Description of Diffusion

Expression (4.3.4) can generalized to describe a continuous system in which μ and T are functions of the position vector $\mathbf{r}$, and S is replaced by entropy density s (the entropy per unit volume):

$$d_i s(\mathbf{r}) = -\nabla\left(\frac{\mu(\mathbf{r})}{T(\mathbf{r})}\right) \cdot d\boldsymbol{\xi}(\mathbf{r}) \tag{A4.1.1}$$

in which the direction of the flow of particles (dN/unit area) is indicated by the vector $d\boldsymbol{\xi}$. From (A4.1.1), the *rate* of entropy production per unit volume due to diffusion can be written in terms of the particle current $\mathbf{J}_N \equiv d\boldsymbol{\xi}/dt$ as

$$\frac{d_i s(\mathbf{r})}{dt} = -\nabla\left(\frac{\mu(\mathbf{r})}{T(\mathbf{r})}\right) \cdot \mathbf{J}_N \tag{A.4.1.2}$$

The particle current $\mathbf{J}_N$ is a response to the gradient $\nabla(\mu(\mathbf{r})/T(\mathbf{r}))$. As we saw in Section 3.4, the entropy production due to each irreversible process in general has the above form of a product of a current or 'flow' $\mathbf{J}_N$ and a 'force', such as the gradient $\nabla(\mu(\mathbf{r})/T(\mathbf{r}))$.

References

1. Leicester, H.M., *The Historical Background of Chemistry*. 1971, New York: Dover.
2. De Donder, T., *L'Affiniteé*. 1927, Paris: Gauthier-Villars.
3. De Donder, T., Van Rysselberghe, P., *Affinity*. 1936, Menlo Park, CA: Stanford University Press.
4. Gibbs, J.W., On the equilibrium of heterogeneous substances. *Trans. Conn. Acad. Sci.*, 1878. **III**: 343–524.
5. Gibbs, J.W., On the equilibrium of heterogeneous substances. *Trans. Conn. Acad. Sci.*, 1875. **III**: 108–248.
6. Gibbs, J.W., *The Scientific Papers of J. Willard Gibbs, Vol. 1: Thermodynamics*. 1961, New York: Dover.
7. Gerhartl, F.J., The A + B = C of chemical thermodynamics. *J. Chem. Ed.*, 1994. **71**: 539–548.
8. Lewis, G.N., Randall, M., *Thermodynamics and Free Energy of Chemical Substances*. 1923, New York: McGraw-Hill.

Examples

Example 4.1 If the change in molar amounts is entirely due to one reaction, show that entropy is a function of V, U and ξ and that

$$\left(\frac{\partial S}{\partial \xi}\right)_{U,V} = \frac{A}{T}$$

Solution Entropy is a function of U, V and N_k: $S(U, V, N_k)$. As shown in Section 4.4 (see Equation (4.4.3)), for change in entropy dS we have

$$dS = \frac{1}{T}dU + \frac{p}{T}dV - \sum_k \frac{\mu_k}{T}dN_k$$

If ξ is the extent of reaction of the single reaction which causes changes in N_k, then

$$dN_k = \nu_k \, d\xi \quad k = 1, 2, \ldots, s$$

in which ν_k is the stoichiometric coefficient of the s species that participate in the reaction. ν_k is negative for the reactants and positive for the products. For the species that do not participate in the reaction $\nu_k = 0$. The change in entropy dS can now be written as

$$dS = \frac{1}{T}dU + \frac{p}{T}dV - \sum_{k=1}^{s} \frac{\mu_k \nu_k}{T}d\xi$$

Now, the affinity of the reaction $A = -\sum_{k=1}^{s}\mu_k \nu_k$ (note that ν_k is negative for the reactants and positive for the products). Hence:

$$dS = \frac{1}{T}dU + \frac{p}{T}dV + A d\xi$$

This shows that S is a function of U, V and ξ and that

$$\left(\frac{\partial S}{\partial \xi}\right)_{U,V} = \frac{A}{T}$$

If N_{10} is the molar amount of the reactant k at time $t = 0$, etc., and if we assume $\xi = 0$ at $t = 0$, then the molar amounts at any time t are $N_{10} + \nu_k\xi(t)$, $N_{20} + \nu_2\xi(t)$, $\ldots$, $N_{s0} + \nu_s\xi(t)$, with all the other molar amounts being constant. Thus, $S = S(U, V, N_{10} + \nu_1\xi(t), N_{20} + \nu_2\xi(t), \ldots, N_{s0} + \nu_s\xi(t))$. Thus, for a given initial molar amounts N_{k0}, the entropy of a closed system with a chemical reaction is a function of U, V and ξ.

Exercises

4.1 In a living cell, which is an open system that exchanges energy and matter with the exterior, the entropy can decrease, i.e. $dS < 0$. Explain how this is possible in terms of d_eS and d_iS. How is the Second Law valid in this case?

4.2 In SI units, what are the units of entropy, chemical potential and affinity?

4.3 Consider a reaction $A \rightarrow 2B$ in the gas phase (i.e. A and B are gases) occurring in a fixed volume V at a fixed temperature T. In the ideal gas approximation, at any time t, if N_A and N_B are molar amounts:

(i) Write an expression for the total entropy.
(ii) Assume that at time $t = 0$, $N_A(0) = N_{A0}$, $N_B(0) = 0$ and the extent of reaction $\xi(0) = 0$. At any time t, express the concentrations $N_A(t)$ and $N_B(t)$ in terms of $\xi(t)$.
(iii) At any time t, write the total entropy as a function of T, V and $\xi(t)$ (and N_{A0} which is a constant).

4.4 Consider the series of reactions:

$$X + Y \rightleftharpoons 2Z \qquad\qquad (1)$$

$$2[Z + W \rightleftharpoons S + T] \qquad\qquad (2)$$

$$\text{Net reaction:} \quad X + Y + 2W \rightleftharpoons 2S + 2T \qquad\qquad (3)$$

Determine the conditions under which the rate of entropy production can be written in terms of the net reaction, i.e. $d_iS/dt = (A_3/T)(d\xi_3/dt)$ in which A_3 and ξ_3 are the affinity and the extent of reaction of the net reaction (3).

4.5 For the reaction scheme

$$Z + T \rightleftharpoons Z^* + D \quad A_6 > 0, \quad A_6(d\xi_6/dt) > 0$$

$$Z^* + W \rightleftharpoons X + Y \quad A_7 > 0, \quad A_7(d\xi_7/dt) > 0$$

(a) Express dN_k/dt for each of the reactants and products, Z, T, Z*, D, etc. in terms of the extents of reaction velocities $d\xi_6/dt$ and $d\xi_7/dt$.
(b) For the steady state of Z*, i.e. $dN_{Z^*}/dt = 0$, show that $d\xi_6/dt = d\xi_7/dt$ and that
(c) the total entropy production d_iS/dt can be written as

$$\frac{d_iS}{dt} = \frac{A_4}{T}\frac{d\xi_4}{dt} + \frac{A_5}{T}\frac{d\xi_5}{dt} \geq 0$$

in which quantities with subscripts 4 and 5 refer to the affinities and extents of reaction of the reactions $X + Y \rightleftharpoons Z + W$ and $T \rightleftharpoons D$ respectively.

4.6   (a) Using the fact that S is a function of U, V and N_k, derive the relation

$$\left(\frac{\partial}{\partial V}\frac{\mu_k}{T}\right)_{U,N_k} + \left(\frac{\partial}{\partial N_k}\frac{p}{T}\right)_{V,U} = 0$$

(b) For an ideal gas, show that

$$\left(\frac{\partial}{\partial V}\frac{\mu_k}{T}\right)_{U,N_k} = -\frac{R}{V}$$

(c) For an ideal gas, show that $(\partial S/\partial V)_{T,N} = nR$ in which n is molar density (moles per unit volume).

5 EXTREMUM PRINCIPLES AND GENERAL THERMODYNAMIC RELATIONS

Extremum Principles in Nature

For centuries we have been motivated by the belief that the laws of Nature are simple, and have been rewarded amply in our search for such laws. The laws of mechanics, gravitation, electromagnetism and thermodynamics can all be stated simply and expressed precisely with a few equations. The current search for a theory that unifies all the known fundamental forces between elementary particles is very much motivated by such a belief. In addition to simplicity, Nature also seems to 'optimize' or 'economize': natural phenomena occur in such a way that some physical quantity is minimized or maximized – or to use one word for both, 'extremized'. The French mathematician Pierre Fermat (1601–1665) noticed that the change of direction of rays of light as they propagate through different media can all be precisely described using one simple principle: *light travels from one point to another along a path that minimizes the time of travel.* Later it was discovered that all the equations of motion in mechanics can be obtained by invoking the *principle of least action*, which states that if a body is at a point x_1 at a time t_1 and at a point x_2 at time t_2, then the motion occurs so as to minimize a quantity called the *action*. (An engaging exposition of these topics can be found in Feynman's Lectures on Physics [1]).

Equilibrium thermodynamics, too, has its extremum principles. In this chapter we will see that the approach to equilibrium under different conditions is such that a **thermodynamic potential** is extremized. Following this, in preparation for the applications of thermodynamics in the subsequent chapters, we will obtain general thermodynamic relations that are valid for all systems.

5.1 Extremum Principles Associated with the Second Law

We have already seen that all isolated systems evolve to the state of equilibrium in which the entropy reaches its maximum value or, equivalently, the rate of entropy production is zero. This is the basic extremum principle of equilibrium thermodynamics. But we do not always deal with isolated systems. In many practical situations, the physical or chemical system under consideration is subject to constant pressure or temperature or both. In these situations, the positivity of entropy change due to irreversible processes, i.e. $d_iS > 0$, implies the evolution of certain thermodynamic functions to their minimum values. Under each *constraint*, such as constant pressure, constant temperature or both, the evolution of the system to the state of

Introduction to Modern Thermodynamics   Dilip Kondepudi
© 2008 John Wiley & Sons, Ltd

equilibrium corresponds to the extremization of a thermodynamic quantity. These quantities are the *Gibbs energy*, the *Helmholtz energy*, and *enthalpy* (which was introduced in Chapter 2), which, as we shall see in this chapter, are functions of state. They are also called **thermodynamic potentials**, in analogy with the potentials associated with forces in mechanics, whose minima are also points of stable mechanical equilibrium. *The systems we consider are either isolated or closed.*

MAXIMUM ENTROPY

As we have seen in Chapter 4, owing to irreversible processes the entropy of an isolated system continues to increase ($d_iS > 0$) until it reaches the maximum possible value. The state thus reached is the state of equilibrium. Therefore, it may be stated that, *when U and V are constant, every system evolves to a state of maximum entropy.*

An equivalent statement is that, when U and V are constant, every system evolves to a state in which the rate of entropy production d_iS/dt vanishes. The later statement refers to irreversible processes, whereas the former refers to the state. When processes are extremely slow, as may be the case for some chemical transformations, the system could be considered to be in 'equilibrium' with respect to all the irreversible processes whose rates have decreased to zero.

MINIMUM ENERGY

The Second Law also implies that, *at constant S and V, every system evolves to a state of minimum energy.* This can be seen as follows. We have seen that, for closed systems, $dU = dQ - p\,dV = T\,d_eS - p\,dV$. Because the total entropy change $dS = d_eS + d_iS$ we may write $dU = T\,dS - p\,dV - T\,d_iS$. Since S and V are constant, $dS = dV = 0$. Therefore, we have

$$dU = -Td_iS \leq 0 \qquad (5.1.1)$$

Thus, in systems whose entropy maintained at a fixed value, driven by irreversible processes, the energy evolves to the minimum possible value.

To keep the entropy constant, the entropy d_iS produced by irreversible processes has to be removed from the system. If a system is maintained at a constant T, V and N_k, the entropy remains constant. The decrease in energy $dU = -T\,d_iS$ is generally due to irreversible conversion of mechanical energy to heat that is removed from the system to keep the entropy (temperature) constant. A simple example is the falling of an object to the bottom of a fluid (Figure 5.1). Here, $dU = -T\,d_iS$ is the heat produced as a result of fluid friction or viscosity. If this heat is removed rapidly so as to keep the temperature constant, the system will evolve to a state of minimum energy. Note that, during the approach to equilibrium, $dU = -T\,d_iS < 0$ for every time interval dt. This represents a continuous conversion of mechanical energy (kinetic energy plus potential energy) into heat; at no time does the conversion occur in the opposite direction.

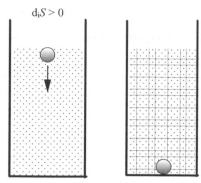

$d_iS > 0$

T and V constant

Figure 5.1 A simple illustration of the principle of minimum energy. In this example, if T and V are constant, then the entropy S is constant. At constant S and V the system evolves to a state of minimum energy

MINIMUM HELMHOLTZ ENERGY

In closed systems maintained at constant T and V, a thermodynamic quantity called the **Helmholtz energy or Helmholtz free energy** evolves to its minimum value. The term 'free energy' has been in use because the Helmholtz energy is the energy that is 'free', available to do work in an idealized reversible process (see Example 5.1). Helmholtz energy, denoted by F, is defined as

$$\boxed{F \equiv U - TS}$$ (5.1.2)

At constant T we have

$$dF = dU - T\,dS = dU - T\,d_eS - T\,d_iS$$
$$= dQ - p\,dV - T\,d_eS - T\,d_iS$$

If V is also kept constant, then $dV = 0$; and for closed systems, $T\,d_eS = dQ$. Thus, at *constant T and V*, we obtain the inequality

$$dF = -Td_iS \leq 0$$ (5.1.3)

as a direct consequence of the Second Law. This tells us that a system whose temperature and volume are maintained constant evolves such that the Helmholtz energy is minimized.

An example of the minimization of F is a reaction, such as $2H_2(g) + O_2(g) \rightleftharpoons 2H_2O(l)$, that takes place at a fixed value of T and V (see Figure 5.2a). To keep T

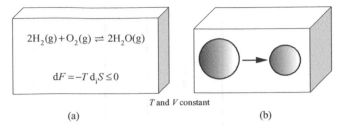

T and V constant

(a)   (b)

Figure 5.2 Examples of minimization of Helmholtz free energy F. (a) If V and T are kept at a fixed value, then a chemical reaction will progress to a state of minimum F (but S is not a constant). In this case the irreversible production of entropy $T d_i S = -\Sigma_k \mu_k dN_k = -dF \geq 0$. (b) Similarly, the contraction of a bubble enclosed in a box of fixed V and T is an example. The contracting force on the bubble's surface decreases the bubble's radius until it reaches a point at which the excess pressure in the bubble balances the contracting force of the surface. In this case we can identify $dF = -T\, d_i S \leq 0$ and determine the excess pressure in the bubble at equilibrium (see Section 5.6)

constant, the heat generated by the reaction has to be removed. In this case, following De Donder's identification of the entropy production in an irreversible chemical reaction (4.1.6), we have $T d_i S = -\Sigma_k \mu_k d_i N_k = -dF \geq 0$.

Another example is the natural evolution of shape of a bubble (Figure 5.2b) enclosed in a box of fixed V and T. In the absence of gravity (or if the bubble is small enough that the gravitational energy is insignificant compared with other energies of the system), regardless of its initial shape, a bubble finally assumes the shape of a sphere of minimal size. The bubble's size decreases irreversibly until the excess pressure inside the bubble balances the contracting force of the surface. During this process, the Helmholtz energy decreases with decreasing surface area. As the area of the bubble decreases irreversibly, the surface energy is transformed into heat which escapes to the surroundings (thus T is maintained constant). The entropy production in this irreversible processes is given by $T d_i S = = -dF$. Generally, Helmholtz energy increases with an increase in surface area (but not always) because molecules at the surface have higher energy than those below the surface. This excess surface energy γ is usually small, of the order of $10^{-2}\,\mathrm{J\,m^{-2}}$. For water, $\gamma = 7.275 \times 10^{-2}\,\mathrm{J\,m^{-2}}$. The thermodynamic drive to minimize the surface energy results in a 'surface tension' (force per unit length) whose numerical value equals γ. We will consider surface energy in more detail at the end of the chapter.

The minimization of Helmholtz energy is a very useful principle. Many interesting features, such as phase transitions and the formation of complex patterns in equilibrium systems [2], can be understood using this principle.

That Helmholtz free energy is a state function follows from its definition (5.1.2). We can show that F is function of T, V and N_k and obtain its derivatives with respect to these variables. From (5.1.2) it follows that $dF = dU - TdS - SdT$. For the change of entropy due to exchange of energy and matter we have $Td_eS = dU + pdV - \Sigma_k\mu_kd_eN_k$. For the change of entropy due to irreversible chemical reaction we have $Td_iS = -\Sigma_k\mu_kd_iN_k$. For the total change in entropy, we have $TdS = Td_eS + Td_iS$. Substituting these expressions for dS in the expression for dF we obtain

$$dF = dU - T\left[\frac{dU + pdV}{T} - \frac{1}{T}\sum_k\mu_kd_eN_k\right] - T\frac{\sum_k\mu_kd_iN_k}{T} - SdT$$
$$= -pdV - SdT + \sum_k\mu_k(d_eN_k + d_iN_k) \qquad (5.1.4)$$

Since $dN_k = d_eN_k + d_iN_k$ we may write Equation (5.1.4) as

$$\boxed{dF = -pdV - SdT + \sum_k\mu_kdN_k} \qquad (5.1.5)$$

This shows that *F is a function of V, T and* N_k. It also leads to the following identification of the derivatives of $F(V, T, N_k)$ with respect to V, T and N_k:[†]

$$\boxed{\left(\frac{\partial F}{\partial V}\right)_{T,N_k} = -p \quad \left(\frac{\partial F}{\partial T}\right)_{V,N_k} = -S \quad \left(\frac{\partial F}{\partial N_k}\right)_{T,V} = \mu_k} \qquad (5.1.6)$$

It is straightforward to include surface or other contributions to the energy (see (2.2.10)–(2.2.11)) into the expression for F and obtain similar derivatives.

If the changes in N_k are only due to a chemical reaction, then F is a function of T, V and the extent of reaction ξ. Then it can easily be shown that (Exercise 5.2)

$$\boxed{\left(\frac{\partial F}{\partial \xi}\right)_{T,V} = -A} \qquad (5.1.7)$$

MINIMUM GIBBS ENERGY

If both the pressure and temperature of a closed system are maintained constant, then the quantity that is minimized at equilibrium is the **Gibbs energy**, also called **Gibbs free energy**. We shall denote this quantity by G. As in the case of Helmholtz free energy, the term 'free energy' is used to note the fact G is the maximum energy available for doing work (through an idealized reversible process). Gibbs energy is defined as the state function

$$\boxed{G \equiv U + pV - TS = H - TS} \qquad (5.1.8)$$

[†] In this and the following chapters, for derivatives with respect to N_k, we assume the subscript $N_{i\neq k}$ is understood and drop its explicit use.

where we have used the definition of enthalpy $H = U + pV$. Just as F evolves to a minimum when T and V are maintained constant, *G evolves to a minimum when the pressure p and temperature T are maintained constant*. When p and T are constant, $dp = dT = 0$ and we can relate dG to d_iS as follows:

$$
\begin{aligned}
dG &= dU + p\,dV + V\,dp - T\,dS - S\,dT \\
&= dQ - p\,dV + p\,dV + V\,dp - T\,d_eS - T\,d_iS - S\,dT \\
&= -T\,d_iS \le 0
\end{aligned}
\tag{5.1.9}
$$

where we have used the fact that $T\,d_eS = dQ$ for closed systems and $dp = dT = 0$.

The Gibbs energy is mostly used to describe chemical processes because the usual laboratory situation corresponds to constant p and T. The irreversible evolution of G to its minimum value can be related to the affinities A_k of the reactions and the reaction velocities $d\xi_k/dt$ (in which the index k identifies different reactions) using (4.1.23)

$$
\frac{dG}{dt} = -T\frac{d_iS}{dt} = -\sum_k A_k \frac{d\xi_k}{dt} \le 0
\tag{5.1.10}
$$

or

$$
dG = -\sum_k A_k d\xi_k \le 0
\tag{5.1.11}
$$

in which the equality on the right-hand side holds at equilibrium. Equation (5.1.11) shows that, at constant p and T, G is a function of the state variables $d\xi_k$, the extent of reaction for reaction k. It also follows that

$$
\boxed{-A_k = \left(\frac{\partial G}{\partial \xi_k}\right)_{p,T}}
\tag{5.1.12}
$$

In view of this relation, calling the affinity the 'Gibbs free energy of reaction', as is commonly done in many texts, is inappropriate. As shown in Figure 5.3b, at constant p and T, the extents of reactions $d\xi_k$ will evolve to a value that minimizes $G(\xi_k, p, T)$.

Note that G evolves to its minimum value monotonically in accordance with the Second Law. Thus, ξ cannot reach its equilibrium value, as a pendulum does, in an oscillatory manner. For this reason, an oscillatory approach to equilibrium in a chemical reaction is impossible. This does not mean that concentration oscillations in chemical systems are not possible (as it was once widely thought). As we will see in later chapters, in systems that are far from equilibrium, concentration oscillations *can* occur.

We showed above that F is function of V, T and N_k. In a similar manner, it is straightforward to show that (Excercise 5.3)

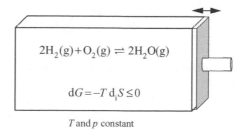

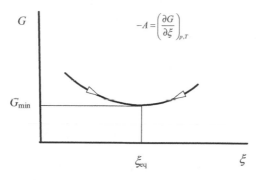

Figure 5.3 Minimization of the Gibbs energy G. (a) Under conditions of constant p and temperature T, irreversible chemical reactions will drive the system to a state of minimum G. (b) The extent of reaction ξ evolves to ξeq, which minimizes G

$$dG = V\,dp - S\,dT + \sum_k \mu_k dN_k \tag{5.1.13}$$

This expression shows that G is function of p, T and N_k and that

$$\left(\frac{\partial G}{\partial p}\right)_{T,N_k} = V \quad \left(\frac{\partial G}{\partial T}\right)_{p,N_k} = -S \quad \left(\frac{\partial G}{\partial N_k}\right)_{T,p} = \mu_k \tag{5.1.14}$$

One very useful property of the Gibbs free energy is its relation to the chemical potential. From a homogeneous system we have shown that (Equation (4.4.10)) $U = TS - pV + \Sigma_k\mu_k N_k$. Substituting this into the definition of G (Equation (5.1.8)) we obtain

$$G = \sum_k \mu_k N_k \tag{5.1.15}$$

For a pure compound, $G = \mu N$. *Therefore, one might think of the chemical potential μ as the Gibbs energy per mole of a pure compound.* For a multicomponent system, dividing (5.1.15) by N, the total molar amount, we see that the molar Gibbs energy

$$G_m \equiv \frac{G}{N} = \sum_k \mu_k x_k \qquad (5.1.16)$$

in which x_k are the mole fractions. Since G must an extensive function, we see that $G(p, T, N_k) = G(p, T, x_k N) = N G_m(p, T, x_k)$ that is G_m is a function of p, T and the mole fractions x_k. From (5.1.16) it then follows that *in a multicomponent system the chemical potential is a function of p, T and the mole fractions x_k: $\mu_k = \mu_k(p, T, x_k)$.* (When we apply these general concepts to particular systems, we will obtain explicit expressions for Gibbs energies and chemical potentials. For example, in Chapter 8 we will see that for mixtures of compounds that interact very weakly with each other, what are called **ideal mixtures**, the chemical potential of a component can be written in the form $\mu_k(p, T, x_k) = \mu_k^*(p, T) + RT \ln x_k$, in which $\mu_k^*(p, T)$ is the chemical potential of the pure compound.)

Furthermore, as shown in Example 5.3, at constant p, and T, we have the differential relation

$$(dG_m)_{p,T} = \sum_k \mu_k dx_k \qquad (5.1.17)$$

In this relation the dx_k are not all independent because $\sum_k x_k = 1$ for mole fractions x_k.

MINIMUM ENTHALPY

In Chapter 2 we introduced the enthalpy

$$\boxed{H \equiv U + pV} \qquad (5.1.18)$$

Like the Helmholtz energy F and the Gibb energy G, the enthalpy is also associated with an extremum principle: *at fixed entropy S and pressure p, the enthalpy H evolves to its minimum value.* This can be seen as before by relating the enthalpy change dH to d_iS. Since we assume that p is constant, we have

$$dH = dU + pdV = dQ \qquad (5.1.19)$$

For a closed system, $dQ = T d_eS = T(dS - d_iS)$. Hence, $dH = T dS - T d_iS$. But because the total entropy S is fixed, $dS = 0$. Therefore, we have the relation

$$dH = -Td_iS \leq 0 \qquad (5.1.20)$$

in accordance with the Second Law. When irreversible chemical reactions take place, we normally do not encounter situations in which the total entropy remains constant. For illustrative purposes, however, it is possible to give an example.

Consider the reaction

$$H_2(g) + Cl_2(g) \rightleftharpoons 2HCl(g)$$

In this reaction, the total number of molecules does not change. As we have seen in Section 3.7, the entropy of an ideal gas $S(V, T, N) = N[s_0 + R\ln(V/N) + C_V\ln(T)]$. Although there is a considerable difference in the heat capacity of molecules with different numbers of atoms, the difference in the heat capacity of two diatomic molecules is relatively small. The difference in the term s_0 is also small for two diatomic molecules. If we ignore these small difference in the entropy between the three species of diatomic molecules, then the entropy, which depends on the N_k, V and T, will essentially remain constant if T and V are maintained constant. At the same time, since the number of molecules does not change, the pressure p remains constant (assuming ideal gas behavior). Since this reaction is exothermic, the removal of heat produced by the reaction is necessary to keep T constant. Under these conditions, both p and S remain constant as the reaction proceeds and the enthalpy reaches its minimum possible value when the system reaches the state of equilibrium. For an arbitrary chemical reaction, V and T have to be adjusted simultaneously so as to keep p and S constant, which is not a simple task.

Just as we derived $dF = -pdV - SdT + \Sigma_k\mu_k dN_k$, it can easily be shown that (Exercise 5.4):

$$dH = T\,dS + V\,dp + \sum_k \mu_k dN_k \tag{5.1.21}$$

This equation shows that H can be expressed as a function of S, p and N_k. The derivatives of H with respect to these variables are

$$\left(\frac{\partial H}{\partial p}\right)_{S,N_k} = V \quad \left(\frac{\partial H}{\partial S}\right)_{p,N_k} = T \quad \left(\frac{\partial H}{\partial N_k}\right)_{S,p} = \mu_k \tag{5.1.22}$$

Once again, if the change in N_k is only due to a chemical reaction, then H is a function of p, S and ξ, and we have

$$\boxed{\left(\frac{\partial H}{\partial \xi}\right)_{p,S} = -A} \tag{5.1.23}$$

EXTREMUM PRINCIPLES AND STABILITY OF EQUILIBRIUM STATE

In thermodynamics, the existence of extremum principles have an important consequence for the behavior of microscopic fluctuations. Since all macroscopic systems are made of a very large number of molecules which are in constant random motion, thermodynamic quantities, such as temperature, pressure and concentration, undergo small fluctuations. Why don't these fluctuations slowly drive the thermodynamic variables from one value to another, just as small random fluctuations in the positions of an object slowly move the object from one location to another (a phenomenon called Brownian motion)? The temperature or concentration of a system in thermodynamic equilibrium fluctuates about a fixed value but does not drift randomly. This is because *the state of equilibrium is stable*. As we have seen, irreversible processes drive the system to the equilibrium state in which one of the potentials is extremized. Thus, whenever a fluctuation drives the system away from the state of equilibrium, irreversible processes restore the state of equilibrium. The tendency of the system to reach and remain at an extremum value of a thermodynamic potential keeps the system stable. In this way the stability of the equilibrium state is related to the existence of thermodynamic potentials.

The state of a system is not always stable. There are situations in which fluctuations can drive a system from one state to another. In this case the initial state is said to be thermodynamically unstable. Some homogeneous mixtures become unstable when the temperature is decreased; driven by fluctuations, they then evolve to a state in which the components separate into two distinct phases, a phenomenon called 'phase separation'. We shall discuss thermodynamic stability more extensively in Chapters 14, 15 and 16.

In addition, when a system is far from thermodynamic equilibrium, the state to which the system will evolve is, in general, not governed by an extremum principle; there is not an identifiable thermodynamic potential that is minimized due to the Second Law. Furthermore, the irreversible processes that assure the stability of the equilibrium state may do just the contrary and make the system unstable. The consequent instability under far-from-equilibrium systems drives the system to states with a high level of organization, such as concentration oscillations and spontaneous formation of spatial patterns. We shall discuss elementary aspects of far-from-equilibrium instability and the consequent 'self-organization' in Chapter 11.

LEGENDRE TRANSFORMATIONS

The relations between the thermodynamic functions $F(T, V, N_k)$, $G(T, p, N_k)$ and $H(S, p, N_k)$ and the total energy $U(S, V, N_k)$, expressed as a function of S, V and N_k (which follows from Equation (4.1.2) introduced by Gibbs), are a particular instances of a general class of relations called Legendre transformations. In a Legendre transformation, a function $U(S, V, N_k)$ is transformed to a function in which one or more of the independent variables S, V, and N_k are replaced by the corresponding partial derivatives of U. Thus, $F(T, V, N_k)$ is a Legendre transform of U in which S is replaced by the corresponding derivative $(\partial U/\partial S)_{V,N_k} = T$. Similarly, $G(T, p, N_k)$ is a Legendre transform of U in which S and V are replaced by their corresponding derivatives $(\partial U/\partial S)_{V,N_k} = T$ and $(\partial U/\partial V)_{V,N_k} = -p$. We thus have the general table of Legendre transforms shown in Table 5.1.

Legendre transforms show us the general mathematical structure of thermodynamics. Clearly, not only are there more Legendre transforms that can be defined of $U(S, V, N_k)$, but also of $S(U, V, N_k)$, and indeed they are used in some situations. A detailed presentation of the Legendre transforms in thermodynamics can be found in the text by Herbert Callen [3]. (Legendre transforms also appear in classical mechanics: the Hamiltonian is a Legendre transform of the Lagrangian).

Table 5.1 Legendre transforms in thermodynamics

$U(S, V, N_k) \rightarrow F(T, V, N_k) = U - TS$	S replaced by $\left(\dfrac{\partial U}{\partial S}\right)_{V,N_k} = T$
$U(S, V, N_k) \rightarrow H(S, p, N_k) = U + pV$	V replaced by $\left(\dfrac{\partial U}{\partial V}\right)_{S,N_k} = -p$
$U(S, V, N_k) \rightarrow G(T, p, N_k) = U + pV - TS$	S replaced by $\left(\dfrac{\partial U}{\partial S}\right)_{V,N_k} = T$
	and V replaced by $\left(\dfrac{\partial U}{\partial V}\right)_{S,N_k} = -p$

5.2 General Thermodynamic Relations

As Einstein noted (see Introduction in Chapter 1), it is remarkable that the two laws of thermodynamics are simple to state, but they relate many different phenomena and have a wide range of applicability. Thermodynamics gives us many general relations between state variables which are valid for *any system in equilibrium*. In this section, we shall present a few important general relations. We will apply these to particular systems in later chapters. As we shall see in Chapter 11, some of these relations can also be extended to nonequilibrium systems that are locally in equilibrium.

THE GIBBS–DUHEM EQUATION

One of the important general relations is the Gibbs–Duhem equation (named after Pierre Duhem (1861–1916) and Josiah Willard Gibbs (1839–1903)). It shows that the intensive variables T, p and μ_k are not all independent. The Gibbs-Duhem equation is obtained from the fundamental relation (4.1.2), through which Gibbs introduced the chemical potential

$$dU = T\,dS - p\,dV + \sum_k \mu_k dN_k \tag{5.2.1}$$

and relation (4.4.10) which can be rewritten as

$$U = TS - pV + \sum_k \mu_k N_k \tag{5.2.2}$$

The latter follows from the assumption that entropy is an extensive function of U, V and N_k and the use of Euler's theorem. The differential of (5.2.2) is

$$dU = T\,dS + S\,dT - V\,dp - p\,dV + \sum_k (\mu_k dN_k + N_k d\mu_k) \tag{5.2.3}$$

This relation can be consistent with (5.2.1) only if

$$\boxed{S\,\mathrm{d}T - V\,\mathrm{d}p + \sum_k N_k\mathrm{d}\mu_k = 0} \tag{5.2.4}$$

This equation is called the **Gibbs–Duhem equation**. It shows that changes in the intensive variables T, p and μ_k cannot all be independent. We shall see in Chapter 7 that the Gibbs–Duhem equation can be used to understand the equilibrium between phases and the variation of boiling point with pressure as described by the Clausius–Clapeyron equation.

At constant T and p, from (5.2.4) it follows that $\Sigma_k N_k(\mathrm{d}\mu_k)_{p,T} = 0$. Since the change in the chemical potential is $(\mathrm{d}\mu_k)_{p,T} = \Sigma_i(\partial\mu_k/\partial N_i)\mathrm{d}N_i$, we can write this expression as

$$\sum_{k,i} N_k\left(\frac{\partial\mu_k}{\partial N_i}\right)_{p,T}\mathrm{d}N_i = \sum_i\left[\sum_k\left(\frac{\partial\mu_k}{\partial N_i}\right)_{p,T}N_k\right]\mathrm{d}N_i = 0 \tag{5.2.5}$$

Since $\mathrm{d}N_i$ are independent and arbitrary variations, (5.2.5) can be valid only if the coefficient of every $\mathrm{d}N_i$ is equal to zero. Thus, we have $\Sigma_k(\partial\mu_k/\partial N_i)_{p,T}N_k = 0$. Furthermore, since

$$\left(\frac{\partial\mu_k}{\partial N_i}\right)_{p,T} = \left(\frac{\partial^2 G}{\partial N_i\partial N_k}\right)_{p,T} = \left(\frac{\partial^2 G}{\partial N_k\partial N_i}\right)_{p,T} = \left(\frac{\partial\mu_i}{\partial N_k}\right)_{p,T}$$

we can write

$$\boxed{\sum_k\left(\frac{\partial\mu_i}{\partial N_k}\right)_{p,T}N_k = 0} \tag{5.2.6}$$

Equation (5.2.6) is an important general result that we will use in later chapters.

THE HELMHOLTZ EQUATION

The Helmholtz equation gives us a useful expression to understand how the total energy U changes with the volume V at constant T. We have seen that the entropy S is a state variable and that it can be expressed as a function of T, V and N_k. The Helmholtz equation follows from the fact that, for a function of many variables, the second 'cross-derivatives' must be equal, i.e.

$$\frac{\partial^2 S}{\partial T\partial V} = \frac{\partial^2 S}{\partial V\partial T} \tag{5.2.7}$$

For closed systems in which no chemical reactions take place, the changes in entropy can be written as

$$dS = \frac{1}{T}dU + \frac{p}{T}dV \qquad (5.2.8)$$

Since U can be expressed as a function of V and T, we have

$$dU = \left(\frac{\partial U}{\partial V}\right)_T dV + \left(\frac{\partial U}{\partial T}\right)_V dT$$

Using this expression in (5.2.8) we obtain

$$\begin{aligned} dS &= \frac{1}{T}\left(\frac{\partial U}{\partial V}\right)_T dV + \frac{1}{T}\left(\frac{\partial U}{\partial T}\right)_V dT + \frac{p}{T}dV \\ &= \left[\frac{1}{T}\left(\frac{\partial U}{\partial V}\right)_T + \frac{p}{T}\right]dV + \frac{1}{T}\left(\frac{\partial U}{\partial T}\right)_V dT \end{aligned} \qquad (5.2.9)$$

The coefficients of dV and dT can now be identified as the derivatives $(\partial S/\partial V)_T$ and $(\partial S/\partial T)_V$ respectively. As expressed in (5.2.7), since the second 'cross-derivatives' must be equal we have

$$\left\{\frac{\partial}{\partial T}\left[\frac{1}{T}\left(\frac{\partial U}{\partial V}\right) + \frac{p}{T}\right]\right\}_V = \left\{\frac{\partial}{\partial V}\left[\frac{1}{T}\left(\frac{\partial U}{\partial T}\right)\right]\right\}_T \qquad (5.2.10)$$

It is matter of simple calculation (Exercise 5.6) to show that (5.2.10) leads to the **Helmholtz equation**:

$$\boxed{\left(\frac{\partial U}{\partial V}\right)_T = T^2\left(\frac{\partial}{\partial T}\frac{p}{T}\right)_V} \qquad (5.2.11)$$

This equation enables us to determine the variation of the energy with volume if the equation of state is known. In particular, it can be used to conclude that, for an ideal gas, the equation $pV = NRT$ implies that, at constant T, the energy U is independent of the volume, i.e. $(\partial U/\partial V)_T = 0$.

THE GIBBS–HELMHOLTZ EQUATION

The Gibbs–Helmholtz equation relates the temperature variation of the Gibbs energy G to the enthalpy H. It is useful for understanding how the state of chemical equilibrium responds to change in temperature; in addition, it provides us a way to determine enthalpies of chemical reactions using data on the variation of Gibbs energy changes with temperature. The Gibbs–Helmholtz equation is obtained as follows. By definition, $G \equiv H - TS$. First, we note that $S = -(\partial G/\partial T)_{p,N_k}$ and write

$$G = H + \left(\frac{\partial G}{\partial T}\right)_{p,N_k} T \tag{5.2.12}$$

It is easy to show (Exercise 5.8) that this equation can be rewritten as

$$\boxed{\frac{\partial}{\partial T}\left(\frac{G}{T}\right) = -\frac{H}{T^2}} \tag{5.2.13}$$

When considering a chemical reaction, this equation can be written in terms of the *changes* in G and H that accompany conversion of reactants to products. If the total Gibbs energy and the enthalpy of the reactants are G_r and H_r respectively and those of the products is G_p and H_p respectively, then the changes due to the reactions will be $\Delta G = G_p - G_r$ and $\Delta H = H_p - H_r$. By applying Equation (5.2.13) to the reactants and the products and subtracting one equation from the other, we obtain

$$\boxed{\frac{\partial}{\partial T}\left(\frac{\Delta G}{T}\right) = -\frac{\Delta H}{T^2}} \tag{5.2.14}$$

In Chapter 8 we will see that a quantity called the 'standard ΔG' of a reaction can be obtained by measuring the equilibrium concentrations of the reactants and products. If the equilibrium concentrations (and hence ΔG) are measured at various temperatures, then the data on the variation of ΔG with T can be used to obtain ΔH, which is the enthalpy of reaction. Equations (5.2.13) and (5.2.14) are referred to as the **Gibbs–Helmholtz equation**.

5.3 Gibbs Energy of Formation and Chemical Potential

Other than heat conduction, every irreversible process – e.g. chemical reactions, diffusion, the influence of electric, magnetic and gravitational fields, ionic conduction, dielectric relaxation – can be described in terms of suitable chemical potentials. Chapter 10 is devoted to some of the processes described using the concept of a chemical potential. All these processes drive the system to the equilibrium state in which the corresponding affinity vanishes. Because of its central role in the description of irreversible processes, we will derive a general expression for the chemical potential in this section.

As we already noted, μ *is the molar Gibbs energy* of *a pure compound*. In general, the Gibbs energy and the chemical potential are related by

$$\left(\frac{\partial G}{\partial N_k}\right)_{p,T} = \mu_k \tag{5.3.1}$$

This equation does not give us a means to relate the chemical potential directly to experimentally measurable quantities such as heat capacities. As we have seen in

Chapter 2, enthalpy can be related to heat capacities; therefore, we seek an expression that relates chemical potential to enthalpy. To this end, we differentiate the Gibbs–Helmholtz equation (5.2.13) with respect to N_k and use (5.3.1) to obtain

$$\frac{\partial}{\partial T}\left(\frac{\mu_k}{T}\right) = -\frac{H_{mk}}{T^2} \quad \text{where} \quad H_{mk} = \left(\frac{\partial H}{\partial N_k}\right)_{p,T,N_{i\neq k}} \tag{5.3.2}$$

in which H_{mk} is called the **partial molar enthalpy** of the compound k.

If the value of the chemical potential $\mu(p_0, T_0)$ at a reference temperature T_0 and pressure p_0 is known, then by integrating Equation (5.3.2) we can obtain the chemical potential at any other temperature T if the partial molar enthalpy $H_{mk}(p_0, T)$ is known as a function of T:

$$\frac{\mu(p_0,T)}{T} = \frac{\mu(p_0,T_0)}{T_0} + \int_{l_0}^{T} \frac{-H_{mk}(p_0,T')}{T'^2}\,dT' \tag{5.3.3}$$

As was shown in Chapter 2 (see Equations (2.4.10) and (2.4.11)), the molar enthalpy of a pure compound $H_{mk}(T)$ can be obtained using the tabulated values of $C_{mp}(T)$, the molar heat capacity at constant pressure. For ideal mixtures, H_{mk} is the same as that of a pure compound. For nonideal mixtures, a detailed knowledge of the heat capacities of the mixture is needed to obtain H_{mk}. As noted earlier, the chemical potential of a component k is not only a function of its amount, mole fraction x_k, but also a function of mole fractions of other components x_j. The chemical potential of a component k depends on how it interacts with other components in the mixture.

For a pure compound, knowing $\mu(p_0, T)$ at a pressure p_0 and temperature T, the value of $\mu(p, T)$ at any other pressure p can be obtained using the expression $d\mu = -S_m\,dT + V_m\,dp$, which follows from Gibbs–Duhem equation (5.2.4), where the molar quantities $S_m = S/N$ and $V_m = V/N$. Since T is fixed, $dT = 0$, and we may integrate this expression with respect to p to obtain

$$\mu(p,T) = \mu(p_0,T) + \int_{p_0}^{p} V_m(p',T)\,dp' \tag{5.3.4}$$

Thus, if the value of the chemical potential $\mu(p_0, T_0)$ is known at a reference pressure p_0 and temperature T_0, Equations (5.3.3) and (5.3.4) tell us that a knowledge of the molar volume $V_m(p, T)$ (or density) and the molar enthalpy $H_m(p, T)$ of a compound will enable us to calculate the chemical potential at any other pressure p and temperature T. An alternative and convenient way of writing (5.3.4) is due to G.N. Lewis (1875–1946), who introduced the concept of **activity** a_k of a compound k [4]. The activity is defined by the expression

$$\boxed{\mu_k(p,T) = \mu_k(p_0,T) + RT\ln a_k = \mu_k^0 + RT\ln a} \tag{5.3.5}$$

in which $\mu_k^0 = \mu_k(p_0, T)$. When we write the chemical potential in this form in terms of activity a_k, it turns out that activity has a direct relation to experimentally measurable quantities such as concentrations and pressure. As an illustration, let us apply (5.3.4) to the case of an ideal gas. Since $V_m = RT/p$, we have

$$\mu(p,T) = \mu(p_0,T) + \int_{p_0}^{p} \frac{RT}{p'}\,\mathrm{d}p'$$
$$= \mu(p_0,T) + RT\ln(p/p_0) = \mu^0 + RT\ln a \tag{5.3.6}$$

which shows that the activity $a = (p/p_0)$ in the ideal gas approximation. In Chapter 6 we will obtain the expression for the activity of gases when the molecular size and molecular forces are taken into account, as in the van der Waals equation.

TABULATION OF THE GIBBS ENERGIES OF COMPOUNDS

The formalism presented above does not give us a way to compute the absolute values of Gibbs energies of compounds. Hence, the convention described in Box 5.1

Box 5.1 Tabulation of Gibbs Free Energies of Compounds.

To compute the *changes* in Gibbs energy in a chemical reaction, the **molar Gibbs energy** $\mu(p_0, T)$ of a compound in its **standard state** (its state at pressure $p_0 = 1$ bar) at temperature T, may be *defined* using the **Gibbs energy of formation**, $\Delta G_f[k, T]$, as follows:

$\Delta G_f^0[k, T] = 0$ for all elements k, at all temperatures T

$\mu_k^0(T) = \mu_k(p_0, T) = \Delta G_f^0[k, T] =$ standard molar Gibbs energy of formation of compound k at temperature T
$=$ Gibbs energy of formation of 1 mol of the compound from its constituent elements, all in their standard states, at temperature T.

Since chemical thermodynamics assumes that there is no interconversion between the elements, the Gibbs energy of formation of elements may be used to define the 'zero' with respect to which the Gibbs energies of all other compounds are measured.

The molar Gibbs energy at any other p and T can be obtained using (5.3.3) and (5.3.4) as shown in the figure below.

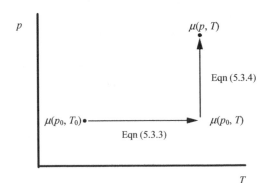

is usually used for tabulating Gibbs energies. Computation of Gibbs energy changes in chemical reactions are based on this convention. Here, the **molar Gibbs energy of formation** of a compound k, denoted by $\Delta G_f^0[k]$, is defined. Since chemical thermodynamics assumes that there is no interconversion between elements, the Gibbs energy of elements may be used to define the 'zero' with respect to which the Gibbs energies of all other compounds are measured. The Gibbs energy of formation of H_2O, written as $\Delta G_f^0[H_2O]$, for example, is the Gibbs energy *change* ΔG in the reaction

$$H_2(g) + \frac{1}{2}O_2(g) \rightarrow H_2O(l)$$

The molar Gibbs energies of formation of compounds $\Delta G_f^0[k] = \mu(p_0, T_0)$ are tabulated generally for $T_0 = 298.15$ K. We shall consider the use of ΔG_f^0 in more detail in Chapter 9 devoted to the thermodynamics of chemical reactions. From these values, the chemical potentials of compounds can be calculated as explained in Box 5.1. We conclude this section by noting that substitution of (5.3.3) into (5.3.4) gives us a general expression for the computation of the chemical potential:

$$\mu(p,T) = \frac{T}{T_0}\mu(p_0, T_0) + \int_{p_0}^{p} V_m(p', T_0)dp' + T\int_{T_0}^{T} \frac{-H_m(p, T')}{T'^2}dT' \qquad (5.3.7)$$

Thus, once the reference chemical potential $\mu(p_0, T_0)$ is defined using some convention, the chemical potential of a compound can be computed using the above formula if the molar volumes V_m and molar enthalpy H_m are known as functions of p and T. These quantities are experimentally measured and tabulated (e.g. *NIST Chemistry Webbook* at http://webbook.nist.gov/chemistry).

5.4   Maxwell Relations

The two laws of thermodynamics establish energy and entropy as functions of state, making them functions of many variables. As we have seen, $U = U(S, V, N_k)$ and $S = S(U, V, N_k)$ are functions of the indicated variables. James Clerk Maxwell (1831–1879) used the rich theory of multivariable functions to obtain a large number of relations between thermodynamic variables. The methods he employed are general, and the relations thus obtained are called the Maxwell relations.

In Appendix 1.1 we introduced the following result: if three variables x, y and z are such that each may be expressed as a function of the other two, $x = x(y, z)$, $y = y(x, z)$ and $z = z(x, y)$, then the theory of multivariable functions gives us the following fundamental relations:

$$\frac{\partial^2 x}{\partial y \partial z} = \frac{\partial^2 x}{\partial z \partial y} \qquad (5.4.1)$$

$$\left(\frac{\partial x}{\partial y}\right)_z = \frac{1}{\left(\dfrac{\partial y}{\partial x}\right)_z} \tag{5.4.2}$$

$$\left(\frac{\partial x}{\partial y}\right)_z \left(\frac{\partial y}{\partial z}\right)_x \left(\frac{\partial z}{\partial x}\right)_y = -1 \tag{5.4.3}$$

Also, if we consider $z = z(x, y)$ and $w = w(x, y)$, two functions of x and y, then the partial derivative $(\partial z/\partial x)_w$, in which the derivative is evaluated at constant w, is given by

$$\left(\frac{\partial z}{\partial x}\right)_w = \left(\frac{\partial z}{\partial x}\right)_y + \left(\frac{\partial z}{\partial y}\right)_x \left(\frac{\partial y}{\partial x}\right)_w \tag{5.4.4}$$

We have already seen how (5.4.1) can be used to derive the Helmholtz equation (5.2.11) in which entropy S was considered a function of T and V. In most cases, (5.4.1)–(5.4.4) are used to write thermodynamic derivatives in a form that can easily be related to experimentally measurable quantities. For example, using the fact that the Helmholtz energy $F(V, T)$ is a function of V and T, (5.4.1) can be used to derive the relation: $(\partial S/\partial V)_T = (\partial p/\partial T)_V$ in which the derivative on the right-hand-side is clearly more easily related to the experiment.

Some thermodynamic derivatives are directly related to properties of materials that can be measured experimentally. Other thermodynamic derivatives are expressed in terms of these quantities. The following are among the most commonly used physical properties in thermodynamics:

isothermal compressibility: $\kappa_T \equiv -\dfrac{1}{V}\left(\dfrac{\partial V}{\partial p}\right)_T$ \hfill (5.4.5)

coefficient of volume expansion: $\alpha \equiv \dfrac{1}{V}\left(\dfrac{\partial V}{\partial T}\right)_p$ \hfill (5.4.6)

Now the **pressure coefficient** $(\partial p/\partial T)_V$, for example, can be expressed in terms of κ_T and α as follows. From (5.4.3), it follows that

$$\left(\frac{\partial p}{\partial T}\right)_V = \frac{-1}{\left(\dfrac{\partial V}{\partial p}\right)_T \left(\dfrac{\partial T}{\partial V}\right)_p}$$

Now, using (5.4.2) and dividing the numerator and the denominator by V we obtain

$$\left(\frac{\partial p}{\partial T}\right)_V = \frac{-\frac{1}{V}\left(\frac{\partial V}{\partial T}\right)_p}{\frac{1}{V}\left(\frac{\partial V}{\partial p}\right)_T} = \frac{\alpha}{\kappa_T} \qquad (5.4.7)$$

GENERAL RELATION BETWEEN C_{mp} AND C_{mV}

As another example of the application of Maxwell's relations, we will derive a general relation between C_{mp} and C_{mV} in terms of α, κ_T, the molar volume V_m and T – all of which can be measured experimentally. We start with the relation we have already derived in Chapter 2, i.e. (2.3.5):

$$C_{mp} - C_{mV} = \left[p + \left(\frac{\partial U_m}{\partial V_m}\right)_T\right]\left(\frac{\partial V_m}{\partial T}\right)_p \qquad (5.4.8)$$

where we have used all molar quantities, as indicated by the subscript 'm'. The first step is to write the derivative $(\partial U/\partial V)_T$ in terms of the derivatives of involving p, V and T, so that we can relate it to α and κ_T. From the Helmholtz equation (5.2.11), it is easy to see that $(\partial U/\partial V)_T + p = T(\partial p/\partial T)_V$. Therefore, we can write (5.4.8) as

$$C_{mp} - C_{mV} = T\left(\frac{\partial p}{\partial T}\right)_V \alpha V_m \qquad (5.4.9)$$

in which we have used the definition (5.4.6) for α. Now, using the Maxwell relation $(\partial p/\partial T)_V = (\alpha/\kappa_T)$ (see (5.4.7)) in (5.4.9) we obtain the general relation

$$C_{mp} - C_{mV} = \frac{T\alpha^2 V_m}{\kappa_T} \qquad (5.4.10)$$

5.5 Extensivity with Respect to N and Partial Molar Quantities

In multicomponent systems, thermodynamic functions such as volume V, Gibbs energy G, and all other thermodynamic functions that can be expressed as functions of p, T and N_k are extensive functions of N_k. This extensivity gives us general thermodynamic relations, some of which we will discuss in this section. Consider the volume of a system as a function of p, T and N_k: $V = V(p, T, N_k)$. At constant p and T, if all the molar amounts were increased by a factor k, the volume V will also increase by the same factor. This is the property of extensivity. In mathematical terms, we have

$$V(p, T, \lambda N_k) = \lambda V(p, T, N_k) \qquad (5.5.1)$$

At constant p and T, using Euler's theorem as was done in Section 4.4, we can arrive at the relation

$$V = \sum_k \left(\frac{\partial V}{\partial N_k} \right)_{p,T} N_k \qquad (5.5.2)$$

It is convenient to define **partial molar volumes** as the derivatives:

$$V_{mk} \equiv \left(\frac{\partial V}{\partial N_k} \right)_{p,T} \qquad (5.5.3)$$

Using this definition, Equation (5.5.2) can be written as

$$\boxed{V = \sum_k V_{mk} N_k} \qquad (5.5.4)$$

Partial molar volumes are intensive quantities. As was done in the case of the Gibbs–Duhem relation, we can derive a relation between the V_{mk} by noting that at constant p and T:

$$(dV)_{p,T} = \sum_k \left(\frac{\partial V}{\partial N_k} \right)_{p,T} dN_k = \sum_k V_{mk} dN_k \qquad (5.5.5)$$

in which we have explicitly noted that the change dV is at constant p and T. Comparing dV obtained from (5.5.4) and (5.5.5), we see that $\Sigma_k N_k (dV_{mk})_{p,T} = 0$. Now $(dV_{mk}) = \Sigma_i (\partial V_{mk}/\partial N_i) dN_i$, so we obtain

$$\sum_k \sum_i N_k \left(\frac{\partial V_{mk}}{\partial N_i} \right) dN_i = 0$$

In this equation, dN_i are arbitrary variations in N_i; consequently, the above equation can be valid only when the coefficient of each $dN_i = 0$, i.e. $\Sigma_k N_k (\partial V_{mk}/\partial N_i) = 0$. Finally, using the property $(\partial V_{mk}/\partial N_i) = (\partial^2 V/\partial N_i \partial N_k) = (\partial V_{mi}/\partial N_k)$ we arrive at the final result:

$$\boxed{\sum_k N_k \left(\frac{\partial V_{mi}}{\partial N_k} \right)_{p,T} = 0} \qquad (5.5.6)$$

Relations similar to (5.5.4) and (5.5.6) can be obtained for all other functions that are extensive in N_k. For Gibbs energy, which is an extensive quantity, the equation corresponding to (5.5.4) is

$$G = \sum_k \left(\frac{\partial G}{\partial N_k} \right)_{p,T} N_k = \sum_k G_{mk} N_k = \sum_k \mu_k N_k \qquad (5.5.7)$$

in which we recognize the **partial molar Gibbs energy** G_{mk} as the chemical potentials μ_k. The equation corresponding to (5.5.6) follows from the Gibbs–Duhem relation (5.2.4) when p and T are constant:

$$\boxed{\sum_k N_k \left(\frac{\partial \mu_i}{\partial N_k}\right)_{p,T} = 0} \tag{5.5.8}$$

Similarly, for the Helmholtz energy F and the enthalpy H we can obtain the following relations:

$$\boxed{F = \sum_k F_{mk} N_k} \quad \boxed{\sum_k N_k \left(\frac{\partial F_{mi}}{\partial N_k}\right)_{p,T} = 0} \tag{5.5.9}$$

$$\boxed{H = \sum_k H_{mk} N_k} \quad \boxed{\sum_k N_k \left(\frac{\partial H_{mi}}{\partial N_k}\right)_{p,T} = 0} \tag{5.5.10}$$

in which the **partial molar Helmholtz energy** $F_{mk} = (\partial F/\partial N_k)_{p,T}$ and **partial molar enthalpy** $H_{mk} = (\partial H/\partial N_k)_{p,T}$. Similar relations can be obtained for entropy S and the total internal energy U.

5.6 Surface Tension

In this section we shall consider some elementary thermodynamic relations involving interfaces [5]. Since molecules at an interface are in a different environment from molecules in the bulk, their energies and entropies are different. Molecules at a liquid–air interface, for example, have larger Helmholtz energy than those in the bulk. At constant V and T, since every system minimizes its Helmholtz energy, the interfacial area shrinks to its minimum possible value, thus increasing the pressure in the liquid (Figure 5.4).

The thermodynamics of such a system can be formulated as follows. Consider a system with two parts, separated by an interface of area Σ (Figure 5.4). For this system we have

$$dU = T\,dS - p''\,dV'' - p'\,dV' + \gamma\,d\Sigma \tag{5.6.1}$$

in which p' and V' are the pressure and the volume of one part and p'' and V'' are the pressure and the volume of the other, Σ is the interfacial area, and the coefficient γ is called the surface tension. Since $dF = dU - T\,dS - S\,dT$, using (5.6.1) we can write dF as

$$dF = -S\,dT - p''\,dV'' - p'\,dV' + \gamma\,d\Sigma \tag{5.6.2}$$

From this it follows that

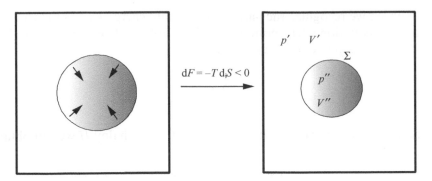

Figure 5.4   To minimize the interfacial Helmholtz energy, a liquid drop shrinks its surface area to the least possible value. As a result, the pressure p'' inside the drop is larger than the external pressure p'. The excess pressure $(p'' - p') = 2\gamma/r$

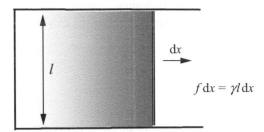

Figure 5.5   Energy is required to enlarge a surface of a liquid. The force per unit length is γ

$$\boxed{\left(\frac{\partial F}{\partial \Sigma}\right)_{T,V',V''} = \gamma}$$
    (5.6.3)

Thus, surface tension γ is the change of F per unit extension of the interfacial area at constant T, V' and V''. This energy is small, usually of the order of $10^{-2}\,\mathrm{J\,m^{-2}}$.

The minimization of Helmholtz energy drives the interface to contract like an elastic sheet. The force per unit length that the interface exerts in its tendency to contract is also equal to γ. This can be seen as follows. Since enlarging an interfacial area increases its Helmholtz energy, work needs to be done. As shown in the Figure 5.5, this means a force f is needed to stretch the surface by an amount dx, i.e. the interface behaves like an elastic sheet. The work done, fdx, equals the increase in the surface energy, $\gamma d\Sigma = (\gamma l\, dx)$, in which l is the width of the surface (Figure 5.5). Thus, $f\, dx = \gamma l\, dx$, or the force per unit length $f/l = \gamma$. For this reason, γ is called the 'surface tension'.

EXCESS PRESSURE IN A LIQUID DROP

In the case of the liquid drop in air shown in Figure 5.4, the difference in the pressures $(p'' - p') = \Delta p$ is the excess pressure inside the liquid drop. An expression for the excess pressure Δp in a spherical liquid drop can be obtained as follows. As shown in Section 5.1, if the total volume of a system and its temperature are constant, then the irreversible approach to equilibrium is described by $-T\,\mathrm{d}_i S = \mathrm{d}F \leq 0$. Now consider an irreversible contraction of the volume V'' of the liquid drop to its equilibrium value when the total volume $V = V' + V''$ and T are constant. Setting $\mathrm{d}T = 0$ and $\mathrm{d}V' = -\mathrm{d}V''$ in (5.6.2) we obtain

$$-T\frac{\mathrm{d}_i S}{\mathrm{d}t} = \frac{\mathrm{d}F}{\mathrm{d}t} = -(p'' - p')\frac{\mathrm{d}V''}{\mathrm{d}t} + \gamma\frac{\mathrm{d}\Sigma}{\mathrm{d}t} \qquad (5.6.4)$$

For a spherical drop of radius r, $\mathrm{d}V'' = (4\pi/3)3r^2\,\mathrm{d}r$ and $\mathrm{d}\Sigma = 4\pi 2r\,\mathrm{d}r$; hence, the above equation can be written as

$$-T\frac{\mathrm{d}_i S}{\mathrm{d}t} = \frac{\mathrm{d}F}{\mathrm{d}t} = [-(p'' - p')4\pi r^2 + \gamma 8\pi r]\frac{\mathrm{d}r}{\mathrm{d}t} \qquad (5.6.5)$$

We see that this expression is a product of a 'thermodynamic force' $-(p'' - p')4\pi r^2 + \gamma 8\pi r$ that causes the 'flow rate' $\mathrm{d}r/\mathrm{d}t$. At equilibrium, both must vanish. Hence, $-(p'' - p')4\pi r^2 + \gamma 8\pi r = 0$. This gives us the well known equation for the excess pressure inside a liquid drop of radius r:

$$\boxed{\Delta p \equiv (p'' - p') = \frac{2\gamma}{r}} \qquad (5.6.6)$$

This result is called the **Laplace equation** because it was first derived by the French mathematician Pierre-Simon Laplace (1749–1827).

CAPILLARY RISE

Another consequence of surface tension is the phenomenon of 'capillary rise': in narrow tubes or capillaries, most liquids rise to a height h (Figure 5.6) that depends on the radius of the capillary. The smaller the radius, the higher the rise. The liquid rises because an increase in the area of the liquid–glass interface lowers the Helmholtz energy. The relation between the height h, the radius r and the surface tension can be derived as follows. As shown in Figure 5.6c, the force of surface tension of the liquid–air interface pulls the surface down while the force at the liquid–glass interface pulls the liquid up. Let the 'contact angle', i.e. the angle at which the liquid is in contact with the wall of the capillary, be θ. When these two forces balance each other along the vertical direction, the force per unit length generated by the liquid–glass interface must be $\gamma\cos\theta$. As the liquid moves up, the liquid–glass interface is increasing while the glass–air interface is decreasing; $\gamma\cos\theta$ is the net force per unit

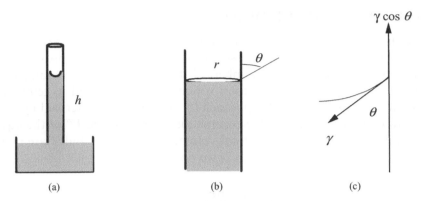

Figure 5.6 Capillary rise due to surface tension. (a) The height h to which the liquid rises depends on the contact angle θ the surface tension γ and the radius r. (b) The contact angle θ specifies the direction in which the force due to the liquid–air interface acts. (c) The vertical component of the force due to the liquid–air interface balances the net force due to the liquid–glass and glass–air interfaces

length due to these two factors. The force per unit length is equal to the interfacial energy per unit area; thus, as the liquid moves up, the decrease in the interfacial energy is $\gamma\cos\theta$ per unit area. Hence, as the liquid moves up and increases the area of the glass–liquid interface, the decrease in Helmholtz energy is $\gamma\cos\theta$ per unit area. On the other hand, as the liquid rises in the capillary, there is an increase in the potential energy of the liquid due to gravity. A liquid layer of thickens dh and density ρ has the mass $(\pi r^2\,dh\,\rho)$ and its potential energy at a height h equal to $(\pi r^2\,dh\,\rho)gh$. For the entire liquid column, this expression has to be integrated from 0 to h. The change in the Helmholtz energy ΔF as the liquid rises is the sum of the potential energy and glass–liquid interfacial energy.

$$\Delta F(h) = \int_0^h gh\rho\pi r^2 dh - 2\pi rh(\gamma\cos\theta) = \frac{\pi\rho gr^2 h^2}{2} - 2\pi rh(\gamma\cos\theta) \qquad (5.6.7)$$

The value of h that minimizes F is obtained by setting $\partial(\Delta F(h)/\partial h) = 0$ and solving for h. This leads to the expression

$$\boxed{h = \frac{2\gamma\cos\theta}{\rho gr}} \qquad (5.6.8)$$

The same result can also be derived by balancing the forces of surface tension and the weight of the liquid column. As shown in Figure 5.6b, the liquid column of height h is held at the surface by the surface tension. The total force due to the surface tension of the liquid along the circumference is $2\pi r\gamma\cos\theta$. Since this force holds the weight of the liquid column, we have

Table 5.2 Examples of surface tension and contact angles

$\gamma/10^{-2}$ J m^{-2} or $\gamma/10^{-2}$ N m^{-1}		Interface	Contact angle/°
Methanol	2.26	Glass–water	0
Benzene	2.89	Glass–many organic liquids[†]	0
Water	7.73	Glass–kerosene	26
Mercury	47.2	Glass–mercury	140
Soap solution	2.3 (approximate)	Paraffin–water	107

[†] Not all organic liquids have a contact angle value of 0°, as is clear in the case of kerosene.
More extensive data may be found in D.R. Lide (ed.) *CRC Handbook of Chemistry and Physics*, 75th edition. 1994, CRC Press: Ann Arbor, MI.

$$2\pi r\gamma\cos\theta = \rho gh\pi r^2 \qquad (5.6.9)$$

from which (5.6.8) follows.

The contact angle θ depends on the interface; see Table 5.2. For a glass–water interface the contact angle is nearly zero, as it is for many, though not all, organic liquids. For a glass–kerosene interface, $\theta = 26°$. The contact angle can be greater than 90°, as in the case of mercury–glass interface, for which θ is about 140°, and for a paraffin–water interface, for which it is about 107°. When θ is greater than 90°, the liquid surface in the capillary is lowered.

References

1. Feynman, R.P., Leighton, R.B. and Sands, M. *The Feynman Lectures on Physics*. 1964, Reading, MA: Addison-Wesley (vol. I: chapter 26; vol. II: chapter 19).
2. Seul, M. and Andelman, D. Domain shapes and patterns: the phenomenology of modulated phases. *Science*, 1995. **267**: 476–483.
3. Callen, H.B. *Thermodynamics*, second edition. 1985, New York: Wiley.
4. Lewis, G.N. and Randall, M. *Thermodynamics and Free Energy of Chemical Substances*. 1925, New York: McGraw-Hill.
5. Defay, R., Prigogine, I. and Bellemans, A. *Surface Tension and Adsorption*. 1966, New York: Wiley.

Examples

Example 5.1 Show that the change in the value of the Helmholtz free energy F corresponds to the work done when T and N_k are constant, thus justifying the name 'free energy' (available for doing work).

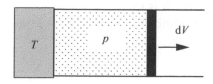

Solution   As a specific example, consider a gas in contact with a thermal reservoir at temperature T. By expanding this gas, work can be done. We can show that the change of F corresponds to the work done at constant T and N_k as follows. $F = U - TS$. From this it follows that (see (5.1.5))

$$dF = -p\,dV - S\,dT + \sum_k \mu_k dN_k$$

At constant T and N_k, $dF = -p\,dV$. Integrating both sides, we see that

$$\int_{F_1}^{F_2} dF = F_2 - F_1 = \int_{V_1}^{V_2} -p\,dV$$

which shows that the change in F is equal to the work done by the gas. The same will be true for any other system.

Example 5.2   For a closed system with one chemical reaction, show that $(\partial F/\partial \xi)_{T,V} = -A$.
Solution   The change in F is given by (see (5.1.5))

$$dF = -p\,dV - S\,dT + \sum_k \mu_k dN_k$$

Since the system is closed, the changes in N_k are due to chemical reaction; hence, we have $dN_k = v_k\,d\xi$ in which v_k are the stoichiometric coefficients (which are negative for the reactants and positive for the products). Thus:

$$dF = -p\,dV - S\,dT + \sum_k v_k\mu_k d\xi$$

Since $\Sigma_k v_k\mu_k = -A$ we have

$$dF = -p\,dV - S\,dT - A\,d\xi$$

When F is considered as a function of V, T and ξ, then

$$dF = \left(\frac{\partial F}{\partial V}\right)_{T,\xi} dV + \left(\frac{\partial F}{\partial T}\right)_{V,\xi} dT + \left(\frac{\partial F}{\partial \xi}\right)_{T,V} d\xi$$

and we see that $(\partial F/\partial \xi)_{T,V} = -A$.

Example 5.3   Using the Gibbs–Duhem relation show that at constant p and T, $(dG_m)_{p,T} = \Sigma_k \mu_k dx_k$ (which is (5.1.17)).
Solution   The molar Gibbs free energy $G_m = \Sigma_k \mu_k x_k$ in which x_k is the mole fraction of component k. Hence:

$$dG_m = \sum_k dx_k\mu_k + \sum_k x_k d\mu_k$$

The Gibbs–Duhem relation is

$$S\,\mathrm{d}T - V\,\mathrm{d}p + \sum_k N_k \mathrm{d}\mu_k = 0$$

Since p and T are constant, $\mathrm{d}T = \mathrm{d}p = 0$. Furthermore, $x_k = N_k/N$ in which N is the total number of moles. Dividing the Gibbs–Duhem equation by N and setting $\mathrm{d}p = \mathrm{d}T = 0$, we have $\Sigma_k x_k \mathrm{d}\mu_k = 0$. Using this result in the expression for $\mathrm{d}G_\mathrm{m}$ above, for constant p and T we see that

$$(\mathrm{d}G_\mathrm{m})_{p,T} = \sum_k \mu_k \mathrm{d}x_k$$

(Note that $\Sigma_k x_k = 1$ and, hence, x_k are not all independent. Hence, the above equation does not imply that $(\partial G_\mathrm{m}/\partial x_k)_{p,T} = \mu_k$.)

Exercises

5.1   Use the expression $T\,\mathrm{d}_i S = -\gamma\mathrm{d}A$ and $T\,\mathrm{d}_e S = \mathrm{d}U + p\,\mathrm{d}V$ in the general expressions for the First and the Second Laws and obtain $\mathrm{d}U = T\,\mathrm{d}S - p\,\mathrm{d}V + \gamma\mathrm{d}A$ (assuming $\mathrm{d}N_k = 0$).

5.2   (a) In an isothermal expansion of a gas from a volume V_i to V_f, what is the change in the Helmholtz free energy F? (b) For a system undergoing chemical transformation at constant V and T, prove (5.1.7).

5.3   Use the relations $\mathrm{d}U = \mathrm{d}Q - p\,\mathrm{d}V$, $T\mathrm{d}_e S = \mathrm{d}Q$ and $T\mathrm{d}_i S = -\Sigma_k \mu_k \mathrm{d}N_k$ to derive

$$\mathrm{d}G = V\,\mathrm{d}p - S\,\mathrm{d}T + \sum_k \mu_k \mathrm{d}N_k$$

which is Equation (5.1.13).

5.4   Use the relations $\mathrm{d}U = \mathrm{d}Q - p\,\mathrm{d}V$, $T\,\mathrm{d}_e S = \mathrm{d}Q$ and $T\mathrm{d}_i S = -\Sigma_k \mu_k \mathrm{d}N_k$ to derive

$$\mathrm{d}H = T\,\mathrm{d}S + V\,\mathrm{d}p + \sum_k \mu_k \mathrm{d}N_k$$

which is Equation (5.1.21).

5.5   For an ideal gas, in an isothermal process, show that the change in the Gibbs energy of a system is equal to the amount of work done by a system in an idealized reversible process.

5.6   Obtain the Helmholtz equation (5.2.11) from (5.2.10).

5.7   (a) Use the Helmholtz equation (5.2.11) to show that, at constant T, the energy of an ideal gas is independent of volume.
(b) Use the Helmholtz equation (5.2.11) to calculate $(\partial U/\partial V)_T$ for N moles of a gas using the van der Waals equation.

5.8   Obtain (5.2.13) from (5.2.12).

5.9   Derive the following general equation, which is similar to the Gibbs–Helmholtz equation:

$$\frac{\partial}{\partial T}\left(\frac{F}{T}\right) = -\frac{U}{T^2}$$

5.10   Assume that ΔH changes little with temperature, integrate the Gibbs–Helmholtz equation (5.2.14) and express ΔG_f at temperature T_f in terms of ΔH, ΔG_i and the corresponding temperature T_i and ΔH.

5.11   Obtain an explicit expression for the Helmholtz energy of an ideal gas as a function T, V and N.

5.12   The variation of Gibbs energy of a substance with temperature is given by $G = aT + b + c/T$. Determine the entropy and enthalpy of this substance as a function of temperature.

5.13   Show that (5.4.10) reduces to $C_{mp} - C_{mV} = R$ for an ideal gas.

5.14   Consider a reaction $X \rightleftharpoons 2Y$ in which X and Y are ideal gases.
(a) Write the Gibbs energy of this system as a function of extent of reaction ξ so that ξ is the deviation from the equilibrium amounts of X and Y, i.e. $N_X = N_{Xeq} - \xi$ and $N_Y = N_{Yeq} + 2\xi$ in which N_{Xeq} and N_{Yeq} are the equilibrium amounts of X and Y.
(b) Through explicit evaluation, show that $(\partial G/\partial \xi)_{p,T} = -A = (2\mu_Y - \mu_X)$.

5.15   (a) By minimizing the free energy $\Delta F(h)$ given by (5.6.1) as a function of h, obtain the expression

$$h = \frac{2\gamma \cos\theta}{\rho g r}$$

for the height of capillary rise due to surface tension.
(b) Assume that the contact angle θ between water and glass is nearly zero and calculate the height of water in a capillary of diameter 0.1 mm.

5.16 (a) Owing to surface tension, the pressure inside a bubble is higher than the outside pressure. Let this excess pressure be Δp. By equating the work done $\Delta p \, dV$ due to an infinitesimal increase dr in the radius r to the increase in surface energy $\gamma \, dA$, show that $\Delta p = 2\gamma/r$
(b) Calculate the excess pressures inside water bubbles of radius 1.0 mm and 1.0 μm.

5.17 What is the minimum energy needed to covert 1.0 mL of water to droplets of diameter 1.0 μm?

5.18 When the surface energy is included we have seen that

$$dU = T \, dS - p \, dV + \mu \, dN + \gamma \, dA$$

in which γ is the surface tension and dA is the change in the surface area. For a small spherical liquid drop of a pure substance, show that the above expression can be written as $dU = T \, dS - p \, dV + \mu'(r) \, dN$ in which $\mu'(r) = \mu + (2\gamma/r)V_m$, a size-dependent chemical potential.

PART II

APPLICATIONS: EQUILIBRIUM AND NONEQUILIBRIUM SYSTEMS

Introduction to Modern Thermodynamics   Dilip Kondepudi
© 2008 John Wiley & Sons, Ltd

PART II

APPLICATIONS: EQUILIBRIUM AND NONEQUILIBRIUM SYSTEMS

6 BASIC THERMODYNAMICS OF GASES, LIQUIDS AND SOLIDS

Introduction

The formalism and general thermodynamic relations that we have seen in the previous chapters have wide applicability. In this chapter we will see how thermodynamic quantities can be calculated for gases, liquids, solids and solutions using general methods.

6.1 Thermodynamics of Ideal Gases

Many thermodynamic quantities, such as total internal energy, entropy, chemical potential, etc., for an ideal gas have been derived in the preceding chapters as examples. In this section, we will bring all these results together and list the thermodynamic properties of gases in the ideal gas approximation. In the following section, we will see how these quantities can be calculated for 'real gases' for which we take into account the size and intermolecular forces.

THE EQUATION OF STATE

Our starting point is the equation of state, *the ideal gas law*:

$$\boxed{pV = NRT} \tag{6.1.1}$$

As we saw in Chapter 1, this approximation is valid for most gases when their densities are less than about $1\,\mathrm{mol\,L^{-1}}$. At this density and temperature of about $300\,\mathrm{K}$, for example, the pressure of $N_2(g)$ obtained using the ideal gas equation is $24.76\,\mathrm{atm}$, whereas that predicted using the more accurate van der Waals equation is $24.36\,\mathrm{atm}$, a difference of only a few percent.

THE TOTAL INTERNAL ENERGY

Through thermodynamics, we can see that the *ideal gas law (6.1.1) implies that the total internal energy U is independent of the volume at fixed T*, i.e. the energy of an ideal gas depends only on its temperature. One arrives at this conclusion using the Helmholtz equation (see (5.2.11)), which is valid for all thermodynamic systems:

$$\left(\frac{\partial U}{\partial V}\right)_T = T^2 \left[\frac{\partial}{\partial T}\left(\frac{p}{T}\right)\right]_V \tag{6.1.2}$$

Introduction to Modern Thermodynamics   Dilip Kondepudi
© 2008 John Wiley & Sons, Ltd

(We remind the reader that the Helmholtz equation is a consequence of the fact that entropy is a state function of V, T and N_k.) Since the ideal gas equation implies that the term $p/T = NR/V$ is independent of T, it immediately follows from (6.1.2) that $(\partial U/\partial V)_T = 0$. Thus, the total internal energy $U(T, V, N)$ of N moles of an ideal gas is independent of the volume at a fixed T. We can get a more explicit expression for U. Since $C_{mV} = (\partial U_m/\partial T)_V$ is found to be independent of T, we can write

$$U_{\text{ideal}} = NU_0 + N\int_0^T C_{mV}\,\mathrm{d}T = N(U_0 + C_{mV}T) \qquad (6.1.3)$$

(The constant U_0 is not defined in classical thermodynamics, but, using the definition of energy that the theory of relativity gives us, we may set $NU_0 = MNc^2$, in which M is the molar mass, N is the molar amount of the substance and c is the velocity of light. In thermodynamic calculations of changes of energy, U_0 does not appear explicitly.)

HEAT CAPACITIES AND ADIABATIC PROCESSES

We have seen earlier that there are two molar heat capacities: C_{mV} and C_{mp}, the former at constant volume and the latter at constant pressure. We have also seen in Chapter 2 that the First Law gives us the following relation between molar heat capacities:

$$\boxed{C_{mp} - C_{mV} = R} \qquad (6.1.4)$$

For an *adiabatic process*, the First Law also gives us the relation

$$\boxed{TV^{\gamma-1} = \text{constant}} \quad \text{or} \quad \boxed{pV^{\gamma} = \text{constant}} \qquad (6.1.5)$$

in which $\gamma = C_{mp}/C_{mV}$. In an adiabatic process, by definition $\mathrm{d}_eS = \mathrm{d}Q/T = 0$. If the process occurs such that $\mathrm{d}_iS \approx 0$, then the entropy of the system remains constant because $\mathrm{d}S = \mathrm{d}_iS + \mathrm{d}_eS$.

ENTROPY AND THERMODYNAMIC POTENTIALS

We have already seen that the entropy $S(V, T, N)$ of an ideal gas is (see (3.7.4))

$$\boxed{S = N[s_0 + C_{mV}\ln(T) + R\ln(V/N)]} \qquad (6.1.6)$$

From the equation of state (6.1.1) and the expressions for U_{ideal} and S it is straightforward to obtain explicit expressions for the enthalpy $H = U + pV$, the Helmholtz energy $F = U - TS$ and the Gibbs energy $G = U - TS + pV$ of an ideal gas (Exercise 6.1).

CHEMICAL POTENTIAL

For the chemical potential of an ideal gas, we obtained the following expression in Section 5.3 (see (5.3.6)):

$$\mu(p,T) = \mu(p_0,T) + RT \ln(p/p_0) \qquad (6.1.7)$$

For a mixture of ideal gases the total energy is the sum of the energies of each component. The same is true for the entropy. The chemical potential of a component k can be expressed in terms of the partial pressures p_k as

$$\mu_k(p_k,T) = \mu_k(p_0,T) + RT \ln(p_k/p_0) \qquad (6.1.8)$$

Alternatively, if x_k is the mole fraction of the component k, since $p_k = x_k p$, the chemical potential can be written as

$$\boxed{\mu_k(p,T,x_k) = \mu_k^0(p,T) + RT \ln(x_k)} \qquad (6.1.9)$$

in which $\mu_k^0(p,T) = \mu_k(p_0,T) + RT \ln(p/p_0)$ is the chemical potential of a pure ideal gas. This form of the chemical potential is generally used in the thermodynamics of multicomponent systems to define an 'ideal mixture'.

ENTROPY OF MIXING AND THE GIBBS PARADOX

Using the expression for the entropy of an ideal gas, we can calculate the increase in the entropy due to irreversible mixing of two gases. Consider two nonidentical gases in chambers of volume V separated by a wall (Figure 6.1). Let us assume that the two chambers contain the same amount, N moles, of the two gases. The total initial entropy of the system is the sum of the entropies of the two gases:

$$S_{init} = N[s_{01} + C_{mV1} \ln(T) + R \ln(V/N)] + N[s_{02} + C_{mV2} \ln(T) + R \ln(V/N)] \quad (6.1.10)$$

Now if the wall separating the two chambers is removed, the two gases will mix irreversibly and the entropy will increase. When the two gases have completely mixed and the system has reached a new state of equilibrium, each gas would be occupying a volume of $2V$. Hence, the total final entropy after the mixing is

$$S_{fin} = N[s_{01} + C_{mV1} \ln(T) + R \ln(2V/N)] + N[s_{02} + C_{mV2} \ln(T) + R \ln(2V/N)] \quad (6.1.11)$$

The difference between (6.1.10) and (6.1.11) is the entropy of mixing $\Delta S_{mix} = S_{fin} - S_{init}$. It is easy to see that

$$\Delta S_{mix} = 2NR \ln 2 \qquad (6.1.12)$$

The generalization of this result to unequal volumes and molar amounts is left as an exercise. It can be shown that if initially the densities of the two gases are

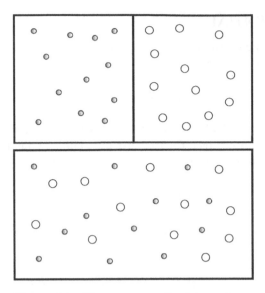

Figure 6.1 The entropy of mixing of two nonidentical gases, however small the difference between the gases, is given by (6.1.12). If the two gases are identical, then there is no change in the entropy

the same, i.e. $(N_1/V_1) = (N_2/V_2)$, then the entropy of mixing can be written as (Exercise 6.2)

$$\Delta S_{\mathrm{mix}} = -RN(x_1 \ln x_1 + x_2 \ln x_2) \tag{6.1.13}$$

where x_1 and x_2 are the mole fractions and $N = N_1 + N_2$.

Gibbs noted a curious aspect of this result. If the two gases were identical, then the states of the gas before and after the removal of the wall are indistinguishable except for the wall; by replacing the wall, the initial state can be restored. This means that there is no irreversible process mixing the two gases. Hence, there is no change in entropy because the initial and final states are the same. But for two nonidentical gases, however small the difference between them is, the change of entropy is given by (6.1.12). Generally, in most physical systems, a small change in one quantity results in a small change in another dependent quantity. Not so with the entropy of mixing; even the smallest difference between two gases leads to an entropy difference of $2NR\ln2$. If the difference between the two gases vanishes, then S_{mix} abruptly drops to zero. This discontinuous behavior of the entropy of mixing is often called **the Gibbs paradox**.

The entropy of mixing (6.1.13) can also be obtained using the statistical formula $S = k_{\mathrm{B}} \ln W$ introduced in Chapter 3 (Box 3.1). Consider a gas containing $(N_1 + N_2)$ moles or $(N_1 + N_2)N_{\mathrm{A}}$ molecules. For this gas, interchange of molecules does not correspond to distinct microstates because the molecules are indistinguishable.

However, if N_2 moles of the gas are replaced by another gas, then an interchange of molecules of the two different gases corresponds to a distinct microstate. Thus, the gas mixture with N_1 moles of one gas and N_2 of another gas has additional microstates in comparison with $(N_1 + N_2)$ moles of one gas. That these additional microstates when used in the formula $S = k_B \ln W$ give the entropy of mixing (6.1.13) can be seen as follows. The number of additional microstates in the mixture is

$$W_{\text{mix}} = \frac{(N_A N_1 + N_A N_2)!}{(N_A N_1)!(N_A N_2)!} \qquad (6.1.14)$$

in which we have introduced the Avogadro number N_A to convert moles to number of molecules. Using the Stirling approximation $N! \approx N \ln N - N$, it can easily be shown that (Exercise 6.2)

$$\Delta S_{\text{mix}} = k_B \ln W_{\text{mix}} = -k_B N_A (N_1 + N_2)(x_1 \ln x_1 + x_2 \ln x_2) \qquad (6.1.15)$$

in which x_1 and x_2 are mole fractions. Equation (6.1.15) is identical to (6.1.13) because $R = k_B N_A$ and $N = N_1 + N_2$. This derivation shows that expression (6.1.13) for the entropy of mixing is not dependent on the interactions between the gas molecules: it is entirely a consequence of distinguishablity of the two components of the system.

6.2   Thermodynamics of Real Gases

Useful though it might be, the idea gas approximation ignores the finite size of the molecules and the intermolecular forces. Consequently, as the gas becomes denser, the ideal gas equation does not predict the relation between the volume, pressure and temperature with good accuracy: one has to use other equations of state that provide a better description. If the molecular size and forces are included in the theory, then one refers to it as a theory of a 'real gas'.

As a result of molecular forces, the total internal energy U, the relation between the molar heat capacities C_{mp} and C_{mV}, the equation for adiabatic processes and other thermodynamic quantities will differ from those for the ideal gas. In this section, we shall see how the thermodynamic quantities of a real gas can be obtained from an equation of state that takes molecular size and forces into account.

The van der Waals equation, which takes into account the intermolecular forces and molecular size, and the critical constants p_c, V_{mc} and T_c were introduced in Chapter 1:

$$\boxed{\left(p + \frac{a}{V_m^2}\right)(V_m - b) = RT} \qquad (6.2.1a)$$

$$p_c = \frac{a}{27b^2} \quad V_{mc} = 3b \quad T_c = \frac{8a}{27bR} \qquad (6.2.1b)$$

in which V_m is the molar volume. Since the van der Waals equation also has its limitations, others have been proposed for gases. Some of the other equations that have been proposed and the corresponding critical constants are as follows:

The Berthelot equation:
$$p = \frac{RT}{V_m - b} - \frac{a}{TV_m^2} \qquad (6.2.2a)$$

$$p_c = \frac{1}{12}\left(\frac{2aR}{3b^3}\right)^{1/2} \quad V_{mc} = 3b \quad T_c = \frac{2}{3}\left(\frac{2a}{3bR}\right)^{1/2} \qquad (6.2.2b)$$

The Dieterici equation:
$$p = \frac{RT\, e^{-a/RTV_m}}{V_m - b} \qquad (6.2.3a)$$

$$p_c = \frac{a}{4e^2 b^2} \quad V_{mc} = 2b \quad T_c = \frac{a}{4Rb} \qquad (6.2.3b)$$

in which a and b are constants similar to the van der Waals constants which can be related to the critical constants as shown. Another equation that is often used is the **virial expansion**, proposed by Kamerlingh Onnes (1853–1926). It expresses the pressure as power series in the molar density $\rho = N/V$:

$$p = RT\frac{N}{V}\left[1 + B(T)\frac{N}{V} + C(T)\left(\frac{N}{V}\right)^2 + \ldots\right] \qquad (6.2.4)$$

in which $B(T)$ and $C(T)$ are functions of temperature, called the **virial coefficients**; they are experimentally measured and tabulated. For example, it is found that experimental data for the virial coefficient can be approximated by the function $B(T) = \alpha - \beta\exp(\gamma/T)$, in which α, β and γ are constants and T is the temperature in kelvin.[*] Values of these constants for a few gases are shown in Table 6.1. It is also

Table 6.1 An empirical function for the second virial coefficient $B(T) = \alpha - \beta \exp(\gamma/T)$

Gas	α/mL mol^{-1}	β/mL mol^{-1}	γ/K	Range of validity/K
Ar	154.2	119.3	105.1	80–1024
N_2	185.4	141.8	88.7	75–700
O_2	152.8	117.0	108.8	90–400
CH_4	206.4	159.5	133.0	110–600
C_2H_6	267.3	191.5	256	210–500

Source: Online Kaye & Laby Tables of Physical and Chemical Constants at the National Physical Laboratory, UK (http://www.kayelaby.npl.co.uk/chemistry).

---

[*]The values of the constants α, β and γ and the ranges of T for which the empirical formula is valid can be found at http://www.kayelaby.npl.co.uk/chemistry/3_5/3_5.html of the National Physical Laboratory, UK.

found that a better fit for experimental data can be obtained by dropping $(N/V)^3$ and higher odd powers from the virial expansion (6.2.4). As expected, (6.2.4) reduces to the ideal gas equation at low densities. The van der Waals constants a and b can be related to the virial coefficients $B(T)$ and $C(T)$ (Exercise 6.4); and, conversely, the virial coefficients can be calculated form the van der Waals constants. Since the ideal gas equation is valid at low pressures, the virial equation may also be written as

$$p = RT\frac{N}{V}[1 + B'(T)p + C'(T)p^2 + \ldots]\qquad(6.2.5)$$

Comparing (6.2.4) and (6.2.5), it can be shown that $B = B'RT$, to a first approximation.

TOTAL INTERNAL ENERGY

For real gases, due to the molecular interaction, the energy is no longer only a function of the temperature. Because the interaction energy of the molecules depends on the distance between the molecules, a change in volume (at a fixed T) causes a change in energy, i.e. the term $(\partial U/\partial V)_T$ does not vanish for a real gas. Molecular forces have a short range. At low densities, since molecules are far apart, the force of interaction is small. As the density approaches zero, the energy of real gas U_{real} approaches the energy of an ideal gas U_{ideal}. We can obtain an explicit expression for U_{real} through the Helmholtz equation, $(\partial U/\partial V)_T = T^2[\partial(p/T)/\partial T]_V$, which is valid for all systems (not only for gases). Upon integration, this equation yields

$$U_{real}(T,V,N) = U_{real}(T,V_0,N) + \int_{V_0}^{V} T^2\left(\frac{\partial}{\partial T}\frac{p}{T}\right)_V dV\qquad(6.2.6)$$

To write this in a convenient form, first we note that, for a fixed N, as the volume $V_0 \to \infty$, the density approaches zero, and, as noted above, U_{real} approaches the energy of an ideal gas U_{ideal} given by (6.1.3). Hence, (6.2.6) can be written as

$$U_{real}(T,V,N) = U_{ideal}(T,N) + \int_{\infty}^{V} T^2\left(\frac{\partial}{\partial T}\frac{p}{T}\right)_V dV\qquad(6.2.7)$$

If $[\partial(p/T)/\partial T]_V$ can be calculated using the equation of state, then explicit expressions for U_{real} could be derived. For example, let us consider the van der Waals equation of state. From (6.2.1) it is easy to see that $p/T = NR/(V - Nb) - a(N/V)^2(1/T)$. Substituting this expression into (6.2.7) we obtain the energy of a van der Waals gas U_{vw}:

$$U_{vw}(T,V,N) = U_{ideal}(T,N) + \int_{\infty}^{V} a\left(\frac{N}{V}\right)^2 dV$$

Evaluation of the integral gives

$$U_{vw}(V, T, N) = U_{ideal} - a\left(\frac{N}{V}\right)^2 V \qquad (6.2.8)$$

Writing the energy in this form shows us that the energy due to molecular interactions is equal to $-a(N/V)^2$ per unit volume. As we expect, as the volume increases U_{vw} approaches U_{ideal}.

MOLAR HEAT CAPACITYS C_{mV} AND C_{mp}

If the molar internal energy U_m of a gas is known, then the molar heat capacity at constant volume $C_{mV} = (\partial U_m/\partial T)_V$ can be calculated. For a real gas, we can use (6.2.7) to obtain the following expression for the molar heat capacity C_{mV}:

$$C_{mV, real} = \left(\frac{\partial U_{m, real}}{\partial T}\right)_V = \left(\frac{\partial U_{m, ideal}}{\partial T}\right)_V + \frac{\partial}{\partial T}\int_\infty^V T^2 \left(\frac{\partial}{\partial T}\frac{p}{T}\right)_V dV$$

which upon explicit evaluation of the derivatives in the integral gives

$$C_{mV, real} = C_{mV, ideal} + \int_\infty^V T\left(\frac{\partial^2 p}{\partial T^2}\right)_V dV \qquad (6.2.9)$$

Given an equation of state, such as the van der Waals equation, the above integral can be evaluated to obtain an explicit expression for C_{mV}. Equation (6.2.9) shows that, *for any equation of state in which p is a linear function of T, $C_{mV,real} = C_{mV,ideal}$.* This is true for the case of the van der Waals equation. The energy due to the molecular interactions depends on the intermolecular distance or density N/V. Because this does not change at constant V, the value of C_{mV} is unaffected by the molecular forces. C_{mV} is the change in kinetic energy of the molecules per unit change in temperature.

Also, given the equation of state, the isothermal compressibility κ_T and the coefficient of volume expansion α (which are defined by Equations (5.4.5) and (5.4.6) respectively) can be calculated. Then, using the general relation

$$\boxed{C_{mp} - C_{mV} = \frac{TV_m\alpha^2}{\kappa_T}} \qquad (6.2.10)$$

(see Equation (5.4.10)) C_{mp} can also be obtained. Thus, using (6.2.9) and (6.2.10), the two molar heat capacities of a real gas can be calculated using its equation of state.

ADIABATIC PROCESSES

For an ideal gas, we have seen in Chapter 2 that in an adiabatic process $TV^{\gamma-1} =$ constant or $pV^\gamma =$ constant (see (2.3.11) and (2.3.12)), in which $\gamma = C_{mp}/C_{mV}$. One can obtain a similar equation for a real gas. An adiabatic process is defined by $dQ = 0$ $= dU + p\, dV$. By considering U as a function of V and T, this equation can be written as

$$\left(\frac{\partial U}{\partial V}\right)_T dV + \left(\frac{\partial U}{\partial T}\right)_V dT + p\, dV = 0 \qquad (6.2.11)$$

Since $(\partial U/\partial T)_V = NC_{mV}$, where N is the molar amount of the gas, this equation becomes

$$\left[\left(\frac{\partial U}{\partial V}\right)_T + p\right]dV = -NC_{mV}dT \qquad (6.2.12)$$

By evaluating the derivative on the right-hand side of the Helmholtz equation (5.2.11), it is easy to see that $[(\partial U/\partial V)_T + p] = T(\partial p/\partial T)_V$. Furthermore, we have also seen in Chapter 5 (see (5.4.7)) that $(\partial p/\partial T)_V = \alpha/\kappa_T$. Using these two relations, (6.2.12) can be written as

$$\frac{T\alpha}{\kappa_T}dV = -NC_{mV}dT \qquad (6.2.13)$$

To write this expression in terms of the ratio $\gamma = C_{mp}/C_{mV}$ we use the general relation:

$$C_{mp} - C_{mV} = \frac{T\alpha^2 V_m}{\kappa_T} \qquad (6.2.14)$$

in which V_m is the molar volume. Combining (6.2.14) and (6.2.13) we obtain

$$N\frac{C_{mp} - C_{mV}}{V\alpha}dV = -NC_{mV}dT \qquad (6.2.15)$$

where we have made the substitution $V_m = V/N$ for the molar volume. Dividing both sides of this expression by C_{mV} and using the definition $\gamma = C_{mp}/C_{mV}$ we obtain the simple expression

$$\frac{\gamma - 1}{V}dV = -\alpha dT \qquad (6.2.16)$$

Generally, γ varies little with volume or temperature, so it may be treated as a constant and Equation (6.2.16) can be integrated to obtain

$$(\gamma - 1)\ln V = -\int \alpha(T)dT + C \qquad (6.2.17)$$

in which we have written α as an explicit function of T. C is the integration constant. An alternative way of writing this expression is

$$\boxed{V^{\gamma-1}\, e^{\int \alpha(T)\,dT} = \text{constant}} \tag{6.2.18}$$

This relation is valid for all gases. For an ideal gas, $\alpha = (1/V)(\partial V/\partial T)_p = 1/T$. When this is substituted into (6.2.18) we obtain the familiar equation $TV^{\gamma-1} = \text{constant}$. If p is a linear function of T, as is the case with the van der Waals equation, since $C_{mV,\text{real}} = C_{mV,\text{ideal}}$, from (6.2.14) it follows that

$$\gamma - 1 = \frac{T\alpha^2 V_m}{C_{mV,\text{ideal}}\kappa_T} \tag{6.2.19}$$

If the equation of state of a real gas is known, then α and γ can be evaluated (numerically, if not analytically) as function of T, and the relation (6.2.18) between V and T can be made explicit for an adiabatic process.

HELMHOLTZ AND GIBBS ENERGIES

The method used to obtain a relation (6.2.7) between U_{ideal} and U_{real} can also be used to relate the corresponding Helmholtz and Gibbs energies. The main idea is that the thermodynamic quantities for a real gas approach those of an ideal gas as $p \to 0$ or $V \to \infty$. Let us consider the Helmholtz energy F. Since $(\partial F/\partial V)_T = -p$ (see (5.1.6)) we have the general expression

$$F(T,V,N) = F(T,V_0,N) - \int_{V_0}^{V} p\,dV \tag{6.2.20}$$

The difference between the Helmholtz energy of a real and an ideal gas at any T, V and N can be obtained as follows. Writing Equation (6.2.20) for a real and an ideal gas, and subtracting one from the other, it is easy to see that

$$F_{\text{real}}(T,V,N) - F_{\text{ideal}}(T,V,N) = F_{\text{real}}(T,V_0,N) - F_{\text{ideal}}(T,V_0,N) - \int_{V_0}^{V} (p_{\text{real}} - p_{\text{ideal}})\,dV$$
$$\tag{6.2.21}$$

Now, since $\lim_{V_0 \to \infty}[F_{\text{real}}(V_0,T,N) - F_{\text{ideal}}(V_0,T,N)] = 0$, we can write the above expression as

$$\boxed{F_{\text{real}}(T,V,N_k) - F_{\text{ideal}}(T,V,N_k) = -\int_{\infty}^{V} (p_{\text{real}} - p_{\text{ideal}})\,dV} \tag{6.2.22}$$

where we have explicitly indicated the fact that this expression is valid for a multicomponent system by replacing N by N_k. Similarly, we can also show that

$$G_{\text{real}}(T, p, N_k) - G_{\text{ideal}}(T, p, N_k) = \int_0^p (V_{\text{real}} - V_{\text{ideal}}) \, \mathrm{d}p \qquad (6.2.23)$$

As an example, let us calculate F using the van der Waals equation. For the van der Waals equation, we have $p_{\text{real}} = p_{\text{vw}} = [NRT/(V - bN)] - (aN^2/V^2)$. Substituting this expression for p_{real} into (6.2.22) and performing the integration one can obtain (Exercise 6.10)

$$F_{\text{vw}}(T, V, N) = F_{\text{ideal}}(T, V, N) - a\left(\frac{N}{V}\right)^2 V - NRT \ln\left(\frac{V - Nb}{V}\right) \qquad (6.2.24)$$

where

$$\begin{aligned} F_{\text{ideal}} &= U_{\text{ideal}} - TS_{\text{ideal}} \\ &= U_{\text{ideal}} - TN[s_0 + C_{\mathrm{m}V} \ln(T) + R\ln(V/N)] \end{aligned} \qquad (6.2.25)$$

Substituting (6.2.25) into (6.2.24) and simplifying we obtain

$$\begin{aligned} F_{\text{vw}} &= U_{\text{ideal}} - a(N/V)^2 V - TN\{s_0 + C_{\mathrm{m}V} \ln(T) + R\ln[(V - Nb)/N] \\ &= U_{\text{vw}} - TN\{s_0 + C_{\mathrm{m}V} \ln(T) + R\ln[(V - Nb)/N]\} \end{aligned} \qquad (6.2.26)$$

where we have used the expression $U_{\text{vw}}(V, T, N) = U_{\text{ideal}} - a(N/V)^2 V$ for the energy of a van der Waals gas (see (6.2.8)). Similarly, the Gibbs energy of a real gas can be calculated using the van der Waals equation.

ENTROPY

The entropy of a real gas can be obtained using expressions (6.2.7) and (6.2.21) for U_{real} and F_{real} because $F_{\text{real}} = U_{\text{real}} - TS_{\text{real}}$. Using the van der Waals equation, for example, the entropy S_{vw} of a real gas can be identified in (6.2.26):

$$S_{\text{vw}}(T, V, N) = N\{s_0 + C_{\mathrm{m}V} \ln(T) + R\ln[(V - Nb)/N]\} \qquad (6.2.27)$$

A comparison of (6.2.27) with the entropy of an ideal gas (6.1.6) shows that, in the van der Waals entropy, the term $(V - Nb)$ takes the place of V in the ideal gas entropy.

CHEMICAL POTENTIAL

The chemical potential for a real gas can be derived from the expression (6.2.23) for the Gibbs free energy. Since the chemical potential of the component k is $\mu_k = (\partial G/\partial N_k)_{p,T}$, by differentiating (6.2.23) with respect to N_k we obtain

$$\mu_{k,\text{real}}(T, p) - \mu_{k,\text{ideal}}(T, p) = \int_0^p (V_{\mathrm{m}k,\text{real}} - V_{\mathrm{m}k,\text{ideal}}) \, \mathrm{d}p \qquad (6.2.28)$$

in which $V_{mk} = (\partial V/\partial N_k)_{p,T}$ is the partial molar volume of the component k by definition. For simplicity, let us consider a single gas. To compare the molar volume of the ideal gas $V_{m,ideal} = RT/p$ with that of a real gas $V_{m,real}$, a **compressibility factor** Z is defined as follows:

$$\boxed{V_{m,real} = ZRT/p}$$ (6.2.29)

For an ideal gas $Z = 1$; a deviation of the value of Z from 1 indicates nonideality. In terms of Z, the chemical potential can be written as

$$\mu_{real}(T, p) = \mu_{ideal}(T, p) + RT \int_0^p \left(\frac{Z-1}{p}\right) dp$$

$$= \mu_{ideal}(p_0, T) + RT \ln\left(\frac{p}{p_0}\right) + RT \int_0^p \left(\frac{Z-1}{p}\right) dp$$ (6.2.30)

in which we have used expression $\mu_{ideal}(p,T) = \mu(p_0,T) + RT \ln(p/p_0)$ for the chemical potential of an ideal gas. The chemical potential is also expressed in terms of a quantity called **fugacity** f, which was introduced by G.N. Lewis, a quantity similar to pressure [1]. To keep the form of the chemical potential of a real gas similar to that of the ideal gases, G.N. Lewis introduced the fugacity f through the definition

$$\mu_{real}(p, T) = \mu_{ideal}(p, T) + RT \ln\left(\frac{f}{p}\right)$$ (6.2.31)

Indeed, we must have $\lim_{p\to 0}(f/p) = 1$ to recover the expression for the ideal gas at a very low pressure. Thus, the deviation of f from the pressure of an ideal gas is a measure of the 'nonideality' of the real gas. Comparing (6.2.30) and (6.2.31), we see that

$$\boxed{\ln\left(\frac{f}{p}\right) = \int_0^p \left(\frac{Z-1}{p}\right) dp}$$ (6.2.32)

It is possible to obtain Z explicitly for various equations such as the van der Waals equation or the virial equation (6.2.5). For example, if we use the virial equation we have

$$Z = \frac{pV_m}{RT} = [1 + B'(T)p + C'(T)p^2 + \ldots]$$ (6.2.33a)

$$= \left[1 + B(T)\left(\frac{N}{V}\right) + C(T)\left(\frac{N}{V}\right)^2 + \ldots\right]$$ (6.2.33b)

Substituting this expression in (6.2.33a) in (6.2.32) we find that, to the second order in p:

G. N. Lewis (1875–1946) (Reproduced courtesy of the AIP Emilio Segre Visual Archive, photo by Francis Simon)

$$\ln\left(\frac{f}{p}\right) = B'(T)p + \frac{C'(T)p^2}{2} + \ldots \tag{6.2.34}$$

Generally, terms of the order p^2 are small and may be ignored. Then, (6.2.34) can be used for the chemical potential of a real gas μ_{real} by noting that (6.2.31) can also be written as

$$\mu_{real}(p,T) = \mu_{ideal}(p,T) + RT\ln\left(\frac{f}{p}\right)$$
$$= \mu_{ideal}(p,T) + RT(B'(T)p + \ldots) \tag{6.2.35}$$

This expression can also be written in terms of the virial coefficients of Equation (6.2.4) by noting the relation $B = B'RT$, to a first approximation. Thus:

$$\mu_{real} = \mu_{ideal}(p,T) + Bp + \ldots \tag{6.2.36}$$

Equations (6.2.35) and (6.2.36) give us the chemical potential of a real gas in terms of its virial coefficients. Similar computation can be performed using the van der Waals equation.

We can also obtain explicit expressions for μ using $(\partial F/\partial N)_{T,V} = \mu$. Using the van der Waals equation, for example, we can write the chemical potential as a function of the molar density $n = N/V$ and temperature T (Exercise 6.9):

$$\mu(n,T) = (U_0 - 2an) + \left(\frac{C_{mV}}{R} + \frac{1}{1-nb} \right) RT - T \left[s_0 + C_{mV} \ln T - R \ln \left(\frac{n}{1-bn} \right) \right]$$
(6.2.37)

CHEMICAL AFFINITIES

Finally, to understand the nature of chemical equilibrium of real gases it is useful to obtain affinities for chemically reacting real gases. The affinity of a reaction $A = -\Sigma_k v_k \mu_k$, in which v_k are the stoichiometric coefficients (which are negative for reactants and positive for products). For a real gas this can be written using the expression (6.2.28) for the chemical potential:

$$A_{real} = A_{ideal} - \sum_k v_k \int_0^p (V_{m,real,k} - V_{m,ideal,k}) \, dp$$
(6.2.38)

This expression can be used to calculate the equilibrium constants for reacting real gases. The partial molar volume $V_{m,ideal,k}$ for all gases is RT/p. Hence, the above expression becomes

$$A_{real} = A_{ideal} - \sum_k v_k \int_0^p \left(V_{m,real,k} - \frac{RT}{p} \right) dp$$
(6.2.39)

With the above quantities, all the thermodynamics of real gases can be described once the real gas parameters, such as the van der Waals constants or the virial coefficients, are known.

6.3 Thermodynamics Quantities for Pure Liquids and Solids

EQUATION OF STATE

For pure solids and liquids, jointly called *condensed phases*, the volume is determined by the molecular size and molecular forces and it does not change much with change in p and T. Since the molecular size and forces are very specific to a compound, the equation of state is specific to that compound. A relation between V, T and p is expressed in terms of the coefficient of thermal expansion α and the isothermal compressibility κ_T defined by (5.4.5) and (5.4.6). If we consider V as a function of p and T, $V(p, T)$, we can write

$$dV = \left(\frac{\partial V}{\partial T} \right)_p dT + \left(\frac{\partial V}{\partial p} \right)_T dp = \alpha V dT - \kappa_T V dp$$
(6.3.1)

The values of α and κ_T are small for solids and liquids. For liquids, the coefficient of thermal expansion α is in the range 10^{-3} to $10^{-4} K^{-1}$ and isothermal compressibility κ_T is about $10^{-5} atm^{-1}$. For solids, α is in the range 10^{-5} to $10^{-6} K^{-1}$ and κ_T is in the range 10^{-6} to $10^{-7} atm^{-1}$. Table 6.2 lists the values of α and κ_T for some liquids and

Table 6.2   List of coefficient of thermal expansion α and isothermal compressibility κ_T for some liquids and solids

Compound	$\alpha/10^{-4}$ K^{-1}	$\kappa_T/10^{-6}$ atm^{-1}
Water	2.1	49.6
Benzene	12.4	92.1
Mercury	1.8	38.7
Ethanol	11.2	76.8
Carbon tetrachloride	12.4	90.5
Copper	0.501	0.735
Diamond	0.030	0.187
Iron	0.354	0.597
Lead	0.861	2.21

solids. Furthermore, the values of α and κ_T are almost constant for T variations of about 100 K and pressure variation of about 50 atm. Therefore, (6.3.1) can be integrated to obtain the following equation of state:

$$V(p,T) = V(p_0, T_0)\exp[\alpha(T - T_0) - \kappa_T(p - p_0)]$$
$$\approx V(p_0, T_0)[1 + \alpha(T - T_0) - \kappa_T(p - p_0)] \tag{6.3.2}$$

THERMODYNAMIC QUANTITIES

Thermodynamically, the characteristic feature of solids and liquids is that μ, S, and H change very little with pressure and, hence, they are essentially functions of T for a given N. If entropy is considered as a function of p and T, then

$$dS = \left(\frac{\partial S}{\partial T}\right)_p dT + \left(\frac{\partial S}{\partial p}\right)_T dp \tag{6.3.3}$$

The first term, $(\partial S/\partial T)_p = NC_{mp}/T$, which relates dS to the experimentally measurable C_{mp}. The second term can be related to α as follows:

$$\left(\frac{\partial S}{\partial p}\right)_T = -\left[\frac{\partial}{\partial p}\left(\frac{\partial G(p,T)}{\partial T}\right)_p\right]_T = -\left[\frac{\partial}{\partial T}\left(\frac{\partial G(p,T)}{\partial p}\right)_T\right]_p = -\left(\frac{\partial V}{\partial T}\right)_p = -V\alpha \tag{6.3.4}$$

With these observations, we can now rewrite (6.3.3) as

$$dS = \frac{NC_{mp}}{T}dT - \alpha V dp \tag{6.3.5}$$

Upon integration, this equation yields

$$S(p,T) = S(0,0) + N\int_0^T \frac{C_{mp}}{T}dT - N\int_0^p \alpha V_m dp \tag{6.3.6}$$

where we have used $V = NV_m$. (That S(0, 0) is well defined is guaranteed by the Nernst theorem.) Since V_m and α do not change much with p, the third term in (6.3.6) can be approximated to $N\alpha V_m p$. For $p = 1$–10 atm, this term is small compared with the second term. For example, in the case of water, $V_m = 18.0 \times 10^{-6} m^3 mol^{-1}$ and $\alpha = 2.1 \times 10^{-4} K^{-1}$. For $p = 10$ bar $= 10 \times 10^5$ Pa, the term $\alpha V_m p$ is about $3.6 \times 10^{-3} J K^{-1} mol^{-1}$. The value of C_{mp}, on the other hand, is about $75 J K^{-1} mol^{-1}$. Though C_{mp} approaches zero so that S is finite as $T \to 0$, the molar entropy of water at $p = 1$ bar and $T = 298$ K is about $70 J K^{-1}$. Thus, it is clear that the third term in (6.3.6) that contains p is insignificant compared with the second term. Since this is generally true for solids and liquids, we may write

$$S(p,T) = S(0,0) + N\int_0^T \frac{C_{mp}(T)}{T} dT \qquad (6.3.7)$$

where we have written C_{mp} explicitly as a function of T. A knowledge of $C_{mp}(T)$ will enable us to obtain the value of entropy of a pure solid or liquid. Note that the integral in (6.3.7) is $\int_0^T d_e S$ because $(NC_{mp} dT/T) = dQ/T = d_e S$.

The chemical potential of condensed phases can be obtained from the Gibbs–Duhem equation $d\mu = -S_m dT + V_m dp$ (see (5.2.4)). Substituting the value of molar entropy into the Gibbs–Duhem equation and integrating, we get

$$\mu(p,T) = \mu(0,0) - \int_0^T S_m(T) dT + \int_0^p V_m dp$$
$$= \mu(T) + V_m p \equiv \mu^0(T) + RT \ln a \qquad (6.3.8)$$

where we assumed that V_m is essentially a constant. Once again, it can be shown that the term containing p is small compared with the first term, which is a function of T. For water, $V_m p = 1.8 J mol^{-1}$ when $p = 1$ atm, whereas the first term is of the order $280 kJ mol^{-1}$. Following the definition of activity a, if we write $V_m p = RT \ln(a)$, then we see that for *liquids and solids the activity is nearly equal to unity*.

In a similar manner, one can obtain other thermodynamic quantities such as enthalpy H and the Helmholtz free energy F.

HEAT CAPACITIES

From the above expressions it is clear that one needs to know the molar heat capacities of a substance as a function of temperature and pressure in order to calculate the entropy and other thermodynamic quantities. A detailed understanding of the theory of molar heat capacities (which requires the statistical mechanics and quantum theory) is beyond the scope of this book. Here we shall only give a brief outline of the Debye's theory of molar heat capacities of solids that provides an approximate general theory. The situation is more complex for liquids because for liquids there is neither complete molecular disorder, as in a gas, nor is there long-range order as the case of a solid.

According to a theory of solids formulated by Peter Debye, the molar heat capacity C_{mV} of a pure solid is of the form

$$C_{mV} = 3RD(T/\theta) \tag{6.3.9}$$

in which $D(T/\theta)$ is a function of the ratio T/θ. The parameter θ depends mainly on the chemical composition of the solid and, to a very small extent, varies with the pressure. As the ratio T/θ increases, the 'Debye function' $D(T/\theta)$ tends to unity, and molar heat capacities of all solids $C_{mV} = 3R$. The fact that the heat capacities of solids tend to have the same value had been observed long before Debye formulated a theory of heat capacities; it is called the law of Dulong and Petit. Debye theory provided an explanation for the law of Dulong and Petit. At very low temperatures, when $T/\theta < 0.1$:

$$D\left(\frac{T}{\theta}\right) \approx \frac{4\pi^4}{5}\left(\frac{T}{\theta}\right)^3 \tag{6.3.10}$$

Thus, Debye's theory predicts that the molar heat capacities at low temperatures will be proportional to the third power of the temperature. Experimentally, this was found to be true for many solids.

Once C_{mV} is known, C_{mp} can be obtained using the general expression $C_{mp} - C_{mV} = TV_m\alpha^2 / \kappa_T$. More detail on this subject can be found in texts on condensed matter. The thermodynamics of liquid and solid mixtures is discussed in Chapters 7 and 8.

Appendix 6.1 Equations of State

In addition to the equations of state presented in Section 6.2, the following equation of state is also used for describing real gases:

Redlich–Kwong equation: $\quad p = \dfrac{RT}{V_m - b} - \dfrac{a}{\sqrt{T}}\dfrac{1}{V_m(V_m + b)} \tag{A6.1.1}$

For this equation, the relations between the critical constants and the constants a and b are

$$a = \frac{0.427\,48R^2T_c^{2.5}}{p_c} \quad b = \frac{0.08664\,RT_c}{p_c}$$

Reference

1. Lewis, G.N., Randall, M. *Thermodynamics and Free Energy of Chemical Substances.* 1925, New York: McGraw-Hill.

Examples

Example 6.1   Show that C_{mV} for a van der Waals gas is the same as that of an ideal gas.

Solution   The relation between C_{mV} for real and ideal gases is given by (6.2.9):

$$C_{mV,\,real} = C_{mV,\,ideal} + \int_{\infty}^{V} T\left(\frac{\partial^2 p}{\partial T^2}\right)_V dV$$

For 1 mol of a van der Waals gas:

$$p = \frac{RT}{V_m - b} - a\frac{1}{V_m^{\,2}}$$

Since this is a linear function of T the derivative $(\partial^2 p/\partial T^2)_V = 0$. Hence, the integral in the expression relating $C_{mV,real}$ and $C_{mV,ideal}$ is zero. Hence, $C_{mV,real} = C_{mV,ideal}$.

Example 6.2   Calculate the total internal energy of a real gas using the Berthelot equation (6.2.2).

Solution   The internal energy of a real gas can be calculated using the relation (6.2.7)

$$U_{real}(T,V,N) = U_{ideal}(T,N) + \int_{\infty}^{V} T^2\left(\frac{\partial}{\partial T}\frac{p}{T}\right)_V dV$$

For the Berthelot equation:

$$p = \frac{RT}{V_m - b} - \frac{a}{TV_m^2}$$

In this case, the integral

$$\int_{\infty}^{V} T^2\left(\frac{\partial}{\partial T}\frac{p}{T}\right)_V dV = -\int_{\infty}^{V} \frac{aN^2}{V^2} T^2 \frac{\partial}{\partial T}\left(\frac{1}{T^2}\right) dV = \int_{\infty}^{V} \frac{2aN^2}{T}\frac{1}{V^2} dV$$

$$= -\frac{2aN^2}{TV}$$

Hence:

$$U_{real}(T,V,N) = U_{ideal}(T,N) - \frac{2aN^2}{TV}$$

Exercises

6.1   For an *ideal gas* obtain the explicit expressions for the following:

   (i) $F(V, T, N) = U - TS$ as a function of V, T and N.
   (ii) $G = U - TS + pV$ as a function of p, T, and N.
   (iii) Use the relation $\mu = (\partial F/\partial N)_{V,T}$ and obtain an expression for μ as function of the number density N/V and T. Also show that $\mu = \mu^0(T) + RT \ln(p/p_0)$, in which $\mu^0(T)$ is a function of T.

6.2   (a) Obtain a general expression for the entropy of mixing of two nonidentical gases of equal molar densities N/V, with molar amounts N_1 and N_2, initially occupying volumes V_1 and V_2. Also show that the entropy of mixing can be written as $\Delta S_{mix} = -RN(x_1 \ln x_1 = x_2 \ln x_2)$, where x_1 and x_2 are the mole fractions and $N - N_1 + N_2$.
   (b) Using the Stirling approximation $N! \approx N \ln N - N$, obtain (6.1.15) from (6.1.14).

6.3   For N_2 the critical values are $p_c = 33.5\,\text{atm}$, $T_c = 126.3\,\text{K}$ and $V_{mc} = 90.1 \times 10^{-3}\,\text{L mol}^{-1}$. Using Equations (6.2.1a)–(6.2.3b), calculate the constants a and b for the van der Waals, Berthelot and Dieterici equations. Plot the p–V_m curves for the three equations at $T = 300\,\text{K}$, $200\,\text{K}$ and $100\,\text{K}$ on the same graph in the range $V_m = 0.1\,\text{L}$ to $10\,\text{L}$ and comment on the differences between the curves.

6.4   Using the van der Waals equation, write the pressure as a function of the density N/V. Assume that the quantity $b(N/V)$ is small and use the expansion $1/(1 - x) = 1 + x + x^2 + x^3 + \ldots$, valid for $x < 1$, to obtain an equation similar to the virial equation

$$p = RT\frac{N}{V}\left[1 + B(T)\frac{N}{V} + C(T)\left(\frac{N}{V}\right)^2 + \ldots\right]$$

Comparing the two series expansions for p, show that van der Waals constants a and b and the virial coefficients $B(T)$ and $C(T)$ are related by $B = b - (a/RT)$ and $C = b^2$.

6.5   The Boyle temperature is defined as the temperature at which the virial coefficient $B(T) = 0$. An empirical function used to fit experimental data is $B(T) = \alpha - \beta \exp(\gamma/T)$, in which α, β and γ are constants tabulated in Table 6.1.

   (a) Using the data in Table 6.1, determine the Boyle temperatures of N_2, O_2 and CH_4.
   (b) Plot $B(T)$ for the three gases on one graph.

6.6   (i) Assume ideal gas energy $U = C_{mV}NT$, where $C_{mV} = 28.46\,\mathrm{J\,K^{-1}}$ for CO_2 and calculate the difference ΔU between U_{ideal} and U_{vw} for $N = 1$, $T = 300\,\mathrm{K}$ at $V = 0.5\,\mathrm{L}$. What percentage of U_{ideal} is ΔU?
(ii) Use Maple/Mathematica to obtain a three-dimensional plot of $\Delta U/U_{ideal}$ for 1 mol of CO_2, in the volume range $V = 22.00\,\mathrm{L}$ to $0.50\,\mathrm{L}$ for $T = 200$ to $500\,\mathrm{K}$.

6.7   Obtain (6.2.9) from (6.2.7) and the definition $C_{mV,real}(\partial U_{real}/\partial T)_V$.

6.8   For CO_2, using the van der Waals equation:

(i) Obtain an expression for the compressibility factor Z. At $T = 300\,\mathrm{K}$ and for $N = 1$, using Mathematica/Maple, plot Z as a function of V from $V = 22.0\,\mathrm{L}$ to $0.5\,\mathrm{L}$.
(ii) Obtain an explicit expression for $(F_{vw} - F_{ideal})$ for 1 mol of CO_2 as a function of T and V in which if T is in kelvin and V is in liters, then $(F_{vw} - F_{ideal})$ is in joules.

6.9   Using the relation $\mu = (\partial F/\partial N)_{V,T}$, show that for a van der Waals gas:

$$\mu(n,T) = (U_0 - 2an) + \left(\frac{C_{mV}}{R} + \frac{1}{1-nb} \right) RT - T\left[s_0 + C_{mV}\ln T - R\ln\left(\frac{n}{1-bn} \right) \right]$$

in which $n = N/V$.

6.10   Obtain (6.2.24) from (6.2.22) and (6.2.26) from (6.2.24) and (6.2.25).

7 THERMODYNAMICS OF PHASE CHANGE

Introduction

Transformations from the liquid to the vapor phases or from the solid to the liquid phases are caused by heat. The eighteenth-century investigations of Joseph Black revealed that these transformations take place at a definite temperature: the boiling point or the melting point. At this temperature the heat absorbed by the substance does not increase its temperature but is 'latent' or concealed; heat's effect is to cause the change from one phase to another, not to increase the substance's temperature. Joseph Black, who clarified this concept, measured the 'latent heat' for the transformation of ice to water.

Under suitable conditions, the phases of a compound can coexist in a state of thermal equilibrium. The nature of this state of thermal equilibrium and how it changes with pressure and temperature can be understood using the laws of thermodynamics. In addition, at the point where the phase transition takes place, some thermodynamic quantities, such as molar entropy, change discontinuously. Based on such discontinuous changes of some thermodynamic quantities, such as molar heat capacity and molar entropy, phase transitions in various materials can be classified into different 'orders'. There are general theories that describe phase transitions of different orders. The study of phase transitions has grown to be a large and interesting subject, and some very important developments occurred during the 1960s and the 1970s. In this chapter, we will only present some of the basic results. For further understanding of phase transitions, we refer the reader to books devoted to this subject [1–3].

7.1 Phase Equilibrium and Phase Diagrams

The conditions of temperature and pressure under which a substance exists in different phases, i.e. gas, liquid and solid, are summarized in a **phase diagram**. A simple phase diagram is shown in Figure 7.1. Under suitable conditions of pressure and temperature, two phases may coexist in thermodynamic equilibrium. The thermodynamic study of phase equilibrium leads to many interesting and useful results. For example, it tells us how the boiling point or freezing point of a substance changes with changes in pressure. We shall see how the thermodynamic formalism developed in the previous chapters enables us to obtain expressions to calculate the boiling point of a liquid at a given pressure.

Introduction to Modern Thermodynamics Dilip Kondepudi
© 2008 John Wiley & Sons, Ltd

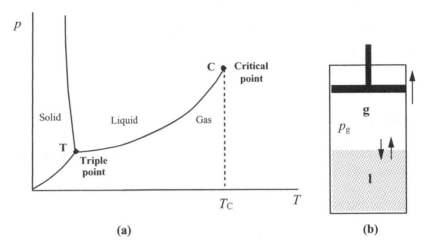

Figure 7.1 (a) Phase diagram for a one-component system showing equilibrium p–T curves (defined by the equality of the chemical potentials), the triple point T and the critical point C. T_c is the critical temperature above which the gas cannot be liquefied by increasing the pressure. (b) A liquid in equilibrium with its vapor. The affinity for the liquid–vapor transformation $A = \mu_l - \mu_g = 0$. An infinitely slow expansion in the system's volume results is a 'reversible' transformation of liquid to gas at an affinity $A \approx 0$

We begin by looking at the equilibrium between liquid and gas phases, as shown in Figure 7.1b. When a liquid is in a closed container, a part of it evaporates and fills the space above it until an equilibrium is reached. The system under consideration is closed and it consists only of the liquid in equilibrium with its vapor at a fixed temperature. In Figure 7.2, the p–V isotherms of a vapor–liquid system are shown. The region of coexistence of the liquid and vapor phases corresponds to the flat portion XY of the isotherms. When $T > T_c$, the flat portion does not exist; there is no distinction between the gas and the liquid phases. The flat portion of each isotherm in Figure 7.2 corresponds to a point on the curve TC in Figure 7.1a; as the temperature approaches T_c, we approach the critical point C.

For a thermodynamic analysis of the equilibrium between liquid and gas phases of a substance let us consider a **heterogeneous system** in which the two phases occupy separate volumes. Under these conditions, the liquid converts irreversibly to vapor, or vice versa, until equilibrium between the two phases is attained. The exchange of matter between the two phases may be considered a 'chemical reaction' which we may represent as

$$l \rightleftharpoons g \tag{7.1.1}$$

Let the chemical potential of the substance k in the two phases be μ_k^g and μ_k^l, with the superscripts 'g' for gas and 'l' for liquid. *At equilibrium, the entropy production due to every irreversible process must vanish.* This implies that the affinity corresponding to liquid–vapor conversion must vanish, i.e.

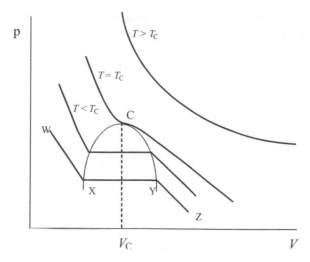

Figure 7.2 p–V isotherms of a gas showing critical behavior. T_c is the critical temperature above which the gas cannot be liquefied by increasing the pressure. In the flat region XY the liquid and the gas phases coexist

$$A = \mu_k^l(p,T) - \mu_k^g(p,T) = 0$$

i.e.

$$\mu_k^l(p,T) = \mu_k^g(p,T) \qquad (7.1.2)$$

in which we have made explicit that the two chemical potentials are functions of pressure and temperature. The pressure of the vapor phase in equilibrium with the liquid phase is called the **saturated vapor pressure**. The equality of the chemical potentials implies that, when a liquid is in equilibrium with its vapor, the pressure and temperature are not independent. This relationship between p and T, as expressed in (7.1.2), gives the **coexistence curve** TC in the phase diagram shown in Figure 7.1a.

A liquid in equilibrium with its vapor is a good system to illustrate the idea of a 'reversible' transformation for which $d_iS = 0$ (Figure 7.1b). Let us assume that initially the system is in equilibrium with $A = 0$. If the volume of the system is increased slowly, the chemical potential of the gas phase will decrease by a small amount, making the affinity for the liquid-to-gas transformation positive. This will result in the conversion of liquid to gas until a new equilibrium is established. In the limit of 'infinitely slow' increase of volume such that the transformation takes place at an arbitrarily small A, i.e. $A \approx 0$, no entropy is produced during this transformation because $d_iS = A \, d\xi \approx 0$. Therefore, it is a reversible transformation. A reversible transformation, of course, is an idealized process taking place at an infinitely slow rate. In any real process that occurs at a nonzero rate, $d_iS = A\,d\xi > 0$; but this change can be made arbitrarily small by slowing the rate of transformation.

Clearly, equality of chemical potentials as in (7.1.2) must be valid between any two phases that are in equilibrium. If there are P phases, then we have the general equilibrium condition:

$$\mu_k^1(p, T) = \mu_k^2(p, T) = \mu_k^3(p, T) = \dots \mu_k^P(p, T) \qquad (7.1.3)$$

The phase diagram Figure 7.1a also shows another interesting feature, viz. the **critical point** C at which the coexistence curve TC (between the liquid and vapor phase) terminates. If the temperature of the gas is above T_c, the gas cannot be liquefied by increasing the pressure. As the pressure increases, the density increases but there is no *transition* to a condensed phase – and no latent heat. In contrast, there is no critical point for the transition between solid and liquid due to the fact that a solid has a definite crystal structure that the liquid does not have. Owing to the definite change in symmetry, the transition between a solid and liquid is always well defined.

A change of phase of a solid is not necessarily a transformation to a liquid. A solid may exist in different phases. Thermodynamically, a phase change is identified by a sharp change in properties such as the heat capacity. In molecular terms, these changes correspond to different arrangements of the atoms, i.e. different crystal structures. For example, at very high pressures, ice exists in different structures, and these are the different solid phases of water. Figure 7.3 shows the phase diagram of water.

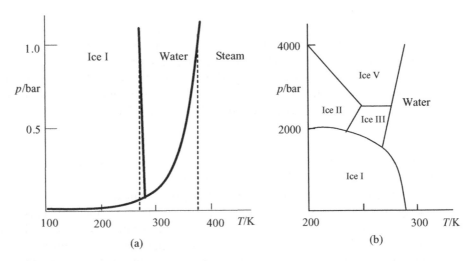

Figure 7.3 (a) The phase diagram of water at ordinary pressures (not to scale). (b) At high pressures, the solid phase (ice) can exist in different phases as shown on the right. The triple point of water is at $p = 0.006$ bar, $T = 273.16$ K. The critical point is at $p_c = 218$ bar, $T_c = 647.3$ K

THE CLAPEYRON EQUATION

The temperature on the coexistence curves corresponds to the temperature at which the transition from one phase to another takes place at a given pressure. Thus, if we obtain an explicit relation between the pressure and the temperature that defines the coexistence curve, then we can know how the boiling point or freezing point changes with pressure. Using the condition for equilibrium (7.1.2), we can arrive at a more explicit expression for the coexistence curve. Let us consider two phases denoted by 1 and 2. Using the Gibbs–Duhem equation, $d\mu = -S_m\, dT + V_m\, dp$, one can derive a differential relation between p and T of the system as follows. From (7.1.3) it is clear that, for a component k, $d\mu_k^1 = d\mu_k^2$. Therefore, we have the equality

$$-S_{m1}dT + V_{m1}dp = -S_{m2}dT + V_{m2}dp \qquad (7.1.4)$$

in which the molar quantities for the two phases are indicated by the subscripts 'm1' and 'm2'. From this it follows that

$$\frac{dp}{dT} = \frac{S_{m1} - S_{m2}}{V_{m1} - V_{m2}} = \frac{\Delta H_{\text{trans}}}{T(V_{m1} - V_{m2})} \qquad (7.1.5)$$

in which we have expressed the difference in the molar entropy between the two phases in terms of the enthalpy of transition: $S_{m1} - S_{m2} = (\Delta H_{\text{trans}}/T)$, where ΔH_{trans} is the *molar enthalpy* of the transition (vaporization, fusion or sublimation). Molar enthalpies of vaporization and fusion of some substances are listed in Table 7.1. More generally, then, we have the equation called the **Clapeyron equation**:

Table 7.1 Enthalpies of fusion of and vaporization at $p = 1$ bar $= 10^5$ Pa $= 0.987$ atm

Substance	T_m/K	$\Delta H_{\text{fus}}/kJ\ mol^{-1}$	T_b/K	$\Delta H_{\text{vap}}/kJ\ mol^{-1}$
He	0.95*	0.021	4.22	0.082
H_2	14.01	0.12	20.28	0.46
O_2	54.36	0.444	90.18	6.820
N_2	63.15	0.719	77.35	5.586
Ar	83.81	1.188	87.29	6.51
CH_4	90.68	0.941	111.7	8.18
C_2H_5OH	156	4.60	351.4	38.56
CS_2	161.2	4.39	319.4	26.74
CH_3OH	175.2	3.16	337.2	35.27
NH_3	195.4	5.652	239.7	23.35
CO_2	217.0	8.33	194.6	25.23
Hg	234.3	2.292	629.7	59.30
CCl_4	250.3	2.5	350	30.0
H_2O	273.15	6.008	373.15	40.66
Ga	302.93	5.59	2676	270.3
Ag	1235.1	11.3	2485	257.7
Cu	1356.2	13.0	2840	306.7

Source: D.R. Lide (ed.) *CRC Handbook of Chemistry and Physics*, 75th edition. 1994, CRC Press: Ann Arbor.
* Under high pressure.

$$\boxed{\frac{dp}{dT} = \frac{\Delta H_{\mathrm{trans}}}{T \Delta V_{\mathrm{m}}}} \qquad (7.1.6)$$

Here, ΔV_{m} is the difference in the *molar volumes* of the two phases. The temperature T in this equation is the transition temperature, i.e. boiling point, melting point, etc. This equation tells us how the transition temperature changes with pressure. For example, for a transition from a solid to a liquid in which there is an increase in the molar volume ($\Delta V > 0$), the freezing point will increase ($dT > 0$) when the pressure is increased ($dp > 0$); if there is a decrease in the molar volume, then the opposite will happen.

THE CLAUSIUS–CLAPEYRON EQUATION

For the case of liquid–vapor transitions, the Clapeyron equation can be further simplified. In this transition $V_{\mathrm{ml}} \ll V_{\mathrm{mg}}$. Therefore, we may approximate $V_{\mathrm{mg}} - V_{\mathrm{ml}}$ by V_{mg}. In this case the Clapeyron equation (7.1.6) simplifies to

$$\frac{dp}{dT} = \frac{\Delta H_{\mathrm{vap}}}{T V_{\mathrm{mg}}} \qquad (7.1.7)$$

As a first approximation, we may use the ideal-gas molar volume $V_{\mathrm{mg}} = RT/p$. Substituting this expression in the place of V_{mg}, and noting that $dp/p = d(\ln p)$, we arrive at the following equation called the **Clausius–Clapeyron equation**:

$$\boxed{\frac{d(\ln p)}{dT} = \frac{\Delta H_{\mathrm{vap}}}{RT^2}} \qquad (7.1.8)$$

This equation is also applicable to a solid in equilibrium with its vapor (e.g. I_2), since the molar volume of the vapor phase is much larger than that of the solid phase. *For a solid in equilibrium with its vapor, ΔH_{sub} takes the place of ΔH_{vap}.* At times, Equation (7.1.8) is also written in its integrated form:

$$\boxed{\ln p_2 - \ln p_1 = \frac{\Delta H_{\mathrm{vap}}}{R}\left(\frac{1}{T_1} - \frac{1}{T_2}\right)} \qquad (7.1.9)$$

As illustrated in Figure 7.4, Equations (7.1.8) and (7.1.9) tell us how the boiling point of a liquid changes with pressure. When a liquid subjected to an external pressure, p_{ext}, is heated, bubbles containing the vapor (in equilibrium with the liquid) can form provided that the vapor pressure $p_{\mathrm{g}} \geq p_{\mathrm{ext}}$. The liquid then begins to 'boil'. If the vapor pressure p is less than p_{ext}, then the bubbles cannot form: they 'collapse'. The temperature at which $p = p_{\mathrm{ext}}$ is what we call the boiling point T_{b}. Hence, in Equations (7.1.8) and (7.1.9) we may interpret p as the pressure to which the liquid is subjected, and T is the corresponding boiling point. It tells us that the boiling point of a liquid decreases with a decrease in pressure p_{ext}.

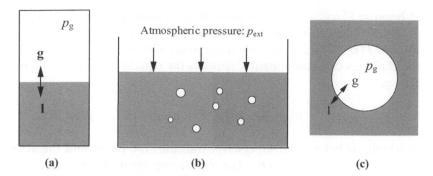

Figure 7.4 Equilibrium between liquid and vapor phases. (a) An isolated system which contains a liquid in equilibrium with its vapor. The pressure of the vapor p_g is called the saturated vapor pressure. (b) When the liquid subject to a pressure p_{ext} (atmospheric pressure) is heated, bubbles of its vapor can form when $p_g \geq p_{ext}$ and the liquid begins to 'boil'. (c) The vapor in the bubble is the saturated vapor in equilibrium with the liquid, as in the case of an isolated system (a)

7.2   The Gibbs Phase Rule and Duhem's Theorem

Thus far we have considered the equilibrium between two phases of a single compound. When many compounds or components and more than two phases are in equilibrium, the chemical potential of each component should be the same in every phase in which it exists. When we have a single phase, such as a gas, its intensive variables, i.e. pressure and temperature, can be varied independently. However, when we consider equilibrium between two phases, such as a gas and liquid, p and T are no longer independent. Since the chemical potentials of the two phases must be equal, $\mu^1(p,T) = \mu^2(p,T)$, which implies that only one of the two intensive variables is independent. In the case of liquid–vapor equilibrium of a single component, p and T are related through relation (7.1.8). The number of independent intensive variables depends on the number of phases in equilibrium and the number of components in the system.

The *independent intensive variables* that specify a state are called its **degrees of freedom**. Gibbs observed that there is general relationship between the number of degrees of freedom f, the number of phases P, and the number of components C:

$$\boxed{f = C - P + 2}$$ (7.2.1)

This can be seen as follows. At a given T, specifying p is equivalent to specifying the density as moles per unit volume (through the equation of state). For a given density, the mole fractions specify the composition of the system. Thus, for each phase, p, T and the C mole fractions x_k^i (in which the superscript indicates the phase and the subscript the component) are the intensive variables that specify the state. Of the C mole fractions in each phase i, there are $(C-1)$ independent mole fractions

x_k^i because $\Sigma_{k=1}^{C} x_k^i = 1$. In a system with C components and P phases, there are a total of $P(C - 1)$ independent mole fractions x_k^i. These, together with p and T, make a total of $P(C-1) + 2$ independent variables. On the other hand, equilibrium between the P phases of a component k requires the equality of chemical potentials in all the phases:

$$\mu_k^1(p, T) = \mu_k^2(p, T) = \mu_k^3(p, T) = \ldots = \mu_k^P(p, T) \qquad (7.2.2)$$

in which, as before, the superscript indicates the phase and the subscript the component. These constitute $(P - 1)$ constraining equation for each component. For the C components we then have a total of $C(P - 1)$ equations between the chemical potentials, which reduces the number of independent intensive variables by $C(P - 1)$. Thus, the total number of independent degrees of freedom is

$$f = P(C - 1) + 2 - C(P - 1) = C - P + 2$$

If a component 'a' does not exist in one of the phases 'b', then the corresponding mole fraction $x_b^a = 0$, thus reducing the number of independent variables by one. However, this also decreases the number of constraining equations by one. Hence, there is no overall change in the number of degrees of freedom.

As an illustration of the Gibbs phase rule, let us consider the equilibrium between the solid, liquid and gas phases of a pure substance, i.e. one component. In this case we have $C = 1$ and $P = 3$, which gives $f = 0$. Hence, for this equilibrium, there are no free intensive variables; there is only one pressure and temperature at which all three phases can coexist. This point is called the **triple point** (see Figure 7.1). At the triple point of H_2O, $T = 273.16\,\mathrm{K} = 0.01\,°\mathrm{C}$ and $p = 611\,\mathrm{Pa} = 6.11 \times 10^{-3}$ bar. This unique condition for the coexistence of the three phases may be used in defining the kelvin temperature scale.

If the various components of the system also chemically react through R independent reactions, then, in addition to (7.2.2), we also have R equations for the chemical equilibrium, viz. the corresponding affinities are zero:

$$A_1 = A_2 = A_3 = \ldots A_R = 0 \qquad (7.2.3)$$

Consequently, the number of degrees of freedom is further decreased by R and we have

$$\boxed{f = C - R - P + 2} \qquad (7.2.4)$$

In older statements of the phase rule, the term 'number of independent components' is used to represent $(C - R)$. In a reaction such as $A \rightleftharpoons B + 2C$, if the amount B and C is entirely a result of decomposition of A, then the amount of B and C is determined by the amount of A that has converted to B and C; in this case the mole fractions of B and C are related, $x_C = 2x_B$. This additional constraint, which depends on the initial preparation of the system, decreases the degrees of freedom by one.

In addition to the phase rule identified by Gibbs, there is another general observation which Pierre Duhem made in his treatise *Traité élémentaire de Mechanique Chimique* which is referred to as **Duhem's theorem**. It states:

Whatever the number of phases, components and chemical reactions, if the initial molar amounts N_k of all the components are specified, the equilibrium state of a closed system is completely specified by two independent variables.

The proof of this theorem is as follows. The state of the entire system is specified by the pressure p, temperature T and the molar amounts N_k^i in which the superscript indicates the P phases and the subscript the C component – a total of CP molar amounts in P phases. Thus, the total number of variables that specify a system is $CP + 2$. Considering the constraints on these variables, for the equilibrium of each component k between the phases we have

$$\mu_k^1(p, T) = \mu_k^2(p, T) = \mu_k^3(p, T) = \ldots = \mu_k^P(p, T) \qquad (7.2.5)$$

a total of $(P - 1)$ equations for each component, a total of $C(P - 1)$ equations. In addition, since the total molar amount, say $N_{k,\text{total}}$, of each component is specified, we have $\Sigma_{i=1}^P N_k^i = N_{k,\text{total}}$ for each component, a total of C equations. Thus, the total number of constraints is $C(P - 1) + C$. Hence, the total number of independent equations is $CP + 2 - C(P - 1) - C = 2$.

The addition of chemical reactions does not change this conclusion because each chemical reaction α adds a new independent variable ξ_α, its extent of reaction, to each phase and at the same time adds the constraint for the corresponding chemical equilibrium $A_\alpha = 0$. Hence, there is no net change in the number of independent variables.

Comparing the Gibbs phase rule and the Duhem theorem, we see the following. The Gibbs phase rule specifies the total number of independent intensive variables regardless of the extensive variables in the system. In contrast, Duhem's equation specifies the total number of independent variables, intensive or extensive, in a closed system.

7.3 Binary and Ternary Systems

Figure 7.1 shows the phase diagram for a single-component system. The phase diagrams for systems with two and three components are more complex. In this section we shall consider examples of two- and three-component systems.

BINARY LIQUID MIXTURES IN EQUILIBRIUM WITH THE VAPOR

Consider a liquid mixture of two components, A and B, in equilibrium with their vapors. This system contains two phases and two components. The Gibbs phase rule tells us that such a system has two degrees of freedom. We may take these degrees of freedom to be the pressure and the mole fraction x_A of component A.

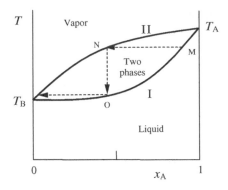

Figure 7.5 The boiling point versus composition of a mixture of two similar liquids, such as benzene and toluene

Thus, if we consider a system subjected to a constant pressure, for each value of the mole fraction x_A there is corresponding temperature at which the two phases are in equilibrium. For example, if the applied pressure is 0.5 bar, for the liquid to be in equilibrium with its vapor the temperature T must be set at an appropriate value (T equals the boiling point at 0.5 bar).

If the applied pressure is the atmospheric pressure, then the temperature corresponds to the boiling point. In Figure 7.5, the curve I is the boiling point as a function of the mole fraction x_A; the boiling points of the two components A and B are T_A and T_B respectively. Curve II shows the composition of the vapor at each boiling temperature. If a mixture with composition corresponding to the point M is boiled, then the vapor will have the composition corresponding to the point N; if this vapor is now collected and condensed, then its boiling point and composition will correspond to the point O. This process enriches the mixture in component B. For such systems, by continuing this process a mixture can be enriched in the more volatile component.

AZEOTROPES

The relation between the boiling point and the compositions of the liquid and the vapor phases shown in Figure 7.5 is not valid for all binary mixtures. For many liquid mixtures the boiling point curve is as shown in Figure 7.6. In this case, there is value of x_A at which the composition of the liquid and the vapor are the same. Such systems are called **azeotropes**. The components of an azeotrope cannot be separated by distillation. For example, in the case of Figure 7.6a, starting at a point to the left of the maximum, if the mixture is boiled and the vapor collected, then the vapor will be enriched in component B while the remaining liquid will be richer in component A and move towards the azeotropic composition. Thus, successive boiling and condensation results in pure B and a mixture with azeotropic composition, not pure A and pure B. The azeotropic composition and the corresponding

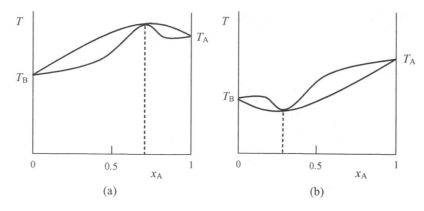

Figure 7.6 The boiling point versus composition of liquid and vapor phases of binary mixtures called azeotropes. Azeotropes have a point at which the vapor and the liquid phases have the same composition. At this point the boiling point is either a maximum (a) or minimum (b)

boiling points for binary mixtures are tabulated. One may notice in Figure 7.6 that the boiling point corresponding to the azeotropic composition occurs at an extremum (maximum or minimum). That this must be so for thermodynamic reasons has been noted by Gibbs and later by Konovolow and Duhem. This observation is called the **Gibbs–Konovalow theorem** [4], which states that:

At constant pressure, in an equilibrium displacement of a binary system, the temperature of coexistence passes through an extremum if the composition of the two phases is the same.

We shall not discuss the proof of this theorem here. An extensive discussion of this and other related theorems may be found in the classic text by Prigogine and Defay [4]. Azeotropes are an important class of solutions whose thermodynamic properties we shall discuss in more detail in Chapter 8. Some examples of azeotropes are given in Table 7.2.

SOLUTION IN EQUILIBRIUM WITH PURE SOLIDS: EUTECTICS

The next example we consider is a solid–liquid equilibrium of two components, A and B, which are miscible in the liquid state but not in the solid state. This system has three phases in all, the liquid with A + B, solid A and solid B.

We can understand the equilibrium of such a system by first considering the equilibrium of two-phase systems, the liquid and one of the two solids, A or B, and then extending it to three phases. In this case, the Gibbs phase rule tells us that, with two components and two phases, the number of degrees of freedom equals two. We can take these two degrees of freedom to be the pressure and composition. Thus, if the mole fraction x_A and the pressure are fixed, then the equilibrium temperature is also fixed. By fixing the pressure at a given value (say the atmospheric pressure) one can obtain an equilibrium curve relating T and x_A. The two curves corresponding to solid A in equilibrium with the liquid and solid B in equilibrium with the

Table 7.2 Examples of azeotropes

	Boiling point (°C)		Azeotropic wt %
	Pure compound	Azeotrope	
Azeotropes formed with water at $p = 1$ bar			
Boiling point of water = 100 °C			
Hydrogen chloride (HCl)	−85	108.58	20.22
Nitric acid (HNO$_3$)	86	120.7	67.7
Ethanol (C$_2$H$_5$OH)	78.32	78.17	96
Azeotropes formed with acetone at $p = 1$ bar			
Boiling point of acetone ((CH$_3$)$_2$CO) = 56.15 °C			
Cyclohexane (C$_6$H$_{12}$)	80.75	53.0	32.5
Methyl acetate (CH$_3$COOCH$_3$)	57.0	55.8	51.7
n-Hexane (C$_6$H$_{14}$)	68.95	49.8	41
Azeotropes formed with methanol at $p = 1$ bar			
Boiling point of methanol (CH$_3$OH) = 64.7 °C			
Acetone ((CH$_3$)$_2$CO)	56.15	55.5	88
Benzene (C$_6$H$_6$)	80.1	57.5	60.9
Cyclohexane (C$_6$H$_{12}$)	80.75	53.9	63.6

Source: D.R. Lide (ed.), *CRC Handbook of Chemistry and Physics*, 75th edition. 1994, CRC Press: Ann Arbor, MI.

liquid are shown in Figure 7.7. In this figure, along the curve EN, the solid A is in equilibrium with the liquid; along the curve EM, solid B is in equilibrium with the solution. The point of intersection of the two curves, E, is called the **eutectic point**, and the corresponding composition and temperature are called the **eutectic composition** and the **eutectic temperature**.

Now, if we consider a three-phase system, the liquid, solid A and solid B, all in equilibrium, then the Gibbs phase rule tells us that there is only one degree of freedom. If we take this degree of freedom to be the pressure and fix it at a particular value, then there is only one point (T, x_A) at which equilibrium can exist between the three phases. This is the eutectic point. This is the point at which the chemical potentials of solid A and solid B are equal to their corresponding chemical potentials in the liquid mixture. Since the chemical potentials of solids and liquids do not change much with changes in pressure, the eutectic composition and temperature are insensitive to variations in pressure.

TERNARY SYSTEMS

As was noted by Gibbs, the composition of a solution containing three components may be represented by points within an equilateral triangle whose sides have a length equal to one. Let us consider a system with components A, B, and C. As shown in Figure 7.8, a point P may be used to specify the mole fractions x_A, x_B and x_C as follows. From the point P, lines are drawn parallel to the sides of the equilateral

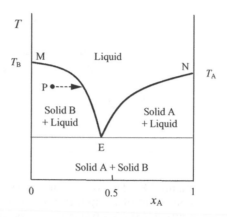

Figure 7.7 The phase diagram of a two-component system with three phases. The system has only
one degree of freedom. For a fixed pressure, the
three phases (the liquid, solid A and solid B) are at
equilibrium at the **eutectic point** E. Along the curve
ME, solid B is in equilibrium with the liquid; and
along the curve NE, solid A is in equilibrium with
the liquid. The point of intersection E specifies the
equilibrium composition and temperature when all
three phases are in equilibrium. At a fixed T, if the
system is initially at point P it will move towards
the equilibrium curve ME. Below the eutectic point
the solid is a mixture of solid A and solid B

triangle. The length of these lines can be used to represent the mole fractions x_A,
x_B and x_C. It is left as an exercise to show that such a construction ensures that
$x_A + x_B + x_C = 1$. In this representation of the composition, we see that:

1. The vertices A, B and C correspond to pure substances.
2. A line parallel to a side of the triangle corresponds to a series of ternary systems
   in which one of the mole fractions remains fixed.
3. A line drawn through one of the apexes to the opposite side represents a set of
   systems in which the mole fractions of two components have a fixed ratio. As the
   apex is approached along this line, the system becomes increasingly richer in the
   component represented by the apex. The variation of some property of a three-
   component solution can be shown in a three-dimensional graph in which the base
   is the above composition triangle; the height will then represent the property.

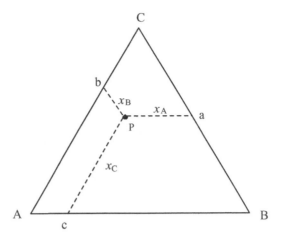

Figure 7.8 The composition of a ternary system consisting of components A, B and C can be represented on a triangular graph because $x_A + x_B + x_C = 1$. The composition is represented as a point P inside an equilateral triangle whose side has a length equal to one. The mole fractions are the lengths of the lines drawn parallel to the sides of the triangle. Showing that Pa + Pb + Pc = 1 for any point P is left as an exercise

As an example, let us consider three components, A, B and C, in two phases: a solution that contains A, B and C, and the other a solid phase of B in equilibrium with the solution. This system has three components and two phases and, hence, has three degrees of freedom, which may be taken as the pressure and the mole fractions x_A and x_B. At constant pressure, every value of x_A and x_B has a corresponding equilibrium temperature. In Figure 7.9a, the point P shows the composition of the solution at a temperature T. As the temperature decreases, the relative values of x_A and x_C remain the same while more of B turns into a solid. According to the observations in point (3) above, this means the point moves along the line BP as shown by the arrow. As the temperature decreases, a point P′ is reached at which the component C begins to crystallize. The system now has two solid phases and one solution phase and, hence, has two degrees of freedom. The composition of the system is now confined to the line P′E. With further decrease in the temperature, component A also begins to crystallize at point E, which corresponds to the eutectic temperature. The system now has only one degree of freedom. At the eutectic temperature and composition, all three components will crystallize out in the eutectic proportions.

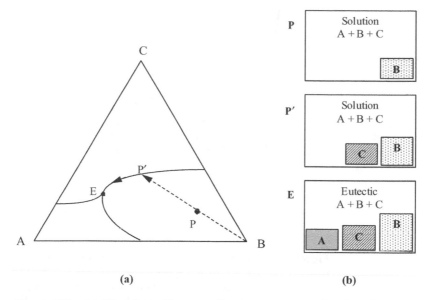

Figure 7.9 (a) The phase diagram of a ternary system showing the composition of the solution as it is cooled. At the point P the system consists of two phases: the solution (A + B + C) in equilibrium with solid B. As the temperature decreases, the composition moves along PP'. At P' the component C begins to crystallize and the composition moves along P'E until it reaches the ternary eutectic point E, at which all components begin to crystallize. (b) The composition of the system at points P, P' and E

7.4 Maxwell's Construction and the Lever Rule

The reader might have noticed that the isotherms obtained from an equation of state, such as the van der Waals equation, do not coincide with the isotherms shown in Figure 7.2 at the part of the curve that is flat, i.e. where the liquid and vapor phases coexist. The flat part of the curve represents what is physically realized when a gas is compressed at a temperature below the critical temperature. Using the condition that the chemical potential of the liquid and the vapor phases must be equal at equilibrium, Maxwell was able to determine the location of the flat part of the curve.

Let us consider a van der Waals isotherm for $T < T_c$ (Figure 7.10). Imagine a steady decrease in volume starting at the point Q. Let the point P be such that, at this pressure, the chemical potentials of the liquid and the vapor phases are equal. At this point the vapor will begin to condense and the volume can be decreased with no change in the pressure. This decrease in volume can continue until all the vapor has condensed to a liquid at the point L. If the volume is maintained at some value has between P and L, then liquid and vapor coexist. Along the line PL the chemical

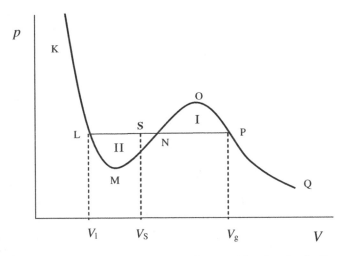

Figure 7.10 Maxwell's construction specifies the physically realized flat part LP with respect to the theoretical isotherm given by an equation of state such as the van der Waals equation. At equilibrium, the chemical potentials at the points L and P must be equal. As shown in the text, this implies that the physically realized states lie on a line LP that makes area I equal area II

potentials of the liquid and the vapor are equal. Thus, the total change in the chemical potential along the curve LMNOP must be equal to zero:

$$\int_{\text{LMNOP}} \mathrm{d}\mu = 0 \qquad (7.4.1)$$

Now, since the chemical potential is a function of T and p, and since the path is an isotherm, it follows from the Gibbs–Duhem relation that $\mathrm{d}\mu = V_{\mathrm{m}}\mathrm{d}p$. Using this relation we may write the above integral as

$$\int_{P}^{O} V_{\mathrm{m}}\mathrm{d}p + \int_{O}^{N} V_{\mathrm{m}}\mathrm{d}p + \int_{N}^{M} V_{\mathrm{m}}\mathrm{d}p + \int_{M}^{L} V_{\mathrm{m}}\mathrm{d}p = 0 \qquad (7.4.2)$$

The area I shown in Figure 7.10 is equal to

$$\int_{P}^{O} V_{\mathrm{m}}\mathrm{d}p - \int_{N}^{O} V_{\mathrm{m}}\mathrm{d}p = \int_{P}^{O} V_{\mathrm{m}}\mathrm{d}p + \int_{O}^{N} V_{\mathrm{m}}\mathrm{d}p$$

which is same as the first two integrals in (7.4.2). Similarly, the sum of the second two terms equals the negative of area II.

Thus, Equation (7.4.2) may be interpreted as

$$(\text{Area I}) - (\text{Area II}) = 0 \qquad (7.4.3)$$

This condition specifies how to locate or construct a flat line on which the chemical potentials of the liquid and the vapor are equal, the one that is physically realized. It is called the **Maxwell construction**.

At point P the substance is entirely in the vapor phase with volume V_g; at the point L it is entirely in the liquid phase with volume V_1. At any point S on the line LP, if a fraction x of substance is in the vapor phase, then the total volume V_S of the system is

$$V_S = x V_g + (1-x)V_1 \qquad (7.4.4)$$

It follows that

$$x = \frac{V_S - V_1}{V_g - V_1} = \frac{SL}{LP} \qquad (7.4.5)$$

From this relation it can be shown that (Exercise 7.10) the mole fraction x of the vapor phase and $(1-x)$ of the liquid phase satisfy

$$\boxed{(SL)x = (SP)(1-x)} \qquad (7.4.6)$$

This relation is called the **lever rule**, in analogy with a lever supported at S, in equilibrium with weights V_1 and V_g attached to either end.

7.5 Phase Transitions

Phase transitions are associated with many interesting and general thermodynamic features. As described below, based on some of these features, phase transitions can be classified into different 'orders'. Thermodynamic behavior in the vicinity of the critical points has been of much interest from the point of view of thermodynamic stability and extremum principles discussed in Chapter 5. A classical theory of phase transitions was developed by Lev Landau; but, in the 1960s, experiments showed that the predictions of this theory were incorrect. This resulted in the development of the modern theory of phase transitions during the 1960s and the 1970s. The modern theory is based on the work of C. Domb, M. Fischer, L. Kadanoff, G.S. Rushbrook, B. Widom, K. Wilson and others. In this section we will only outline some of the main results of the thermodynamics of phase transitions. A detailed description of the modern theory of phase transitions, which uses the mathematically advanced concepts of renormalization-group theory, is beyond the scope of this book. For a better understanding of this rich and interesting subject we refer the reader to books on this topic [1–3].

GENERAL CLASSIFICATION OF PHASE TRANSITIONS

When transition from a solid to a liquid or from a liquid to vapor takes place, there is a discontinuous change in the entropy. This can clearly be seen (see Figure 7.11) if we plot molar entropy $S_m = -(\partial G_m/\partial T)_p$ as function of T, for a fixed p and N. The same is true for other derivatives of G_m, such as $V_m = (\partial G_m/\partial p)_T$. The chemical potential changes continuously, but its derivative is discontinuous. At the transition temperature, because of the existence of latent heat, the molar heat capacities ($\Delta Q/\Delta T$) have a 'singularity' in the sense they become infinite, i.e. heat absorbed ΔQ causes no change in temperature, i.e. $\Delta T = 0$. Transitions of this type are classified as **first-order** phase transitions.

The characteristic features of **second-order phase transitions** are shown in Figure 7.12. In this case the changes in the thermodynamic quantities are not so drastic: changes in S_m and V_m are continuous, but their derivatives are discontinuous. Similarly, for the chemical potential it is the second derivative that is discontinuous; the molar heat capacity does not have a singularity, but it has a discontinuity. Thus, depending on the order of the derivatives that are discontinuous, phase transitions are classified as transitions of first or second order.

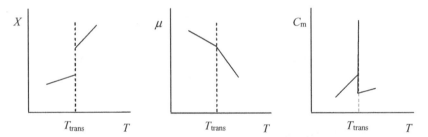

Figure 7.11 The change of thermodynamic quantities in a first-order phase transition that occurs at the temperature T_{trans}. X is a molar extensive quantity such as S_m or V_m that changes discontinuously

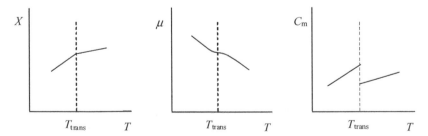

Figure 7.12 The change of thermodynamic quantities in a second-order phase transition that occurs at the temperature T_{trans}. X is a molar extensive quantity such as S_m or V_m whose derivative changes discontinuously

BEHAVIOR NEAR THE CRITICAL POINT

The classical theory of phase transitions was developed by Lev Landau to explain the coexistence of phases and the critical point at which the distinction between the phases disappears. Landau's theory explains the critical behavior in terms of the minima of the Gibbs free energy. According to this theory, as shown in Figure 7.13, in the coexistence region for a given p and T, G as a function of V has two minima. As the critical point is approached, the minima merge into one broad minimum. The classical theory of Landau makes several predictions regarding the behavior of systems near the critical point. The predictions of the theory are, in fact, quite general, valid for large classes of systems. Experiments done in the 1960s did not support these predictions. We shall list below some of the discrepancies between theory and experiments using the liquid–vapor transition as an example, but the experimental values are those obtained for many similar systems. Also, all the classical predictions can be verified using the van der Waals equation of state as an example.

* For the liquid–vapor transition, as the critical temperature was approached from below ($T < T_c$), the theory predicted that

$$V_{mg} - V_{ml} \propto (T_c - T)^\beta \quad \beta = 1/2 \tag{7.5.1}$$

However, experiments showed that β was in the range 0.3–0.4, not equal to 0.5.

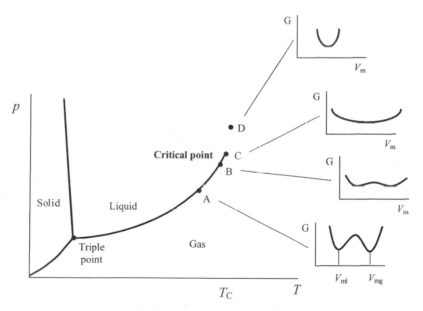

Figure 7.13 Classical theory of phase transitions is based on the shape of the Gibbs energy. The Gibbs energies associated with the points A, B, C and D are shown in the insets. As the system moves from A to D, the Gibbs energy changes from a curve with two minima to a curve with one minimum, as shown in the small figures

- Along the critical isotherm, as the critical pressure p_c is approached from above, the theory predicted

$$V_{mg} - V_{ml} \propto (p - p_c)^{1/\delta} \quad \delta = 3 \tag{7.5.2}$$

Experiments place the value of δ in the range 4.0–5.0.

- When the gas can be liquefied, it is easy to see that the isothermal compressibility $\kappa_T = -(1/V)(\partial V/\partial p)_T$ diverges during the transition (the flat part of the p–V isotherm). Above the critical temperature, since there is no transition to liquid there is no divergence. According to classical theory, as the critical temperature is approached from above, the divergence of κ_T should be according to

$$\kappa_T \propto (T - T_c)^{-\gamma} \quad \gamma = 1 \tag{7.5.3}$$

Experimental values of γ were found to be in the range 1.2–1.4.

- We have seen in Chapter 6 that the values of molar heat capacity C_{mV} for real and ideal gases are the same if the pressure is a linear function of the temperature. This means the value of C_{mV} does not diverge (though the value of C_p diverges). Thus, according to classical theory, if

$$C_{mV} \propto (T - T_c)^{-\alpha} \quad \text{then} \quad \alpha = 0 \tag{7.5.4}$$

Experimentally, the value of α found was in the range –0.2 to +0.3.

The failure of the classical or Landau theory initiated a reexamination of the critical behavior. The main reason for the discrepancy was found to be the role of fluctuations. Near the critical point, due to the flat nature of the Gibbs energy, large long-range fluctuations arise in the system, but they were not properly included in Landau's theory. Kenneth Wilson incorporated these fluctuations into the theory through the development of new mathematical techniques and the theory of the **renormalization group**. The modern theory of critical behavior not only predicts the experimental values of the exponents α, β, γ and δ more successfully than the classical theory, but it also relates these exponents. For example, the modern theory predicts that

$$\beta = \frac{2 - \alpha}{1 + \delta} \quad \text{and} \quad \gamma = \frac{(\alpha - 2)(1 - \delta)}{(1 + \delta)} \tag{7.5.5}$$

Since a detailed presentation of the theory of the renormalization group is beyond the scope of this book, we will leave the reader with only this brief outline of the limitations of the classical theory and accomplishments of the modern theory.

References

1. Stanley, H.E. *Introduction to Phase Transitions and Critical Phenomena*. 1971, Oxford: Oxford University Press.
2. Ma, S.-K. *Modern Theory of Critical Phenomena*. 1976, New York: Addison-Wesley.
3. Pfeuty, P. and Toulouse, G. *Introduction to the Renormalization Group and Critical Phenomena*. 1977, New York: Wiley.
4. Prigogine, I. and Defay, R. *Chemical Thermodynamics*, fourth edition. 1967, London: Longmans.

Examples

Example 7.1   A chemical reaction occurs in CCl_4 at room temperature, but it is very slow. To increase its speed to a desired value, the temperature needs to be increased to 80 °C. Since CCl_4 boils at 77 °C at $p = 1.00$ atm, the pressure has to be increased so that CCl_4 will boil at a temperature higher than 80 °C. Using the data in Table 7.1, calculate the pressure at which CCl_4 will boil at 85 °C.

Solution   From the Clausius–Clapeyron equation we have

$$\ln p - \ln(1.00 \text{ atm}) = \frac{30.0 \times 10^3}{8.314}\left(\frac{1}{350} - \frac{1}{358}\right) = 0.230$$

$$p = (1.00 \text{ atm})e^{0.23} = 1.26 \text{ atm}$$

Example 7.2   If a system contains two immiscible liquids (such as CCl_4 and CH_3OH), how many phases are there?

Solution   The system will consist of three layers. A layer rich in CCl_4, a layer rich in CH_3OH and a layer of vapor of CCl_4 and CH_3OH. Thus, there are three phases in this system.

Example 7.3   Determine the number of degrees of freedom of a two-component liquid mixture in equilibrium with its vapor.

Solution   In this case $C = 2$, $P = 2$. Hence, the number of degrees of freedom $f = 2 - 2 + 2 = 2$. These two degrees of freedom can be, for example, T and the mole fraction x_1 of one of the components. The pressure of the system (vapor phase in equilibrium with the liquid) is completely specified by x_1 and T.

Example 7.4   How many degrees of freedom does an aqueous solution of the weak acid CH_3COOH have?

Solution   The acid decomposition is

$$CH_3COOH \rightleftharpoons CH_3COO^- + H^+$$

The number of components is $C = 4$ (water, CH_3COOH, CH_3COO^- and H^+). The number of phases is $P = 1$. There is one chemical reaction in equilibrium; hence $R = 1$. However, since the concentrations of CH_3COO^- and H^+ are equal, the degrees of freedom is reduced by one. Hence, the number of degrees of freedom $f = C - R - P + 2 - 1 = 4 - 1 - 1 + 2 - 1 = 3$.

Exercises

7.1 The heat of vaporization of hexane is $30.8 \, kJ \, mol^{-1}$. The boiling point of hexane at a pressure of 1.00 atm is 68.9 °C. What will the boiling point be at a pressure of 0.50 atm?

7.2 The atmospheric pressure decreases with height. The pressure at a height h above sea level is given approximately by the barometric formula $p = p_0 e^{-Mgh/RT}$, in which $M = 0.0289 \, kg \, mol^{-1}$, and $g = 9.81 \, m \, s^{-2}$. Assume that the enthalpy of vaporization of water is $\Delta H_{vap} = 40.6 \, kJ \, mol^{-1}$ and predict at what temperature water will boil at a height of 3.5 km.

7.3 At atmospheric pressure, CO_2 turns from solid to gas, i.e. it sublimates. Given that the triple point of CO_2 is at $T = 216.58 \, K$ and $p = 518.0 \, kPa$, how would you obtain liquid CO_2?

7.4 In a two-component system, what is the maximum number of phases that can be in equilibrium?

7.5 Determine the number of degrees of freedom for the following systems:
(a) solid CO_2 in equilibrium with CO_2 gas;
(b) an aqueous solution of fructose;
(c) $Fe(s) + H_2O(g) \rightleftharpoons FeO(s) + H_2(g)$.

7.6 Draw qualitative figures of T versus x_A curves (Figure 7.6) for the azeotropes in Table 7.2.

7.7 In Figure 7.8, show that $PA + PB + PC = 1$ for any point P.

7.8 In the triangular representation of the mole fractions of ternary solution, show that along the line joining an apex and a point on the opposite side, the ratio of two of the mole fractions remain constant while mole fraction of the third component changes.

7.9 On triangular graph paper, mark points representing the following compositions:
(a) $x_A = 0.2$, $x_B = 0.4$, $x_C = 0.4$
(b) $x_A = 0.5$, $x_B = 0$, $x_C = 0.5$

(c) $x_A = 0.3$, $x_B = 0.2$, $x_C = 0.5$

(d) $x_A = 0$, $x_B = 0$, $x_C = 1.0$.

7.10 Obtain the lever rule (7.4.6) from (7.4.5).

7.11 When the van der Waals equation is written in terms of the reduced variables p_r, V_r and T_r (see 1.4.6), the critical pressure, temperature and volume are equal to one. Consider small deviations from the critical point, $p_r = 1 + \delta p$ and $V_r = 1 + \delta V$ on the critical isotherm. Show that δV is proportional to $(\delta p)^{1/3}$. This corresponds to the classical prediction (7.5.2).

8 THERMODYNAMICS OF SOLUTIONS

8.1 Ideal and Nonideal Solutions

Many properties of solutions can be understood through thermodynamics. For example, we can understand how the boiling and freezing points of a solution change with composition, how the solubility of a compound changes with temperature and how the osmotic pressure depends on the concentration.

We begin by obtaining the chemical potential of a solution. As noted in Chapter 5 (Equation (5.3.5)), the general expression for the chemical potential of a substance may be written as $\mu(p,T) = \mu^0(p_0,T) + RT\ln a$ in which a is the activity and μ^0 is the chemical potential of the standard state in which $a = 1$. For an ideal gas mixture, in Equation (6.1.9) we saw that the chemical potential of a component can be written in terms of its mole fraction x_k in the form $\mu_k(p,T,x_k) = \mu_k^0(p,T) + RT\ln x_k$. We shall see in this section that properties of many dilute solutions can be described by a chemical potential of the same form. This has led to the following definition of an **ideal solution** as a solution for which

$$\mu_k(p,T,x_k) = \mu_k^0(p,T) + RT\ln x_k \qquad (8.1.1)$$

where $\mu_k^0(p,T)$ is the chemical potential of a reference state which is independent of x_k. We stress that the similarity between ideal gas mixtures and ideal solutions is only in the dependence of the chemical potential on the mole fraction; the dependence on the pressure, however, is entirely different, as can be seen from the general expression for the chemical potential of a liquid (6.3.8).

In (8.1.1), if the mole fraction of the 'solvent' x_s is nearly equal to one, i.e. for dilute solutions, then for the chemical potential of the solvent the reference state $\mu_k^0(p,T)$ may be taken to be $\mu_k^*(p,T)$, the chemical potential of the pure solvent. For the other components, $x_k \ll 1$; as we shall see below, (8.1.1) is still valid in a small range, but in general the reference state is not $\mu_k^*(p,T)$. Solutions for which (8.1.1) is valid for all values of x_k are called **perfect solutions**. When $x_k = 1$, since we must have $\mu_k(p,T) = \mu_k^*(p,T)$, it follows that for perfect solutions

$$\mu_k(p,T,x_k) = \mu_k^*(p,T) + RT\ln x_k \quad \forall x_k \qquad (8.1.2)$$

The activity of **nonideal solutions** is expressed as $a_k = \gamma_k x_k$ in which γ_k is the **activity coefficient**, a quantity introduced by G.N. Lewis. Thus, the chemical potential of nonideal solutions is written as

Introduction to Modern Thermodynamics   Dilip Kondepudi
© 2008 John Wiley & Sons, Ltd

$$\begin{aligned}\mu_k(p,T,x_k) &= \mu_k^0(p,T) + RT\ln a_k\\ &= \mu_k^0(p,T) + RT\ln(\gamma_k x_k)\end{aligned} \tag{8.1.3}$$

The activity coefficient $\gamma_k \to 1$ as $x_k \to 1$.

Let us now look at conditions under which ideal solutions are realized. We consider a solution with many components, whose mole fractions are x_i, in equilibrium with its vapor (see Figure 8.1). At equilibrium, the affinities for the conversion of liquid to the gas phase are zero for each component i, i.e. for each component, the chemical potentials in the two phases are equal. If we use the ideal gas approximation for the component in the vapor phase we have

$$\mu_{i,l}^0(p_0,T) + RT\ln a_i = \mu_{i,g}^0(p_0,T) + RT\ln(p_i/p_0) \tag{8.1.4}$$

in which the subscripts l and g indicate the liquid and gas phases. The physical meaning of the activity a_i can be seen as follows. Consider a pure liquid in equilibrium with its vapor. Then $p_i = p_i^*$, the vapor pressure of a pure liquid in equilibrium with its vapor. Since a_i is nearly equal to one for a pure liquid, $\ln(a_i) \approx 0$. Hence, (8.1.4) can be written as

$$\mu_{i,l}^0(p_0,T) = \mu_{i,g}^0(p_0,T) + RT\ln(p_i^*/p_0) \tag{8.1.5}$$

Subtracting (8.1.5) from (8.1.4) we find that

$$\boxed{RT\ln a_i = RT\ln(p_i/p_i^*) \quad \text{or} \quad a_i = \frac{p_i}{p_i^*}} \tag{8.1.6}$$

i.e. the activity is the ratio of the partial vapor pressure p_i of the component when in a solution and its vapor pressure $p*$ when it is a pure substance. By measuring the vapor pressure of a substance, its activity can be determined.

For an ideal solution, Equation (8.1.4) takes the form

$$\mu_{i,l}^0(p,T) + RT\ln(x_i) = \mu_{i,g}^0(p_0,T) + RT\ln(p_i/p_0) \tag{8.1.7}$$

From this equation it follows that the partial pressure in the vapor phase and the mole fraction of a component can be written as

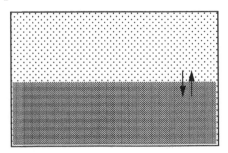

Figure 8.1 Equilibrium between a solution and its vapor

$$\boxed{p_i = K_i x_i} \qquad (8.1.8)$$

in which

$$K_i(p,T) = p_0 \exp\left(\frac{\mu_{i,l}^0(p,T) - \mu_{i,g}^0(p_0,T)}{RT} \right) \qquad (8.1.9)$$

As indicated, the term $K_i(p,T)$ is, in general, a function of p and T; but since the chemical potential of the liquid $\mu_{i,l}^0(p,T)$ changes little with p, it is essentially a function of T. K_i has the dimensions of pressure. For any component, when $x_i = 1$, we must have $K(p^*,T) = p^*$, the vapor pressure of the pure substance (Figure 8.2). (This is consistent with (8.1.9) because when we set $p = p_0 = p^*$, the exponent $\mu_l^0(T,p^*) - \mu_g^0(T,p^*) = 0$ because the vapor and the liquid are in equilibrium.) At a given temperature T, if $x_1 \approx 1$ for a particular component, which is called the 'solvent', then since the change of K_i is small for changes in pressure we may write

$$\boxed{p_1 = p_1^* x_1} \qquad (8.1.10)$$

Experiments conducted by François-Marie Raoult (1830–1901) in the 1870s showed that if the mole fraction of the solvent is nearly equal to unity, i.e. for dilute solutions, then (8.1.10) is valid. For this reason, (8.1.10) is called the **Raoult's law**. The chemical potential of the vapor phase of the solvent $\mu_{s,g} = \mu_{s,g}(p_0,T) + RT \ln(p_s/p_0)$ can now be related to its mole fraction in the solution by using Raoult's law and by setting $p_0 = p^*$:

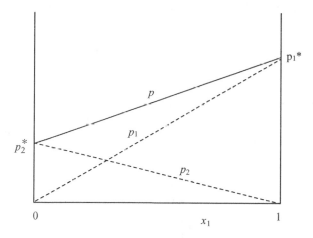

Figure 8.2 The vapor pressure diagram of a **perfect binary solution** for which (8.1.1) is valid for all values of the mole fraction x_1. p_1^* and p_2^* are the vapor pressures of the pure substances; p_1 and p_2 are the partial pressures of the two components in the mixture and p is the total vapor pressure

$$\mu_{s,g}(p, T, x_s) = \mu_{s,g}(p^*, T) + RT \ln x_s \qquad (8.1.11)$$

For a minor component of a solution, when its mole fraction $x_k \ll 1$, (8.1.10) is not valid but (8.1.8) is still valid. This relation is called **Henry's law** after William Henry (1774–1836), who studied this aspect for the solubility of gases [1]:

$$\boxed{p_i = K_i x_i \quad x_i \ll 1} \qquad (8.1.12)$$

The constant K_i is called the Henry's constant. Some values of Henry's constants are given in Table 8.1. In the region where Henry's law is valid, K_i is not equal to the vapor pressure of the pure substance. The graphical representation of the Henry's constant is shown in Figure 8.3. (Also, where Henry's law is valid, in general the chemical potential of the reference state μ_i^0 is not the same as the chemical potential of the pure substance.) Only for a perfect solution do we have $K_i = p_i^*$ when $x_i \ll 1$, but such solutions are very rare. Many dilute solutions obey Raoult's and Henry's laws to a good approximation.

When the solution is not dilute, the **nonideal** behavior is described using the activity coefficients γ_i in the chemical potential:

$$\mu_i(p, T, x_i) = \mu_i^0(p, T) + RT \ln(\gamma_i x_i) \qquad (8.1.13)$$

The deviation from Raoult's or Henry's law is a measure of γ_i. For nonideal solutions, as an alternative to the activity coefficient, an **osmotic coefficient** ϕ_i is defined through

$$\mu_i(p, T, x_i) = \mu_i^0(p, T) + \phi_i RT \ln(x_i) \qquad (8.1.14)$$

As we will see in the following section, the significance of the osmotic coefficient lies in the fact that it is the ratio of the osmotic pressure to that expected for ideal solutions. From (8.1.13) and (8.1.14) it is easy to see that

$$\phi_k - 1 = \frac{\ln \gamma_k}{\ln x_k} \qquad (8.1.15)$$

Table 8.1 Henry's law constants at 25 °C for atmospheric gases

Gas	$K/10^4$ atm	Volume in the atmosphere/ppm
$N_2(g)$	8.5	780 900
$O_2(g)$	4.3	209 500
$Ar(g)$	4.0	9 300
$CO_2(g)$	0.16	380
$CO(g)$	5.7	–
$He(g)$	13.1	5.2
$H_2(g)$	7.8	0.5
$CH_4(g)$	4.1	1.5
$C_2H_2(g)$	0.13	–

Source: D.R. Lide (ed.) *CRC Handbook of Chemistry and Physics*, 75th edition. 1994, CRC Press: Ann Arbor, MI.

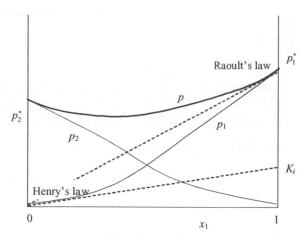

Figure 8.3 The vapor pressure diagram of a binary solution. When the mole fraction is very small or nearly equal to one, we have ideal behavior. The minor component obeys Henry's law, while the major component obeys Raoult's law. p_1^* and p_2^* are the vapor pressures of the pure substances; p_1 and p_2 are the partial pressures of the two components in the mixture and p is the total vapor pressure. The deviation from the partial pressure predicted by Henry's law or Raoult's law can be used to obtain the activity coefficients

8.2 Colligative Properties

By using the chemical potential of ideal solutions we can derive several properties of ideal solutions that depend on the *total number of the solute particles* and not on the chemical nature of the solute. (For example, a 0.2 M solution of NaCl will have colligative concentration of 0.40 M due to the dissociation into Na^+ and Cl^-.) Such properties are collectively called **colligative properties**.

CHANGES IN BOILING AND FREEZING POINTS

Equation (8.1.11) could be used to obtain an expression for the increase in the boiling point and the decrease in the freezing point of solutions (Figure 8.4). As we noted in Chapter 7, a liquid boils when its vapor pressure $p = p_{\text{ext}}$, the atmospheric (or applied external) pressure. Let T^* be the boiling temperature of the pure solvent and T the boiling temperature of the solution. *We assume that the mole fraction of the solvent is x_2 and that of the solute is x_1.* We assume that the solute is nonvolatile so that the gas phase of the solution is pure solvent. At equilibrium, the chemical potentials of the liquid and gas phases of the solvent must be equal:

$$\mu_{2\text{g}}^*(p_{\text{ext}}, T) = \mu_{2,\text{l}}(p_{\text{ext}}, T, x_2) \tag{8.2.1}$$

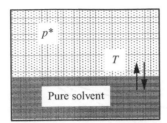 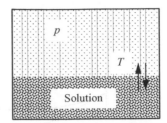

Figure 8.4 The vapor pressure of a solution with a non-volatile solute is less than that of a pure solvent. Consequently, the boiling point of a solution increases with the solute concentration

Using (8.1.11) we can now write this equation as

$$\mu_{2,g}^*(p_{ext}, T) = \mu_{2,l}(p_{ext}, T, x_2) = \mu_{2,l}^*(T) + RT \ln x_2 \qquad (8.2.2)$$

Since the chemical potential of a pure substance $\mu = G_m$, the molar Gibbs energy, we have

$$\frac{\mu_{2,g}^*(p_{ext}, T) - \mu_{2,l}^*(T)}{RT} = \frac{\Delta G_m}{RT} = \frac{\Delta H_m - T\Delta S_m}{RT} = \ln x_2 \qquad (8.2.3)$$

in which Δ denotes the difference between the liquid and the gas phase. Generally, ΔH_m does not vary much with temperature. Therefore, $\Delta H_m(T) = \Delta H_m(T^*) = \Delta H_{vap}$. Also, $\Delta S_m = \Delta H_{vap}(T)/T^*$, and $x_2 = (1 - x_1)$, in which $x_1 \ll 1$ is the mole fraction of the solute. With these observations we can write Equation (8.2.3) as

$$\ln(1 - x_1) = \frac{\Delta H_{vap}}{R}\left(\frac{1}{T} - \frac{1}{T^*}\right) \qquad (8.2.4)$$

If the difference $T - T^* = \Delta T$ is small, then it is easy to show that the terms containing T and T^* can be approximated to $-\Delta T/T^{*2}$. Furthermore, since $\ln(1 - x_1) \approx -x_1$ when $x_1 \ll 1$, we can approximate (8.2.4) by the relation

$$\boxed{\Delta T = \frac{RT^{*2}}{\Delta H_{vap}} x_1} \qquad (8.2.5)$$

which relates the change in boiling point to the mole fraction of the solute. In a similar way, by considering a pure solid in equilibrium with the solution, one can derive the following relation for the decrease in freezing point ΔT in terms of the enthalpy of fusion ΔH_{fus}, the mole fraction x_k of the solute and the freezing point T^* of the pure solvent:

$$\boxed{\Delta T = \frac{RT^{*2}}{\Delta H_{fus}} x_1} \qquad (8.2.6)$$

The changes in the boiling point and the freezing point are often expressed in terms of **molality**, i.e. *moles of solute/kilogram of solvent*, instead of mole fraction. For

Table 8.2 Ebullioscopic and cryoscopic constants

Compound	K_b/°C kg mol^{-1}	T_b/°C	K_f/°C kg mol^{-1}	T_f/°C
Acetic acid, CH_3COOH	3.07	118	3.90	16.7
Acetone, $(CH_3)_2CO$	1.71	56.3	2.40	−95
Benzene, C_6H_6	2.53	80.10	5.12	5.53
Carbon disulfide, CS_2	2.37	46.5	3.8	−111.9
Carbon tetrachloride, CCl_4	4.95	76.7	30	−23
Nitrobenzene, $C_6H_5NO_2$	5.26	211	6.90	5.8
Phenol, C_6H_5OH	3.04	181.8	7.27	40.92
Water, H_2O	0.51	100.0	1.86	0.0

Source: G.W.C. Laye and T.H. Laby (eds) *Tables of Physical and Chemical Constants*. 1986, London: Longmans.

dilute solutions, the conversion from mole fraction x to molality m is easily done. If M_s is the molar mass of the solvent in kilograms, then the mole fraction of the solute

$$x_1 = \frac{N_1}{N_1 + N_2} \approx \frac{N_1}{N_2} = M_s\left(\frac{N_1}{M_s N_2}\right) = M_s m_1$$

Equations (8.2.5) and (8.2.6) are often written as

$$\Delta T = K(m_1 + m_2 + \ldots + m_s) \tag{8.2.7}$$

in which the molalities of all the 's' species of solute particles is shown explicitly. The constant K is called the **ebullioscopic constant** for changes in boiling point and the **cryoscopic constant** for changes in freezing point. The values of ebullioscopic and cryoscopic constants for some liquids are given in Table 8.2.

OSMOTIC PRESSURE

When a solution and pure solvent are separated by a semi-permeable membrane (see Figure 8.5a), which is permeable to the solvent but not the solute molecules, the solvent flows into the chamber containing the solution until equilibrium is reached. This process is called **osmosis** and was noticed in the mid eighteenth century. In 1877, a botanist named Pfeffer made a careful quantitative study of it. Jacobus Henricus van't Hoff (1852–1911), who was awarded the first Nobel Prize in chemistry in 1901 for his contributions to thermodynamics and chemistry [1], found that a simple equation, similar to that of an ideal gas, could be used to describe the observed data.

As shown in Figure 8.5, let us consider a solution and a pure solvent separated by a membrane that is permeable to the solvent but not the solute. Initially, the chemical potentials of the solvent on the two sides of the

Jacobus van't Hoff (1852–1911) (Reproduced courtesy of the AIP Emilio Segre Visual Archive, Brittle Book Collection)

membrane may not be equal, the chemical potential on the solution side being smaller. Unequal chemical potentials will cause a flow of the solvent from higher to a lower chemical potential, i.e. a flow of pure solvent towards the solution. The affinity driving this solvent flow is

$$A = \mu^*(p, T) - \mu(p', T, x_2) \tag{8.2.8}$$

in which x_2 is the mole fraction of the solvent, p' is the pressure of the solution and p the pressure of the pure solvent. Equilibrium is reached when the chemical potentials become equal and the corresponding affinity (8.2.8) equals zero. Using Equation (8.1.1) for an ideal solution, the affinity of this system can be written as

$$A = \mu^*(p, T) - \mu^*(p', T) - RT \ln x_2 \tag{8.2.9}$$

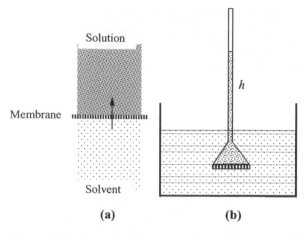

Figure 8.5 Osmosis: the pure solvent flows toward the solution through a semi-permeable membrane until its chemical potentials in both chambers are equal

in which $\mu^0 = \mu*$, the chemical potential of the pure solvent. When $p = p'$, the affinity takes the following simple form:

$$A = -RT \ln x_2 \qquad (8.2.10)$$

The flow of the solvent into the solution can generate a pressure difference between the solvent and the solution (Figure 8.5b). The flow continues until the difference between solvent pressure p and solution pressure p' makes $A = 0$. When $A = 0$, the pressure difference $(p' - p) = \pi$, is called the **osmotic pressure**. In the experimental set-up shown in Figure 8.5b, the liquid level in the solution rises to a height h above the pure-solvent level when equilibrium is reached. The excess pressure in the solution $\pi = h\rho g$, in which ρ is the solution density and g is the acceleration due to gravity. At equilibrium, from (8.2.9) it follows that

$$A = 0 = \mu^*(p, T) - \mu^*(p + \pi, T) - RT \ln x_2 \qquad (8.2.11)$$

At constant temperature, the change in the chemical potential with pressure $d\mu = (\partial\mu/\partial p)_T \, dp = V_m \, dp$, where V_m is the partial molar volume. Since the partial molar volume of a liquid changes very little with pressure, we may assume it to be a constant equal to V_m^*, the solvent molar volume (because when $\mu = \mu^*$, $V_m = V_m^*$). Hence, we may write

$$\mu^*(p + \pi, T) \approx \mu^*(p, T) + \int_p^{p+\pi} d\mu = \mu^*(p, T) + \int_p^{p+\pi} V_m^* dp$$

$$= \mu^*(p, T) + V_m^* \pi \qquad (8.2.12)$$

Also, as we noted before, for dilute solutions, $\ln(x_2) = \ln(1 - x_1) \approx -x_1$. If N_1 is the molar amount of the solute and N_2 is the molar amount of the solvent, then, since $N_2 \gg N_1$, we see that $x_1 = N_1/(N_1 + N_2) \approx N_1/N_2$. Hence, we see that $\ln(x_2) \approx -N_1/N_2$. Using (8.2.12) and the fact that $\ln(x_2) \approx -N_1/N_2$, Equation (8.2.11) can be written as

$$RT\frac{N_1}{N_2} = V_m^* \pi$$

i.e.

$$RTN_1 = N_2 V_m^* \pi = V\pi \qquad (8.2.13)$$

in which $V = N_2 V_m^*$ is nearly the volume of the solution (the correction due to the solute being small). This shows that the osmotic pressure π obeys an ideal-gas-like equation:

$$\boxed{\pi = \frac{N_{\text{solute}}RT}{V_{\text{solution}}} = [S]RT} \qquad (8.2.14)$$

in which $[S]$ is the molar concentration of the solution. This is the **van't Hoff equation** for the osmotic pressure. The osmotic pressure is as if an ideal gas consisting of the solute particles is occupying a volume equal to the solution volume. By measuring the osmotic pressure one can determine the molar amount N_{solute} of a solute. Thus, if the mass of the solute is known, then its molar mass can be calculated. The measurement of osmotic pressure is used to determine the molar mass or molecular weight of large biomolecules for which semi-permeable membranes can be easily found (Exercise 8.10).

Table 8.3 shows a comparison between experimentally measured osmotic pressures and those calculated using the van't Hoff equation (8.2.14) for an aqueous

Table 8.3 Comparison between theoretical osmotic pressure calculated using van't Hoff's equation and the experimentally observed osmotic pressure for an aqueous solution of sucrose at two temperatures

$T = 273$ K			$T = 333$ K		
Concentration/mol L^{-1}	π (atm)		Concentration/mol L^{-1}	π (atm)	
	Experiment	Theory		Experiment	Theory
0.029 22	0.65	0.655	0.098	2.72	2.68
0.058 43	1.27	1.330	1.923	5.44	5.25
0.131 5	2.91	2.95	0.370 1	10.87	10.11
0.273 9	6.23	6.14	0.533	16.54	14.65
0.532 8	14.21	11.95	0.685 5	22.33	18.8
0.876 6	26.80	19.70	0.827 3	28.37	22.7

Source: I. Prigogine and R. Defay, *Chemical Thermodynamics*, 4th edition. 1967, London: Longmans.

solution of sucrose. We see that for concentrations up to about 0.2 M the van't Hoff equation agrees with experimental values. Deviation from the van't Hoff equation is not necessarily due to deviation from ideality. In deriving the van't Hoff equation, we also assumed a dilute solution. Using (8.1.11) and (8.2.12), it is easy to see that the osmotic pressure can also be written as

$$\pi_{\text{ideal}} = \frac{-RT \ln x_2}{V_m^*} \tag{8.2.15}$$

where x_2 is the mole fraction of the solvent. Here we have indicated explicitly that the osmotic pressure in this expression is valid for ideal solutions. This formula was obtained by J.J. van Larr in 1894.

For nonideal solutions, instead of an activity coefficient γ, an osmotic coefficient ϕ is defined through

$$\mu(p, T, x_2) = \mu^*(p, T) + \phi RT \ln x_2 \tag{8.2.16}$$

in which μ^* is the chemical potential of the pure solvent. At equilibrium, when the affinity vanishes and osmotic pressure is π we have the equation

$$\mu^*(p, T) = \mu^*(p + \pi, T) + \phi RT \ln x_2 \tag{8.2.17}$$

Following the same procedure as above, we arrive at the following expression for the osmotic pressure of a nonideal solution:

$$\boxed{\pi = \frac{-\phi RT \ln x_2}{V_m^*}} \tag{8.2.18}$$

Equation (8.2.18) was proposed by Donnan and Guggenheim in 1932. From Equations (8.2.15) and (8.2.18) it follows that $\phi = \pi/\pi_{\text{ideal}}$. Hence, the name 'osmotic coefficient' for ϕ. We can also relate the affinity to the osmotic pressure. When the solution and the pure solvent are at the same pressure the affinity is $A = \mu^*(p,T) - \mu^*(p,T) - \phi RT \ln x_2 = -\phi RT \ln x_2$. Using this in (8.2.18) we see that

$$\pi - \frac{A}{V_m^*} \quad \text{when } p_{\text{solution}} = p_{\text{solvent}}. \tag{8.2.19}$$

Another approach for the consideration of nonideal solutions is similar to that used to obtain the virial equation for real gases. In this case the osmotic pressure is written as

$$\pi = [S]RT(1 + B(T)[S] + \ldots) \tag{8.2.20}$$

in which $B(T)$ is a constant that depends on the temperature. The experimental data on the osmotic pressure of polymer solutions (such as polyvinyl chloride in cyclo-hexanone) shows a fairly linear relation between $\pi/[S]$ and $[S]$. Also, the value of $B(T)$ changes sign from negative to positive as the temperature increases. The temperature at which B equals zero is called the **theta temperature**. If the concentration is expressed in grams per liter, which we shall denote by $[C]$, then (8.2.20) can be written as

$$\pi = \frac{[C]RT}{M_s}\left\{1 + B(T)\frac{[C]}{M_s} + \dots\right\} \tag{8.2.21}$$

in which M_s is molar mass of the solute. With this equation, a plot of $\pi/[C]$ versus $[C]$ is expected to be linear with an intercept equal to RT/M_s. From the intercept, the molar mass can be determined. From the slope, equal to BRT/M_s^2, the 'virial constant' B can be obtained.

8.3   Solubility Equilibrium

The solubility of a solid in a solvent depends on the temperature. **Solubility** is the concentration when the solid solute is in equilibrium with the solution: it is the concentration at saturation. Thermodynamics gives us a quantitative relation between solubility and temperature. In considering the solubilities of solids one must distinguish between ionic solutions and nonionic solutions. When ionic solids, such as NaCl, dissolve in polar solvents, such as water, the solutions contain ions, Na$^+$ and Cl$^-$. Since ions interact strongly even in dilute solutions, the activities cannot be approximated well by mole fractions. For nonionic solutions, such as sugar in water or naphthalene in acetone, the activity of dilute solution can be approximated by the mole fraction.

NONIONIC SOLUTIONS

For dilute nonionic solutions, we may assume ideality and use the expression (8.1.1) for the chemical potential to analyze the conditions for thermodynamic equilibrium. Solutions of higher concentrations require a more detailed theory (as can be found for instance in the classic text by Prigogine and Defay [2]). Recall that, like it does for liquids, the chemical potential of a solid varies very little with pressure and so it is essentially a function only of the temperature. If $\mu_s^*(T)$ is the chemical potential of the pure solid in equilibrium with the solution, then we have (using (8.1.1))

$$\mu_s^*(T) = \mu_1(T) = \mu_1^*(T) + RT\ln(x_1) \tag{8.3.1}$$

in which the μ_1 the chemical potential of the solute in the solution phase (liquid phase), μ_1^* is the chemical potential of the pure solute in the liquid phase, and x_1 is the mole fraction of the solute. If $\Delta G_{fus}(T) = \mu_1^* - \mu_s^*$ is the molar Gibbs energy of fusion at temperature T, then the above equation can be written in the form

$$\ln x_1 = -\frac{1}{R}\frac{\Delta G_{fus}}{T} \tag{8.3.2}$$

In this form the temperature dependence of the solubility is not explicit because ΔG_{fus} is itself a function of T. This expression can also be written in terms of the enthalpy of fusion ΔH_{fus} by differentiating this expression with respect to T and using the Gibbs–Helmholtz equation $d(\Delta G/T)dT = -(\Delta H/T^2)$ (5.2.14); this equation can be written as

$$\boxed{\frac{d\ln(x_1)}{dT} = \frac{1}{R}\frac{\Delta H_{fus}}{T^2}} \qquad (8.3.3)$$

Since ΔH_{fus} does not change much with T, this expression can be integrated to obtain a more explicit dependence of solubility with temperature.

IONIC SOLUTIONS

Ionic solutions, also called **electrolytes**, are dominated by electrical forces which can be very strong. To get an idea of the strength of electrical forces, it is instructive to calculate the force of repulsion between two cubes of copper of side 1 cm in which one in a million Cu atoms is a Cu^+ ion, when the two cubes are 10 cm apart. The force is sufficient to lift a weight of 16×10^6 kg (Exercise 8.13).

Owing to the enormous strength of electrical forces, there is almost no separation between positive and negative ions in a solution; positive and negative charges aggregate to make the net charge in every macroscopic volume nearly zero, i.e. every macroscopic volume is electrically neutral. Solutions, and indeed most matter, maintain *electroneutrality* to a high degree. Thus if c_k (mol L^{-1}) are the concentrations of positive and negative ions with *ion numbers* (number of electronic charges) z_k, the total charge carried by an ion per unit volume is $Fz_k c_k$, in which $F = eN_A$ is the Faraday constant, equal to the product of the electronic charge $e = 1.609 \times 10^{-19}$ C and the Avogadro number N_A. Since electroneutrality implies that the net charge is zero, we have

$$\sum_k Fz_k c_k = 0 \qquad (8.3.4)$$

Let us consider the solubility equilibrium of a sparingly soluble electrolyte AgCl in water:

$$AgCl(s) \rightleftharpoons Ag^+ + Cl^- \qquad (8.3.5)$$

At equilibrium:

$$\mu_{AgCl} = \mu_{Ag^+} + \mu_{Cl^-} \qquad (8.3.6)$$

In ionic systems, since the positive and negative ions always come in pairs, physically it is not possible to measure the chemical potentials μ_{Ag^+} and μ_{Cl^-} separately; only their sum can be measured. A similar problem arises for the definition of enthalpy and Gibbs energy of formation. For this reason, for ions, these two quantities are defined with respect to a new reference state based on the H$^+$ ions, as described in Box 8.1. For the chemical potential, a **mean chemical potential** is defined by

---

Box 8.1 Enthalpy and Gibbs Free Energy of Formation of Ions.

When ionic solutions form, the ions occur in pairs; therefore, it is not possible to isolate the enthalpy of formation of a positive or negative ion. Hence, we cannot obtain the heats of formation of ions with the usual elements in their standard state as the reference state. For ions, the enthalpy of formation, is tabulated by defining the ΔH_f of formation of H^+ as zero at all temperatures. Thus

$$\Delta H_f^0[H^+(aq)] = 0 \quad \text{at all temperatures}$$

With this definition it is now possible to obtain the ΔH_f of all other ions. For example, to obtain the heat of formation of $Cl^-(aq)$, at a temperature T, the enthalpy of solution of HCl is measured. Thus, $\Delta H_f^0[Cl^-(aq)]$ is the heat of solution at temperature T:

$$HCl \rightarrow H^+(aq) + Cl^-(aq)$$

The tabulated values of enthalpies are based on this convention. Similarly, for the Gibbs energy:

$$\Delta G_f^0[H^+(aq)] = 0 \quad \text{at all temperatures}$$

For ionic systems, it has become customary to use the **molality scale** (mol/kg solvent). This scale has the advantage that the addition of another solute does not change the molality of a given solute. The values of ΔG_f^0 and ΔH_f^0 for the formation of ions in water at $T = 298.15\,K$ are tabulated for the **standard state** of an ideal dilute solution at a concentration of $1\,\text{mol}\,kg^{-1}$. This standard state is given the subscript 'ao'. Thus, the chemical potential or the activity of an ion is indicated by 'ao'. The chemical potential of an ionized salt, $\mu_{salt} \equiv v_+\mu_+ + v_-\mu_-$, and the corresponding activity are denoted with the subscript '**ai**'.

$$\mu_\pm = \frac{1}{2}(\mu_{Ag^+} + \mu_{Cl^-}) \tag{8.3.7}$$

so that (8.3.6) becomes

$$\mu_{AgCl} = 2\mu_\pm \tag{8.3.8}$$

In general, for the decomposition of a neutral compound W into positive and negative ions, A^{Z+} and B^{Z-} respectively (with ion numbers $Z+$ and $Z-$), we have

$$W \rightleftharpoons v_+A^{Z+} + v_-B^{Z-} \tag{8.3.9}$$

in which v_+ and v_- are the stoichiometric coefficients. The mean chemical potential in this case is defined as

$$\boxed{\mu_\pm = \frac{v_+\mu_+ + v_-\mu_-}{v_+ + v_-} = \frac{\mu_{salt}}{v_+ + v_-}} \tag{8.3.10}$$

in which $\mu_{salt} \equiv \nu_+\mu_+ + \nu_-\mu_-$. Here, we have written the chemical potential of the positive ion A^{Z+} as μ_+ and that of the negative ion B^{Z-} as μ_-.

The activity coefficients γ of electrolytes are defined with respect to ideal solutions. For example, the mean chemical potential for AgCl is written as

$$\mu_\pm = \frac{1}{2}[\mu_{Ag^+}^0 + RT\ln(\gamma_{Ag^+}x_{Ag^+}) + \mu_{Cl^-}^0 + RT\ln(\gamma_{Cl^-}x_{Cl^-})] \qquad (8.3.11)$$
$$= \mu_\pm^0 + RT\ln\sqrt{\gamma_{Ag^+}\gamma_{Cl^-}x_{Ag^+}x_{Cl^-}}$$

where $\mu_\pm^0 = 1/2(\mu_{Ag^+}^0 + {}_{Cl^-}^0)$. Once again, since the activity coefficients of the positive and negative ions cannot be measured individually, a mean activity coefficient $\gamma_\pm$ is defined by

$$\gamma_\pm = (\gamma_{Ag^+}\gamma_{Cl^-})^{1/2} \qquad (8.3.12)$$

In the more general case of (8.3.9), the **mean ionic activity coefficient** is defined as

$$\gamma_\pm = (\gamma_+^{\nu_+}\gamma_-^{\nu_-})^{1/(\nu_++\nu_-)} \qquad (8.3.13)$$

where we have used γ_+ and γ_- for the activity coefficients of the positive and negative ions.

The chemical potentials of dilute solutions may be expressed in terms of **molality** m_k (moles of solute per kilogram of solvent) or molarities c_k (moles of solute per liter of solution*) instead of mole fractions x_k. In electrochemistry, it is more common to use molality m_k. For dilute solutions, since $x_k = N_k/N_{solvent}$, we have the following conversion formulas for the different units:

$$x_k = m_k M_s \quad \text{and} \quad x_k = V_{ms}c_k \qquad (8.3.14)$$

in which M_s is the molar mass of the solvent in kilograms and V_{ms} the molar volume of the solvent in liters. The corresponding chemical potentials then are written as

$$\mu_k^x = \mu_k^{x0} + RT\ln(\gamma_k x_k) \qquad (8.3.15)$$

$$\mu_k^m = \mu_k^{x0} + RT\ln M_s + RT\ln(\gamma_k m_k)$$
$$= \mu_k^{m0} + RT\ln\left(\frac{\gamma_k m_k}{m^0}\right) \qquad (8.3.16)$$

$$\mu_k^c = \mu_k^{x0} + RT\ln V_{ms} + RT\ln(g_k c_k)$$
$$= \mu_k^{c0} + RT\ln\left(\frac{g_k c_k}{c^0}\right) = \mu_k^{c0} + RT\ln\left(\frac{r_k[k]}{[k]^0}\right) \qquad (8.3.17)$$

---

* Molarity of k is also expressed as $[k]$.

in which the definitions of the reference chemical potentials μ_k^{m0} and μ_k^{c0} in each concentration scale are self-evident. The activity in the molality scale is written in the dimensionless form as $a_k = \gamma_k m_k / m^0$, in which m^0 is standard value of molality equal to (1 mol of solute per kilogram of solvent). Similarly, the activity in the molarity scale is written as $a_k = \gamma_k c_k / c^0$, in which c^0 equals (1 mol per liter of solution). For electrolytes the mean chemical potential $\mu_{\pm}$ is usually expressed in the molality scale; the tabulation of ΔG_f^0 and ΔH_f^0 for the formation of ions in water at $T = 298.15\,\text{K}$ is usually for the standard state of an ideal dilute solution at a concentration of $1\,\text{mol}\,\text{kg}^{-1}$. This standard state is given the subscript 'ao'.

In the commonly used molality scale, the solution equilibrium of AgCl expressed in (8.3.8) can now be written as

$$\mu_{\text{AgCl}}^0 + RT \ln a_{\text{AgCl}} = 2\mu_{\pm}^{m0} + RT \ln \left[\frac{\gamma_{\pm}^2 m_{\text{Ag}^+} m_{\text{Cl}^-}}{(m^0)^2} \right] \qquad (8.3.18)$$

Since the activity of a solid is nearly equal to one, $a_{\text{AgCl}} \approx 1$. Hence, we obtain the following expression for the **equilibrium constant**[†] for solubility in the molality scale:

$$K_{\text{m}}(T) \equiv \frac{\gamma_{\pm}^2 m_{\text{Ag}^+} m_{\text{Cl}^-}}{(m^0)^2} = a_{\text{Ag}^+} a_{\text{Cl}^-} = \exp \left(\frac{\mu_{\text{AgCl}}^0 - 2\mu_{\pm}^{m0}}{RT} \right) \qquad (8.3.19)$$

The equilibrium constant for electrolytes is also called the **solubility product** K_{SP}. For sparingly soluble electrolytes such as AgCl, even at saturation the solution is very dilute and $\gamma_{\pm} \approx 1$. In this limiting case, the solubility product

$$K_{\text{SP}} \approx m_{\text{Ag}^+} m_{\text{Cl}^-} \qquad (8.3.20)$$

in which we have not explicitly included m^0, which has a value equal to one.

ACTIVITY, IONIC STRENGTH AND SOLUBILITY

A theory of ionic solutions developed by Peter Debye and Erich Hückel in 1923 (which is based on statistical mechanics and is beyond the scope of this text) provides an expression for the activity. We shall only state the main result of this theory, which works well for dilute electrolytes. The activity depends on a quantity called the **ionic strength** I defined by

$$I = \frac{1}{2} \sum_k z_k^2 m_k \qquad (8.3.21)$$

The activity coefficient of an ion k in the molality scale is given by

$$\log_{10}(\gamma_k) = -A z_k^2 \sqrt{I} \qquad (8.3.22)$$

in which

$$A = \frac{N_A^2}{2.3026} \left(\frac{2\pi \rho_s}{R^3 T^3} \right)^{1/2} \left(\frac{e^2}{4\pi \varepsilon_0 \varepsilon_r} \right)^{3/2} \qquad (8.3.23)$$

[†] A general definition of the equilibrium constant is discussed in Chapter 9.

where ρ_s is the density of the solvent, e is the electronic charge, $\varepsilon_0 = 8.854 \times 10^{-12}\,C^2$ $N^{-1}\,m^{-2}$ is the permittivity of vacuum, and ε_r the relative permittivity of the solvent (ε_r = 78.54 for water). For ions in water, at $T = 298.15\,K$, we find $A = 0.509\,kg^{1/2}\,mol^{-1/2}$. Thus, at $25\,°C$ the activity of ions in dilute solutions can be approximated well by the expression

$$\log_{10}(\gamma_k) = -0.509 z_k^2 \sqrt{I} \qquad (8.3.24)$$

The Debye–Hückel theory enables us to understand how solubility is influenced by ionic strength. For example, let us look at the solubility of AgCl. If the $m_{Ag^+} = m_{Cl^-} = S$, the *solubility*, we may write the equilibrium constant K_m as

$$K_m(T) \equiv \gamma_{\pm}^2 m_{Ag^+} m_{Cl^-} = \gamma_{\pm}^2 S^2 \qquad (8.3.25)$$

The ionic strength depends not only on the concentration of Ag^+ and Cl^- ions, but also on all the other ions. Thus, for example, the addition of nitric acid, HNO_3, which adds H^+ and NO_3^- ions to the system, will change the activity coefficient $\gamma_{\pm}$. But the equilibrium constant, which is a function of T only (as is evident from (8.3.19)) remains unchanged if T is constant. As a result, the value of m (or solubility in molal) will change with the ionic strength I. If the concentration of HNO_3 (which dissociates completely) is m_{HNO_3}, then the ionic strength will be

$$\begin{aligned} I &= \frac{1}{2}(m_{Ag^+} + m_{Cl^-} + m_{H^+} + m_{NO_3^-}) \\ &= S + m_{HNO_3} \end{aligned} \qquad (8.3.26)$$

Using (8.3.12) for $\gamma_{\pm}$ for AgCl and substituting (8.3.24) in (8.3.25) we can obtain the following relation between the solubility S of AgCl and the concentration of HNO_3:

$$\log_{10}(S) = \frac{1}{2}\log_{10}(K_m(T)) + 0.509\sqrt{S + m_{HNO_3}} \qquad (8.3.27)$$

If $S \ll m_{HNO_3}$ then the above relation can be approximated by

$$\log_{10}(S) = \frac{1}{2}\log_{10}(K_m(T)) + 0.509\sqrt{m_{HNO_3}} \qquad (8.3.28)$$

Thus, a plot of $\log S$ versus $\sqrt{m_{HNO_3}}$ should yield a straight line, which is indeed found to be the case experimentally. In fact, such plots can be used to determine the equilibrium constant K_m and the activity coefficients.

8.4 Thermodynamic Mixing and Excess Functions

PERFECT SOLUTIONS

A perfect solution is one for which the chemical potential of the form $\mu_k(p, T, x_k) = \mu_k^*(p, T) + RT \ln(x_k)$ is valid for all values of the mole fraction x_k. The molar Gibbs energy of such a solution is

$$G_m = \sum_k x_k \mu_k = \sum_k x_k \mu_k^* + RT \sum_k x_k \ln x_k \tag{8.4.1}$$

If each of the components were separated, then the total Gibbs energy for the components is the sum $G_m^* = \Sigma_k x_k G_{mk}^* = \Sigma_k x_k \mu_k^*$ in which we have used the fact that, for a pure substance, G_{mk}^*, the molar Gibbs energy of k, is equal to the chemical potential μ_k^*. Hence, the change (decrease) in the *molar* Gibbs energy due to the *mixing* of the components in the solution is

$$\Delta G_{mix} = RT \sum_k x_k \ln x_k \tag{8.4.2}$$

and

$$G_m = \sum_k x_k G_{mk}^* + \Delta G_{mix} \tag{8.4.3}$$

Since the molar entropy $S_m = -(\partial G_m / \partial T)_p$, it follows from (8.4.2) and (8.4.3) that

$$S_m = \sum_k x_k S_{mk}^* + \Delta S_{mix} \tag{8.4.4}$$

$$\Delta S_{mix} = -R \sum_k x_k \ln x_k \tag{8.4.5}$$

where ΔS_{mix} is the molar entropy of mixing. This shows that, during the formation of a perfect solution from pure components at a fixed temperature, the decrease in G is $\Delta G_{mix} = -T\Delta S_{mix}$. Since $\Delta G = \Delta H - T\Delta S$, we can conclude that, for the formation of a perfect solution at a fixed temperature, $\Delta H = 0$. This can be verified explicitly by noting that the Helmholtz equation (5.2.11) can be used to evaluate the enthalpy. For G given by (8.4.2) and (8.4.3) we find

$$H_m = -T^2 \left(\frac{\partial}{\partial T} \frac{G_m}{T} \right) = \sum_k x_k H_{mk}^* \tag{8.4.6}$$

Thus, the enthalpy of the solution is the same as the enthalpy of the pure components and there is no change in the enthalpy of a perfect solution due to mixing. Similarly, by noting that $V_m = (\partial G_m / \partial p)_T$, it is easy to see (Exercise 8.16) that there is no change in the molar volume due to mixing, i.e. $\Delta V_{mix} = 0$; the total volume is the sum of the volumes of the components in the mixture. Furthermore, since $\Delta U = \Delta H - p\Delta V$, we see also that $\Delta U_{mix} = 0$. Thus, for a perfect solution, the *molar quantities* for mixing are

$$\Delta G_{mix} = RT \sum_k x_k \ln x_k \tag{8.4.7}$$

$$\Delta S_{mix} = -R \sum_k x_k \ln x_k \tag{8.4.8}$$

$$\Delta H_{mix} = 0 \tag{8.4.9}$$

$$\Delta V_{\mathrm{mix}} = 0 \qquad (8.4.10)$$

$$\Delta U_{\mathrm{mix}} = 0 \qquad (8.4.11)$$

In a perfect solution, the irreversible process of mixing of the components at constant p and T is entirely due to the increase in entropy; no heat is evolved or absorbed.

IDEAL SOLUTIONS

Dilute solutions may be ideal over a small range of mole fractions x_i. In this case the molar enthalpy H_{m} and the molar volume V_{m} may be a linear function of the partial molar enthalpies H_{mi} and partial molar volumes V_{mi}. Thus:

$$H_{\mathrm{m}} = \sum_i x_i H_{mi} \quad \text{and} \quad V_{\mathrm{m}} = \sum_i x_i V_{mi} \qquad (8.4.12)$$

However, the partial molar enthalpies H_{mi} may not be equal to the molar enthalpies of pure substances if the corresponding mole fractions are small. The same is true for the partial molar volumes. On the other hand, if x_i is nearly equal to one, then H_{mi} will be nearly equal to the molar enthalpy of the pure substance. A dilute solution for which (8.4.12) is valid will exhibit ideal behaviour, but it may have a nonzero enthalpy of mixing. To see this more explicitly, consider a dilute ($x_1 \gg x_2$) binary solution for which H_{m1}^* and H_{m2}^* are the molar enthalpies of the two pure components. Then, before mixing, the molar enthalpy is

$$H_{\mathrm{m}}^* = x_1 H_{\mathrm{m1}}^* + x_2 H_{\mathrm{m2}}^* \qquad (8.4.13)$$

After mixing, since for the major component (for which $x_1 \approx 1$) we have $H_{\mathrm{m1}}^* = H_{ml}$, the molar enthalpy will be

$$H_{\mathrm{m}} = x_1 H_{\mathrm{m1}}^* + x_2 H_{\mathrm{m2}} \qquad (8.4.14)$$

The molar enthalpy of mixing is then the difference between the above two enthalpies:

$$\Delta H_{\mathrm{mix}} = H_{\mathrm{m}} - H_{\mathrm{m}}^* = x_2(H_{\mathrm{m2}} - H_{\mathrm{m2}}^*) \qquad (8.4.15)$$

In this way, an ideal solution may have a nonzero enthalpy of mixing. The same may be true for the volume of mixing.

EXCESS FUNCTIONS

For nonideal solutions, the molar Gibbs energy of mixing is

$$\Delta G_{\mathrm{mix}} = RT \sum_i x_i \ln(\gamma_i x_i) \qquad (8.4.16)$$

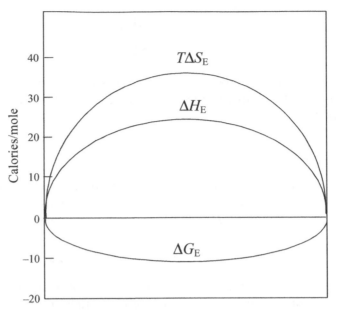

Figure 8.6 Thermodynamic excess function for a solution of *n*-heptane (component 1) and *n*-hexadecane (component 2) at 20°C. The graph shows molar excess functions as a function of the mole fraction x_2 of *n*-hexadecane

The difference between the Gibbs energies of mixing of perfect and nonideal solutions is called the **excess Gibbs energy**, which we shall denote by ΔG_E. From (8.4.7) and (8.4.16) it follows that

$$\Delta G_E = RT \sum_i x_i \ln \gamma_i \qquad (8.4.17)$$

Other excess functions, such as excess entropy and enthalpy, can be obtained from ΔG_E. For example:

$$\Delta S_E = -\left(\frac{\partial \Delta G_E}{\partial T}\right)_p = -RT \sum_i x_i \frac{\partial \ln \gamma_i}{\partial T} - R \sum_i x_i \ln x_i \qquad (8.4.18)$$

Similarly ΔH_E can be obtained using the relation

$$\Delta H_E = -T^2 \left(\frac{\partial}{\partial T} \frac{\Delta G_E}{T}\right) \qquad (8.4.19)$$

These excess functions can be obtained experimentally through measurements of vapor pressure and heats of reaction (Figure 8.6).

REGULAR AND ATHERMAL SOLUTIONS

Nonideal solutions may be classified into two limiting cases. In one limiting case, called **regular solutions**, $\Delta G_E \approx \Delta H_E$, i.e. most of the deviation from ideality is due to the excess enthalpy of mixing. Since $\Delta G_E = \Delta H_E - T\Delta S_E$, it follows that for regular solutions $\Delta S_E \approx 0$. Furthermore, since $\Delta S_E = -(\partial \Delta G_E/\partial T)_p$, from (8.4.18) it follows that the activity coefficients

$$\ln \gamma_i \propto \frac{1}{T} \tag{8.4.20}$$

For regular binary solutions, the activities may be approximated by the function $\gamma_k = \alpha x_k^2/RT$.

The other limiting case of nonideal solutions is when $\Delta G_E \approx -T\Delta S_E$, in which case the deviation from ideality is mostly due to the excess entropy of mixing and $\Delta H_E \approx 0$. In this case, using (8.4.17) in (8.4.19), we see that $\ln \gamma_i$ are independent of T. Such solutions are called **athermal solutions**. Solutions in which the component molecules are of nearly the same size but differ in intermolecular forces generally behave like regular solutions. Solutions whose component molecules have very different sizes but do not differ significantly in their intermolecular forces, as in the case of monomers and polymers, are examples of athermal solutions.

8.5 Azeotropy

In Chapter 7 we discussed azeotropes briefly. We shall apply the thermodynamics of solutions that was presented in the previous sections of this chapter to azeotropes. For an azeotrope in equilibrium with its vapor, the composition of the liquid and the vapor phases are the same. At a fixed pressure, a liquid mixture is an azeotrope at a particular composition called the **azeotropic composition**. In a closed system, an **azeotropic transformation** is one in which there is an exchange of matter between two phases without change in composition. In this regard, an azeotrope is similar to the vaporization of a pure substance. This enables us to obtain the activity coefficients of azeotropes just as can be done for a pure substance.

Let us consider a binary azeotrope. As we have seen in Section 8.1, the chemical potentials of the components can be written in the form $\mu_k(T, p, x_k) = \mu_k^0(T, p) + RT \ln(\gamma_k x_k)$ in which activity coefficient γ_k is a measure of the deviation from ideality. If $\gamma_{k,l}$ and $\gamma_{k,g}$ are the activity coefficients of component k in the liquid and gas phases respectively, then by considering an azeotropic transformation it can be shown that (Exercise 8.17)

$$\ln \left(\frac{\gamma_{k,g}}{\gamma_{k,l}} \right) = \int_{T_k^*}^{T} \frac{\Delta H_{vap,k}}{RT^2} dT - \frac{1}{RT} \int_{p^*}^{p} \Delta V_{mk}^* dp \tag{8.5.1}$$

in which $\Delta H_{vap,k}$ is the heat of vaporization of component k, and ΔV_{mk}^* is the change in the molar volume of the pure component between the liquid and the vapor phases. $T*$ is the boiling point of the pure solvent at pressure $p*$. If we consider an azeotropic transformation at a fixed pressure, e.g. $p = p* = 1$ atm, then since ΔH_{vap} generally does not change much with T, we obtain

$$\ln\left(\frac{\gamma_{k,g}}{\gamma_{k,l}}\right) = \frac{-\Delta H_{vap,k}}{R}\left(\frac{1}{T} - \frac{1}{T^*}\right) \qquad (8.5.2)$$

For the activity coefficient of the vapor phase, if we use the ideal gas approximation $\gamma_{k,g} = 1$. This gives us an explicit expression for the activity coefficient of the liquid phase:

$$\ln(\gamma_{k,l}) = \frac{\Delta H_{vap,k}}{R}\left(\frac{1}{T} - \frac{1}{T^*}\right) \qquad (8.5.3)$$

With this expression, the activity coefficient of a component of an azeotrope, can be calculated, and it gives a simple physical meaning to the activity coefficient. Molecular theories of solutions give us more insight into the relation between the intermolecular forces and the thermodynamics of azeotropes [3].

References

1. Laidler, K.J., *The World of Physical Chemistry*. 1993, Oxford: Oxford University Press.
2. Prigogine, I. and Defay, R., *Chemical Thermodynamics*, fourth edition. 1967, London: Longmans.
3. Prigogine, I., *Molecular Theory of Solutions*. 1957, New York: Interscience Publishers.

Examples

Example 8.1 In the oceans, to a depth of about 100 m the concentration of O_2 is about $0.25 \times 10^{-3}\,mol\,L^{-1}$. Compare this value with the value obtained using Henry's law assuming equilibrium between the atmospheric oxygen and the dissolved oxygen.

Solution The partial pressure of O_2 in the atmosphere is $p_{O_2} \approx 0.2\,atm$. Using Henry's law constant in Table 8.1 we have for the mole fraction of the dissolved oxygen x_{O_2}

$$p_{O_2} = K_{O_2}x_{O_2}$$

Hence:

$$x_{O_2} = \frac{p_{O_2}}{K_{O_2}} = \frac{0.2\,atm}{4.3 \times 10^4\,atm} = 4.6 \times 10^{-6}$$

i.e. there are $4.6 \times 10^{-6}\,mol$ of O_2 per mole of H_2O. Noting that 1 L of H_2O is equal to 55.5 mol, the above mole fraction of O_2 can be converted a concentration in moles per liter:

$$c_{O_2} = 4.6 \times 10^{-6} \times 55.5\,mol\,L^{-1} = 2.5 \times 10^{-4}\,mol\,L^{-1}$$

which is equal to the measured concentration of O_2 in the oceans.

Example 8.2 In an aqueous solution of NH_3 at 25.0°C, the mole fraction of NH_3 is 0.05. For this solution, calculate the partial pressure of water vapor assuming ideality. If the vapor pressure is found to be 3.40 kPa, what is the activity a of water, and what is its activity coefficient γ?

Solution If $p*$ is the vapor pressure of pure water at 25.0°C, then, according to Raoult's law (8.1.10), the vapor pressure of the above solution is given by $p = x_{H_2O}p*$ $= 0.95p*$. The value of $p*$ can be obtained as follows. Since water boils at 373.0 K at $p = 1.0\,atm = 101.3\,kPa$, we know that its vapor pressure is 101.3 kPa at 373.0 K. Using the Clausius–Clapeyron equation, we can calculate the vapor pressure at 25.0°C = 298.0 K:

$$\ln p_1 - \ln p_2 = \frac{\Delta H_{vap}}{R}\left(\frac{1}{T_2} - \frac{1}{T_1}\right)$$

With $p_2 = 1\,atm$, $T_2 = 373.0$, $T_1 = 298.0$, $\Delta H_{vap} - 40.66\,kJ\,mol^{-1}$ (see Table 7.1), the vapor pressure, p_1 (atm) can be computed:

$$\ln(p_1/atm) = -3.299$$

i.e.

$$p_1 = \exp(-3.299)\,atm = 0.0369\,atm - 101.3 \times 0.0369\,kPa$$
$$= 3.73\,kPa = p*$$

Hence, the vapor pressure $p*$ of pure water at 25°C is 3.738 kPa. For the above solution in which the mole fraction of water is 0.95, the vapor pressure for an ideal solution according to Raoult's law should be

$$p = 0.95 \times 3.73\,kPa = 3.54\,kPa$$

For an ideal solution, the activity a is the same as the mole fraction x_1. As shown in (8.1.6), in the general case the activity $a = p/p*$. Hence, if the measured vapor pressure is 3.40 kPa, then the activity

$$a_1 - 3.40/3.738 - 0.909$$

The activity coefficient is defined by $a_k = \gamma_k x_k$. Hence, $\gamma_1 = a_1/x_1 = 0.909/0.95 = 0.956$.

Example 8.3 Living cells contain water with many ions. The osmotic pressure corresponds to that of an NaCl solution of about 0.15 M. Calculate the osmotic pressure at $T = 27°C$.

Solution Osmotic pressure depends on the 'colligative concentration', i.e. the number of particles per unit volume. Since NaCl dissociates into Na^+ and Cl^- ions, the colligative molality of the above solution is 0.30 M. Using the van't Hoff equation (8.2.14), we can calculate the osmotic pressure π:

$$\pi = RT[S] = (0.0821\,\text{L atm K}^{-1}\text{mol}^{-1})(300.0\,\text{K})(0.30\,\text{mol L}^{-1}) = 7.40\,\text{atm}$$

If an animal cell is immersed in water, then the water flowing into the cell due to osmosis will exert a pressure of about 7.4 atm and causes the cell to burst. Plant cell walls are strong enough to withstand this pressure.

Example 8.4   At $p = 1$ atm, the boiling point of an azeotropic mixture of C_2H_5OH and CCl_4 is 338.1 K. The heat of vaporization of C_2H_5OH is 38.58 kJ mol^{-1} and its boiling point is 351.4 K. Calculate the activity coefficient of ethanol in the azeotrope.

Solution   This can be done by direct application of Equation (8.5.3), where $\Delta H_{1,\text{vap}}$ = 38.58 kJ mol^{-1}, $T = 338.1$ K and $T* = 351.4$ K:

$$\ln(\gamma_{1k}) = \frac{38.5 \times 10^3}{8.314}\left(\frac{1}{338.1} - \frac{1}{351.4}\right) = 0.519$$

i.e.

$$\gamma_{1,k} = 1.68$$

Exercises

8.1   Obtain equation (8.1.8) from (8.1.7).

8.2   14.0 g of NaOH is dissolved in 84.0 g of H_2O. The solution has a density of 1.114×10^3 kg m^{-3}. For the two components, NaOH and H_2O, in this solution, obtain (a) the mole fractions, (b) the molality and (c) molarity.

8.3   The composition of the atmosphere is shown in Table 8.1. Using the Henry's law constants, calculate the concentrations of N_2, O_2 and CO_2 in a lake.

8.4   Obtain (8.2.5) from (8.2.4) for small changes in the boiling point of a solution.

8.5   (a) The solubility of $N_2(g)$ in water is about the same as in blood serum. Calculate the concentration (in mol L^{-1}) of N_2 in the blood.
(b) The density of seawater is 1.01 g mL^{-1}. What is the pressure at a depth of 100 m? What will the blood serum concentration (in mol L^{-1}) of N_2 be at this depth? If divers rise too fast, then any excess N_2 can form bubbles in the blood, causing pain, paralysis and distress in breathing.

8.6   Assuming Raoult's law holds, predict the boiling point of a 0.5 M aqueous solution of sugar. Do the same for NaCl, but note that the number of particles (ions) per molecule is twice that of a nonionic solution. Raoult's law is a colligative property that depends on the number of solute particles.

8.7   Ethylene glycol ($OH—CH_2—CH_2—OH$) is used as an antifreeze. (Its boiling point is 197 °C and freezing point is −17.4 °C.)
(a) Look up the density of ethylene glycol in the *CRC Handbook* or other tables and write a general formula for the freezing point of a mixture of X mL of ethylene glycol in 1.00 L of water for X in the range 0–100 mL.
(b) If the lowest expected temperature is about −10 °C, what is the minimum amount (in milliliters per liter of H_2O) of ethylene glycol do you need in your coolant?
(c) What is the boiling point of the coolant that contains 300 mL of ethylene glycol per liter of water?

8.8   What will be the boiling point of a solution of 20.0 g of urea (($NH_2)_2CO$) in 1.25 kg of nitrobenzene (use Table 8.2).

8.9   A 1.89 g pellet of an unknown compound was dissolved to 50 mL of acetone. The change in the boiling point was found to be 0.64 °C. Calculate the molar mass of the unknown. Density of acetone is 0.7851 g/mL and the value of ebullioscopic constant K_b may be found in Table 8.2.

8.10   A solution of 4.00 g hemoglobin in 100 mL was prepared and its osmotic pressure was measured. The osmotic pressure was found to be 0.0130 atm at 280 K. (a) Estimate the molar mass of hemoglobin. (b) If 4.00 g of NaCl is dissolved in 100 mL of water, what would the osmotic pressure be?

(Molecular weights of some proteins: ferricytochrome c 12 744; myoglobin 16 951; lysozyme 14 314; immunoglobulin G 156 000; myosin 570 000.)

8.11   The concentration of the ionic constituents of seawater are:

Ion	Cl^-	Na^+	SO_4^{2-}	Mg^{2+}	Ca^{2+}	K^+	HCO_3^-
Concentration/M	0.55	0.46	0.028	0.054	0.010	0.010	0.0023

Many other ions are present in much lower concentrations. Estimate the osmotic pressure of seawater due to its ionic constituents.

8.12   The concentration of NaCl in seawater is approximately 0.5 M. In the process of reverse osmosis, seawater is forced through a membrane impermeable to the ions to obtain pure water. The applied pressure has to be larger than the osmotic pressure.

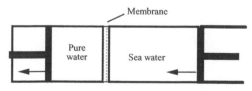

(a) At 25 °C, what is the minimum pressure needed to achieve reverse osmosis? What is the work done in obtaining 1.0 L of pure water from seawater?

(b) If the cost of 1 kWh of electrical power is about $0.15, what would be the energy cost of producing 100 L of water from seawater through reverse osmosis if the process is 50% efficient in using the electrical power to obtain pure water?

(c) Suggest another process to obtain pure water from seawater.

8.13 Consider two cubes of copper of side 1 cm. In each cube, assume that one out of million Cu atoms is Cu^+. Using Coulomb's law, calculate the force between the two cubes if they are placed at a distance of 10 cm.

8.14 Calculate the ionic strength and the activity coefficients of a 0.02 M solution of $CaCl_2$.

8.15 The solubility product of AgCl is 1.77×10^{-10}. Calculate the concentration of Ag^+ ions in equlibrium with solid AgCl.

8.16 Show that for a perfect solution the molar volume of mixing $\Delta V_{mix} = 0$.

8.17 Consider a binary azeotrope. The chemical potentials of a component, say 2, in the gas and the liquid phases can be written as:

$$\mu_{2,g}(T, p, x) = \mu^*_{2,g}(T, p) + RT \ln(\gamma_{2,g} x_2)$$

and

$$\mu_{2,l}(T, p, x) = \mu^*_{2,l}(T, p) + RT \ln(\gamma_{2,l} x_2)$$

in which μ^* is the chemical potential of the pure substance. Note that the mole fraction is the same in the two phases. Use Equation (5.3.7) to derive the relation (8.5.1).

8.18 A regular solution is one for which the excess entropy $\Delta S_E = 0$. Show that this implies that $\ln \gamma_i \propto 1/T$ in which γ_i is the activity coefficient.

9 THERMODYNAMICS OF CHEMICAL TRANSFORMATIONS

9.1 Transformations of Matter

Transformations of matter take place through chemical, nuclear and elementary particle reactions. We shall speak of 'chemical transformations' in this broader sense. Though thermodynamics was founded in our daily experience, its reach is vast, ranging from the most simple changes like the melting of ice to the state of matter during the first few minutes after the big bang, to the radiation that fills the entire universe today.

Let us begin by looking at the transformation that matter undergoes at various temperatures. Box 9.1 gives an overview of the reactions that take place at various temperatures ranging from those during the first few minutes after the big bang [1] to terrestrial and interstellar temperatures. All these transformations or reactions can be associated with enthalpies of reaction and an equilibrium characterized by the vanishing of the corresponding affinities.

Our present knowledge of the universe is based on the radiation emitted by galaxies that we can detect and on the motion of galaxies due to gravitational forces exerted by matter that is visible and invisible. Astrophysical data on observable gravitational effects indicate that only about 4% of the energy density in the universe is in the form of the protons, neutrons and electrons that make up ordinary matter in all the visible galaxies. Of the rest, 74% is in an unknown form spread diffusely throughout the universe; this is called **dark energy**. The remaining 22% is matter in galaxies that is not visible and it is called **dark matter**; its presence is inferred through the gravitational effects it has on visible matter. The universe is also filled with thermal radiation* at a temperature of about 2.73 K (usually called **cosmic microwave background**) and particles called neutrinos, which interact only very weakly with protons, neutrons and electrons.

The small amount of matter which is in the form of stars and galaxies is not in thermodynamic equilibrium. The affinities for the reactions that are currently occurring in the stars are not zero. The nuclear reactions in the stars produce all the known elements from hydrogen [2–4]. Hence, the observed properties, such as the abundance of elements in stars and planets, cannot be described using the theory of chemical equilibrium. A knowledge of the rates of reactions and the history of the star or planet is necessary to understand the abundance of elements.

---

*The precise thermodynamic nature of thermal radiation is discussed in Chapter 12.

Introduction to Modern Thermodynamics   Dilip Kondepudi
© 2008 John Wiley & Sons, Ltd

Box 9.1 Transformation of Matter at Various Temperatures.

Temperature>10^{10}K. This was the temperature during the first few minutes of the universe after the big bang. At this temperature, the thermal motion of the protons and neutrons is so violent that even the strong nuclear forces cannot contain them as nuclei of elements. Electron–positron pairs appear and disappear spontaneously and are in thermal equilibrium with radiation. (The threshold for electron–positron pair production is about 6×10^9 K.)

Temperature range 10^9–10^7 K. At about 10^9 K, nuclei begin to form and nuclear reactions occur in this range. Temperatures as high as 10^9 are reached in stars and supernova, where heavier elements are synthesized from H and He. The *binding energy per nucleon* (proton or neutron) is in the range $(1.0–1.5) \times 10^{-12}$ J $\approx (6.0–9.0) \times 10^6$ eV, which corresponds to $(6.0–9.0) \times 10^8$ kJ mol^{-1}.

Temperature range 10^6–10^4 K. In this range, electrons bind to nuclei to form atoms, but the bonding forces between atoms are not strong enough to form stable molecules. At a temperature of about 1.5×10^5 K, hydrogen atoms begin to ionize. The ionization energy of 13.6 eV corresponds to 1310 kJ mol^{-1}. Heavier atoms require larger energies for complete ionization. To ionize a carbon atom completely, for example, requires 490 eV of energy, which corresponds to 47187 kJ mol^{-1}.* Carbon atoms will be completely dissociated at T $\approx 5 \times 10^6$ K into electrons and nuclei. In this temperature range, matter exists as free electrons and nuclei, a state of matter called *Plasma*.

Temperature range 10–10^4 K. Chemical reactions take place in this range. The chemical bond energies are of the order of 10^2 kJ mol^{-1}. The C—H bond energy is about 414 kJ mol^{-1}. At a temperature of about 5×10^4 K, chemical bonds will begin to break. The intermolecular forces, such as hydrogen bonds, are of the order 10 kJ mol^{-1}. The enthalpy of vaporization of water, which is essentially the breaking of hydrogen bonds, is about 40 kJ mol^{-1}.

*1 eV = 1.6×10^{-19} J = 96.3 kJ mol^{-1}; T = (Energy in J mol^{-1})/R = (Energy in J)/k_B.

When a system reaches thermodynamic equilibrium, however, its history is of no importance. Regardless of the path leading to equilibrium, the state of equilibrium can be described by general laws. In this chapter we shall first look at the nature of chemical reactions and equilibrium, then we study the relation between entropy production and the rates chemical reactions that drive the system to equilibrium.

9.2 Chemical Reaction Rates

In studying chemical reactions and the approach to equilibrium, it is also our purpose to look explicitly at the entropy production while the reactions are in progress. In other words, we would like to obtain explicit expressions for the entropy production d$_i$S/dt in terms of the rates of reactions. The introduction of reaction rates takes us beyond the classical thermodynamics of equilibrium states that was formulated by Gibbs and others.

In general, the laws of thermodynamics cannot specify reaction rates (which depend on many factors, such as the presence of catalysts); but, as we shall see in later chapters, close to thermodynamic equilibrium – called the 'linear regime' – thermodynamic formalism can be used to show that rates are linearly related to the affinities. But the general problem of specifying the rates of chemical reactions has become a subject in itself and it goes by the name of 'chemical kinetics'. Some basic aspects of chemical kinetics will be discussed in this section.

We have already seen that the entropy production due to a chemical reaction may be written in the form (see (4.1.16))

$$\frac{d_i S}{dt} = \frac{A}{T} \frac{d\xi}{dt} \tag{9.2.1}$$

in which ξ is the extent of reaction introduced in Section 2.5 and A is the affinity, expressed in terms of the chemical potentials. The time derivative of ξ is related to the rate of reaction. The precise definition of the rate of reaction is given in Box 9.2. For the following simple reaction*:

$$Cl(g) + H_2(g) \rightleftharpoons HCl(g) + H(g) \tag{9.2.2}$$

the affinity A and the extent of reaction ξ are defined by

$$A = \mu_{Cl} + \mu_{H_2} - \mu_{HCl} - \mu_H \tag{9.2.3}$$

$$d\xi = \frac{dN_{Cl}}{-1} = \frac{dN_{H_2}}{-1} = \frac{dN_{HCl}}{1} = \frac{dN_H}{1} \tag{9.2.4}$$

As explained in Box 9.2, the forward reaction rate is $k_f[Cl][H_2]$, in which the square brackets indicate concentrations and k_f is the forward rate constant, which depends on temperature. Similarly, the reverse reaction rate is $k_r[HCl][H]$. The time derivative of ξ is the *net rate of conversion* of reactants Cl and H_2 to the products HCl and H due to the forward and reverse reactions. Since the reaction rates are generally expressed as functions of concentrations, it is more convenient to define this net rate per unit volume. Accordingly, we define a **reaction velocity** v as

$$v = \frac{d\xi}{Vdt} = k_f[Cl][H_2] - k_r[HCl][H] \tag{9.2.5}$$

Note that this equation follows from (9.2.4) and the definition of the forward and reverse rates. For example, in a homogeneous system, the rate of change of the concentration of Cl is $dN_{Cl}/V\,dt = -k_f[Cl][H_2] + k_r[HCl][H]$. More generally, if R_f and R_r are the forward and reverse reaction rates, we have

---

* For a detailed study of this reaction, see *Science* **273**, 1519 (1996).

Box 9.2 Reaction Rate and Reaction Velocity.

The reaction rate is defined as the number of reactive events per second per unit volume. It is usually expressed as $mol\,L^{-1}\,s^{-1}$. Chemical reactions depend on collisions. In most reactions, only a very small fraction of the collisions result in a chemical reaction. For each reacting species, since the number of collisions per unit volume is proportional to its concentration, the rates are proportional to the product of the concentrations. A **reaction rate** refers to conversion of the reactants to the products or vice versa. Thus, for the reaction

$$Cl(g) + H_2(g) \rightleftharpoons HCl(g) + H(g)$$

the **forward rate** $R_f = k_f[Cl][H_2]$ and the **reverse rate** $R_r = k_r[HCl][H]$. In a reaction, both forward and reverse reactions take place simultaneously. For thermodynamic considerations, we define the **velocity** of a reaction as the **rate of net conversion** of the reactants to products. Thus:

Reaction velocity $v =$ **Forward rate** $-$ **Reverse rate**
$$= k_f[Cl][H_2] - k_r[HCl][H]$$
$$= R_f - R_r$$

In a homogeneous system, the reaction velocity v in terms of the extent of reaction is given by

$$\boxed{v = \frac{d\xi}{V\,dt} = R_f - R_r}$$

in which V is the volume of the system. In practice, monitoring the progress of a reaction by noting the change in some property (such as refractive index or spectral absorption) of the system generally amounts to monitoring the change in the extent of reaction ξ.

$$\boxed{v = \frac{d\xi}{V\,dt} = R_f - R_r} \qquad (9.2.6)$$

The reaction velocity units are $mol\,L^{-1}\,s^{-1}$.

In the above example, the rate of reaction bears a direct relation to the stoichiometry of the reactants, but this is not always true. In general, for a reaction such as

$$2X + Y \rightarrow products, \quad rate = k[X]^a[Y]^b \qquad (9.2.7)$$

in which k is a temperature-dependent **rate constant**, and the exponents a and b are not necessarily integers. The rate is said to be of *order a in [X]* and of *order b in [Y]*. The sum of all the orders of the reactants $a + b$ is called the **order of the reac-**

tion. Reaction rates can take complex forms because they may be the result of many intermediate steps with widely differing rates that depend on the presence of catalysts. If all the intermediate steps are known, then each step is called an **elementary step**. Rates of elementary steps do bear a simple relation to the stoichiometry: the exponents equal the stoichiometric coefficients. If reaction (9.2.7) were an elementary step, for example, then its rate would be $k[X]^2[Y]$.

In many cases, the temperature dependence of the rate constant is given by the **Arrhenius equation**:

$$\boxed{k = k_0\, e^{-E_a/RT}} \tag{9.2.8}$$

Svante Arrhenius (1859–1927) proposed it in 1889 and showed its validity for a large number of reactions [5,6]. The term k_0 is called the **pre-exponential factor** and E_a the **activation energy**. For the forward reaction of (9.2.2), $Cl + H_2 \rightarrow HCl + H$, we have, for example, $k_0 = 7.9 \times 10^{10}\,L\,mol^{-1}\,s^{-1}$ and $E_a = 23\,kJ\,mol^{-1}$. For large variations in temperature, the Arrhenius equation was found to be inaccurate in predicting the variation of the rate constant, though it is quite useful for many reactions.

A more recent theory that is based on statistical mechanics and quantum theory was developed in the 1930s by Wigner, Pelzer, Eyring, Polyani and Evans. According to this theory, the reaction occurs through a **transition state** (see Box 9.3). We shall discuss transition state theory in some detail later in this chapter. The concept of a transition state leads to the following expression for the rate constant:

$$k = \kappa\left(\frac{kT}{h}\right)e^{-(\Delta H^\dagger - T\Delta S^\dagger)/RT} = \kappa\left(\frac{kT}{h}\right)e^{-\Delta G^\dagger/RT} \tag{9.2.9}$$

in which $k_B = 1.381 \times 10^{-23}\,J\,K^{-1}$ is the Boltzmann constant and h is the Planck's constant. The terms $\Delta H^\dagger$ and $\Delta S^\dagger$ are the transition-state enthalpy and entropy respectively, as explained briefly in Box 9.3. The term κ is small, of the order of unity that is characteristic of the reaction. *A catalyst increases the rate of reaction by altering the transition state such that $(\Delta H^\dagger - T\Delta S^\dagger) = \Delta G^\dagger$ decreases.*

RATE EQUATIONS USING THE EXTENT OF REACTIONS

Reaction rates are generally determined empirically. The mechanisms of reactions, which detail all the elementary steps, are usually a result of long and detailed study. Once the reaction rate laws are known, the time variation of the concentration can be obtained by integrating the rate equations, which are coupled differential equations. (Box 9.4 lists elementary first- and second-order reactions.) For example, if we have an elementary reaction of the form

$$X \underset{k_r}{\overset{k_f}{\rightleftharpoons}} 2Y \tag{9.2.10}$$

then the concentrations are governed by the following differential equations:

Box 9.3 Arrhenius Equation and Transition State Theory.

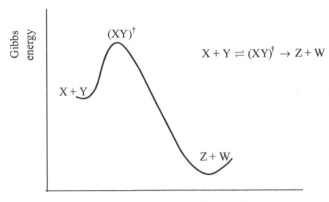

$$X + Y \rightleftharpoons (XY)^\dagger \rightarrow Z + W$$

Reaction coordinate

According to the Arrhenius equation, the rate constant of a chemical reaction is of the form

$$k = k_0 e^{-E_a / RT}$$

The rate constant k has this form because, for the reactants to convert to products, the collisions must have sufficient energy to overcome an energy barrier. As shown in the above figure, the transformation from the reactants to the products is schematically represented with a 'reaction coordinate' and the energy of the molecules undergoing the reaction.

According to the transition state theory, the reactants X and Y reversibly form a **transition state** $(XY)^\dagger$. The transition state then irreversibly transforms to the products. The difference in the enthalpy and entropy between the free molecules X and Y and the transition state are denoted by $\Delta H^\dagger$ and $\Delta S^\dagger$ respectively. The main result of the transition state theory (which is obtained using principles of statistical mechanics) is that the rate constant is of the form

$$k = \kappa (k_B T / h)\exp[-(\Delta H^\dagger - T\Delta S^\dagger) / RT] = \kappa (k_B T / h)\exp[(-\Delta G^\dagger / RT)]$$

in which $k = 1.381 \times 10^{23}\,\mathrm{J\,K^{-1}}$ is the Boltzmann constant and $h = 6.626 \times 10^{-34}\,\mathrm{J\,s}$ is Planck's constant. κ is a term of the order of unity that is characteristic of the reaction.

A catalyst increases the rate of reaction by altering the transition state such that $(\Delta H^\dagger - T\Delta S^\dagger) = \Delta G^\dagger$ *decreases.*

$$-\frac{1}{V}\frac{d\xi}{dt} = \frac{d[X]}{dt} = -k_f[X] + k_r[Y]^2 \tag{9.2.11}$$

$$2\frac{1}{V}\frac{d\xi}{dt} = \frac{d[Y]}{dt} = 2k_f[X] - 2k_r[Y]^2 \tag{9.2.12}$$

Box 9.4 Elementary Reactions.

To obtain an explicit analytic expression for the concentrations of the reactants and products as a function of time, we must solve differential equations such as (9.2.11) and (9.2.12). Generally, this is possible only in the case of simple reactions. For more complex reactions, one can obtain numerical solutions using a computer. Two elementary reactions for which we can obtain explicit expressions for the concentrations as functions of time are given below.

First-order reaction. For a decomposition reaction X → (products), in which the reverse reaction rate is so small that it can be neglected, we have the differential equation

$$\frac{d[X]}{dt} = -k_f[X]$$

It is easy to see that solution of this equation is

$$[X](t) = [X]_0 e^{-k_f t}$$

in which $[X]_0$ is the concentration at time $t = 0$. This is the well-known exponential decay; in a given amount of time, $[X]$ decreases by the same fraction. In particular, the time it takes for any initial value of $[X]$ to decrease by a factor of 1/2 is called the **half-life**. It is usually denoted by $t_{1/2}$. The half-life can be computed by noting that $\exp(-k_f t_{1/2}) = 1/2$, i.e.

$$t_{1/2} = \frac{\ln(2)}{k_f} = \frac{0.6931}{k_f}$$

Second-order reaction. For the elementary reaction 2X → (products), if the reverse reaction can be neglected, the rate equation is

$$\frac{d[X]}{dt} = -2k_f[X]^2$$

The solution is obtained by evaluating

$$\int_{[X]_0}^{[X]} \frac{d[X]}{[X]^2} = -\int_0^t 2k_f dt$$

which give us

$$\frac{1}{[X]} - \frac{1}{[X]_0} = 2k_f t$$

Given k_f and $[X]_0$ at $t = 0$, this expression gives us the value $[X]$ at any time t.

Without loss of generality, we may assume $V = 1$ and simplify the notation. These two equations are not independent. *In fact, there is only one independent variable ξ for every independent reaction.* If $[X]_0$ and $[Y]_0$ are the values of the concentrations at $t = 0$, then by assigning $\xi(0) = 0$ and using $d\xi = -d[X]$ and $2d\xi = d[Y]$ it is easy to see that $[X] = [X]_0 - \xi$ and $[Y] = [Y]_0 + 2\xi$. Substituting these values into (9.2.1) we obtain

$$\frac{d\xi}{dt} = k_f([X]_0 - \xi) - k_r([Y]_0 + 2\xi)^2 \tag{9.2.13}$$

In this equation, the initial concentrations $[X]_0$ and $[Y]_0$ appear explicitly and $\xi(0) = 0$ for all initial concentrations. The solution $\xi(t)$ of such an equation can be used to obtain the rate of entropy production, as will be shown explicitly in Section 9.5. Differential equations such as these, and more complicated systems, can be solved numerically on a computer, e.g. using software such as Mathematica or Maple (sample programs are provided in the appendix). Furthermore, in describing reactions involving solid phases, concentration cannot be used to describe the change in the amount of a solid phase; the extent of reaction ξ, which represents the change in the total amounts of a reactant or product, is a convenient variable for this purpose.

When many reactions are to be considered simultaneously, we will have one ξ for each independent reaction, denoted by ξ_k, and the entire system will be described by a set of coupled differential equations in ξ_k. Only in a few cases can we find analytical solutions to such equations, but they can be readily solved numerically using Mathematica, Maple or other similar software.

REACTION RATES AND ACTIVITIES

Though reaction rates are generally expressed in terms of concentrations, one could equally well express them in terms of activities. In fact, we shall see in the following sections that the connection between affinities and reaction rates can be made more easily if the reaction rates are expressed in terms of activities. For example, for the elementary reaction

$$X + Y \rightleftharpoons 2W \tag{9.2.14}$$

the forward rate R_f and the reverse rate R_r may be written as

$$R_f = k_f a_X a_Y \quad \text{and} \quad R_r = k_r a_W^2 \tag{9.2.15}$$

The rate constants k_f and k_r in (9.2.15) will have units of $mol\,L^{-1}\,s^{-1}$; their numerical values and units differ from those of the rate constants when R_f and R_r are expressed in terms of concentrations (Exercise 9.11).

Experimentally we know that reaction rates do depend on the activities; they are not specified by concentrations alone. For example, at fixed values of temperature and concentrations of the reactants, it is well known that the rates of ionic reactions can be altered by changing the ionic strength of the solution (usually known as the 'salt effect'). This change in the rate is due to a change in the activities. It has become general practice, however, to express the reaction rates in terms of the concentrations and to include the effects of changing activities in the rate constants. Thus, the rate constants are considered functions of the ionic strength when rates are expressed in terms of concentrations. Alternatively, if the rates are expressed in terms of activities, then the rate constant is independent of the ionic strength; a change in rate due to a change in ionic strength would be because activity depends on ionic strength.

9.3 Chemical Equilibrium and the Law of Mass Action

In this section we shall study chemical equilibrium in detail. At equilibrium, the pressure and temperature of all components and phases are equal; the affinities and the corresponding velocities of reactions vanish. For example, for a reaction such as

$$X + Y \rightleftharpoons 2Z \tag{9.3.1}$$

at equilibrium we have

$$A = \mu_X + \mu_Y - 2\mu_Z = 0 \quad \text{and} \quad \frac{d\xi}{dt} = 0 \tag{9.3.2}$$

or

$$\mu_X + \mu_Y = 2\mu_Z \tag{9.3.3}$$

The condition that the 'thermodynamic force', affinity A, equals zero implies that the corresponding 'thermodynamic flow', i.e. the reaction velocity $d\xi/dt$, also equals zero. The condition $A = 0$ means that at equilibrium the 'stoichiometric sum' of the chemical potentials of the reactants and products are equal, as in (9.3.3). It is easy to generalize this result to an arbitrary chemical reaction of the form

$$a_1 A_1 + a_2 A_2 + a_3 A_3 + \ldots + a_n A_n \rightleftharpoons b_1 B_1 + b_2 B_2 + b_3 B_3 + \ldots + b_m B_m \tag{9.3.4}$$

in which the a_k are the stoichiometric coefficients of the reactants A_k and the b_k are the stoichiometric coefficients of the products B_k. The corresponding condition for chemical equilibrium will then be

$$a_1 \mu_{A_1} + a_2 \mu_{A_2} + a_3 \mu_{A_3} + \ldots + a_n \mu_{A_n} = a_1 \mu_{B_1} + a_2 \mu_{B_2} + a_3 \mu_{B_3} + \ldots + a_m \mu_{B_m} \tag{9.3.5}$$

Such equalities of chemical potentials are valid for all reactions: changes of phase, and chemical, nuclear and elementary particle reactions. Just as a difference in temperature drives the flow of heat until the temperatures difference vanishes, a nonzero affinity drives a chemical reaction until the affinity vanishes.

To understand the physical meaning of the mathematical conditions such as (9.3.3) or (9.3.5), we express the chemical potential in terms of experimentally measurable quantities. We have seen in Section 5.3 (Equation (5.3.5)) that the chemical potential in general can be expressed as

$$\mu_k(p,T) = \mu_k^0(T) + RT \ln a_k \qquad (9.3.6)$$

in which a_k is the activity and $\mu_k^0(T_0) = \Delta G_f^0[k]$ is the standard molar Gibbs energy of formation (Box 5.1), the values of which are tabulated. This being a general expression, for gases, liquids and solids we have the following explicit expressions:

- *ideal gas*: $a_k = p_k/p_0$, where p_k is the partial pressure;
- *real gases*: expressions for activity can be derived using (6.2.30), as was shown in Section 6.2;
- *pure solids and liquids*: $a_k \approx 1$;
- *solutions*: $a_k \approx \gamma_k x_k$, where γ_k is the activity coefficient and x_k is the mole fraction.

For ideal solutions, $\gamma_k = 1$. For nonideal solutions, γ_k is obtained by various means, depending on the type of solution. The chemical potential can also be written in terms of the concentrations by appropriately redefining μ_k^0.

We can now use (9.3.6) to express the condition for equilibrium (9.3.3) in terms of the activities, which are experimentally measurable quantities:

$$\mu_X^0(T) + RT \ln(a_{X,eq}) + \mu_Y^0(T) + RT \ln(a_{Y,eq}) = 2[\mu_Z^0(T) + RT \ln(a_{Z,eq})] \qquad (9.3.7)$$

where the equilibrium values of the activities are indicated by the subscript 'eq'.

This equation can be rewritten as

$$\boxed{\frac{a_{Z,eq}^2}{a_{X,eq} a_{Y,eq}} = \exp\left[\frac{\mu_X^0(T) + \mu_Y^0(T) - 2\mu_Z^0(T)}{RT}\right] \equiv K(T)} \qquad (9.3.8)$$

$K(T)$, as defined above, is called the **equilibrium constant**. It is a function only of temperature. That the equilibrium constant as defined above is a function of T only is an important thermodynamic result. It is called the **law of mass action**. $\mu_k^0(T) = \Delta G_f^0[k,T]$ is the standard molar Gibbs energies of formation of compound k at a temperature T. The 'Standard Thermodynamic Properties' table at the end of the book lists this quantity at $T = 298.15\,K$. It is convenient and conventional to define the **Gibbs energy of reaction** ΔG_r as

$$\begin{aligned}
\Delta G_r(T) &= -[\mu_X^0(T) + \mu_Y^0(T) - 2\mu_Z^0(T)] \\
&= 2\Delta G_f^0[Z,T] - \Delta G_f^0[X,T] - \Delta G_f^0[Y,T]
\end{aligned} \qquad (9.3.9)$$

The equilibrium constant is then written as

$$K(T) = \exp(-\Delta G_r^0/RT)$$
$$= \exp[-(\Delta H_r^0 - T\Delta S_r^0)/RT] \qquad (9.3.10)$$

in which ΔG_r^0, ΔH_r^0 and ΔS_r^0 are respectively the standard Gibbs energy, enthalpy and entropy of the reaction at temperature T, though their temperature dependence is usually not explicitly indicated. The activities in (9.3.8) can be written in terms of partial pressures p_k or mole fractions x_k. If (9.3.1) were an ideal-gas reaction, then $a_k = p_k/p_0$. With $p_0 = 1$ bar and p_k measured in bars, the equilibrium constant takes the form

$$\frac{p_{Z,eq}^2}{p_{X,eq}\, p_{Y,eq}} = K_p(T) = \exp(-\Delta G_r^0/RT) \qquad (9.3.11)$$

At a given temperature, regardless of the initial partial pressures, the chemical reaction (9.3.1) will irreversibly proceed towards the state of equilibrium in which the partial pressures will satisfy equation (9.3.11). This is one form of the *law of mass action*. K_p is the *equilibrium constant* expressed in terms of the partial pressures. Since in an ideal gas mixture $p_k = (N_k/V)RT = [k]RT$ (in which R is in the units of bar L mol^{-1} K^{-1}), the law of mass action can also be expressed in terms of the concentrations of the reactants and products:

$$\frac{[Z]_{eq}^2}{[X]_{eq}[Y]_{eq}} = K_c(T) \qquad (9.3.12)$$

in which K_c is the equilibrium constant expressed in terms of the concentrations. In general, for a reaction of the form $aX + bY \rightleftharpoons cZ$, it is easy to obtain the relation $K_c = (RT)^\alpha K_p$, where $\alpha = a + b - c$ (Exercise 9.14). In the particular case of reaction (9.3.1) α happens to be zero.

If one of the reactants were a pure liquid or a solid, then the equilibrium constant will not contain corresponding 'concentration' terms. For example, let us consider the reaction

$$O_2(g) + 2C(s) \rightleftharpoons 2CO(g) \qquad (9.3.13)$$

Since $a_{C(s)} \approx 1$ for the solid phase, the equilibrium constant in this case is written as

$$\frac{a_{CO,eq}^2}{a_{O_2,eq}\, a_{C,eq}^2} = \frac{p_{CO,eq}^2}{p_{O_2,eq}} = K_p(T) \qquad (9.3.14)$$

Equations (9.3.9) and (9.3.10) provide us with means of calculating the equilibrium constant $K(T)$ using the tabulated values of $\Delta G_f^0[k]$. If the activities are expressed in terms of partial pressures, then we have K_p. Some examples are shown in Box 9.5.

Box 9.5 The Equilibrium Constant.

A basic result of equilibrium chemical thermodynamics is that the rate constant $K(T)$ is a function of temperature only. It can be expressed in terms of the standard Gibbs energy of reaction ΔG_r^0 (Equations (9.3.9) and (9.3.10)):

$$K(T)=\exp(-\Delta G_r^0/RT)$$

For a reaction such as $O_2(g) + 2C(s) \rightleftharpoons 2CO(g)$ the equilibrium constant at 298.15 K can be calculated using the tabulated values of standard Gibbs energy of formation ΔG_f^0 at $T = 298.15$ K:

$$\Delta G_r^0 = 2\Delta G_f^0[CO] - 2\Delta G_f^0[C] - \Delta G_f^0[O_2]$$
$$= -2(137.2)\,kJ\,mol^{-1} - 2(0) - (0) = -274.4\,kJ\,mol^{-1}$$

Using this value in the expression $K(T) = \exp(-\Delta G_r^0 / RT)$ we can calculate $K(T)$ at $T = 298.15$ K:

$$K(T) = \exp(-\Delta G_r^0/RT)$$
$$= \exp[274.4 \times 10^3 / (8.314 \times 298.15)]$$
$$= 1.19 \times 10^{48}$$

Similarly, for the reaction $CO(g) + 2H_2(g) \rightleftharpoons CH_3OH(g)$ at $T = 298.15$ K:

$$\Delta G_r^0 = 2\Delta G_f^0[CH_3OH] - \Delta G_f^0[CO] - 2\Delta G_f^0[H_2]$$
$$= -161.96\,kJ\,mol^{-1} - (-137.2\,kJ\,mol^{-1}) - 2(0) = -24.76\,kJ\,mol^{-1}$$

for which the equilibrium constant is

$$K(T) = \exp(-\Delta G_r^0 / RT)$$
$$= \exp[24.76 \times 10^3 / (8.1314 \times 298.15)]$$
$$= 2.18 \times 10^4$$

RELATION BETWEEN THE EQUILIBRIUM CONSTANTS AND THE RATE CONSTANTS

Chemical equilibrium can also be described as a state in which the forward rate of every reaction equals its reverse rate. If the reaction $X + Y \rightleftharpoons 2Z$ is an elementary step, and if we express the reaction rates in terms of the activities, then when the velocity of the reaction is zero we have

$$k_f a_X a_Y = k_r a_Z^2 \qquad (9.3.15)$$

From a theoretical viewpoint, writing the reaction rates in terms of activities rather than concentrations is better because the state of equilibrium is directly related to activities, not concentrations.

Comparing (9.3.15) and the equilibrium constant (9.3.8), we see that

$$K(T) = \frac{a_Z^2}{a_X a_Y} = \frac{k_f}{k_r} \qquad (9.3.16)$$

Thus, the equilibrium constant can also be related to the rate constants k_r and k_f when the rates are expressed in terms of the activities. It must be emphasized that Equation (9.3.8) is valid even if the forward and reverse rates do not have the form shown in (9.3.15); in other words, (9.3.8) is valid whether the reaction X + Y $\rightleftharpoons$ 2Z is an elementary reaction step or not. *The relation between the activities and the equilibrium constant is entirely a consequence of the laws of thermodynamics; it is independent of the kinetic rates of the forward and reverse reactions.*

THE VAN'T HOFF EQUATION

Using (9.3.10), the temperature variation of the equilibrium constant $K(T)$ can be related to the enthalpy of reaction ΔH_r. From (9.3.10) it follows that

$$\frac{d \ln K(T)}{dT} = -\frac{d}{dT} \frac{\Delta G_r}{RT} \qquad (9.3.17)$$

But, according to the Gibbs–Helmholtz equation (5.2.14), the variation of ΔG with temperature is related to ΔH by $(\partial / \partial T)(\Delta G / T) = -\Delta H / T^2$. Using this in the above equation we have

$$\boxed{\frac{d \ln K(T)}{dT} = \frac{\Delta H_r}{RT^2}} \qquad (9.3.18)$$

This relation enables us to deduce how the equilibrium constant $K(T)$ depends on the temperature. It is called the **van't Hoff equation**. In many situations of interest, the heat of reaction ΔH_r changes very little with temperature and may be assumed to be a constant equal to the standard enthalpy of reaction at 298.15 K, which we denote by ΔH_r^0. Thus, we may integrate (9.3.18) and obtain

$$\ln K(T) = \frac{-\Delta H_r^0}{RT} + C \qquad (9.3.19)$$

Experimentally, the equilibrium $K(T)$ constant can be obtained at various temperatures. According to (9.3.19), a plot of $\ln K(T)$ versus $1/T$ should result in a straight line with a slope equal to $-\Delta H_r^0 / R$. This method can be used to obtain the values of ΔH_r^0.

RESPONSE TO PERTURBATION FROM EQUILIBRIUM: THE LE CHATELIER–BRAUN PRINCIPLE

When a system is perturbed from its state of equilibrium, it will relax to a new state of equilibrium. Le Chatelier and Braun noted in 1888 that a simple principle may be used to predict the direction of the response to a perturbation from equilibrium. Le Chatelier stated this principle thus:

Any system in chemical equilibrium undergoes, as a result of a variation in one of the factors governing the equilibrium, a compensating change in a direction such that, had this change occurred alone it would have produced a variation of the factors considered in the *opposite* direction.

To illustrate this principle, let us consider the reaction

$$N_2 + 3H_2 \rightleftharpoons 2NH_3$$

in equilibrium. In this reaction, the total molar amount of all components decreases when the reactants convert to products. If the pressure of this system is suddenly increased, then the system's response will be the production of more NH_3, which decreases the total molar amount and thus the pressure. The compensating change in the system is in a direction *opposite* to that of the perturbation. The new state of equilibrium will contain more NH_3. Similarly, if a reaction is exothermic, if heat is supplied to the system, then the product will be converted to reactants, which has the effect of opposing the increase in temperature. Though this principle has its usefulness, it does not always give unambiguous results. For this reason, a more general approach under the name 'theorems of moderation' has been developed [7]. This approach provides a very precise and accurate description of the response of a system in equilibrium to a perturbation from this state, which is always the evolution to another state of equilibrium.

Le Chatelier's principle only describes the response of a system in thermodynamic equilibrium; it says nothing about the response of a system that is maintained in a nonequilibrium state. Indeed, the response of a nonequilibrium system to small changes in temperature could be extraordinarily complex. This is obviously evident in living organisms, which are nonequilibrium systems. A small change in temperature could denature an enzyme and result in profound changes in the state of an organism. Nonequilibrium systems can be extraordinarily sensitive to small perturbations.

9.4   The Principle of Detailed Balance

There is an important aspect of the state of chemical equilibrium, and the state of thermodynamic equilibrium in general, that must be noted, namely the **principle of detailed balance**.

We noted earlier that, for a given reaction, the state of equilibrium depends only on the stoichiometry of the reaction, not its actual mechanism. For example, in the reaction $X + Y \rightleftharpoons 2Z$ considered above, if the forward and reverse reaction rates were given by

$$R_f = k_f a_X a_Y \quad \text{and} \quad R_r = k_r a_Z^2 \tag{9.4.1}$$

respectively, then result that $a_Z^2/a_X a_Y = K(T)$ at equilibrium can be interpreted as the balance between forward and reverse reactions:

$$R_f = k_f a_X a_Y = R_r = k_r a_Z^2$$

so that

$$\frac{a_Z^2}{a_X a_Y} = K(T) = \frac{k_f}{k_r} \tag{9.4.2}$$

However, the equilibrium relation $a_Z^2/a_X a_Y = K(T)$ was not obtained using any assumption regarding the kinetic mechanism of the reaction. It remains valid even if there was a complex set of intermediate reactions that result in the overall reaction $X + Y \rightleftharpoons 2Z$. This feature could be understood through the **principle of detailed balance**, according to which:

In the state of equilibrium, every transformation is balanced by its exact opposite or reverse.

That the principle of detailed balance implies that $a_Z^2 / a_X a_Y = K(T)$ regardless of the mechanism can be seen through the following example. Assume that the reaction really consists of two steps:

(a) $$X + X \rightleftharpoons W \tag{9.4.3}$$

(b) $$W + Y \rightleftharpoons 2Z + X \tag{9.4.4}$$

which results in the net reaction $X + Y \rightleftharpoons 2Z$. According to the principle of detailed balance, at equilibrium we must have

$$\frac{a_W}{a_X^2} = \frac{k_{fa}}{k_{ra}} \equiv K_a \qquad \frac{a_Z^2 a_X}{a_W a_Y} = \frac{k_{fb}}{k_{rb}} \equiv K_b \tag{9.4.5}$$

in which the subscripts a and b stand for reactions (9.4.3) and (9.4.4) respectively. The thermodynamic equation for equilibrium $a_Z^2 / a_X a_Y = K(T)$ can now be obtained as the product of these two equations:

$$\frac{a_W a_Z^2 a_X}{a_X^2 a_W a_Y} = \frac{a_Z^2}{a_X a_Y} = K_a K_b = K \tag{9.4.6}$$

From this derivation it is clear that this result will be valid for an arbitrary set of intermediate reactions.

The principle of detailed balance is a very general principle, valid for all transformations. It is in fact valid for the exchange of matter and energy between any two volume elements of a system in equilibrium. The amount of matter and energy transferred from volume element X to volume element Y exactly balances the energy and matter transferred from volume element Y to volume element X (see Figure 9.1). The same can be said of the interaction between the volume elements Y and Z and X and Z. One important consequence of this type of balance is that the removal or isolation of one of the volume elements from the system, say Z, does not alter the states of X or Y or

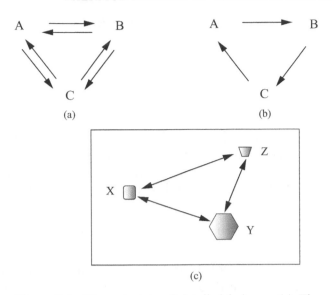

Figure 9.1 The principle of detailed balance. (a) The equilibrium between three interconverting compounds A, B and C is a result of 'detailed balance' between each pair of compounds. (b) Though a conversion from one compound to another as shown can also result in concentrations that remain constant in time, such a state is not the equilibrium state. (c) The principle of detailed balance has a more general validity. The exchange of matter (or energy) between any two regions of a system is balanced in detail; the amount of matter going from X to Y is balanced by exactly the reverse process

the interaction between them. This is another way of saying that there is no long-range correlation between the various volume elements. As we shall see in later chapters, the principle of detailed balance is not valid in nonequilibrium systems that make a transition to organized dissipative structures. Consequently, the removal or isolation of a volume element at one part will alter the state of a volume element located elsewhere. It is then said to have long-range correlations. We can see this clearly if we compare a droplet of water that contains carbon compounds in thermal equilibrium and a living cell that is in an organized state far from thermodynamic equilibrium. Removal of a small part of the water droplet does not change the state of other parts of the droplet, whereas removing a small part of a living cell is likely to have a drastic influence on other parts of the cell.

9.5   Entropy Production due to Chemical Reactions

The formalism of the previous sections can now be used to relate entropy production to reaction rates more explicitly. In Chapter 4 we saw that the entropy production rate due to a chemical reaction is given by

$$\frac{d_i S}{dt} = \frac{A}{T}\frac{d\xi}{dt} \geq 0 \qquad (9.5.1)$$

Our objective is to relate the affinity A and $d\xi/dt$ to the reaction rates, so that the entropy production is written in terms of the reaction rates. In order to do this, let us consider the reaction that we have considered before:

$$X + Y \rightleftharpoons 2Z \qquad (9.5.2)$$

Assuming that this is an elementary step, we have for the forward and reverse rates that

$$R_f = k_f a_X a_Y \quad \text{and} \quad R_r = k_r a_Z^2 \qquad (9.5.3)$$

Since the forward and reverse rates must be equal at equilibrium, we have seen from (9.4.2) that

$$K(T) = \frac{k_f}{k_r} \qquad (9.5.4)$$

The velocity of reaction v, which is simply the difference between the forward and reverse reaction rates, is related to $d\xi/dt$ as shown in (9.2.6). The reaction rates R_f and R_r can themselves be expressed as functions of the extent of reaction ξ, as was shown in Section 9.2:

$$\frac{1}{V}\frac{d\xi}{dt} = [R_f(\xi) - R_r(\xi)] \qquad (9.5.5)$$

To obtain the velocity of reaction as a function of time, this differential equation has to be solved. An example is presented below.

Turning now to the affinity A, we can relate it to the reaction rates in the following manner. By definition, the affinity of the reaction (9.5.2) is

$$\begin{aligned}
A &= \mu_X + \mu_Y - 2\mu_Z \\
&= \mu_X^0(T) + RT\ln(a_X) + \mu_Y^0(T) + RT\ln(a_Y) - 2[\mu_Z^0(T) + RT\ln(a_Z)] \qquad (9.5.6) \\
&= [\mu_X^0(T) + \mu_Y^0(T) - 2\mu_Z^0(T)] + RT\ln(a_X) + RT\ln(a_Y) - 2RT\ln(a_Z)
\end{aligned}$$

Since $[\mu_X^0(T) + \mu_Y^0(T) - 2\mu_Z^0(T)] = -\Delta G_r^0 = RT\ln K(T)$, the above equation can be written as

$$A = RT\ln K(T) + RT\ln\left(\frac{a_X a_Y}{a_Z^2}\right) \qquad (9.5.7)$$

This is an alternative way of writing the affinity. At equilibrium, $A = 0$. To relate A to the reaction rates, we use (9.5.4) and combine the two logarithm terms:

$$A = RT \ln \frac{k_\text{f}}{k_\text{r}} + RT \ln\left(\frac{a_\text{X} a_\text{Y}}{a_\text{Z}^2}\right) = RT \ln\left(\frac{k_\text{f} a_\text{X} a_\text{Y}}{k_\text{r} a_\text{Z}^2}\right) \tag{9.5.8}$$

This leads us to the relations we are seeking if we use the expressions in (9.5.3) to write this expression in terms of the reaction rates:

$$\boxed{A = RT \ln\left(\frac{R_\text{f}}{R_\text{r}}\right)} \tag{9.5.9}$$

Clearly, this equation is valid for any elementary step because the rates of elementary steps are directly related to the stoichiometry. Now we can substitute (9.5.5) and (9.5.9) in the expression for the entropy production rate (9.5.1) and obtain

$$\boxed{\frac{1}{V}\frac{d_i S}{dt} = \frac{1}{V}\frac{A}{T}\frac{d\xi}{dt} = R(R_\text{f} - R_\text{r})\ln(R_\text{f}/R_\text{r}) \geq 0} \tag{9.5.10}$$

which is an expression that relates *entropy production rate per unit volume* to the reaction rates. (Note that R is the gas constant.) Also, as required by the Second Law, the right-hand side of this equation is positive, whether $R_\text{f} > R_\text{r}$ or $R_\text{f} < R_\text{r}$. Another point to note is that in (9.5.10) the forward and reverse rates R_f and R_r can be expressed in terms of concentrations, partial pressures or any other convenient variables of the reactants; the reaction rates need not be expressed only in terms of activities, as in (9.5.3).

The above equation can be generalized to several simultaneous reactions, each indexed by the subscript k. The rate of total entropy production per unit volume is the sum of the rates at which entropy is produced in each reaction:

$$\boxed{\frac{1}{V}\frac{d_i S}{dt} = \frac{1}{V}\sum_k \frac{A_k}{T}\frac{d\xi_k}{dt} = R\sum_k (R_{kf} - R_{kr})\ln(R_{kf}/R_{kr})} \tag{9.5.11}$$

in which R_{kf} and R_{kr} are the forward and reverse reaction rates of the kth reaction. This expression is useful for computing the entropy production in terms of the reaction rates, but *it is valid only for elementary steps whose reaction rates are specified by the stoichiometry*. This is not a serious limitation, however, because every reaction is ultimately the result of many elementary steps. If the detailed mechanism of a reaction is known, then an expression for the entropy production can be written for any chemical reaction.

AN EXAMPLE

As an example of entropy production due to an irreversible chemical reaction, consider the simple reaction:

$$L \rightleftharpoons D \tag{9.5.12}$$

which is the interconversion or 'racemization' of molecules with mirror-image structures. Molecules that are not identical to their mirror image are said to be *chiral* and
the two mirror-image forms are called *enantiomers*. Let [L] and [D] be the concentrations of the enantiomers of a chiral molecule. If at time $t = 0$ the concentrations are
[L] = L_0 and [D] = D_0, and $\xi(0) = 0$, then we have the following relations:

$$\frac{d[L]}{-1} = \frac{d[D]}{+1} = \frac{d\xi}{V} \tag{9.5.13}$$

$$[L] = L_0 - (\xi/V) \quad [D] = D_0 + (\xi/V) \tag{9.5.14}$$

Relations (9.5.14) are obtained by integrating (9.5.13) and using the initial conditions. For notational convenience we shall assume $V = 1$. At the end of the calculation we can reintroduce the V factor. Racemization can be an elementary first-order
reaction for which the forward and reverse reactions are

$$R_f = k[L] = k(L_0 - \xi) \quad R_r = k[D] = k(D_0 + \xi) \tag{9.5.15}$$

Note that the rate constants for the forward and reverse reactions are the same due
to symmetry: L must convert to D with the same rate constant as D to L. Also, from
(9.5.15) and (9.5.9) one can see that the affinity is a function of the state variable ξ
for a given set of initial concentrations.

To obtain the entropy production as an explicit function of time, we must obtain
R_f and R_r as functions of time. This can be done by solving the differential equation
defining the velocity of this reaction:

$$\frac{d\xi}{dt} = R_f - R_r = k(L_0 - \xi) - k(D_0 + \xi)$$

i.e.

$$\frac{d\xi}{dt} = 2k\left(\frac{L_0 - D_0}{2} - \xi\right) \tag{9.5.16}$$

This first-order differential equation can be easily solved by defining $x = [(L_0 - D_0)
/2] - \xi$ so that the equation reduces to $dx/dt = -2kx$. The solution is

$$\xi(t) = \frac{L_0 - D_0}{2}(1 - e^{-2kt}) \tag{9.5.17}$$

With this expression for $\xi(t)$, the rates (9.5.15) can be written as explicit functions
of time:

$$R_f = \frac{k(L_0 + D_0)}{2} + \frac{k(L_0 - D_0)}{2}e^{-2kt} \tag{9.5.18}$$

$$R_r = \frac{k(L_0 + D_0)}{2} - \frac{k(L_0 - D_0)}{2}e^{-2kt} \qquad (9.5.19)$$

With (9.5.18) and (9.5.19), we can now also write the rate of entropy production (9.5.10) as an explicit function of time:

$$\frac{1}{V}\frac{d_iS}{dt} = R(R_f - R_r)\ln(R_f/R_r)$$

$$\frac{1}{V}\frac{d_iS}{dt} = R[k(L_0 - D_0)e^{-2kt}]\ln\left[\frac{(L_0 + D_0) + (L_0 - D_0)e^{-2kt}}{(L_0 + D_0) - (L_0 - D_0)e^{-2kt}}\right] \qquad (9.5.20)$$

As $t \to \infty$, the system reaches equilibrium, at which

$$\xi_{eq} = \frac{L_0 - D_0}{2} \quad \text{and} \quad [L]_{eq} = [D]_{eq} = \frac{L_0 + D_0}{2} \qquad (9.5.21)$$

The volume term can be reintroduced by replacing ξ_{eq} by ξ_{eq}/V.

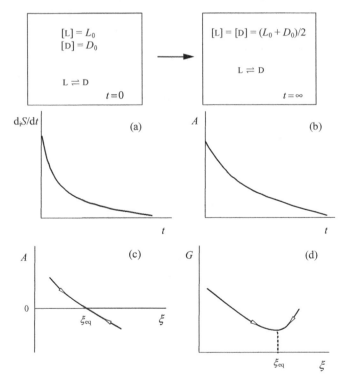

Figure 9.2   Racemization of enantiomers as an example of a chemical reaction. The associated entropy production, the time variation of A are shown in (a) and (b). State functions A and G as functions of ξ are shown in (c) and (d)

In Chapter 5 (see (5.1.12)) we noted the relation between affinity A and the Gibbs energy G: $A = -(\partial G/\partial \xi)_{p,T}$. Both A and G are functions of state, which can be expressed as functions of ξ and the initial molar amounts of reactants and products. As ξ approaches its equilibrium value ξ_{eq}, the Gibbs energy reaches its minimum value and the affinity A goes to zero, as shown in Figure 9.2.

The entropy production for more complex reactions can be obtained numerically using computers. Mathematica code for the above example is given in Appendix 9.1. The student is encouraged to alter this code to develop codes for more complex reactions.

9.6   Elementary Theory of Chemical Reaction Rates

The rates of chemical reactions depend on several factors. In previous sections we discussed the dependence of rates on concentrations and introduced the Arrhenius and transition-state rate constants. According to the Arrhenius theory, the rate constant has the form $k_0 \exp(-E_a/RT)$, whereas transition-state theory gives a rate constant of the form $k_0 \exp(-\Delta G^\dagger/RT)$. In this section we will introduce the reader to the theoretical basis that leads to these expressions.

ARRHENIUS THEORY OF RATES

When the molecular nature of compounds became established, theories of rates of chemical reactions began to emerge. That molecules were in incessant and rapid chaotic motion was established by the kinetic theory of gases. A natural consequence was a view that chemical reactions were a consequence of molecular collisions. When molecules collide somehow, an atomic rearrangement occurs and the products are formed. But not every collision between reacting molecules results in the formation of products. In fact, quantitative estimates indicated that only a very small fraction of the collisions were 'reactive collisions'. This naturally raised the question as to why only certain collisions between reactant molecules resulted in the formation of products.

One of the first successful theories of reaction rates is due to the Swedish chemist Svante Arrhenius (1859–1927), but it is noted that others, especially van't Hoff, also made important contributions to this theory [5,6]. The success of Arrhenius theory is mainly in explaining the temperature dependence of reaction rates. To explain why only a small fraction of molecular collisions resulted in reactions, the concept of 'activation energy' was introduced. This is the idea that the colliding molecules must have sufficient energy to activate the reaction, i.e. the breaking of bonds and formation of new bonds. Only a small fraction of molecules have the required activation energy (an idea proposed by the German chemist L. Pfundler). To compute the probability that the collision has the required activation energy, the Boltzmann principle is taken as a guide. We recall that, according to the Maxwell–Boltzmann probability distribution, the probability that a molecule has energy E is proportional to $\exp(-E/RT)$. Using this principle, it could be argued that if a certain activation energy E_a is required in a collision between reacting molecules to generate the product, this will happen with a probability proportional to $\exp(-E_a/RT)$. Thus, the reaction rate must be proportional to a factor $\exp(-E_a/RT)$; that is, of all the collisions that occur in a unit volume in unit time, a fraction $k_0\exp(-E_a/RT)$ will be reactive collisions. Thus, the Arrhenius rate constant

$$k = k_0 e^{-E_a/RT}$$

where k_0 is called the **pre-exponential factor**.

The next step is to compute the number of collisions that occur in unit time in a unit volume. For gases, this can be done using the Maxwell–Boltzmann distribution (Section 1.6). Let us consider the reaction A + B → (products). Let r_A and r_B be the radii of the A and B molecules respectively. For small molecules, radii can be estimated from tabulated bond lengths. Figure 9.3 shows the path of a molecule of A as it undergoes collisions with molecules in its path. An observer located on the molecule A will observe a stream of molecules; collisions with molecules of B occur when the distance between the center of A and the center of a streaming B is equal to or less than the sum $r_A + r_B$. Consider a cylinder of radius $r_A + r_B$ with the path of the molecule A as its axis. Molecule A will collide with all B molecules in such a cylinder. From the viewpoint of an observer on A, molecules will be streaming at an average speed v_r, which is equal to the average relative velocity between A and B molecules. Thus, in unit time, on the average, a molecule of A will collide with

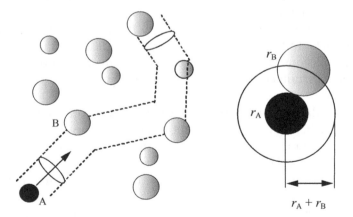

Figure 9.3   The elementary bimolecular reaction A + B → (products) is a result of collisions between the molecules of A and B. Approximating the molecule's shape to be spherical, we assume the radii of molecules of A and B are r_A and r_B respectively. As shown, on average, in unit time, a molecule of A (filled sphere) will collide with all molecules in the cylinder of cross-section $\pi(r_A + r_B)^2$ and length v_r

all B molecule in the volume $\pi(r_A + r_B)^2 v_r$. The term $\pi(r_A + r_B)^2$ is called the **collision cross-section**. If n_B is the moles of B molecules per unit volume, then a single A molecule will collide with $\pi(r_A + r_B)^2 v_r n_B N_A$ vrmolecules of B (N_A is the Avogadro number). Thus, the average total number of collisions between A and B molecules per unit volume per unit time, called the **collision frequency** z_{AB}, equals

$$z_{AB} = \pi(r_A + r_B)^2 v_r n_B n_A N_A^2 \tag{9.6.1}$$

in which n_A is the moles of A molecules per unit volume. Using the Maxwell–Boltzmann distribution it can be shown that the average relative velocity between A and B molecules is given by

$$v_r = \left(\frac{8k_B T}{\pi\mu}\right)^{1/2} \quad \text{in which} \quad \mu = \frac{m_A m_B}{m_A + m_B} \tag{9.6.2}$$

where m_A and m_B are the masses of molecules A and B respectively. The factor μ is called the **reduced mass**. Of all the collisions, only a fraction $\exp(-E_a/RT)$ are reactive collisions that result in the formation of products. Hence, the reaction rate (number of reactive collisions per unit time per unit volume) equals

$$
\begin{aligned}
\text{Rate} &= z_{AB}\exp(-E_a/RT)\\
&= N_A^2 \pi(r_A + r_B)^2 \left(\frac{8k_B T}{\pi\mu}\right)^{1/2} n_B n_A \exp(-E_a/RT)
\end{aligned} \tag{9.6.3}
$$

If the rate measured in moles per unit volume per unit time, then

$$\text{Rate/mol}\,m^{-3}\,s^{-1} = N_A (r_A + r_B)^2 \left(\frac{8\pi k_B T}{\mu}\right)^{1/2} \exp(-E_a/RT)\, n_A n_B \tag{9.6.4}$$

in which all quantities are in SI units. If the unit of length is taken to be decimeters, then the concentrations will be molarities [A] and [B] and the rate will be in the units of moles per liter. We can now identify the pre-exponential factor k_0 in the Arrhenius rate:

$$\boxed{k_0 = N_A (r_A + r_B)^2 \left(\frac{8\pi k_B T}{\mu}\right)^{1/2}} \tag{9.6.5}$$

At $T = 300\,K$, the value of k_0 is of the order of $10^8\,m^3\,mol^{-1}\,s^{-1} = 10^{11}\,L\,mol^{-1}\,s^{-1}$. The changes in k_0 due to changes in T are small compared with the corresponding changes in the exponential factor in the rate constant.

A number of other expressions were also suggested to explain the temperature dependence of reaction rates, as Laidler notes [6], but they found less and less support as experimental data were gathered. In addition, the expression suggested by Arrhenius had a strong theoretical basis that the other expressions lacked.

TRANSITION STATE THEORY

Transition state theory postulates the existence of a transition state which is in equilibrium with the reactants. The transition state has an unstable mode which results in a conversion to products. For a reaction $X + Y \rightarrow Z + W$, the mechanism is

$$X + Y \underset{k_{1r}}{\overset{k_{1f}}{\rightleftharpoons}} (XY)^{\dagger} \xrightarrow{k_{2f}} Z + W \tag{9.6.6}$$

The rate of product formation is $k_{2f}[XY^{\dagger}]$. The assumption that the transition state is in equilibrium with the reactants implies that

$$[(XY)^{\dagger}]/[X][Y] = k_{1f}/k_{1r} = K_1(T) = \exp(-\Delta G^{\dagger}/RT) \tag{9.6.7}$$

in which $K_1(T)$ is the equilibrium constant and $\Delta G^{\dagger}$ is the Gibbs energy of reaction. The reaction rate can be written as:

$$\text{Rate} = k_{2f} K_1(T)[X][Y] \tag{9.6.8}$$

The use of statistical thermodynamics to calculate the rate constant gives $k_{2f} = \kappa(k_B T/h)$, in which κ is a term of the order of unity. Therefore, the rate constant has the form

$$k = \kappa \left(\frac{k_B T}{h} \right) \exp(-\Delta G^{\dagger}/RT) \tag{9.6.9}$$

In contrast to Arrhenius theory, the transition state theory has a thermodynamic basis and predicts the existence of a transition state. The pre-exponential factor it predicts is proportional to T; this is in contrast to Arrhenius theory, which predicts a $T^{1/2}$ dependence. Transition state theory predicts a change in the rate of reaction due to a factor that might change $\Delta G^{\dagger}$. One such factor is the effect of solvents. In solutions, if the reactants are ionic then it is observed that the reaction rate depends on the dielectric constant of the solvent. This effect, called the 'solvent effect', can be explained by noting that a change in dielectric constant changes the value of $\Delta G^{\dagger}$. In general, transition state theory gives more insight into the nature of a chemical reaction than Arrhenius theory does and it is widely used.

9.7   Coupled Reactions and Flow Reactors

In the previous sections we discussed some basic aspects of chemical kinetics. In this supplementary section we shall look at more complex reactions. Box 9.4 summarizes the main aspects of first- and second-order reactions. In these cases, the reverse reactions were not considered, but in many cases the reverse reaction cannot be ignored. We shall now consider some examples below.

ZERO-ORDER REACTIONS

In certain conditions, the rate of a reaction can be essentially independent of the concentration of the initial reactants. For example, a reaction such as

$$X \rightarrow Y \tag{9.7.1}$$

could have a rate of product formation such as

$$\frac{d[Y]}{dt} = k \tag{9.7.2}$$

in which k is a constant. Such a reaction may be said to be of zero order in the reactant X. Such a rate law clearly indicates that the reaction mechanism that controls the conversion of X to Y depends on the concentration of another compound and that increasing the amount of X does not increase the rate of conversion to Y. For example, let us assume the formation of Y depends on X binding to a catalyst C to form a complex CX and that the complex CX converts to C and Y:

$$X + C \xrightarrow{k_{1f}} CX \xrightarrow{k_{2f}} Y + C \tag{9.7.3}$$

The rate of product formation depends on the amount of the complex CX. If all the catalyst is bound to the reactant X, then increasing the amount of X does not increase the rate of product formation. $[C]_T$ is the total amount of C, then the rate of reaction when the complex is saturated is

$$\frac{d[Y]}{dt} = k_{2f}[C]_T \tag{9.7.4}$$

Such rates can be observed in reactions catalyzed by solid catalysts and in enzymes. The solution to Equation (9.7.2) is $[Y] = [Y]_0 + kt$.

REVERSIBLE FIRST-ORDER REACTION

In general, if the forward and the reverse rate constants are not equal, then rate equations are of the form

$$A \underset{k_r}{\overset{k_f}{\rightleftharpoons}} B \tag{9.7.5}$$

$$\frac{d[A]}{dt} = -R_f + R_r = -k_f[A] + k_r[B] \tag{9.7.6}$$

in which R_f and R_r are the forward and reverse reactions rates. Let $[A]_0$ and $[B]_0$ be the initial concentrations. In the above reaction, the total concentration, which we shall denote as $T = [A] + [B] = [A]_0 + [B]_0$ remains constant. Hence, the above rate equation can be rewritten as

$$\frac{d[A]}{dt} = -k_f[A] + k_r(T - [A]) = -(k_f + k_r)[A] + k_r T \qquad (9.7.7)$$

The solution to this equation is

$$[A] = \frac{k_r}{k_f + k_r}T + \left([A]_0 - \frac{k_r}{k_f + k_r}T\right)e^{-(k_f + k_r)t} \qquad (9.7.8)$$

The reaction could also be described in terms of the extent of reaction ξ, as was done in Section 9.5 for the racemization reaction $L \rightleftharpoons D$. This is left as an exercise for the student.

CONSECUTIVE FIRST-ORDER REACTIONS

Sequential conversion of compounds is quite common in natural and industrial processes. Sequential transformations in nature more often than not are cyclical. Let us consider a very simple example: conversion of A to B to C, in which we ignore the reverse reactions.

$$A \xrightarrow{k_{1f}} B \xrightarrow{k_{2f}} C \qquad (9.7.9)$$

We assume that all the rates are first order and that, at $t = 0$, $[A] = [A]_0$, $[B] = 0$ and $[C] = 0$. The kinetic equations for the concentrations of A, B and C are

$$R_{1f} = k_{1f}[A] \quad R_{2f} = k_{2f}[B] \qquad (9.7.10)$$

$$\frac{d[A]}{dt} = -R_{1f} = -k_{1f}[A] \qquad (9.7.11)$$

$$\frac{d[B]}{dt} = R_{1f} - R_{2f} = k_{1f}[A] - k_{2f}[B] \qquad (9.7.12)$$

$$\frac{d[C]}{dt} = R_{2f} = k_{2f}[B] \qquad (9.7.13)$$

This set of coupled equation can be solved analytically. The solution to (9.7.11) is

$$[A] = [A]_0 \exp(-k_{1f}t) \qquad (9.7.14)$$

This solution can be substituted into the equation for [B], (9.7.12); we get

$$\frac{d[B]}{dt} + k_{2f}[B] = k_{1f}[A]_0 \exp(-k_{1f}t) \qquad (9.7.15)$$

This is a first-order differential equation of the form $(dX/dt) + cX = f(t)$ in which c is a constant and $f(t)$ is a function of time. The general solution to such an equation is

$$X(t) = X(0)e^{-ct} + e^{-ct} \int_0^t e^{ct'} f(t') dt' \qquad (9.7.16)$$

Using this general solution we can write the solution to (9.7.15) and show that

$$[B] = \frac{k_{1f}[A]_0}{k_{2f} - k_{1f}} (e^{-k_{1f}t} - e^{-k_{2f}t}) \qquad (9.7.17)$$

in which we have used $[B]_0 = 0$. If the initial concentration $[C]_0 = 0$, then the total amount $[A] + [B] + [C] = [A]_0$. Using this relation, one can obtain the time variation of $[C]$:

$$\begin{aligned}
[C] &= [A]_0 - [A] - [B] \\
&= [A]_0 \left[1 - e^{-k_{1f}t} - \frac{k_{1f}}{k_{2f} - k_{1f}} (e^{-k_{1f}t} - e^{-k_{2f}t}) \right]
\end{aligned} \qquad (9.7.18)$$

Alternatively, the rate equations can be written and solved in terms of the extents of reaction ξ_1 and ξ_2 of the two reactions (9.7.9). For simplicity, and without loss of generality, we shall assume the system volume $V = 1$ so that concentrations and ξ values could be related without explicitly including V. The extent of reaction for the two reactions and the corresponding changes in concentrations are related by

$$\frac{d[A_1]}{-1} = \frac{d[B_1]}{+1} = d\xi_1 \qquad \frac{d[B_2]}{-1} = \frac{d[C_2]}{+1} = d\xi_2 \qquad (9.7.19)$$

in which the subscripts indicate changes due to the first and second reactions in (9.7.9). The total change in the concentration of A is only due to the reaction A $\rightarrow$ B and that of C is only due to B $\rightarrow$ C, i.e.

$$d[A] = d[A_1] = -d\xi_1 \quad \text{or} \quad [A] = [A]_0 - \xi_1 \qquad (9.7.20)$$

and

$$d[C] = d[C_2] = +d\xi_2 \quad \text{or} \quad [C] = [C]_0 + \xi_2 \qquad (9.7.21)$$

where we have assumed $\xi = 0$ at $t = 0$ (the subscript 0 indicates values at $t = 0$). Since the change in the intermediate $[B]$ is due to both reactions, we write:

$$d[B] = d[B_1] + d[B_2] = d\xi_1 - d\xi_2 \quad \text{or} \quad [B] = [B]_0 + \xi_1 - \xi_2 \qquad (9.7.22)$$

The velocities of the two reactions are

$$\frac{d\xi_1}{dt} = R_{1f} - R_{1r} = k_{1f}[A] = k_{1f}([A]_0 - \xi_1) \qquad (9.7.23)$$

$$\frac{d\xi_2}{dt} = R_{2f} - R_{2r} = k_{2f}[B] = k_{2f}([B]_0 + \xi_1 - \xi_2) \qquad (9.7.24)$$

These two first-order linear differential equations could be solved using the methods outlined above. By substituting the solutions $\xi_1(t)$ and $\xi_2(t)$ into (9.7.20)–(9.7.22), the time variation of concentrations [A], [B] and [C] can be obtained (Exercise 9.20). Describing the kinetics of reactions using extents of reaction has some notable aspects:

- Each extent of reaction is an independent variable and the number of independent variables in a set of reactions is equal to the number of extents of reaction. The time variations of all reacting species are expressed in terms of these independent variables.
- The initial values of all reactants appear explicitly in the equations and the initial values of all extents of reaction may be assumed to be zero.
- The rate of entropy production is expressed in terms of the velocities $d\xi_k/dt$ and the chemical potentials of the reacting species.

THE STEADY-STATE ASSUMPTION

In many chemical reactions, the concentration of an intermediate compound or complex may be approximated to be constant. Take, for example, the following **Michaelis–Menten mechanism**, which describes enzyme reactions:

$$E + S \underset{k_{1r}}{\overset{k_{1f}}{\rightleftarrows}} ES \xrightarrow{k_{2f}} P + E \qquad (9.7.25)$$

Enzyme E complexes with the substrate S to form the complex ES, which in turn transforms to product P and the enzyme. The complexation of E and S to form ES occurs very rapidly and reversibly. In contrast, the conversion of ES to P and E happens relatively slowly. The rapidity of the reaction $E + S \rightleftharpoons ES$ keeps the concentration of ES essentially a constant close to its equilibrium value; any decrease in [ES] due to product formation is quickly compensated by the production of ES. Hence, we can assume that [ES] is in a *steady state*, i.e. its time derivative is zero. Taking the two steps of the reaction (9.7.19), the steady state assumption can be expressed as

$$\frac{d[ES]}{dt} = k_{1f}[E][S] - k_{1r}[ES] - k_{2f}[ES] = 0 \qquad (9.7.26)$$

In the above reaction, the total concentration of enzyme [E_0] in the free and complex form is a constant:

$$[E] + [ES] = [E_0] \qquad (9.7.27)$$

Combining (9.7.26) and (9.7.27) we can write $k_{1f}([E_0] - [ES])[S] - k_{1r}[ES] - k_{2f}[ES] = 0$. From this, it follows that

$$[ES] = \frac{k_{1f}[E_0][S]}{k_{1f}[S] + (k_{1r} + k_{2f})} \qquad (9.7.28)$$

The rate of formation of the product P is $k_{2f}[ES]$ and is usually written in the following form:

$$R = \frac{d[P]}{dt} = k_{2f}[ES] = \frac{k_{2f}[E_0][S]}{[S] + (k_{1r} + k_{2r})/k_{1f}} = \frac{R_{max}[S]}{[S] + K_m} \qquad (9.7.29)$$

in which $R_{max} = k_{2f}[E_0]$ is the maximum rate of product formation and $K_m = (k_{1r} + k_{2r})/k_{1f}$. It can be seen from (9.7.29) that the rate at which the P is generated has the following properties:

- when $[S] \ll K_m$, the rate is proportional to $[S]$;
- when $[S] \gg K_m$, the rate reaches its maximum value and is independent of $[S]$;
- when $[S] = K_m$, that rate reaches half the maximum value.

FLOW REACTORS

Many industrial chemical reactions take place in a flow reactor into which reactants flow and from which products are removed. The kinetic equations for such systems must consider the inflow and outflow. To see how the kinetic equations are written for a flow reactor, let us consider the following reaction, which we assume requires a catalyst:

$$A \xrightarrow{k_{1f}} B \xrightarrow{k_{2f}} C \qquad (9.7.30)$$

We assume that the reaction takes place in a solution. The solution containing A flows into the reactor (Figure 9.4) of volume V. In the reactor, activated by a catalyst, the conversion A to B and C takes place. The fluid in the reactor is rapidly stirred so that we may assume that it is homogeneous. The outflow is a solution containing B, C and unconverted A. If the objective is to produce B and C, then the reaction should be rapid enough so that very little A is in the outflow. We consider a flow rate of f liters per second of a solution of concentration $[A]_{in}$ in mol L^{-1}. The amount of A flowing into the reactor per second equals $[A]_{in}f$. Hence, the rate at which the concentration of A increases due to the inflow into the reactor of volume V is $[A]_{in}f/V$. Similarly, the rate of decrease in concentrations of A, B and C due to the outflow are $[A]f/V$, $[B]f/V$ and $[C]f/V$ respectively. The term f/V has units of s^{-1}. Its inverse, $V/f \equiv \tau$, is called the **residence time** (because it roughly corresponds to the time the flowing fluid resides in the reactor before it flows out). Taking the flow into consideration, the kinetic equations for the reactor can be written as

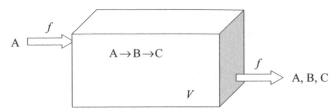

Figure 9.4 A flow reactor into which fluid containing A flows. Owing to a catalyst in the reactor, conversion of A → B → C take place in the reactor. The outflow consists of unconverted A and the products B and C. The amount of fluid flowing into the reactor per unit time is f. Inflow rate equals the outflow rate at steady state

$$\frac{d[A]}{dt} = [A]_{in}(f/V) - k_{1f}[A] - [A](f/V) \qquad (9.7.31)$$

$$\frac{d[B]}{dt} = k_{1f}[A] - k_{2f}[B] - [B](f/V) \qquad (9.7.32)$$

$$\frac{d[C]}{dt} = k_{2f}[B] - [C](f/V) \qquad (9.7.33)$$

This set of linear coupled equations can be solved for steady states by setting $d[A]/dt = d[B]/dt = d[C]/dt = 0$. If, initially, the reactor contains no A, B or C, then the flow will result in an initial increase in the concentration of the three reactants and then the reactor will approach a steady state in which the concentrations are constant. The steady states, which we identify with a subscript 's', are easily calculated:

$$[A]_s = \frac{[A]_{in}(f/V)}{k_{1f} + (f/V)} \qquad (9.7.34)$$

$$[B]_s = \frac{k_{1f}[A]_s}{k_{2f} + (f/V)} \qquad (9.7.35)$$

$$[C]_s = \frac{k_{2f}[B]_s}{f/V} \qquad (9.7.36)$$

If the rate constants k_{1f} and k_{2f} are large compared with f/V, then the steady-state concentrations $[A]_s$ and $[B]_s$ will be small and $[C]_s$ will be large. This corresponds to almost complete conversion of A into product C, which will flow out of the reactor. On the other hand, if the flow rate is high, then the conversion in the reactor will only be partial. Because they are coupled linear equations, (9.7.31)–(9.7.33) can also be solved analytically; generally, however, chemical kinetics leads to coupled non-linear equations, which cannot be solved analytically. They can, of course, be investigated numerically.

The above simple example illustrates how kinetic equations for a reactor can be written. Generalizing it to reactions more complex than (9.7.30) is straightforward. The purpose of some reactors is to combust fuel and generate heat. At the steady state, heat is generated at a constant rate. If the enthalpies of the reactions are known, then at a steady state, the rate at which heat is generated in the reactor can be calculated.

Appendix 9.1 Mathematica Codes

In Mathematica, numerical solutions to the rate equation can be obtained using the 'NDSolve' command. Examples of the use of this command in solving simple rate equations are given below. The results can be plotted using the 'Plot' command. Numerical output can be exported to graphing software using the 'Export' command.

CODE A: LINEAR KINETICS X → PRODUCTS

```
(* Linear Kinetics *)
k=0.12;
Soln1=NDSolve[{X'[t]== -k*X[t],  X[0]==2.0},X,{t,0,10}]

{{X→InterpolatingFunction[{{0.,10.}},<>]}}
```

The above output indicates that the solution has been generated as an interpolating function.

The solution can be plotted using the following command. Here, '/.Soln1' specifies that the values of $X[t]$ are to be calculated using the interpolation function generated by Soln1.

```
Plot[Evaluate[X[t]/.Soln1],{t,0,10}]
```

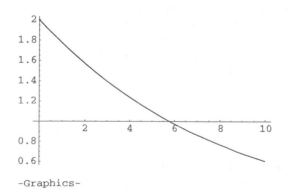

```
-Graphics-
```

To write output files for spreadsheets use the 'Export' command and the file format 'List'. For more detail see the Mathematica help file for the 'Export' command. In the command below, the output filename is data1.txt. This file can be read by most spreadsheets and graphing software.

The command '$X[t]$/.Soln1' specifies that $X[t]$ is to be evaluated using Soln1 defined above. TableForm outputs data in a convenient form.

```
Export["data1.txt",
    Table[{t,X[t] /.Soln1}, {t,1,10}] //TableForm, "List"]
```

```
data1.txt
```

To obtain a table of t and $X[t]$, the following command can be used:

```
Table[{t,X[t] /.Soln1},{t,1,5}] //TableFrom
```

```
    1        1.77384
    2        1.57326
    3        1.39536
    4        1.23757
    5        1.09763
```

CODE B: MATHEMATICA CODE FOR THE REACTION X + 2Y ⇌ 2Z

In writing codes for kinetic equations, we shall define the forward and reverse rates, R_f and R_r respectively, and use these in the rate equations. Thus, we avoid typing the same expression many times.

```
(* Reaction X+2Y ⇌ 2Z *)

kf=0.5;kr=0.05;
Rf:=kf*X[t]*(Y[t]^2); Rr:=kr*Z[t]^2;

Soln2=NDSolve[{ X'[t]== -Rf+Rr,
               Y'[t]== 2*(-Rf+Rr),
               Z'[t]== 2*(Rf-Rr),
               X[0]==2.0,Y[0]==3.0,Z[0]==0.0},
               {X,Y,Z},{t,0,3}]
```

```
{{X → InterpolatingFunction[{{0.,3.}},<>],
  Y → InterpolatingFunction[{{0.,3.}},<>],
  Z → InterpolatingFunction[{{0.,3.}},<>]}}
```

The above output indicates that the solution as an interpolating function has been generated. The solution can be plotted using the following command:

```
Plot[Evaluate[{X[t],Y[t],Z[t]}/.Soln2],{t,0,3}]
```

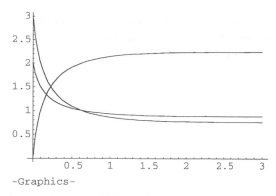

-Graphics-

As shown in Code A, the data could be written to an output file that graphing software can read the Export command.

CODE C: MATHEMATICA CODE FOR RACEMIZATION REACTION L $\rightleftharpoons$ D AND CONSEQUENT ENTROPY PRODUCTION

```
(* Racemization Kinetics:L ⇌ D *)

kf=1.0;kr=1.0;
Rf:=kf*XL[t];  Rr:=kr*XD[t];

Soln3=NDSolve[{ XL'[t]== -Rf+Rr,
               XD'[t]== Rf-Rr,
           XL[0]==2.0,XD[0]==0.001},
           {XL,XD},{t,0,3}]

{{XL → InterpolatingFunction[{{0.,3.}},<>],
  XD → InterpolatingFunction[{{0.,3.}},<>]}}
```

The output indicates an interpolating function has been generated. As before, the solution can be plotted.

```
Plot[Evaluate[XL[t],XD[t]}/.Soln3],{t,0,3}]
```

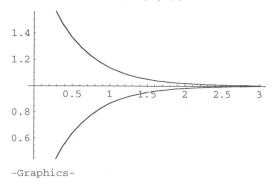

-Graphics-

The rate of entropy production can be obtained from the interpolating functions and expression (9.5.20). Note: in Mathematica, Log is ln.

```
(*Calculation of entropy production "sigma"*)
R=8.1314; sigma=R*(Rf-Rr)*Log[Rf/Rr];
Plot[Evaluate[sigma/.Soln3],{t,0,0.5}]
```

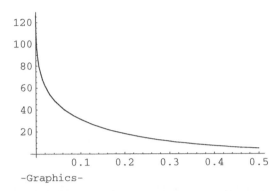

-Graphics-

References

1. Weinberg, S., *The First Three Minutes*. 1980, New York: Bantam.
2. Taylor, R.J., *The Origin of the Chemical Elements*. 1975, London: Wykeham Publications.
3. Norman, E.B., Stellar alchemy: the origin of the chemical elements. *J. Chem. Ed.*, **71** (1994) 813–820.
4. Clayton, D.D., *Principles of Stellar Evolution and Nucleosynthesis*. 1983, Chicago: University of Chicago Press.
5. Laidler, K.J., *The World of Physical Chemistry*, 1993, Oxford: Oxford University Press.
6. Laidler, K.J., The development of Arrhenius equation. *J. Chem. Ed.*, **61** (1984) 494–498.
7. Prigogine, I. and Defay, R., *Chemical Thermodynamics*, fourth edition. 1967, London: Longmans.

Examples

Example 9.1 At a temperature T, the average energy $h\nu$ of a thermal photon is roughly equal to kT. As discussed in Chapter 2, at high temperatures electron-positron pairs will be spontaneously produced when the energy of photons is larger than rest energy $2mc^2$ of an electron positron pair (where m is the mass of the electron). Calculate temperature at which electron-positron pair production occurs.
Solution For pair production:

$$h\nu = k_B T = 2mc^2 = (2 \times 9.10 \times 10^{-31}\,\text{kg})(3.0 \times 10^8\,\text{m s}^{-1})^2 = 1.64 \times 10^{-13}\,\text{J}$$

Hence, the corresponding $T = (1.64 \times 10^{-13}\,\text{J})/(1.38 \times 10^{-23}\,\text{J K}^{-1}) = 1.19 \times 10^{10}\,\text{K}$.

Example 9.2 Consider a second-order reaction $2X \rightarrow$ (products) whose rate equation is $d[X] / dt = -2k_f[X]^2 = -k[X]^2$ in which we set $k = 2k_f$. (a) Show that the half-life $t_{1/2}$ for this reaction depends on the initial value of $[X]$ and is equal to $1/([X]_0 k)$. (b) Assume that $k = 2.3 \times 10^{-1} M^{-1} s^{-1}$ and obtain the value of $[X]$ at a time $t = 60.0 s$ if the initial concentration $[X]_0 = 0.50 M$.

Solution (a) As shown in Box 9.4, the solution to the rate equation is

$$\frac{1}{[X]} - \frac{1}{[X]_0} = kt$$

Multiplying both sides by $[X]_0$ we obtain

$$\frac{[X]_0}{[X]} = 1 + [X]_0 kt$$

Since at $t = t_{1/2}$ the ratio $[X]_0/[X] = 2$, we must have $[X]_0 k t_{1/2} = 1$ or $t_{1/2} = 1/([X]_0 k)$.

(b) If the initial concentration $[X]_0 = 0.50 M$, $k = 0.23 M^{-1} s^{-1}$ and $t = 60.0 s$ we have:

$$\frac{1}{[X]} - \frac{1}{0.50} = 0.23 \times 60 \, mol \, L^{-1}$$

Solving for $[X]$ we get $[X] = 0.063 \, mol \, L^{-1}$.

Example 9.3 For the water dissociation reaction $H_2O \rightleftharpoons OH^- + H^+$ the enthalpy of reaction $\Delta H_r = 55.84 \, kJ \, mol^{-1}$. At 25 °C, the value of the equilibrium constant $K = 1.00 \times 10^{-14}$ and pH is 7.0. At 50 °C, what will the pH be?

Solution Given $K(T)$ at one temperature T_1, its value at another temperature T_2 can be obtained using the van't Hoff equation (9.3.19):

$$\ln K(T_1) - \ln K(T_2) = \frac{-\Delta H_1}{R} \left(\frac{1}{T_1} - \frac{1}{T_2} \right)$$

For this example, we have

$$\ln K = \ln(1.0 \times 10^{-14}) - \frac{55.84 \times 10^3}{8.314} \left(\frac{1}{323} - \frac{1}{298} \right) = -30.49$$

Hence, K at 50 °C is equal to $\exp(-30.49) = 5.73 \times 10^{-14}$. Since the equilibrium constant $K = [OH^-][H^+]$ and because $[OH^-] = [H^+]$, we have

$$pH = -\log[H^+] = -\log[\sqrt{K}] = -\frac{1}{2}\log[5.73 \times 10^{-14}] = 6.62$$

Exercises

9.1   When the average kinetic energy of molecules is nearly equal to the bond energy, molecular collisions will begin to break the bonds. (a) The C—H bond energy is about $414 \, kJ \, mol^{-1}$. At what temperature will the C—H bonds in methane begin to break? The average binding energy per nucleon (neutron or proton) is in the range $(6.0–9.0) \times 10^6 \, eV$ or $(6.0–9.0) \times 10^8 \, kJ \, mol^{-1}$. At what temperature do you expect nuclear reactions to take place?

9.2   For the reaction $Cl + H_2 \rightarrow HCl + H$, the activation energy $E_a = 23.0 \, kJ \, mol^{-1}$ and $k_0 = 7.9 \times 10^{10} \, mol^{-1} \, L \, s^{-1}$. What is the value of rate constant at $T = 300.0 \, K$? If $[Cl] = 1.5 \times 10^{-4} \, mol \, L^{-1}$ and $[H_2] = 1.0 \times 10^{-5} \, mol \, L^{-1}$, what is the forward reaction rate at $T = 350.0 \, K$?

9.3   For the decomposition of urea in an acidic medium, the following data were obtained for rate constants at various temperatures:

Temperature/°C	50	55	60	65	70
Rate constant $k/10^{-8} \, s^{-1}$	2.29	4.63	9.52	18.7	37.2

(a) Using an Arrhenius plot, obtain the activation energy E_a and the pre-exponential factor k_0.
(b) Apply the transition state theory to the same data, plot $\ln(k/T)$ versus $1/T$ and obtain $\Delta H^{\dagger}$ and $\Delta S^{\dagger}$ of the transition state.

9.4   Consider the dimerization of the triphenylmethyl radical Ph_3C, which can be written as the reaction

$$A \rightleftharpoons 2B$$

The forward and reverse rate constants for this reaction at $300 \, K$ are found to be $k_f = 0.406 \, s^{-1}$ and $k_r = 3.83 \times 10^2 \, mol^{-1} \, L \, s^{-1}$. Assume that this reaction is an elementary step. At $t = 0$ the initial concentration of A and B are $[A]_0 = 0.041 \, M$ and $[B]_0 = 0.015 \, M$.

(a) What is the velocity of the reaction at $t = 0$?
(b) If ξ_{eq} is the extent of reaction at equilibrium ($\xi = 0$ at $t = 0$), write the equilibrium concentrations of A and B in terms of $[A]_0$, $[B]_0$ and ξ_{eq}.
(c) Use (b) to obtain the value of ξ_{eq} by solving the appropriate quadratic equation (you may use Maple) and obtain the equilibrium concentrations of [A] and [B].

9.5   (a) Write the rate equations for the concentrations of X, Y and Z in the following reaction:

$$X + Y \rightleftharpoons 2Z$$

(b) Write the rate equation for the extent of reaction ξ.

(c) When the system reaches thermal equilibrium, $\xi = \xi_{eq}$. If $[X]_0$, $[Y]_0$ and $[Z]_0$ are the initial concentrations, write the equilibrium concentrations in terms of the initial concentrations and ξ_{eq}.

9.6 Radioactive decay is a first-order reaction. If N is the number of radioactive nuclei at any time, then $dN/dt = -kN$. ^{14}C is radioactive with a half-life of 5730 yrs. What is the value of k? For this process, do you expect k to change with temperature?

9.7 If $d[A]/dt = -k[A]^\alpha$, show that the half-life is

$$t_{1/2} = \frac{2^{\alpha-1} - 1}{(\alpha - 1)k[A]_0^{\alpha-1}}$$

9.8 Find an analytical solution to the reversible reaction $[L] \underset{k_r}{\overset{k_f}{\rightleftharpoons}} [D]$, in which L and D are enantiomers. Enantiomeric excess (EE) is defined as

$$EE \equiv \frac{|[L] - [D]|}{[L] + [D]}$$

If the initial EE = 1.0, how long does it take for it to reach 0.5? (Amino acid racemization is used in dating of biological samples.)

9.9 (a) For the bimolecular reaction $A + B \overset{k_f}{\longrightarrow} P$ the rate equation is

$$\frac{d[A]}{dt} = -k_f[A][B]$$

Show that

$$\frac{1}{[B]_0 - [A]_0} \ln\left(\frac{[A][B]_0}{[B][A]_0}\right) = -k_f t$$

(b) Write the above rate equation in terms of the extent of reaction ξ and solve it.

9.10 The chirping rate of crickets depends on temperature. When the chirping rate is plotted against $1/T$ it is observed to follow the Arrhenius law (see K.J. Laidler, *J. Chem. Ed.*, **49** (1972) 343). How would you explain this observation?

9.11 Consider the reaction $X + Y \rightleftharpoons 2Z$ in the gas phase. Write the reaction rates in terms of the concentrations $[X]$, $[Y]$ and $[Z]$ as well as in terms of the activities. Find the relation between the rate constants in the two ways of writing the reaction rates.

9.12 When atmospheric CO_2 dissolves in water it produces carbonic acid H_2CO_3 (which causes natural rain to be slightly acidic). At 25.0°C the equilibrium constant K_a for the reaction $H_2CO_3 \rightleftharpoons HCO_3^- + H^+$ is specified by $pK_a = 6.63$. The enthalpy of this reaction $\Delta H_r = 7.66 \, \text{kJ} \, \text{mol}^{-1}$. Calculate the pH at 25°C and at 35°C. (Use Henry's law to obtain $[H_2CO_3]$.)

9.13 Equilibrium constants can vary over an extraordinary range, as the following examples demonstrate. Obtain the equilibrium constants for the following reactions at $T = 298.15 \, \text{K}$, using the tables for $\mu(p_0, T_0) = \Delta G_f^0$:
(a) $2NO_2(g) \rightleftharpoons N_2O_4 (g)$
(b) $2CO(g) + O_2(g) \rightleftharpoons 2CO_2(g)$
(c) $N_2(g) + O_2(g) \rightleftharpoons 2NO(g)$

9.14 (a) For a reaction of the form $aX + bY \rightleftharpoons cZ$, show that the equilibrium constants K_c and K_p are related by $K_c = (RT)^\alpha K_p$ where $\alpha = a + b - c$.
(b) Using the definition of enthalpy $H = U + pV$, show that the van't Hoff equation for a gas-phase reaction can also be written as

$$\frac{d \ln K_c}{dT} = \frac{\Delta U_r}{RT^2}$$

in which K_c is the equilibrium constant expressed in terms of concentrations.

9.15 Ammonia may be produced through the reaction of $N_2(g)$ with $H_2(g)$:

$$N_2(g) + 3H_2(g) \rightleftharpoons 2NH_3(g)$$

(a) Calculate the equilibrium constant of this reaction at 25°C using the thermodynamic tables.
(b) Assuming that there is no significant change in the enthalpy of reaction ΔH_r, use the van't Hoff equation to obtain the approximate ΔG_r and the equilibrium constant at 400°C.

9.16 2-Butene is a gas that has two isomeric forms, cis and trans. For the reaction:

$$cis\text{-}2\text{-butene} \rightleftharpoons trans\text{-}2\text{-butene} \quad \Delta G_r^0 = -2.41 \, \text{kJ} \, \text{mol}^{-1}$$

calculate the equilibrium constant at $T = 298.15 \, \text{K}$. If the total amount of butene is 2.5 mol, then, assuming ideal gas behavior, determine the number of moles of each isomer.

9.17 Determine if the introduction of a catalyst will alter the affinity of a reaction or not.

9.18   For the reaction $A \underset{k_r}{\overset{k_f}{\rightleftharpoons}} B$, write the equation for the velocity of reaction $d\xi/dt$ in terms of the initial values $[A_0]$ and $[B_0]$.

9.19   For the reaction $X + 2Y \rightleftharpoons 2Z$, write explicitly the expression for the entropy production in terms of the rates and as a function of ξ.

9.20   As shown in Section 9.7, for the reaction $A \xrightarrow{k_{1f}} B \xrightarrow{k_{2f}} C$ the extents of reaction obey the equations

$$\frac{d\xi_1}{dt} = R_{1f} - R_{1r} = k_{1f}[A] = k_{1f}([A]_0 - \xi_1)$$

$$\frac{d\xi_2}{dt} = R_{2f} - R_{2r} = k_{2f}[B] = k_{2f}([B]_0 + \xi_1 - \xi_2)$$

Solve these equations with initial conditions $\xi_1 = \xi_2 = 0$ at $t = 0$. Assume $[A] = [A]_0$, $[B] = 0$ and $[C] = 0$ and show that

$$[C] = [A]_0 \left[1 - e^{-k_{1f}t} - \frac{k_{1f}}{k_{2f} - k_{1f}} (e^{-k_{1f}t} - e^{-k_{2f}t}) \right]$$

9.21   Write the complete set of rate equations for all the species in the Michaelis–Menten reaction mechanism:

$$E + S \underset{k_{1r}}{\overset{k_{1f}}{\rightleftharpoons}} ES \xrightarrow{k_{2f}} P + E$$

Write Mathematica/Maple code to solve them numerically with the following numerical values for the rate constants and initial values (assuming all quantities are in appropriate units): $k_{1f} = 1.0 \times 10^2$, $k_{1r} = 5.0 \times 10^3$, $k_{2f} = 2.0 \times 10^3$; and at $t = 0$, $[E] = 3.0 \times 10^{-4}$, $[S] = 2 \times 10^{-2}$, $[ES] = 0$, $[P] = 0$. Using the numerical solutions, check the validity of the steady-state assumption.

9.22   Calculate k_0 for the reaction between H_2 and O_2 at $T = 298\,K$ using the bond lengths 74 pm for H—H and 121 pm for O—O.

10 FIELDS AND INTERNAL DEGREES OF FREEDOM

The Many Faces of Chemical Potential

The concept of chemical potential is very general, applicable to almost any transformation of matter as long as there is a well-defined temperature. We have already seen how the condition for thermodynamic equilibrium for chemical reactions leads to the law of mass action. We shall now see how particles in a gravitational or electric field, electrochemical reactions, and transport of matter through diffusion can all be viewed as 'chemical transformations' with associated chemical potential and affinity.

10.1 Chemical Potential in a Field

The formalism for the chemical potential presented in Chapter 9 can be extended to electrochemical reactions and to systems in a field, such as a gravitational field. When a field is present, the energy due to a field must be included in the expression for a change in energy. As a result, the energy of a constituent depends on it location.

We start with a simple system: the transport of chemical species which carry electrical charge from a position where the potential is ϕ_1 to a position where the potential is ϕ_2. For simplicity, we shall assume that our system consists of two parts, each with a well-defined potential, while the system as a whole is closed (see Figure 10.1). The situation is as if the system consists of two phases and transport of particles dN_k is a 'chemical reaction'. For the corresponding extent of reaction $d\xi_k$ we have

$$-dN_{1k} = dN_{2k} = d\xi_k \qquad (10.1.1)$$

in which dN_{1k} and dN_{2k} are the changes in the molar amount in each part. The change in energy due to the transport of the ions is given by

$$dU = TdS - pdV + F\phi_1 \sum_k z_k dN_{1k} + F\phi_2 \sum_k z_k dN_{2k} + \sum_k \mu_{1k} dN_{1k} + \sum_k \mu_{2k} dN_{2k} \qquad (10.1.2)$$

in which z_k is the charge of ion k and F is the Faraday constant (the product of the electronic charge e and the Avogadro number N_A: $F = eN_A = 9.6485 \times 10^4\,\mathrm{C\,mol^{-1}}$). Using (10.1.1), the change in the entropy dS can now be written as

Introduction to Modern Thermodynamics   Dilip Kondepudi
© 2008 John Wiley & Sons, Ltd

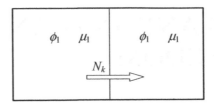

Figure 10.1 A simple situation illustrating the thermodynamics of a system in the presence of an electric field. We consider two compartments, one with associated potential ϕ_1 and the other ϕ_2. It is as if there are two phases; ions will be transported from one to the other until the electrochemical potentials are equal

$$T\,\mathrm{d}S = \mathrm{d}U + p\,\mathrm{d}V - \sum_k [(F\phi_1 z_k + \mu_{1k}) - (F\phi_2 z_k + \mu_{2k})]\mathrm{d}\xi_k \qquad (10.1.3)$$

Thus, we see that the introduction of a potential ϕ associated with a field is equivalent to adding a term to the chemical potential. This makes it possible to extend the definition of the chemical potential to include the field. Thus, the **electrochemical potential** $\tilde{\mu}$, which was introduced by Guggenheim [1] in 1929 is defined as

$$\boxed{\tilde{\mu}_k = \mu_k + Fz_k\phi} \qquad (10.1.4)$$

Clearly, this formalism can be extended to any field to which a potential may be associated. If ψ is the potential associated with the field, then the energy of interaction *per mole* of the component k may be written in the form $\tau_k\psi$. For the electric field $\tau_k = Fz_k$; for the gravitational field, $\tau_k = M_k$, where M_k is the molar mass. The corresponding chemical potential which includes the potential is

$$\boxed{\tilde{\mu}_k = \mu_k + \tau_k\psi} \qquad (10.1.5)$$

The affinity $\tilde{A}_k$ for electrochemical reactions can be written just as was done for other chemical reactions:

$$\tilde{A}_k = \tilde{\mu}_{1k} - \tilde{\mu}_{2k} = [(F\phi_1 z_k + \mu_{1k}) - (F\phi_2 z_k + \mu_{2k})] \qquad (10.1.6)$$

The increase in the entropy due to transfer of charged particles from one potential to another can now be written as

$$\mathrm{d}_i S = \sum_k \frac{\tilde{A}_k}{T}\mathrm{d}\xi_k \qquad (10.1.7)$$

At equilibrium:

$$\tilde{A}_k = 0 \quad \text{or} \quad \mu_{1k} - \mu_{2k} = -z_k F(\phi_1 - \phi_2) \tag{10.1.8}$$

The basic equations of equilibrium electrochemistry follow from (10.1.8).

Because electrical forces are very strong, in ionic solutions the electrical field produced by even small changes in charge density results in very strong forces between the ions. Consequently, in most cases the concentrations of positive and negative ions are such that net charge density is virtually zero, i.e. **electroneutrality** is maintained to a high degree. In a typical electrochemical cell, most of the potential difference applied to the electrodes appears in the vicinity of the electrodes and only a small fraction of the total potential difference occurs across the bulk of the solution. The solution is electrically neutral to an excellent approximation. As a result, an applied electric field does not separate positive and negative charges and so does not create an appreciable concentration gradient.

When we consider the much weaker gravitational field, however, an external field can produce a concentration gradient. As noted above, for a gravitational field, the coupling constant τ_k is the molar mass M_k. For a gas in a uniform gravitational field, for example, $\psi = gh$, where g is the strength of the field and h is the height; from (10.1.8) we see that

$$\mu_k(h) = \mu_k(0) - M_k gh \tag{10.1.9}$$

For an ideal-gas mixture, using $\mu_k(h) = \mu_k^0(T) + RT\ln[p_k(h)/p_0]$ in the above equation we obtain the well-known **barometric formula**:

$$\boxed{p_k(h) = p_k(0)e^{-M_k gh/RT}} \tag{10.1.10}$$

Note how this formula is derived assuming that the temperature T is uniform, i.e. the system is assumed to be in thermal equilibrium. The temperature of the Earth's atmosphere is not uniform; in fact, as shown in Figure 10.2, it varies between $-60\,^{\circ}\text{C}$ and $+20\,^{\circ}\text{C}$ in the troposphere and stratosphere, the two layers in which almost all of the atmospheric gases reside.

ENTROPY PRODUCTION IN CONTINUOUS SYSTEM

In considering thermodynamic systems in a field, we often have to consider continuous variation of the thermodynamic fields. In this case, $\tilde{\mu}$ is a function of position and entropy has to be expressed in terms of entropy density $s(\mathbf{r})$, i.e. entropy per unit volume, which depends on position $\mathbf{r}$. For simplicity, let us consider a one-dimensional system, i.e. a system in which the entropy and all other variables, such as μ, change only along one direction, say x (Figure 10.3). Let $s(x)$ be the entropy density per unit length. We shall assume that the temperature is constant throughout the system. Then the entropy in a small volume element between x and $x + \delta$ is equal to $s(x)\delta$. An expression for affinity in this small volume element can be written as

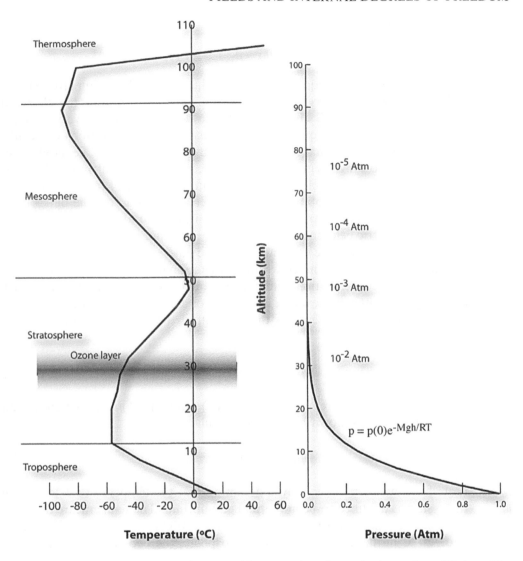

Figure 10.2 The actual state of the Earth's atmosphere is not in thermal equilibrium. The temperature varies with height as shown. At thermal equilibrium, the concept of a chemical potential that includes a field leads to the well-known barometric formula $p(h) = p(0)e^{-Mgh/RT}$

$$\tilde{A} = \tilde{\mu}(x) - \tilde{\mu}(x + \delta) = \tilde{\mu}(x) - \left(\tilde{\mu}(x) + \frac{\partial \tilde{\mu}}{\partial x} \delta \right) = -\frac{\partial \tilde{\mu}}{\partial x} \delta \qquad (10.1.11)$$

The velocity of the reaction $d\xi_k / dt$ for this elemental volume is the flow of particles of component k, i.e. the particle current of k. We shall denote this particle current of k by J_{Nk}. Then by writing expression (10.1.7) for this elemental volume we obtain

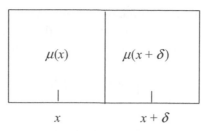

Figure 10.3 An expression for the entropy production in a continuous system can be obtained by considering two adjacent cell separated by a small distance δ. The entropy in the region between x and $x + \delta$ is equal to $s(x)\delta$. The affinity, which is the difference in the chemical potential is given by (10.1.11)

$$\frac{d_i(s(x)\delta)}{dt} = \sum_k \frac{\tilde{A}_k}{T}\frac{d\xi_k}{dt} = -\sum_k \frac{1}{T}\left(\frac{\partial \tilde{\mu}_k}{\partial x}\right)\delta\frac{d\xi_k}{dt} \tag{10.1.12}$$

Simplifying this expression, and using the definition $J_{Nk} = d\xi_k / dt$, the following expression for *entropy production per unit length* due to particle flow is obtained:

$$\boxed{\frac{d_i s(x)}{dt} = -\sum_k \frac{1}{T}\left(\frac{\partial \tilde{\mu}_k}{\partial x}\right)J_{Nk}} \tag{10.1.13}$$

ENTROPY PRODUCTION DUE TO ELECTRICAL CONDUCTION AND OHM'S LAW

To understand the physical meaning of expression (10.1.13), let us consider the flow of electrons in a conductor. In a conductor in which the electron density and temperature are uniform, the chemical potential of the electron μ_e (which is a function of the electron density and T) is constant. Therefore, the derivative of the electrochemical potential is

$$\frac{\partial \tilde{\mu}_e}{\partial x} = \frac{\partial}{\partial x}(\mu_e - Fe\phi) = -\frac{\partial}{\partial x}(Fe\phi) \tag{10.1.14}$$

Since the electric field $E = -\partial\phi / \partial x$ and the conventional electric current $I = -eFJ_e$, using (10.1.14) in expression (10.1.13) we obtain the following expression for the entropy production:

$$\frac{d_i s}{dt} = eF\left(\frac{\partial\phi}{\partial x}\right)\frac{J_e}{T} = \frac{EI}{T} \tag{10.1.15}$$

Since the electric field is the change of potential per unit length, it follows that the integral of E over the entire length L of the system is the potential difference V across the entire system. The total entropy production from $x = 0$ to $x = L$ is

$$\frac{\mathrm{d}S}{\mathrm{d}t} = \int_0^L \left(\frac{\mathrm{d}_i s}{\mathrm{d}t}\right) \mathrm{d}x = \int_0^L \frac{EI}{T}\, \mathrm{d}x = \frac{VI}{T} \tag{10.1.16}$$

Now it is well known that the product VI, of potential difference and the current, is the heat generated per unit time, called the **ohmic heat**. The flow of an electric current through a resistor is a dissipative process that converts electrical energy into heat. For this reason we may write $VI = \mathrm{d}Q / \mathrm{d}t$. Thus, for a flow of electric current, we have

$$\boxed{\frac{\mathrm{d}_i S}{\mathrm{d}t} = \frac{VI}{T} = \frac{1}{T}\frac{\mathrm{d}Q}{\mathrm{d}t}} \tag{10.1.17}$$

This shows that the entropy production is equal to the dissipative heat divided by the temperature.

We noted in Chapter 3 that the entropy production due to each irreversible process is a product of a thermodynamic force and the flow which it drives (see (3.4.7)). In the above case, the flow is the electric current; the corresponding force is the term V/T. Now it is generally true that, when a system is close to thermodynamic equilibrium, the flow is proportional to the force. Hence, based on thermodynamic reasoning, we arrive at the conclusion

$$I = L_e \frac{V}{T} \tag{10.1.18}$$

in which L_e is a constant of proportionality for the electron current. L_e is called the **linear phenomenological coefficient**. Relations such as (10.1.18) are the basis of linear nonequilibrium thermodynamics, which we shall consider in more detail in Chapter 11. We see at once that this corresponds to the familiar **Ohm's law**, $V = IR$, where R is the resistance, if we identify

$$L_e = \frac{T}{R} \tag{10.1.19}$$

This is an elementary example of how the expression for entropy production can be used to obtain linear relations between thermodynamic forces and flows, which often turn out to be empirically discovered laws such as Ohm's law. In Section 10.3 we shall see that a similar consideration of entropy production due to diffusion leads to another empirically discovered law called Fick's law of diffusion. *Modern thermodynamics enables us to incorporate many such phenomenological laws into one unified formalism.*

10.2   Membranes and Electrochemical Cells

MEMBRANE POTENTIALS

Just as equilibrium with a semi-permeable membrane resulted in a difference in pressure (the osmotic pressure) between the two sides of the membrane, equilibrium of ions across a membrane that is permeable to one ion but not another results in an electric potential difference. As an example, consider a membrane separating two solutions of KCl of *unequal* concentrations (Figure 10.4). We assume that the membrane is permeable to K^+ ions but is impermeable to the larger Cl^- ions. Since the concentrations of the K^+ ions on the two sides of the membrane are unequal, K^+ ions will begin to flow to the region of lower concentration from the region of higher concentration. Such a flow of positive charge, without a counterbalancing flow of negative charge, will cause a build up in a potential difference that will oppose the flow. Equilibrium is reached when the electrochemical potentials of K^+ on the two sides become equal, at which point the flow will stop. We shall denote the two sides with superscripts α and β. Then the equilibrium of the K^+ ion is established when

$$\tilde{\mu}_{K^+}^{\alpha} = \tilde{\mu}_{K^+}^{\beta} \tag{10.2.1}$$

Since the electrochemical potential of an ion k is $\tilde{\mu}_k = \mu_k + z_k F\phi = \mu_k^0 + RT\ln a_k + z_k F\phi$, in which a_k is the activity and z_k the ion number (which is $+1$ for K^+), the above equation can be written as

$$\mu_{K^+}^0 + RT\ln a_{K^+}^{\alpha} + F\phi^{\alpha} = \mu_{K^+}^0 + RT\ln a_{K^+}^{\beta} + F\phi^{\beta} \tag{10.2.2}$$

From this equation it follows that the potential difference, i.e. the **membrane potential** $\phi^{\alpha} - \phi^{\beta}$ across the membrane, can now be written as

$$\phi^{\alpha} - \phi^{\beta} = \frac{RT}{F}\ln\left(\frac{a_{K^+}^{\beta}}{a_{K^+}^{\alpha}}\right) \tag{10.2.3}$$

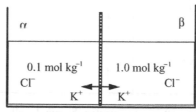

Figure 10.4   A membrane potential is generated when a membrane permeable to K^+ but not to Cl^- separates two solutions of KCl of unequal concentrations. In this case, the flow of the permeable K^+ ions is counterbalanced by the membrane potential

In electrochemistry, the concentrations are generally measured using the molality scale, as was discussed in Chapter 8. In the simplest approximation, the activities may be replaced by molalities m_{K^+}, i.e. the activity coefficients are assumed to be unity. Hence, one may estimate the membrane potential with the formula $\phi^\alpha - \phi^\beta = (RT/F)\ln(m_{K^+}^\beta/m_{K^+}^\alpha)$.

ELECTROCHEMICAL AFFINITY AND ELECTROMOTIVE FORCE

In an electrochemical cell, the reactions at the electrodes that transfer electrons can generate an electromotive force (EMF). An electrochemical cell generally has different phases that separate the two electrodes (Figure 10.5). By considering entropy production due to the overall reaction and the electric current flowing through the system we can derive a relationship between the activities and the EMF. In an electrochemical cell, the reactions at the two electrodes can be generally written as:

$$X + ne^- \rightarrow X_{red} \quad \text{'reduction'} \tag{10.2.4}$$

$$Y \rightarrow Y_{ox} + ne^- \quad \text{'oxidation'} \tag{10.2.5}$$

each is called a **half-reaction**; the overall reaction is

$$X + Y \rightarrow X_{red} + Y_{ox} \tag{10.2.6}$$

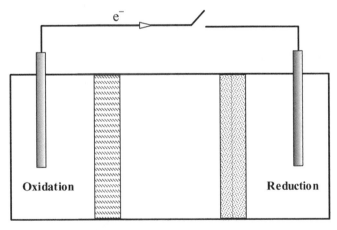

Figure 10.5 An electrochemical cell consisting of many phases that generate an EMF due to half-reactions at the electrodes. The electrode reactions are as given in (10.2.4) and (10.2.5). Upon closing the circuit, chemical reactions occurring within the cell will generate an EMF that will drive a current. Cells such as this are represented by a cell diagram denoting the various phases and junctions. In a cell diagram, the reduction reaction is on the right:

Electrode|Y|. . .|. . .|. . .|X|Electrode

For example, the half-reactions

$$Cu^{2+} + 2e^- \rightarrow Cu(s)$$
$$Zn(s) \rightarrow Zn^{2+} + 2e^-$$

at the two electrodes results in the overall reaction

$$Cu^{2+} + Zn(s) \rightarrow Zn^{2+} + Cu(s)$$

(Thus, a zinc rod placed in an aqueous solution of $CuSO_4$ will dissolve and metallic copper will be deposited.)

Reactions at the electrodes may be more complicated than those indicated above, but the main idea is the same: at one electrode, electrons are transferred *from* the electrode; at the other electrode, electrons are transferred *to* the electrode. In representing electrochemical cells diagramatically, it has become a convention to place the **'reduction' half-reaction on the right**. Thus, the electrode on the right-hand side of the diagram supplies the electrons that reduce the reactants.

Since the reactions at the electrodes may occur at different electrical potentials, we must use the electrochemical affinity to formulate the thermodynamics of an electrochemical cell. If $\tilde{A}$ is the electrochemical affinity and ξ is the extent of reaction, the entropy production due to such reaction is

$$\frac{d_i S}{dt} = \frac{\tilde{A}}{T} \frac{d\xi}{dt} \qquad (10.2.7)$$

Since each mole of reacting X transfers n moles of electrons (see (10.2.4)), and since $d\xi/dt$ is the velocity of the reaction, the relation between the current I (which is the amount of charge transferred per second) is

$$I = nF \frac{d\xi}{dt} \qquad (10.2.8)$$

in which F is the Faraday constant, i.e. the amount of charge carried by a mole of electrons. Substituting (10.2.8) in (10.2.7) we find

$$\frac{d_i S}{dt} = \frac{1}{T} \frac{\tilde{A}}{nF} I \qquad (10.2.9)$$

Comparing this expression with (10.1.17) we obtain the following relation between the voltage and the associated electrochemical affinity:

$$\boxed{V = \frac{\tilde{A}}{nF}} \qquad (10.2.10)$$

in which n is the number of electrons transferred in the oxidation–reduction reaction. For a given $\tilde{A}$, the larger the number of electrons transferred, the smaller the potential difference.

Using the electrode reactions (10.2.4) and (10.2.5), the above expression can be more explicitly written in terms of the chemical potentials:

$$X + ne^- \to X_{red} \text{ (right)} \quad \tilde{A}^R = (\mu_X^R + n\mu_e^R - nF\phi^R) - \mu_{X_{red}}^R \tag{10.2.11}$$

$$Y \to Y_{ox} + ne^- \text{ (left)} \quad \tilde{A}^L = \mu_Y^L - (n\mu_e^L - nF\phi^L + \mu_{Y_{ox}}^L) \tag{10.2.12}$$

in which the superscripts indicate the reactions at the right and left electrodes. The electrochemical affinity of the electron in the left electrode is written as $\tilde{\mu}_e = \mu_e^L - F\phi^L$, and similarly for the electrons in the right electrode. The overall electrochemical affinity $\tilde{A}$, which is the sum of the two affinities, can now be written as

$$\tilde{A} = \tilde{A}^R + \tilde{A}^L = (\mu_X^R + \mu_Y^L - \mu_{X_{red}}^R - \mu_{Y_{ox}}^L) + n(\mu_e^R - \mu_e^L) - nF(\phi^R - \phi^L) \tag{10.2.13}$$

If the two electrodes are identical, then $\mu_e^R = \mu_e^L$ and the only difference between the two electrodes is in their electrical potential ϕ. By virtue of (10.2.10), we can now write the voltage V associated with the electrochemical affinity as

$$V = \frac{\tilde{A}}{nF} = \frac{1}{nF}(\mu_X^R + \mu_Y^L - \mu_{X_{red}}^R - \mu_{Y_{ox}}^L) - (\phi^R - \phi^L) \tag{10.2.14}$$

Now let us consider the 'terminal voltage' $V_{cell} = \phi^R - \phi^L$, the potential difference between the terminals for which $\tilde{A} = 0$. It is the open-circuit condition with zero current, similar to the osmotic pressure difference at zero affinity. This terminal voltage V_{cell} is called the **EMF of the cell**. From (10.2.14) we see that

$$\boxed{V_{cell} = \frac{1}{nF}(\mu_X^R + \mu_Y^L - \mu_{X_{red}}^R - \mu_{Y_{ox}}^L)} \tag{10.2.15}$$

For a nonzero $\tilde{A}$, i.e. for nonzero current, the terminal voltage is less than the EMF. On the other hand, if the potentials of the two electrodes are equalized by shorting the two terminals, then the flow of current $I = nF(d\xi/dt)$ is limited only by the rate of electron transfer at the electrodes. Under these conditions the voltage $V = \tilde{A}/nF$ is also equal to the right-hand side of (10.2.15).

It is more convenient to write cell EMF (10.2.15) in terms of the activities by using the general expression $\mu_k = \mu_k^0 + RT\ln a_k$ for the reactants and products. This leads to

$$\boxed{V = V_0 - \frac{RT}{nF}\ln\left(\frac{a_{X_{red}}^R a_{Y_{ox}}^L}{a_X^R a_Y^L}\right)} \tag{10.2.16}$$

where

$$V_0 = \frac{1}{nF}(\mu_{X0}^{R} + \mu_{Y0}^{L} - \mu_{X_{red}0}^{R} - \mu_{Y_{ox}0}^{L}) = \frac{-\Delta G_{rxn}^{0}}{nF} \qquad (10.2.17)$$

Equation (10.2.16) relates the cell potential to the activities of the reactants; it is called the **Nernst equation**. As we expect, V is zero at equilibrium and the equilibrium constant of the electrochemical reaction can be written as

$$\boxed{\ln K = \frac{-\Delta G_{rxn}^{0}}{RT} = \frac{nFV_0}{RT}} \qquad (10.2.18)$$

Box. 10.1 Electrochemical Cells and Cell Diagrams.

When there is an external flow of current, there must be a compensating current within the cell. This can be accomplished in many ways, each defining a type of electrochemical cell. The choice of electrodes is also decided by the experimental conditions and the need to use an electrode without undesirable side reactions. Electrochemical cells often incorporate **salt bridges** and **liquid junctions**.

Liquid junctions. When two different liquids are in contact, usually through a porous wall, it is called a liquid junction. The concentrations of ions and their electrochemical potentials on either side of a liquid junction are generally not equal; the electrochemical potential difference causes a diffusional flow of ions. If the rates of flow of the different ions are unequal, then a potential difference will be generated across the liquid junction. Such a potential is called the **liquid junction potential**. The liquid junction potential may be reduced by the use of a **salt bridge**, in which the flows of the positive and negative ions are nearly equal.

Salt bridge. A commonly used salt bridge consists of a solution of KCl in agarose jelly. In this medium, the flow of K^+ and Cl^- are nearly equal.

Cell diagrams. An electrochemical cell diagram is drawn adopting the following conventions:

• Reduction reaction occurs at the electrode on the right.
• The symbol '|' indicates a phase boundary, such as the boundary between a solid electrode and a solution.
• The symbol ':' indicates a liquid junction, such as a porous wall separating a solution of $CuSO_4$ and CuCl.
• The symbol '||' or ': :' indicates a salt bridge, such as KCl in agarose jelly.

For example, the cell in Figure 10.6 is represented by the following cell diagram:

$$Zn(s)|Zn^{2+}||H^+|Pt(s)$$

GALVANIC AND ELECTROLYTIC CELLS

A cell in which a chemical reaction generates an electric potential difference is called a **galvanic cell**; if an external source of electric voltage drives a chemical reaction, then it is called an **electrolytic cell**.

Let us consider a simple reaction. When Zn reacts with an acid, H_2 is evolved. This reaction is a simple electron-transfer reaction:

$$Zn(s) + 2H^+ \rightarrow Zn^{2+} + H_2 \qquad (10.2.19)$$

The reason why the electrons migrate from one atom to another is a difference in electrical potential; that is, in the above reaction, when an electron moves from a Zn atom to an H^+ ion, it is moving to a location of lower potential energy. An interesting possibility now arises: if the reactants are placed in a 'cell' such that the only way an electron transfer can occur is through a conducting wire, then we have a situation in which a chemical affinity drives an electric current. Such a cell would be a **galvanic cell**, as shown in Figure 10.6, in which the sum of the electrode reactions' half-reactions is (10.2.19) and the flow of electrons occurs through an external circuit. Conversely, through an external EMF, the electron transfer can be reversed, which is the case in an **electrolytic cell**.

The EMF generated by a galvanic cell, as shown above, is given by the Nernst equation. In the above example the cell EMF is given by

$$V = V_0 - \frac{RT}{nF} \ln \left(\frac{a_{H_2} a_{Zn^{2+}}}{a_{Zn(s)} a_{H^+}^2} \right) \qquad (10.2.20)$$

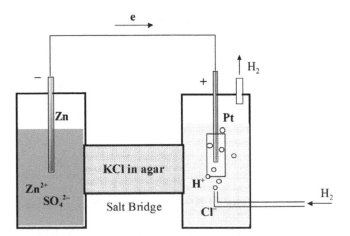

Figure 10.6 An example of a galvanic cell that is driven by the reaction $Zn(s) + 2H^+ \rightarrow Zn^{2+} + H_2$. The two electrode chambers are connected through a salt bridge that allows for the flow of current without introducing a liquid junction potential

CONCENTRATION CELL

The affinity generated by a concentration difference can also generate an EMF. A simple example of an $AgNO_3$ concentration cell in which a concentration-driven EMF can be realized is shown in Figure 10.7. The two beakers linked by KNO_3 salt bridge (a gel containing KNO_3 solution). A silver electrode is placed in each beaker. If the two electrodes are connected by a wire, the difference in electrochemical potential of Ag^+ ions causes a flow of electrons from one silver electrode to another, absorbing Ag^+ in the beaker that has a higher concentration and releasing them in the beaker that has lower concentration.

The reactions at the two electrodes are

$$Ag^+ + e^- \rightarrow Ag(s)\,(\beta) \quad \text{and} \quad Ag(s) \rightarrow Ag^+ + e^-(\alpha) \tag{10.2.21}$$

which amounts to transfer of Ag^+ ions from a higher concentration to a lower concentration. The electroneutrality is maintained in both beakers by the migration of K^+ and NO_3^- ions through the salt bridge. For such a cell V_0 in the Nernst equation equals zero because the reaction at one electrode is the reverse of the reaction at the other and the standard states of reactants and products are the same. Thus, for a concentration cell:

$$V_{\text{cell}} = -\frac{RT}{nF} \ln \left(\frac{a^{\beta}_{Ag^+}}{a^{\alpha}_{Ag^+}} \right) \tag{10.2.22}$$

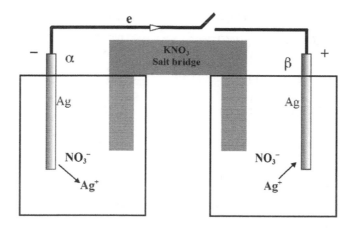

Figure 10.7 A concentration difference can generate an EMF. Two beakers containing $AgNO_3$ solutions at different concentrations are connected by a KNO_3 salt bridge. A silver electrode is placed in each cell. The difference in concentrations generates an EMF

STANDARD ELECTRODE POTENTIALS

Just as the tabulation of the Gibbs energies of formation facilitates the computation of equilibrium constants, the tabulation of 'standard electrode potentials' facilitates the computation of equilibrium constants for electrochemical reactions. A voltage is assigned to each electrode half-reaction with the convention that the voltage of the hydrogen-platinum electrode, $H^+|Pt$, is zero. That is, the electrode reaction $H^+ + e^- \rightarrow \frac{1}{2}H_2(g)$ at a Pt electrode is taken to be the reference and the voltages associated with all other electrode reactions are measured with respect to it. *The standard electrode potentials are the potentials when activities of all the reactants and products equal one at $T = 298.15\,K$.* For any cell, the voltages of the corresponding standard potentials are added to obtain the cell potentials. Since these potential correspond to the situation when all the activities are equal to one, it follows from the Nernst equation that the standard cell voltage is equal to V_0.

Example 10.3 shows how an equilibrium constant may be computed using the standard electrode potentials. A list of some of the commonly used standard electrode potentials is given in Table 10.1. In using the standard potentials, one must

Table 10.1   Standard electrode potentials

Electrode reaction	V_0/V	Electrode
$\frac{1}{3}Au^{3+} + e^- \rightarrow \frac{1}{3}Au$	1.50	$Au^{3+}|Au$
$\frac{1}{2}Cl_2(g) + e^- \rightarrow Cl^-$	1.360	$Cl^-|Cl_2(g)|Pt$
$Ag^+ + e^- \rightarrow Ag(s)$	0.799	$Ag^+|Ag$
$Cu^+ + e^- \rightarrow Cu(s)$	0.521	$Cu^+|Cu$
$\frac{1}{2}Cu^{2+} + e^- \rightarrow \frac{1}{2}Cu(s)$	0.339	$Cu^{2+}|Cu$
$AgCl + e^- \rightarrow Ag + Cl^-$	0.222	$Cl^-|AgCl(s)|Ag$
$Cu^{2+} + e^- \rightarrow Cu^+$	0.153	$Cu^{2+},Cu^+|Pt$
$H^+ + e^- \rightarrow \frac{1}{2}H_2(g)$	0.0	$H^+|H_2|Pt$
$\frac{1}{2}Pb^{2+} + e^- \rightarrow \frac{1}{2}Pb(s)$	−0.126	$Pb^{2+}|Pb(s)$
$\frac{1}{2}Sn^{2+} + e^- \rightarrow \frac{1}{2}Sn(s)$	−0.140	$Sn^{2+}|Sn(s)$
$\frac{1}{2}Ni^{2+} + e^- \rightarrow \frac{1}{2}Ni(s)$	−0.250	$Ni^{2+}|Ni(s)$
$\frac{1}{2}Cd^{2+} + e^- \rightarrow \frac{1}{2}Cd(s)$	−0.402	$Cd^{2+}|Cd(s)$
$\frac{1}{2}Zn^{2+} + e^- \rightarrow \frac{1}{2}Zn(s)$	−0.763	$Zn^{2+}|Zn(s)$
$Na^+ + e^- \rightarrow Na(s)$	−2.714	$Na^+|Na(s)$
$Li^+ + e^- \rightarrow Li(s)$	−3.045	$Li^+|Li(s)$

Note. (a) Changing the stoichiometry does not change V_0. (b) If the reaction is reversed, the sign of V_0 also reverses.

note that: (a) changing the stoichiometry does not change V_0 (b) if the reaction is reversed, then the sign of V_0 also reverses.

10.3 Isothermal Diffusion

We have already seen in Section 4.3 that the flow of particles from a region of high concentration to a region of lower concentration is a flow driven by unequal chemical potentials. For a discrete system consisting of two parts of equal temperature T, one with chemical potential μ_1 and molar amount N_1 and the other with chemical potential μ_2 and molar amount N_2 we have the following relation:

$$-dN_1 = dN_2 = d\xi \tag{10.3.1}$$

The entropy production that results from unequal chemical potentials is

$$d_iS = -\left(\frac{\mu_2 - \mu_1}{T}\right)d\xi = \frac{A}{T}d\xi > 0 \tag{10.3.2}$$

The positivity of this quantity, required by the Second Law, implies that particle transport is from a region of higher chemical potential to a region of lower chemical potential. It is the diffusion of particles from a region of higher chemical potential to a region of lower chemical potential. In many situations this is a flow of a component from a higher concentration to a lower concentration. At equilibrium the concentrations become uniform, but this need not be so in every case. For example, when a liquid is in equilibrium with its vapor or when a gas reaches equilibrium in the presence of a gravitational field, the chemical potentials becomes uniform, not the concentrations. *The tendency of the thermodynamic forces that drive matter flow is to equalize the chemical potential, not the concentrations.*

DIFFUSION IN A CONTINUOUS SYSTEM AND FICK'S LAW

Expression (10.3.2) can be generalized to describe a continuous system as was done for the general case of a field in Section 10.1 (Figure 10.3). Let us consider a system in which the variation of the chemical potential is along one direction only, say x. We shall also assume that T is uniform and does not change with position. Then, as in Equation (10.1.13), we have for diffusion that

$$\frac{d_is(x)}{dt} = -\sum_k \frac{1}{T}\left(\frac{\partial \mu_k}{\partial x}\right)J_{Nk} \tag{10.3.3}$$

For simplicity, let us consider the flow of a single component k:

$$\frac{d_is(x)}{dt} = -\frac{1}{T}\left(\frac{\partial \mu_k}{\partial x}\right)J_{Nk} \tag{10.3.4}$$

We note, once again, that the entropy production is the product of a thermodynamic flow J_{Nk} and the force, $-(1/T)(\partial\mu_k/\partial x)$, that drives it. The identification of a thermodynamic force and the corresponding flow enables us to relate the two. Near equilibrium, the flow is linearly proportional to the force. In the above case we can write this linear relation as

$$J_{Nk} = -L_k \frac{1}{T}\left(\frac{\partial\mu_k}{\partial x}\right)$$

(10.3.5)

The constant of proportionality, L_k, is the linear phenomenological coefficient for diffusional flow. We saw earlier that in an ideal fluid mixture the chemical potential can be written as $\mu(p,T,x_k) = \mu(p,T) + RT\ln x_k$, in which x_k is the mole fraction, which in general is a function of position. If n_{tot} is the total molar density and n_k is the molar density of component k, then the mole fraction $x_k = n_k/n_{tot}$. We shall assume that the change of n_{tot} due to diffusion is insignificant, so that $\partial\ln x_k/\partial x = \partial\ln n_k/\partial x$. Then, substituting $\mu(p,T,x_k) = \mu(p,T) + RT\ln x_k$ into (10.3.5), we obtain the following thermodynamic relation between the diffusion current J_{Nk} and the concentration:

$$J_{Nk} = -L_k R \frac{1}{n_k}\frac{\partial n_k}{\partial x}$$

(10.3.6)

Empirical studies of diffusion have led to what is called Fick's law. According to **Fick's law**:

$$\boxed{J_{Nk} = -D_k \frac{\partial n_k}{\partial x}}$$

(10.3.7)

in which D_k is the diffusion coefficient of the diffusing component k. Typical values of the diffusion coefficients for gases and liquids are given in Table 10.2. Clearly, this expression is the same as (10.3.6) if we make the identification

$$\boxed{D_k = \frac{L_k R}{n_k}}$$

(10.3.8)

Table 10.2 Diffusion coefficients of molecules in gases and liquids

Compound, in air ($p = 101.325$ kPa, $T = 293.15$ K)	$D/10^{-4}$ m^2 s^{-1}	Solute, in water ($T = 298.15$ K)	$D/10^{-9}$ m^2 s^{-1}
CH_4	0.106	Sucrose	0.52
Ar	0.148	Glucose	0.67
CO_2	0.160	Alanine	0.91
CO	0.208	Ethylene glycol	1.16
H_2O	0.242	Ethanol	1.24
He	0.580	Acetone	1.28
H_2	0.627		

Source: D.R. Lide (ed.) *CRC Handbook of Chemistry and Physics*, 75th edition. 1994, CRC Press: Ann Arbor.

This gives us a relation between the thermodynamic phenomenological coefficient L_k and the empirical diffusion coefficient.

An important point to note is that the thermodynamic relation (10.3.5) is valid in all cases, whereas Fick's law (10.3.7) is not. For example, in the case of a liquid in equilibrium with its vapor, since the chemical potential is uniform, $(\partial\mu_k/\partial x) = 0$ and (10.3.5) correctly predicts $J_{Nk} = 0$; but (10.3.7) does not predict $J_{Nk} = 0$ because $(\partial n_k/\partial x) \neq 0$. In general, if we write (10.3.5) as $J_{Nk} = -(L_k/T)(\partial\mu_k/\partial n_k)(\partial n_k/\partial x)$, then we see that, depending on the sign of $(\partial\mu_k/\partial n_k)$, J_{Nk} can be positive or negative when $(\partial n_k/\partial x) > 0$. Thus, the flow is toward the region of lower concentration when $(\partial\mu_k/\partial n_k) > 0$, but the flow can be to the region of higher concentration when $(\partial\mu_k/\partial n_k) < 0$. The latter situation arises when a mixture of two components is separating into two phases: each component flows from a region of lower concentration to a region of higher concentration. As we shall see in later chapters, the system is 'unstable' when $(\partial\mu_k/\partial n_k) < 0$.

THE DIFFUSION EQUATION

In the absence of chemical reactions, the only way the molar density $n_k(x, t)$ can change with time is due to the flow J_{Nk}. Consider a small cell of size δ at a location x (Figure 10.8). The molar amount in this cell is equal to $n_k(x, t)\delta$. The rate of change of the molar amount in this cell is $\partial(n_k(x,t)\delta)/\partial t$. This change is due the net flow, i.e.

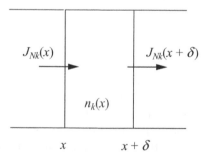

Figure 10.8 In the absence of chemical reactions, the change in the molar amount of a substance in a small cell of size δ, at a location x, equals the net flow, the difference in the flow J_{Nk} into and out of the cell. The number of moles in the cell of size δ is $n_k\delta$. The net flow into the cell of size δ is given by (10.3.9). This difference in the flow will cause a net rate of change in the mole amount $\partial(n_k(x,t)\delta)/\partial t$. On equating the net flow to the rate of change of the molar amount, we obtain the equation $\partial n_k(x,t)/\partial = -\partial J_{Nk}/\partial x$

the difference between the inflow and the outflow of component k in the cell. The net flow into the cell of size δ is equal to

$$J_{Nk}(x) - J_{Nk}(x+\delta) = J_{Nk}(x) - \left(J_{Nk}(x) + \frac{\partial J_{Nk}}{\partial x}\delta \right) = -\frac{\partial J_{Nk}}{\partial x}\delta \qquad (10.3.9)$$

Equating the net flow to the rate of change of the molar amount, we obtain the equation

$$\frac{\partial n_k(x,t)}{\partial t} = -\frac{\partial J_{Nk}}{\partial x} \qquad (10.3.10)$$

Using Fick's law (10.3.7), we can write this equation entirely in terms of $n_k(x, t)$ as

$$\boxed{\frac{\partial n_k(x,t)}{\partial t} = D_k \frac{\partial^2 n_k(x,t)}{\partial x^2}} \qquad (10.3.11)$$

This partial differential equation for $n_k(x)$ is the **diffusion equation** for the component k. It is valid in a homogeneous system. In a homogeneous system, diffusion tends to eliminate concentration differences and equalize the concentrations throughout the system. But it must be borne in mind that, in general, the thermodynamic force tends to equalize the chemical potential, not the concentrations.

THE STOKES–EINSTEIN RELATION

The viscous force on a particle in a fluid and its diffusive motion are both results of random molecular collisions. A particle diffuses due to random collisions it undergoes with the fluid molecules, and it can also transfer its momentum to the fluid molecules during these collisions. The latter process appears as the viscous force on a macro level. Through thermodynamics one can see that the diffusion coefficient and the coefficient of viscous force or 'friction' must be related – a reflection of the fact that both are the result molecular collisions. This relation is called the **Stokes– Einstein relation**.

Fick's law gives us the diffusion current in the presence of a concentration gradient. In the presence of a field, there is also a current which is proportional to the strength of the field. For example, in the presence of an electric field **E**, an ion carrying a charge ez_k will drift at constant speed proportional to the magnitude of the force $ez_k|\mathbf{E}|$. This happens because the force due to the field F_{field} (whose magnitude equals $ez_k|\mathbf{E}|$ for ions) accelerates the ion till the opposing viscous or frictional force, which is proportional to the velocity, balances F_{field}. When the ion moves at a speed v, the viscous force equals $\gamma_k v$, in which γ_k is the coefficient of viscous force. When the two forces balance, $\gamma_k v = F_{field}$ and the ion will drift with a **terminal velocity** v. Hence, the terminal or **drift velocity** can be written as

$$v = \frac{F_{\text{field}}}{\gamma_k} \qquad (10.3.12)$$

Since the number of ions that drift is proportional to the concentration n_k, the ionic drift gives rise to the following particle current density I_k due to the component k:

$$I_k = v n_k = \frac{e z_k}{\gamma_k} n_k |\mathbf{E}| = -\Gamma_k n_k \frac{\partial \phi}{\partial x} \qquad (10.3.13)$$

in which the constant $\Gamma_k = e z_k / \gamma_k$ is called the **ionic mobility** of the ion k. (Note that the total electric current density due to all the ions $I = \sum_k e z_k I_k$.) Similarly, a molecule of mass m_k, falling freely in the atmosphere, or any fluid, will reach a 'terminal velocity' $v = g m_k / \gamma_k$, where g is the acceleration due to gravity. In general, for any potential ψ associated with a conservative field, the mobility a component k is defined by

$$J_{\text{field}} = -\Gamma_k n_k \frac{\partial \psi}{\partial x} \qquad (10.3.14)$$

Linear phenomenological laws of nonequilibrium thermodynamics lead to a general relation between mobility Γ_k and the diffusion coefficient D_k, This relation can be obtained as follows. The general expression for the chemical potential in a field with potential ψ is given by $\tilde{\mu}_k = \mu_k + \tau_k \psi$, in which τ_k is the interaction energy per mole due to the field (10.1.5). In the simplest approximation of an ideal system, if we write the chemical potential in terms of the concentration n_k, then we have

$$\tilde{\mu}_k = \mu_k^0 + RT \ln(n_k) + \tau_k \psi \qquad (10.3.15)$$

A gradient in this chemical potential will result in a thermodynamic flow

$$J_{Nk} = -L_k \frac{1}{T}\left(\frac{\partial \tilde{\mu}_k}{\partial x}\right) = -\frac{L_k}{T}\left(\frac{RT}{n_k}\frac{\partial n_k}{\partial x} + \tau_k \frac{\partial \psi}{\partial x}\right) \qquad (10.3.16)$$

where we have used $\partial \ln x_k / \partial x = \partial \ln n_k / \partial x$. In (10.3.16), the first term on the right-hand side is the familiar diffusion current and the second term is the drift current due to the field. Comparing this expression with Fick's law (10.3.7) and expression (10.3.14) that defines mobility, we see that:

$$\frac{L_k R}{n_k} = D_k \qquad \frac{L_k \tau_k}{T} = \Gamma_k n_k \qquad (10.3.17)$$

From these two relations it follows that the diffusion coefficient D_k and the mobility Γ_k have the following general relation:

$$\boxed{\frac{\Gamma_k}{D_k} = \frac{\tau_k}{RT}} \qquad (10.3.18)$$

This relation was first obtained by Einstein and is sometimes called the **Einstein relation**. For ionic systems, as we have seen in Section 10.1 (see (10.1.5)), $\tau_k = Fz_k = eN_A z_k$ and $\Gamma_k = ez_k/\gamma_k$. Since $R = k_B N_A$, in which k_B is the Boltzmann constant and N_A the Avogadro number, Equation (10.1.16) for **ionic mobility** Γ_k becomes

$$\frac{\Gamma_k}{D_k} = \frac{ez_k}{\gamma_k D_k} = \frac{z_k F}{RT} = \frac{ez_k}{k_B T} \tag{10.3.19}$$

which leads to the following general relation between the diffusion coefficient D_k and the friction coefficient γ_k of a molecule or ion k, called the **Stokes–Einstein relation**:

$$\boxed{D_k = \frac{k_B T}{\gamma_k}} \tag{10.3.20}$$

Reference

1. Guggenheim, E.A., *Modern Thermodynamics*. 1933, London: Methuen.

Examples

Example 10.1 Use the barometric formula to estimate the pressure at an altitude of 3.0 km. The temperature of the atmosphere is not uniform (so it is not in equilibrium). Assume an average temperature $T = 270.0$ K.
Solution The pressure at an altitude h is given by the barometric formula $p(h) = p(0)e^{-gMh/RT}$. For the purpose of estimating, since 78% of the atmosphere consists of N_2, we shall use the molar mass of N_2 for M. The pressure at an altitude of 3.0 km will be

$$p(3\,\text{km}) = (1\,\text{atm})\exp\left[-\frac{(9.8\,\text{m s}^{-2})(28.0\times10^{-3}\,\text{kg mol}^{-1})3.0\times10^{3}\,\text{m}}{(8.314\,\text{J K}^{-1}\,\text{mol}^{-1})(270\,\text{K})}\right]$$
$$= (1\,\text{atm})\exp(-0.366)$$
$$= 0.69\,\text{atm}$$

Example 10.2 Calculate the membrane potential for the setup shown in Figure 10.4.
Solution In this case, the expected potential difference across the membrane is

$$V = \phi^\alpha - \phi^\beta = \frac{RT}{F}\ln\left(\frac{1.0}{0.1}\right) = 0.0257\ln(10)$$
$$= 0.0592\,\text{V}$$

Example 10.3   Calculate the standard cell potential V_0 for the cell shown in Figure 10.6. Also calculate the equilibrium constant for the reaction $Zn(s) + 2H^+ \rightarrow H_2(g) + Zn^{2+}$.

Considering the two electrode reactions, we have

$$2H^+ + 2e^- \rightarrow H_2(g) \quad 0.00\,V$$
$$Zn(s) \rightarrow Zn^{2+} + 2e^- \quad +0.763\,V$$

The total cell potential is

$$V_0 = 0 + 0.763\,V = 0.763\,V$$

and the equilibrium constant is

$$K = \exp\left(\frac{2FV_0}{RT}\right)$$
$$= \exp\left(\frac{2 \times 9.648 \times 10^4 \times 0.763}{8.314 \times 298.15}\right)$$
$$= 6.215 \times 10^{25}$$

Exercises

10.1   Use the chemical potential of an ideal gas in (10.1.9) and obtain the barometric formula (10.1.10). Use the barometric formula to estimate the boiling point of water at an altitude of 2.50 km above sea level. Assume an average $T = 270\,K$.

10.2   A heater coil is run at a voltage of 110 V and it draws 2.0 A current. If its temperature is equal to 200 °C, what is the rate of entropy production due to this coil?

10.3   Calculate the equilibrium constants at $T = 25.0\,°C$ for the following electrochemical reactions using the standard potentials in Table 10.1:

(i) $Cl_2(g) + 2Li(s) \rightarrow 2Li^+ + 2Cl^-$
(ii) $Cd(s) + Cu^{2+} \rightarrow Cd^{2+} + Cu(s)$
(iii) $2Ag(s) + Cl_2(g) \rightarrow 2Ag^+ + 2Cl^-$
(iv) $2Na(s) + Cl_2(g) \rightarrow Na^+ + Cl^-$.

10.4   If the reaction $Ag(s) + Fe^{3+} + Br^- \rightarrow AgBr(s) + Fe^{2+}$ is not in equilibrium it can be used generate a EMF. The 'half-cell' reactions that correspond to the oxidation and reduction in this cell are

$$Ag(s) + Br^- \rightarrow AgBr(s) + e^- \quad V_0 = -0.071\,V$$
$$Fe^{3+} + e^- \rightarrow Fe^{2+} \quad\quad V_0 = 0.771\,V$$

(a) Calculate V_0 for this reaction.

(b) What is the EMF for the following activities at $T = 298.15\,K$: $a_{Fe^{3+}} = 0.98$; $a_{Br^-} = 0.30$; $a_{Fe^{2+}} = 0.01$.

(c) What will be the EMF at $T = 0.0\,°C$?

10.5 The K^+ concentration inside a nerve cell is much larger than the concentration outside it. Assume that the potential difference across the cell membrane is $90\,mV$. Assuming that the system is in equilibrium, estimate the ratio of concentration of K^+ inside and outside the cell.

10.6 Verify that

$$n(x) = \frac{n(0)}{2\sqrt{\pi Dt}} e^{-x^2/4Dt}$$

is the solution of the diffusion equation (10.3.11). Using Mathematica or Maple, plot this solution for various values of t for one of the gases listed in Table 10.2, assuming $n(0) = 1$. This gives you an idea of how far a gas will diffuse in a given time. Obtain a simple expression to estimate the distance a molecule will diffuse in a time t, given its diffusion coefficient D.

10.7 Compute the diffusion current corresponding to the barometric distribution $n(x) = n(0)e^{-gMx/RT}$.

11 INTRODUCTION TO NONEQUILIBRIUM SYSTEMS

Introduction

We live in a world that is not in thermodynamic equilibrium, a world that is constantly evolving, a world that creates order with one hand and destroys it with another. The knowledge we acquired during the twentieth century makes this all too clear. The radiation filling the universe, which is at 2.73 K, is not in equilibrium with the matter, and matter itself has been evolving through the production of elements in the birth-and-death cycle of stars and planets. On our planet, the atmosphere is not in thermodynamic equilibrium: its nonuniform temperature, chemical composition (which includes the chemically reactive O_2) and the many cycles that characterize its dynamic nature are all driven by the flux of solar energy. Nonequilibrium conditions are also common in industry and the laboratory. In contrast, the world of nineteenth-century classical thermodynamics is confined to equilibrium states; processes that occur at a finite rate are not a part of the theory. In classical thermodynamics, changes in quantities such as entropy and Gibbs energy are calculable only for infinitely slow 'quasi-static' processes. It is a subject that describes a static world, so much so that some had called it 'thermostatics' [1].* The birth of an expanded formalism of twentieth-century thermodynamics, capable of describing nonequilibrium systems and irreversible processes that take place at a finite rate, was an inevitability. In this chapter we present the reader with a brief overview of nonequilibrium thermodynamics, a subject that is still evolving with new hypotheses regarding rates of entropy production and nonequilibrium states [3].

Nonequilibrium systems can be broadly classified as near-equilibrium systems, in which there is a linear relation between forces and flows, also called the **linear regime**, and far-from-equilibrium systems, in which the relationship between forces and flows is nonlinear, also called the **nonlinear regime**. In the near-equilibrium linear regime we will discuss Onsager reciprocal relations. In the far-from-equilibrium nonlinear regime, we will encounter spontaneous self-organization and the concept of **dissipative structures**.

---

* See also the following comment on classical thermodynamics in Ref. [2]: '. . . in view of the emphasis upon equilibrium, or static, states, it is surely more appropriate to talk in this context of thermostatics and reserve the name thermodynamics for the . . . detailed examination of dynamic, or off-equilibrium, situations'.

Introduction to Modern Thermodynamics   Dilip Kondepudi
© 2008 John Wiley & Sons, Ltd

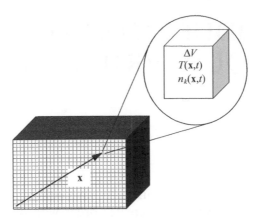

Figure 11.1 The concept of local equilibrium. A nonequilibrium system could be considered as an ensemble of elemental cells of volume ΔV, each in equilibrium, exchanging energy and matter with the neighboring cells. Temperature T and molar density n_k vary from cell to cell and in time; they are functions of position and time

11.1   Local Equilibrium

Though classical thermodynamics is based on equilibrium states, it remains very important in that it acts as the foundation for the thermodynamics of nonequilibrium systems. The formulation of the thermodynamics of irreversible processes is based on the concept of **local equilibrium**. It is a concept that considers a nonequilibrium system as an ensemble of elemental cells of volume ΔV, each in equilibrium, exchanging energy and matter with neighboring cells (Figure 11.1). It assumes that each elemental cell is large enough to have a well-defined temperature, concentration, etc., but which is small enough to describe spatial variations of thermodynamic quantities. For almost every macroscopic system (whose size typically is of the order of a centimeter or larger) we can meaningfully assign a temperature, and other thermodynamic variables to every such elemental cell. The size of ΔV of a cell could be, for example, of the order $(\mu m)^3 = 10^{-15}$ L. Detailed studies have shown the validity of the concept of local equilibrium [4, 5] and that *equilibrium thermodynamic relations are valid for the thermodynamic variables assigned to an elemental cell*. Thus, all intensive thermodynamic variables T, p, μ, become functions of position $\mathbf{x}$ and time t:

$$T = T(\mathbf{x},\, t), \quad p = p(\mathbf{x},\, t) \quad \mu = \mu(\mathbf{x},\, t)$$

The extensive variables are replaced by densities s, u and n_k which are now function of position $\mathbf{x}$ and time t:

$s(T(\mathbf{x}), n_k(\mathbf{x}), \ldots) = s(\mathbf{x}, t)$ the entropy per unit volume

$u(T(\mathbf{x}), n_k(\mathbf{x}), \ldots) = u(\mathbf{x}, t)$ the energy per volume   (11.1.1)

$n_k(\mathbf{x}, t)$ the molar amount of component k per unit volume

where the two fundamental thermodynamic quantities s and u are functions of T, n_k and other variables that can be measured experimentally. (In some formulations, the extensive quantities are replaced by entropy, energy and volume per unit mass). The Gibbs relation $dU = T\,dS - p\,dV + \Sigma_k \mu_k\,dN_k$ is assumed to be valid for small volume elements. With $U = uV$ and $S = sV$, it follows that relations such as

$$\left(\frac{\partial u}{\partial s}\right)_{n_k} = T \quad T\,ds = du - \sum_k \mu_k dn_k \qquad (11.1.2)$$

are valid for the densities in every cell at location $\mathbf{x}$ and time t (Exercise 11.1). In these equations, the volume does not appear because s, u and n_k are densities.

Now let us look at the reasons which make local equilibrium a valid assumption. First, we must look at the concept of temperature. In Chapter 1 we saw that the **Maxwell distribution of velocities** describes a gas in equilibrium. The probability $P(v)$ that a molecule has a velocity v is given by

$$\boxed{P(v)\,d^3v = \left(\frac{\beta}{\pi}\right)^{3/2} e^{-\beta v^2} d^3v} \qquad (11.1.3)$$

$$\beta = \frac{m}{2k_B T} \qquad (11.1.4)$$

The temperature is identified through the relation (11.1.4), in which m is the mass of the molecule and k_B is the Boltzmann constant. *For velocity distributions that significantly deviate from the Maxwell distribution, temperature is not well defined.* On a microscopic scale, in every elemental volume ΔV, we could look at the deviations from the Maxwell distribution to assess the validity of the assumption that T is well defined in every elemental cell. Only under very extreme conditions do we find significant deviations from the Maxwell distribution. Any initial distribution of velocities quickly becomes Maxwellian due to molecular collisions. Molecular-dynamics computer simulations have revealed that the Maxwell distribution is reached in less than 10 times the average collision time, which in a gas at a pressure of 1 atm is about 10^{-8} s. Consequently, physical processes that perturb the system significantly from the Maxwell distribution have to be very rapid, on a timescale of 10^{-8} s. A detailed statistical mechanical analysis of the assumption of local equilibrium was done by Ilya Prigogine [4] in 1949, from which one can obtain a very precise understanding of its wide-ranging validity.

Chemical reactions are of particular interest to us. In almost all chemical reactions, only a very small fraction of molecular collisions result in a chemical reaction, as we have seen in the Arrhenius theory of rates. Collisions between molecules that result in a chemical reaction are called **reactive collisions**. For a gas at a pressure of 1 atm, the collision frequency is about $10^{31}\,L\,s^{-1}$. If nearly every collision resulted in a chemical reaction, then the resulting rate would be of the order of $10^8\,mol\,L^{-1}\,s^{-1}$! Reaction rates that approach such a large value are extremely rare. Most of the reaction rates we encounter indicate that reactive collision rates are many orders of magnitude smaller than the collision rates. Between reactive collisions, the system quickly relaxes to equilibrium, redistributing the change in energy due to the chemical reaction. In other words, any perturbation of the Maxwell distribution due to a chemical reaction quickly relaxes back to the Maxwellian with a slightly different temperature. Hence, on the timescale of most chemical reactions, temperature is locally well defined. All these aspects can be seen well in molecular dynamics simulations with modern computers [6].

Next, let us look at the sense in which thermodynamic variables, such as molar density, and quantities such as entropy that are functions of these variables, may be considered functions of position. Every thermodynamic quantity undergoes fluctuations. For a small elemental volume ΔV we can meaningfully associate a value for a thermodynamic quantity Y only when the size of the fluctuations (**root-mean-square (RMS)** value, for example) δY is very small compared with Y. Clearly, this condition will not be satisfied if ΔV is too small. The thermodynamic theory of fluctuations tells us that, if $\tilde{N}$ is the number of particles in ΔV, the RMS value of the fluctuations $\delta \tilde{N} = \sqrt{\delta \tilde{N}}$. It is easy to calculate the relative value of fluctuations in $\tilde{N}$, $\delta \tilde{N}/\tilde{N}$, in a cell of volume $\Delta V = (1\,\mu m)^3 = 10^{-15}\,L$, filled with an ideal gas at $T = 298\,K$ and $p = 1$ atm and see that $\delta \tilde{N}/\tilde{N} \approx 2 \times 10^{-4}$. For liquids and solids, the same value of $\delta \tilde{N}/\tilde{N}$ will correspond to an even smaller volume. Hence, it is meaningful to assign a molar density $n(\mathbf{x}, t)$ to a volume with a characteristic size of a micrometer. The same is generally true for other thermodynamic variables.

There is one more point to note regarding the concept of local equilibrium. If we are to assign a molar density to a volume ΔV, then the molar density in this volume should be nearly uniform. This means that the variation of molar density with position on the scale of micrometer should be very nearly uniform, a condition satisfied by most macroscopic systems. Thus, we see that a theory based on local equilibrium is applicable to a wide range of macroscopic systems. For almost all systems that we encounter, thermodynamics based on local equilibrium has excellent validity.

EXTENDED THERMODYNAMICS

In the above approach, an implicit assumption is that the thermodynamic quantities do not depend on the gradients in the system, i.e. it is postulated that entropy s is a function of the temperature T and the mole number density n_k, but not their gradients. Nevertheless, flows represent a level of organization, however small. This implies that the local entropy in a nonequilibrium system may be a little smaller than the equilibrium entropy. In the recently developed formalism of **extended irreversible thermodynamics**, gradients are included in the basic formalism and small corrections to the local entropy due to the flows appear [7]. This formalism is needed for systems such as shock waves, where very large gradients of $n_k(\mathbf{x})$ are encountered.

11.2 Local Entropy Production, Thermodynamic Forces and Flows

As we noted in the previous section, the Second Law of thermodynamics must be a local law. If we divide a system into r parts, then not only is

$$d_i S = d_i S^1 + d_i S^2 + \ldots + d_i S^r \geq 0 \tag{11.2.1}$$

in which the $d_i S^k$ is the entropy production in the kth part, but also

$$d_i S^k \geq 0 \tag{11.2.2}$$

for every k. Clearly, this statement that the entropy production due to irreversible processes is positive in every part is stronger than the classical statement of the Second Law that the entropy of an isolated system can only increase or remain unchanged.* It must also be noted that Second Law as stated by (11.2.2) does not require that the system be isolated. *It is valid for all systems, regardless of the boundary conditions.*

The local increase in entropy in continuous systems can be defined by using the entropy density $s(\mathbf{x}, t)$. As was the case of the total entropy, $ds = d_i s + d_e s$, with $d_i s \geq 0$. We define the rate of local entropy production thus:

$$\boxed{\sigma(x,t) \equiv \frac{d_i s}{dt}} \tag{11.2.3}$$

The total rate of entropy production in the systems is

$$\frac{d_i S}{dt} = \int_V \sigma(\mathbf{x},t)\,dV \tag{11.2.4}$$

---

*One general point to note about the First and Second Laws is that both must be *local laws*. In fact, to be compatible with the principle of relativity, and to be valid regardless of the observer's state of motion, these laws *must* be local. Nonlocal laws of conservation of energy or of entropy production are inadmissible because the notion of simultaneity is relative. Consider two parts of a system spatially separated by some nonzero distance. If changes in energy δu_1 and δu_2 occur in these two parts *simultaneously* in one frame of reference so that $\delta u_1 + \delta u_2 = 0$, the energy is conserved. However, in another frame of reference that is in motion with respect to the first, the two changes in energy *will not occur simultaneously*. Thus, during the time between one change of u and the other, the law of conservation of energy will be violated. Similarly, the entropy changes in a system δS_1 and δS_2 at two spatially separated parts of a system must be independently positive. The simultaneous decrease of one and increase of the other so that their sum is positive is inadmissible.

Table 11.1   Table of thermodynamic forces and flows

Irreversible process	Force F_k^*	Flow J_k	Units
Heat conduction	$\dfrac{\partial}{\partial x}\dfrac{1}{T}$	Heat flow J_q	J m^{-2} s^{-1}
Diffusion	$-\dfrac{\partial}{\partial x}\dfrac{\mu_k}{T}$	Diffusion current J_k	mol m^{-2} s^{-1}
Electrical conduction	$\dfrac{-(\partial\phi/\partial x)}{T}=\dfrac{E}{T}$	Electric current I_k	C m^{-2} s^{-1}
Chemical reactions	$\dfrac{A_j}{T}$	Velocity of reaction $v_j=\dfrac{1}{V}\dfrac{d\xi_j}{dt}$	mol m^{-3} s^{-1}

* In three dimensions, the derivatives $\partial/\partial x$ is replaced by the gradient operator ∇ and corresponding flows are vectors.

Table 11.2   Empirical laws

Fourier's law of heat conduction	$J_q=-\kappa\dfrac{\partial}{\partial x}T(x)$	κ = heat conductivity
Fick's law of diffusion	$J_k=-D_k\dfrac{\partial}{\partial x}n_k(x)$	D_k = diffusion coefficient
Ohm's law	$I=V/R$ or $I=E/\rho$	I = current, V = voltage, E = electric field, ρ = resistivity (resistance/unit length/unit cross-sectional area)
Reaction rate laws X + Y $\rightarrow$ Products	Rate = $k[X]^a[Y]^b$	[X] = concentration (mol L^{-1})

Modern thermodynamics formulated by De Donder, Onsager, Prigogine and others is founded on the explicit expression for σ in terms of the irreversible processes that we can identify and study experimentally.

Using conservation laws, the rate of entropy production per unit volume can be written as a bilinear product of terms that we can identify as **thermodynamic forces** and **thermodynamic flows**. We shall not present the derivation of this result, but refer the reader to more detailed texts [5, 8]. We use the symbols F_k and J_k to represent thermodynamic forces and flows. Table 11.1 lists some of the most common forces and flows encountered in nonequilibrium systemsand Table 11.2 lists various empirical laws that are unified under the thermodynamic formalism. The rate of entropy production per unit volume σ can be written as

$$\sigma=\sum_k F_k J_k \tag{11.2.5}$$

The forces drive the flows. *In this chapter, for simplicity, we will consider systems in which spatial variations of thermodynamic quantities are only along the x coordinate.*

The gradient of $1/T$, i.e. $\partial(1/T)/\partial x$, for example, is the force that drives the heat flow J_q; similarly, the gradient $\partial(-\mu_k/T)/\partial x$ drives diffusion flow J_k (μ_k is the chemical potential of component k and J_k is its flow). At equilibrium, all the forces and the corresponding flows vanish, i.e. the flows J_k are functions of forces F_k such that they vanish when $F_k = 0$.

11.3 Linear Phenomenological Laws and Onsager Reciprocal Relations

At equilibrium, there are no thermodynamic flows because the forces that drive them are zero. While it is clear that the flows are function of the forces, there is no general theory that enables us to specify the flow for a given force. However, as is the case with many physical variables, we may assume that the flows are analytic functions of the forces, such that both equal zero at equilibrium. Then, when a system is close to the state of equilibrium, the flows can be expected to be linear functions of the forces. Accordingly, the following relation between the flows and the forces is postulated close to equilibrium:

$$\boxed{J_k = \sum_j L_{kj} F_j}$$

(11.3.1)

Here, L_{kj} are constants called **phenomenological coefficients** or **Onsager coefficients**. Note that (11.3.1) implies that a force such as gradient of $1/T$ not only can cause the flow of heat, but can also drive other flows, such as a flow of matter or an electrical current. The thermoelectric effect is one such **cross-effect**, in which a thermal gradient drives not only a heat flow but also an electrical current, and vice versa (Figure 11.2). The thermoelectric phenomenon was investigated in the 1850s by William Thomson [9] (Lord Kelvin), who gave theoretical explanations for the observed **Seebeck** and **Peltier** effects (Figure 11.2) (but Kelvin's reasoning was later found to be incorrect). Another example is 'cross-diffusion', in which a gradient in the concentration of one compound drives a diffusion current of another.

Phenomenological laws and the cross-effects between the flows were independently studied, but, until the formalism presented here was developed in the 1930s, there was no *unified* theory of all the cross-effects. Relating the entropy production to the phenomenological laws ((11.2.5) and (11.3.1)) is the first step in developing a unified theory. For conditions under which the linear phenomenological laws (11.3.1) are valid, entropy production (11.2.5) takes the quadratic form:

$$\boxed{\sigma = \sum_{jk} L_{jk} F_j F_k > 0}$$

(11.3.2)

Here, the forces F_k can be positive or negative. The coefficients L_{jk} form a matrix. A matrix that satisfies the condition (11.3.2) is said to be **positive definite**. The properties of positive definite matrices are well characterized. For example, a two-dimensional matrix L_{ij} is positive definite only when the following conditions are satisfied (Exercise 11.3):

$$L_{11} > 0 \quad L_{22} > 0 \quad (L_{12} + L_{21})^2 < 4L_{11}L_{22} \tag{11.3.3}$$

In general, the diagonal elements of a positive definite matrix must be positive. More general properties of positive definite matrices can be found in Ref. [8]. Thus, according to the Second Law, the 'proper coefficients' L_{kk} should be positive (the 'cross-coefficients' L_{ik} ($i \neq k$) can be positive or negative).

ONSAGER RECIPROCAL RELATIONS

One of the most important results of linear nonequilibrium thermodynamics is that L_{jk} obey the **Onsager reciprocal relations**: $L_{jk} = L_{kj}$. These relations imply there is an equivalence between the cross-effects. The effect of heat flow in generating electric current is equal to the opposite cross-effect, but this equivalence can only be seen when the Onsager coefficients L_{kj} have been correctly identified. This point will be made clear when we discuss Onsager relations for the thermoelectric effect later in this section. That reciprocal relations $L_{jk} = L_{kj}$ were associated with cross-effects had already been noticed by Lord Kelvin and others in the nineteenth century. The early explanations of the reciprocal relations were based on thermodynamic reasoning that was not on a firm footing. For this reason, Kelvin and others regarded the reciprocal relations only as conjectures. A well-founded theoretical explanation for these relations was developed by Onsager in 1931 [10]. Onsager's theory is based on the *principle of detailed balance* or *microscopic reversibility* that is valid for systems at equilibrium. Discussions of this principle and the derivation of these reciprocal relations are beyond the scope of this introductory chapter; they can be found in Refs [5] and [8].

ONSAGER RECIPROCAL RELATIONS IN THERMOELECTRIC PHENOMENA

As an illustration of the theory presented in the last two sections, let us consider thermoelectric effects, which involve the flow of heat J_q and electric current I_e in conducting wires (in which the subscript indicates that the flow corresponds to the flow of electrons). We shall consider a one-dimensional case in which all the gradients are along the x axis. For a one-dimensional system, such as a conducting wire, the vectorial aspect of $\mathbf{J}_q$ and $\mathbf{I}_e$ is unimportant and they may be treated as scalars. The entropy production per unit volume due to these two irreversible processes and the associated linear phenomenological laws are

Lars Onsager (1903–1976) (Reproduced courtesy of the AIP Emilio Segre Visual Archive)

$$\sigma = J_q \frac{\partial}{\partial x}\left(\frac{1}{T}\right) + \frac{I_e E}{T} \tag{11.3.4}$$

$$\boxed{J_q = L_{qq} \frac{\partial}{\partial x}\left(\frac{1}{T}\right) + L_{qe} \frac{E}{T}} \tag{11.3.5}$$

$$\boxed{I_e = L_{ee} \frac{E}{T} + L_{eq} \frac{\partial}{\partial x}\left(\frac{1}{T}\right)} \tag{11.3.6}$$

In the above equations, E is the electric field (volts per unit length). To relate the coefficients L_{qq} and L_{ee} with the heat conductivity κ and resistance R, we compare these equations with those in Table 11.2. Fourier's law of heat conduction of heat conduction is valid when the electric field $E = 0$. Comparing the heat conduction term $J_q = -(1/T^2)L_{qq}(\partial T/\partial x)$ with Fourier's law leads to the identification

$$\kappa = \frac{L_{qq}}{T^2} \tag{11.3.7}$$

We can now specify more precisely what is meant by the **near-equilibrium linear regime**; by this we mean that L_{qq}, L_{ee}, etc. may be treated as constants. This is a valid assumption only when κT^2 can be treated as constant independent of position. Since $T(x)$ is a function of position, such an assumption is strictly not valid. It is valid only in the approximation that the change in T from one end of the system to another is small compared with the average T, i.e. if the average temperature is T_{avg}, then $|T(x) - T_{avg}|/T_{avg} \ll 1$ for all x. Hence, we may approximate $T^2 \approx T_{avg}^2$ and use κT_{avg}^2 in place of κT^2.

To find the relation between L_{ee} and the resistance R, we note that $V = -\Delta\phi = \int_0^l E\, dx$, in which l is the length of the system. The current I_e is a constant in the entire system. At constant temperature ($\partial T/\partial x = 0$), the current is due entirely to electrical potential difference. With $\partial T/\partial x = 0$, integrating (11.3.6) over the length l of the system we obtain

$$\int_0^l I_e\, dx = \frac{L_{ee}}{T}\int_0^l E\, dx \quad \text{or} \quad I_e l = \frac{L_{ee}}{T}V \tag{11.3.8}$$

Comparing this equation with Ohm's law (Table 11.2), we make the identification

$$L_{ee} = \frac{T}{R/l} = \frac{T}{r} \tag{11.3.9}$$

in which r is the *resistance per unit length*. As noted in Table 11.2, Ohm's law can also be stated in general as $I_e = E/\rho$, in which ρ is the *specific resistance*, I is the current density and E the electric field. Comparing this expression with (11.3.9), we have the general relation

$$L_{ee} = \frac{T}{\rho} \tag{11.3.10}$$

When we consider a one-dimensional system, ρ is replaced by r, the resistance per unit length.

The Seebeck effect

Let us now consider relating L_{qe} and L_{eq} to experimentally measured quantities. In the Seebeck effect (Figure 11.2), a temperature difference between two junctions of dissimilar metals produces an EMF. For this system, Equations (11.3.5) and (11.3.6) may be used. The EMF is measured at zero current. Setting $I_e = 0$ in (11.3.6) we obtain

$$0 = L_{ee}ET - L_{eq}\frac{\partial}{\partial x}T \tag{11.3.11}$$

Table 11.3 Some experimental data confirming Onsager reciprocal relations

Thermocouple	$T/°C$	$\pi/T /\mu V\ K^{-1}$	$-\Delta\phi/\Delta T /\mu V\ K^{-1}$	L_{qe}/L_{eq}
Cu–Al	15.8	2.4	3.1	0.77
CuNi	0	18.6	20.0	0.930
CuNi	14	20.2	20.7	0.976
CuFe	0	−10.16	−10.15	1.000
CuBi	20	−71	−66	1.08
FeNi	16	33.1	31.2	1.06
FeHg	18.4	16.72	16.66	1.004

Source: Miller, D.G., *Chem. Rev.*, **60** (1960) 15–37.

This equation may be integrated to obtain a relation between the temperature difference ΔT and the EMF generated due to this temperature difference $\Delta\phi = -\int_0^l E\,dx$. In doing this integration, we shall assume that the total variation $\int_0^l dT = \Delta T$ is small and make the approximation $\int_0^l TE\,dx \approx T\int_0^l E\,dx = -T\Delta\phi$. This gives us the relation

$$L_{eq} = -L_{ee}T\left(\frac{\Delta\phi}{\Delta T}\right)_{I=0} \tag{11.3.12}$$

Experimentally, the ratio $-(\Delta\phi/\Delta T)_{I=0}$, called the **thermoelectric power**, is measured. Some typical values of this quantity are shown in Table 11.3. As can be seen from this table, the thermoelectric power may be of either sign. Using (11.3.12), the coefficient L_{eq} can be related to the measured quantities.

The Peltier effect

In the Peltier effect, the two junctions are maintained at a constant temperature while a current I is passed through the system (Figure 11.2). This causes a flow of heat from one junction to another. The two junctions are maintained at the same temperature only by removing heat from one of the junctions and thus maintaining a steady heat flow J_q. Under these conditions, the ratio

$$\Pi = \frac{J_q}{I_e} \tag{11.3.13}$$

which can be measured, is called the **Peltier heat**. Some typical values of Π/T are shown in Table 11.3. The phenomenological coefficient L_{qe} can be related to the Peltier heat as follows. Since there is no temperature difference between the two junctions, $\partial T/\partial x = 0$, and Equations (11.3.5) and (11.3.6) become

$$J_q = L_{qe}\frac{E}{T} \tag{11.3.14}$$

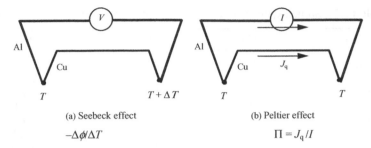

(a) Seebeck effect   (b) Peltier effect

$-\Delta\phi/\Delta T$ $\Pi = J_q/I$

Fig. 11.2 An example of a 'cross effect' between thermodynamic forces and flows is the thermoelectric effect. (a) In the Seebeck effect, two dissimilar metal wires are joined and the junctions are maintained at different temperatures. As a result, an EMF is generated. The EMF generated is generally of the order of 10^{-5} V per kelvin of temperature difference and it may vary from sample to sample. (b) In the Peltier effect, the two junctions are maintained at the same temperature and an electric current is passed through the system. The electric current I drives a heat flow J_q from one junction to the other. The Peltier heat current is generally of the order of 10^{-5} J s^{-1} A^{-1}.

$$I_e = L_{ee} \frac{E}{T} \tag{11.3.15}$$

Dividing (11.3.14) by (11.3.15) we see $J_q/I_e = L_{qe}/L_{ee}$. Using (11.3.13) and (11.3.9) we obtain

$$L_{qe} = \Pi L_{ee} = \Pi \frac{T}{R/l} = \Pi \frac{T}{r} \tag{11.3.16}$$

In this manner, all the phenomenological coefficients L_{qe} and L_{eq} can be related to the experimental parameters of the cross-effects.

Having identified all the linear phenomenological coefficients in terms of the experimentally measured quantities, we can now turn to the reciprocal relations according to which one must find

$$L_{qe} = L_{eq} \tag{11.3.17}$$

Upon using (11.3.12) for L_{eq} and (11.3.16) for L_{qe} we find

$$-L_{ee}T\left(\frac{\Delta\phi}{\Delta T}\right) = \Pi L_{ee} \quad \text{or} \quad \boxed{-\left(\frac{\Delta\phi}{\Delta T}\right) = \frac{\Pi}{T}} \tag{11.3.18}$$

Experimental data verifying this prediction for some pairs of conductors is shown in Table 11.3. Other reciprocal relations predicted by Onsager have also been verified experimentally [8, 11]. Onsager reciprocal relations are excellent demonstrations of the usefulness of the modern formulation of thermodynamics to irreversible processes.

11.4   Symmetry-Breaking Transitions and Dissipative Structures

The increase of entropy has often been associated with increase of disorder, the destruction of patterns and establishment of uniformity. In many instances this is indeed true: diffusion destroys patterns of inhomogeneous distributions of matter, and heat conduction tends to make the temperature uniform. For this reason, it is sometimes claimed that the evolution of an organized state, such as living cells, is in violation of the Second Law. But it is not true that irreversible processes *always* destroy structure and order or organized states. In fact, *when a system is far from thermodynamic equilibrium, irreversible processes can drive a system to evolve spontaneously to organized states*. Irreversible processes that make life possible produce entropy just as much as those that destroy it; both are in accord with the Second Law. In this section we will consider some examples of nonequilibrium systems which spontaneously make transitions to organized states or organized structures. Since the nonequilibrium organized structures are a result of irreversible processes that dissipate free energy and produce entropy, they are called **dissipative structures**, a concept introduced by Ilya Prigogine. In open systems, dissipative structures can be maintained indefinitely through a flow of matter and energy.

Ilya Prigogine (1917–2003) (Reproduced courtesy of the University of Texas at Austin)

GENERAL FEATURES OF DISSIPATIVE STRUCTURES

All dissipative structures have some general features that could be seen through a simple example, viz., the convection patterns that were briefly described in Chapter 2 (Figure 11.3). These patterns arise in a fluid placed between two horizontal plates. The two plates are maintained at different temperatures by appropriate heating and

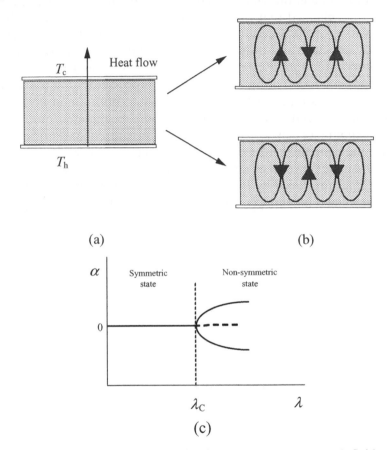

(a)   (b)

(c)

Figure 11.3   A simple example of a **dissipative structure**. A fluid is placed between two plates and heated from below. The temperature difference $\Delta T = T_h - T_c$ between the two plates drives a heat flow. (a) When ΔT is small, the heat flow is due to conduction and the fluid is static. (b) When ΔT exceeds a critical value, organized convection patterns emerge spontaneously. (c) General features of dissipative structures. α is an **order parameter**, which for the convective pattern could be the speed of convective flow at the center. λ is the **critical parameter**, which can be taken as ΔT for the convective instability. When $\lambda < \lambda_c$ the system is in a state that has the symmetries of the equilibrium state, such as spatial homogeneity or time translation invariance. When $\lambda > \lambda_c$, the system makes a transition to a state which does not have the symmetries of the equilibrium state. These states are generally organized sates, such as the convection pattern in (b)

cooling. The lower plate is maintained at a higher temperature T_h and the upper plate is at a lower temperature T_c. The temperature difference $\Delta T = T_h - T_c$, is a measure of the nonequilibrium conditions of this system. $\Delta T = 0$ is the equilibrium state. When ΔT is close to zero, heat flow from the lower plate to the upper plate is through heat conduction; the fluid is static. As ΔT increases, there comes a point at which the static state of the fluid becomes unstable and convection begins; the whole fluid then becomes organized into convection cells (Figure 11.3b). The convection cells are organized into symmetric patterns. The establishment of these symmetric patterns is entirely a consequence of irreversible processes in a fluid in the presence of a gravitational field. Entropy-producing irreversible processes create the organized convection cells.

The transition to organized states when the value of a parameter, such as ΔT, is above a threshold or a critical value is a general feature of all transitions to dissipative structures. It is somewhat similar to phase transitions that take place at a critical temperature.

There is one other general feature we can also see in this example: *a change of symmetry of the system*. Below the convection threshold, the fluid is the same at all points on a horizontal plane (ignoring the obvious effects at the boundary), i.e. the system has a translational symmetry. Above the threshold, this symmetry is 'broken' by the convection patterns. Furthermore, there are two possible convection patterns (Figure 11.3b) corresponding to clockwise or counterclockwise flow of a particular convection cell. Of the two possible convection patterns, the system will evolve to one depending on random fluctuations and other influences that might favor one of the two patterns. This is the phenomenon of **spontaneous symmetry breaking** in nonequilibrium systems.

To describe these features precisely, two variables are defined. The first variable is the **critical parameter** λ. It is generally a measure of distance from equilibrium. In many situations the critical parameter can be defined so that $\lambda = 0$ corresponds to the equilibrium state; when $\lambda \geq \lambda_c$, the system makes a transition to a dissipative structure. λ_c is called the **critical value** of the parameter λ. The second variable is called the **order parameter** or **amplitude of the dissipative structure** α. It is a measure of the organized state. It can be defined in such a way that $\alpha = 0$ when $\lambda \leq \lambda_c$, and $\alpha \neq 0$ when $\lambda \geq \lambda_c$, as shown in Figure 11.3c. The two branches representing $\alpha \neq 0$ correspond to the two possible convection patterns in Figure 11.3b. The symmetry of the system changes when α changes from zero to a nonzero value. Figure 11.3c summarizes these general features of dissipative structures.

SPONTANEOUS CHIRAL SYMMETRY BREAKING

The concept of dissipative structures and the breaking of symmetry in nonequilibrium systems give us insight into a very intriguing asymmetry in nature. The chemistry of life as we know it is founded on a remarkable asymmetry. A molecule whose geometrical structure cannot be superimposed on its mirror image is said to possess **chirality**, or handedness. Mirror-image structures of a chiral molecule are called **enantiomers**. Just as we distinguish the left and the right hands, the two mirror-image

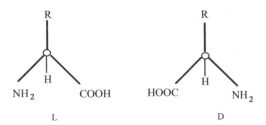

Figure 11.4 The L and D forms of amino
acids. With only very rare exceptions, the
amino acids in living cells are of the L form

structures are identified as L- and D-enantiomers (L for 'levo' and D for 'dextro'; R and S is another convention for identifying the two enantiomers). Amino acids, the building blocks of proteins, and sugars in DNA are chiral molecules. From bacteria to man, nearly all amino acids that take part in the chemistry of life are L-amino acids and the sugars in DNA and RNA are D-sugars (Figure 11.4). As Francis Crick notes [12]:

The first great unifying principle of biochemistry is that the key molecules have the same hand in all organisms.

Biomolecular asymmetry is all the more remarkable because chemical reactions show equal preference for the two mirror-image forms (except for very small differences due to parity nonconserving electroweak interactions [13–15]).

Biochemistry's hidden asymmetry was discovered by Louis Pasteur in 1857. A century-and-half later, its true origin remains an unsolved problem, but we can see how such a state might be realized in the framework of dissipative structures. First, we note that such an asymmetry can arise only under far-from-equilibrium conditions; at equilibrium, the concentrations of the two enantiomers will be equal. The maintenance of this asymmetry requires constant catalytic production of the preferred enantiomer in the face of interconversion between enantiomers, called **racemization**. (Racemization drives the system to the equilibrium state in which the concentrations of the two enantiomers will become equal.) Second, the transition from a symmetric state, which contains equal amounts of L- and D-enantiomers, to a state of broken symmetry or an asymmetric state, in which the two enantiomers are present in unequal amounts, must happen when the value of an appropriately defined critical parameter exceeds a threshold value. These features can be illustrated using a model reaction.

In 1953, F.C. Frank [16] devised a simple model reaction scheme with chiral autocatalysis that could amplify a small initial asymmetry. We shall modify this reaction scheme so that its nonequilibrium aspects, instability and transition to an asymmetric state can be clearly seen. The reaction scheme we consider, which includes chirally autocatalytic reactions, is shown below.

$$S + T \underset{k_{1r}}{\overset{k_{1f}}{\rightleftharpoons}} X_L \tag{11.4.1}$$

$$S + T + X_L \underset{k_{2r}}{\overset{k_{2f}}{\rightleftharpoons}} 2X_L \tag{11.4.2}$$

$$S + T \underset{k_{1r}}{\overset{k_{1f}}{\rightleftharpoons}} X_D \tag{11.4.3}$$

$$S + T + X_D \underset{k_{2r}}{\overset{k_{2f}}{\rightleftharpoons}} 2X_D \tag{11.4.4}$$

$$X_L + X_D \overset{k_{3f}}{\longrightarrow} P \tag{11.4.5}$$

Each enantiomer of X is produced directly from the achiral* reactants S and T, as shown in (11.4.1) and (11.4.3), and autocatalytically, as shown in (11.4.2) and (11.4.4). In addition, the two enantiomers react with one another and turn into an inactive P. Owing to the chiral symmetry of chemical processes, the rate constants for the direct reactions, (11.4.1) and (11.4.3), as well as the autocatalytic reactions, (11.4.2) and (11.4.4), must be equal. It is easy to see that at equilibrium the system will be in a *symmetric state*, i.e. $[X_L] = [X_D]$ (Exercise 11.4).

Now let us consider an open system into which S and T are pumped and P removed. For mathematical simplicity, we assume that the pumping is done in such a way that the concentrations [S] and [T] are maintained at a fixed level and that due to removal of P the reverse reaction in (11.4.5) may be ignored. The kinetic equations of this system are

$$\frac{d[X_L]}{dt} = k_{1f}[S][T] - k_{1r}[X_L] + k_{2f}[X_L][S][T] - k_{2r}[X_L]^2 - k_3[X_L][X_D] \tag{11.4.6}$$

$$\frac{d[X_D]}{dt} = k_{1f}[S][T] - k_{1r}[X_D] + k_{2f}[X_D][S][T] - k_{2r}[X_D]^2 - k_3[X_L][X_D] \tag{11.4.7}$$

To make the symmetric and asymmetric states explicit, it is convenient to define the following variables:

$$\lambda = [S][T] \quad \alpha = \frac{[X_L] - [X_D]}{2} \quad \beta = \frac{[X_L] + [X_D]}{2} \tag{11.4.8}$$

When Equations (11.4.6) and (11.4.7) are rewritten in terms of α and β we have (Exercise 11.5)

$$\frac{d\alpha}{dt} = -k_{1r}\alpha + k_{2f}\lambda\alpha - 2k_{2r}\alpha\beta \tag{11.4.9}$$

---

*Objects that do not possess a sense of handedness are called achiral. The molecule NH_3 is an example of an achiral molecule.

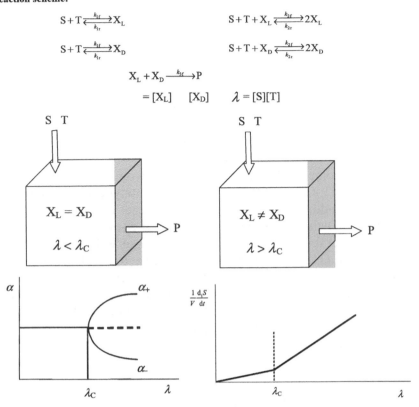

Reaction scheme:

$$S + T \underset{k_{1r}}{\overset{k_{1f}}{\rightleftharpoons}} X_L \qquad\qquad S + T + X_L \underset{k_{2r}}{\overset{k_{2f}}{\rightleftharpoons}} 2X_L$$

$$S + T \underset{k_{1r}}{\overset{k_{1f}}{\rightleftharpoons}} X_D \qquad\qquad S + T + X_D \underset{k_{2r}}{\overset{k_{2f}}{\rightleftharpoons}} 2X_D$$

$$X_L + X_D \overset{k_{3f}}{\longrightarrow} P$$

$$= [X_L] \qquad [X_D] \qquad \lambda = [S][T]$$

S   T   S   T

$$X_L = X_D \qquad\qquad\qquad X_L \neq X_D$$

$$\lambda < \lambda_C \qquad\qquad\qquad \lambda > \lambda_C$$

→ P   → P

Figure 11.5 A simple autocatalytic reaction scheme in which X_L and X_D are produced with equal preference. However, in an open system this leads to a dissipative structure in which $X_L \neq X_D$, a state of broken symmetry. At the bottom is a diagram showing some general features of transitions to dissipative structures and the way the rate of entropy production σ behaves as a function of the critical parameter λ

$$\frac{d\beta}{dt} = k_{1f}\lambda - k_{1r}\beta + k_{2f}\lambda\beta - k_{2r}(\beta^2 + \alpha^2) - k_3(\beta^2 - \alpha^2) \qquad (11.4.10)$$

The above system has the reaction mechanism needed to make a transition to a dissipative structure in which the chiral symmetry between the two enantiomers is broken. The breaking of chiral symmetry is in the following sense: though the reaction mechanism and the rates for the production of the two enantiomers are identical, the system reaches a state in which the concentrations of the two enantiomers are unequal. The overall features of this system are similar to those in the previous example of convection patterns. For this open system, whose kinetics are described by (11.4.6) and (11.4.7), the critical parameter λ and the order parameter α can be defined as shown in (11.4.8). The behavior of this system is summarized in Figure 11.5. The symmetric state, in which the amounts of X_L and X_D are equal, is charac-

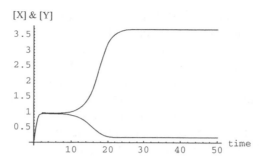

Figure 11.6 Time evolution of X_L and X_D, obtained using Mathematica Code A given in Appendix 11.1, showing how, for $\lambda > \lambda_c$, a small initial fluctuation in X_L grows to establish a state of broken symmetry in which the concentrations of X_L and X_D are unequal. Initial values: $[X_L]_0 = 0.002$ and $[X_D]_0 = 0.0$

terized by $\alpha = 0$; the asymmetric state is one in which $\alpha \neq 0$. There exists a critical value of λ beyond which the symmetric state becomes unstable and the system makes a transition to an asymmetric state. In the asymmetric state, $\alpha > 0$ or $\alpha < 0$, each corresponding to the dominance of one of the enantiomers, X_L or X_D. Using the Mathematica code provided for the numerical solutions of (11.4.6) and (11.4.7) in Appendix 11.1, these features can be studied and the behavior of this system explored (Figure 11.6). A full mathematical analysis and the determination of the critical value λ_c is beyond the scope of this introductory chapter, but it can be found in Ref. [8]. The rate of entropy production in this system can also be computed using the general formula (9.5.11)

$$\sigma = \frac{1}{V}\frac{d_i S}{dt} = \frac{1}{V}\sum_k \frac{A_k}{T}\frac{d\xi_k}{dt} = R\sum_k (R_{kf} - R_{kr})\ln(R_{kf}/R_{kr}) \qquad (11.4.11)$$

in which R_{kf} and R_{kr} are the forward and reverse reaction rates of reactions (11.4.1)–(11.4.5) and R is the universal gas constant. The behavior of σ as a function of λ is shown in Figure 11.5; there is a sharp change in its slope at the transition point.

The above example gives us insight into how the chiral asymmetry we see in the molecules of life might have arisen under prebiotic conditions. It is also clear that the maintenance of this asymmetric state requires continuous production of entropy. The 'first great unifying principle of biochemistry' is a clear indication that the entire edifice of life stands as a dissipative structure.

11.5 Chemical Oscillations

Oscillation of a pendulum and the orbital motion of planets are by far the most commonly discussed periodic phenomena in physics. Their description is a triumph of Newtonian mechanics. But there is one fundamental aspect that must be noted about these quintessential 'clocks': they are governed by time-reversible laws of mechanics and, as such, make no fundamental distinction between the system's evolution into the future and into the past. The equations of motion remain invariant when time t is replaced by $-t$. This means that by looking at such periodic motion it is impossible to say whether we are seeing evolution of the system into the past or into the future. If we imagine the motion of a pendulum or a planet running backwards in time, we see another possible motion that is consistent with all the laws of mechanics. The same is true for all periodic phenomena described using time-reversible laws. But there is a wide range of periodic phenomena that we see all around us that is fundamentally different from pendulums and planets: the beating of the heart, the flashing of a firefly, and the chirping of a cricket are governed by irreversible processes that are not time symmetric. As a consequence, if we imagine these periodic phenomena running backwards in time then we see processes that are impossible, i.e. processes that cannot be realized because they violate the Second Law. Their description is in the realm of thermodynamics, not mechanics. In this section we will see examples of periodic phenomena in chemical systems; irreversible chemical kinetic equations describe their behavior. In these systems, the concentrations of the reactants oscillate with clocklike periodicity and the process continuously generates entropy.

Some early reports on observations of concentration oscillations were discounted because it was widely believed that such behavior was not consistent with thermodynamics. It was for this reason that the report on oscillating reactions by Bray in 1921 and Belousov in 1958 were met with skepticism [17]. While it is true that oscillations of the extent of reaction ξ about its equilibrium value violate the Second Law, oscillation of concentrations can occur in nonequilibrium systems with continuous increase in ξ without any violation of the Second Law. A continuous increase in entropy accompanies such oscillations. When it was realized that systems far from thermodynamic equilibrium could exhibit oscillations, interest in these and other oscillating reactions rose sharply and gave rise to a rich study of dissipative structures in chemical systems.

THE BRUSSELATOR

In 1968, Prigogine and Lefever [18] developed a simple model chemical reaction which demonstrated clearly how a nonequilibrium system can become unstable and make a transition to an oscillatory state. This model also proved to be a rich source for theoretical understanding of propagating waves and almost every other phenomenon observed in real chemical systems that are generally extremely complex. Owing to its enormous impact on the study of dissipative structures, it often called the **Brusselator** (after its place of origin, the Brussels school of thermodynamics) or the 'tri-molecular model' due to the tri-molecular autocatalytic step in the reaction scheme. Because of its theoretical simplicity, we shall first discuss this reaction. The reaction scheme of the Brusselator (in which the reverse reactions rates are assumed to be very small and, hence, can be ignored) is the following:

$$A \xrightarrow{k_{1f}} X \tag{11.5.1}$$

$$B + X \xrightarrow{k_{2f}} Y + D \tag{11.5.2}$$

$$2X + Y \xrightarrow{k_{3f}} 3X \tag{11.5.3}$$

$$X \xrightarrow{k_{4f}} E \tag{11.5.4}$$

The net reaction of this scheme is A + B → D + E. We assume that the concentrations of the reactants A and B are maintained at a desired nonequilibrium value through appropriate flows. The products D and E are removed as they are formed. We also assume that the reaction occurs in a solution that is well stirred and, hence, homogeneous and write the following rate equations for the species X and Y:

$$\frac{d[X]}{dt} = k_{1f}[A] - k_{2f}[B][X] + k_{3f}[X]^2[Y] - k_{4f}[X] \tag{11.5.5}$$

$$\frac{d[Y]}{dt} = k_{2f}[B][X] - k_{3f}[X]^2[Y] \tag{11.5.6}$$

The behavior of the intermediates X and Y depends on the relative values of [A] and [B] in the system. The **steady-state solutions** to (11.5.5) and (11.5.6) are defined as those for which $d[X]/dt = 0$ and $d[Y]/dt = 0$. One can easily verify that the stationary solutions are (Exercise 11.6)

$$[X]_s = \frac{k_{1f}}{k_{4f}}[A] \quad [Y]_s = \frac{k_{4f}k_{2f}}{k_{3f}k_{1f}}\frac{[B]}{[A]} \tag{11.5.7}$$

When the steady state is close to equilibrium it is stable, i.e. if the system is momentarily perturbed by a random fluctuation it returns to the steady state. But as the system is driven away from equilibrium through an appropriate inflow of A and B and an outflow of D and E, the steady state (11.5.7) becomes unstable and the system makes a transition to a periodic state in which the concentrations of [X] and [Y] oscillate. The threshold for the transition to the oscillatory state is given by

$$[B] > \frac{k_{4f}}{k_{2f}} + \frac{k_{3f}k_{1f}^2}{k_{2f}k_{4f}^2}[A] \tag{11.5.8}$$

Using the Mathematica code provided in Appendix 11.1, numerical solutions to (11.5.5) and (11.5.6) can be obtained. With these solutions the steady states and the transition to oscillations can be easily investigated and the behavior of the system under various conditions can be explored. Figure 11.7 shows the oscillation in [X] and [Y] obtained using this code. The mathematical analysis of the

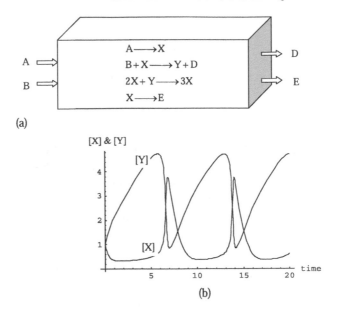

(a)

(b)

Figure 11.7 (a) The Brusselator as an open system with an inflow of A and B and outflow of D and E. When condition (11.5.8) is satisfied, the concentrations of X and Y oscillate. (b) Oscillations of X and Y of the Brusselator obtained using the Mathematica code given in Appendix 11.1, Code **B**

instability of the steady state and the transition to oscillating state can be found in Ref. [8].

THE BELOUSOV–ZHABOTINSKY REACTION

Once it became clear that concentration oscillations are not inconsistent with the laws of thermodynamics (as the theoretical models of oscillating reactions showed), the neglected 1958 report by Belousov and the later experiments of Zhabotinsky reported in a 1964 article [19] gained interest. The experimental studies of Belousov and Zhabotinsky on oscillating chemical reactions, which were conducted in the Soviet Union, were made known to the Western world through the Brussels School of thermodynamics headed by Ilya Prigogine. In the USA, the study of the 'Belousov–Zhabotinsky' oscillations was taken up by Field, Körös and Noyes [20], who performed a through study of the reaction mechanism in the early 1970s. This was an important landmark in the study of oscillating reactions. Field, Körös and Noyes identified the key steps in the rather complex **Belousov–Zhabotinsky reaction** and developed a model, which we shall refer to as the **FKN model**, consisting of only three variables that showed how the oscillations might arise.

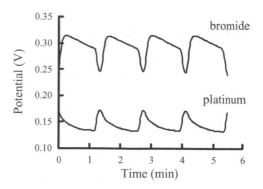

Figure 11.8 Experimentally observed oscillations in the Belousov–Zhabotinsky reaction. The concentrations are measured using electrodes. The oscillations shown are of [Br⁻]. (Courtesy: John Pojman)

The Belousov–Zhabotinsky reaction is basically catalytic oxidation of an organic compound such as malonic acid ($CH_2(COOH)_2$). The reaction occurs in an aqueous solution and is easily performed in a beaker by simply adding the following reactants in the concentrations shown:

$$[H^+] = 2.0\,\text{M} \quad [CH_2(COOH)_2] = 0.28\,\text{M} \quad [BrO_3^-] = 6.3 \times 10^{-2}\,\text{M} \quad [Ce^{4+}] = 2.0 \times 10^{-3}\,\text{M}$$

After an initial 'induction' period, the oscillatory behavior can be seen in the variation of the concentration of the Ce^{4+} ion, due to which there is a change in color from colorless to yellow. Many variations of this reaction (with more dramatic variations of color) are known today.

Box 11.1 contains a simplified version of the reaction mechanism based on which the FKN model was developed. Later models of the Belousov–Zhabotinzky reactions have included as many as 22 reactions steps. The FKN model of the Belousov–Zhabotinsky reaction makes the following identification: $A = [BrO_3^-]$, $X = [HBrO_2]$, $Y = [Br^-]$, $Z = [Ce^{4+}]$, $P = [HBrO]$ and $B = [Org]$, the organic species that is oxidized. In modeling the reaction, $[H^+]$ is absorbed in the definition of the rate constant. The reaction scheme consists of the following steps:

			Rate	
Generation of $HBrO_2$:	$A + Y \to X + P$		$k_{1f}[A][Y]$	(11.5.9)
Autocatalytic production of $HBrO_2$:	$A + X \to 2X + 2Z$		$k_{2f}[A][X]$	(11.5.10)
Consumption of $HBrO_2$:	$X + Y \to 2P$		$k_{3f}[X][Y]$	(11.5.11)
	$2X \to A + P$		$k_{4f}[X]^2$	(11.5.12)
Oxidation of the organic reactants:	$B + Z \to (f/2)Y$		$k_{5f}[B][Z]$	(11.5.13)

Box 11.1 The Belousov-Zhabotinsky Reaction and The FKN Model.

The Field–Körös–Noyes (FKN) model of the Belousov–Zhabotinsky reaction consists of the following steps, with $A = [BrO_3^-]$, $X = [HBrO_2]$, $Y = [Br^-]$, $Z = [Ce^{4+}]$, $P = [HBrO]$ and $B = [Org]$ is the organic species that is oxidized . In modeling the reaction, $[H^+]$ is absorbed in the definition of the rate constant.

- Generation of $HBrO_2$: $A + Y \rightarrow X + P$

$$BrO_3^- + Br^- + 2H^+ \rightarrow HBrO_2 + HBrO \qquad (BZ1)$$

- Autocatalytic production of $HBrO_2$: $A + X \rightarrow 2X + 2Z$

$$BrO_3^- + HBrO_2 + H^+ \rightarrow 2BrO_2 + H_2O \qquad (BZ2)$$

$$BRO_2 + Ce^{3+} + H^+ \rightarrow HBrO_2 + Ce^{4+} \qquad (BZ3)$$

The net reaction, (BZ2) + 2(BZ3), is autocatalytic in $HBrO_2$. Since the rate-determining step is (BZ2), the reaction is modeled as $BrO_3^- + HBrO_2 \xrightarrow{H^+, Ce^{3+}} 2Ce^{4+} + 2HBrO_2$

- Consumption of $HBrO_2$: $X + Y \rightarrow 2P$ and $2X \rightarrow A + P$

$$HBrO_2 + Br^- + H^+ \rightarrow 2HBrO \qquad (BZ4)$$

$$2HBrO_2 \rightarrow BrO_3^- + HBrO + H^+ \qquad (BZ5)$$

- Oxidation of the organic reactants: $B + Z \rightarrow (f/2)Y$

$$CH_2(COOH)_2 + Br_2 \rightarrow BrCh(COOH)_2 + H^+ + Br^- \qquad (BZ6)$$

$$Ce^{4+} + \frac{1}{2}[CH_2(COOH)_2 + BrCH(COOH)_2] \rightarrow \frac{f}{2}Br^- + Ce^{3+} + Products \qquad (BZ7)$$

The oxidation of the organic species is a complex reaction. It is approximated by a single reaction in which (BZ7) is the rate-limiting step. In the FKN model, concentration '[B]' of the organic species is assumed to be constant. The value of the effective stoichiometric coefficient f is a variable parameter. Oscillations occur if f is in the range 0.5–2.4.

The corresponding rate equations for [X], [Y] and [Z] are:

$$\frac{d[X]}{dt} = k_{1f}[A][Y] + k_{2f}[A][X] - k_{3f}[X][Y] - 2k_{4f}[X]^2 \qquad (11.5.14)$$

$$\frac{d[Y]}{dt} = -k_{1f}[A][Y] - k_{3f}[X][Y] + \frac{f}{2}k_{5f}[B][Z] \qquad (11.5.15)$$

$$\frac{d[Z]}{dt} = 2k_{2f}[A][X] - k_{5f}[B][Z] \qquad (11.5.16)$$

As in the previous example, the oscillatory behavior of these equations may be studied numerically quite easily using the Mathematica code (code C) provided in

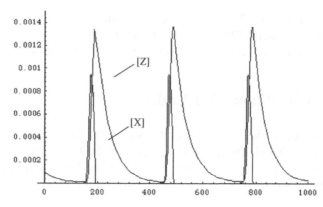

Figure 11.9 Oscillatory solutions obtained numerically using the FNK model of the Belousov–Zhabotinsky reaction. In the FKN model, [X] = [HBrO₂] and [Z] = [Ce⁴⁺]. For the Mathematica code used to obtain these oscillations, see Appendix 11.1, Mathematica Code C

Appendix 11.1 (Figure 11.9). For numerical solutions of the above reaction, one may use the following data [21]:

$$k_{1f} = 1.28\,\mathrm{mol^{-1}\,L\,s^{-1}} \qquad k_{2f} = 8.0\,\mathrm{mol^{-1}\,L\,s^{-1}} \qquad k_{3f} = 8.0\times10^{-5}\,\mathrm{mol^{-1}\,L\,s^{-1}}$$
$$k_{4f} = 2.0\times10^{3}\,\mathrm{mol^{-1}\,L\,s^{-1}} \qquad k_{5f} = 1.0\,\mathrm{mol^{-1}\,L\,s^{-1}} \tag{11.5.17}$$
$$[B] = [\mathrm{Org}] = 0.02\,\mathrm{M} \qquad A = [\mathrm{BrO_3^-}] = 0.06\,\mathrm{M} \qquad 0.5 < f < 2.4$$

The Belousov–Zhabotinsky reaction shows oscillations of great variety and complexity; it even exhibits chaos. In chaotic systems arbitrarily close initial conditions diverge exponentially; the system exhibits aperiodic behavior. A review by Epstein and Showalter summarizes these features [22]. It also produces propagating waves and multistability. A large number of very interesting phenomena have been observed in this reaction [21, 23].

OTHER OSCILLATING REACTIONS

During the last three decades, many more oscillating chemical reactions have been discovered. Indeed, Irving Epstein and coworkers have developed a systematic way of designing oscillating chemical reactions· [24, 25]. In biochemical systems, one of the most interesting oscillating behaviors is found in the glycolytic reaction. A recent monograph by Albert Goldbeter [26] summarizes the vast amount of work done on oscillatory biochemical systems.

The above examples of dissipative structures are just the tip of the iceberg of the rich and complex world of far-from-equilibrium systems. Chemical oscillations, propagating waves, and chaotic unpredictable behavior emerging from perfectly deterministic kinetic equation have all been identified in physico-chemical systems. The text by

Epstein and Pojman [27] is an excellent introduction to this subject. An earlier book by Nicolis and Prigogine [28] gives a more advanced overview. For further reading in this subject, a list of titles is provided in the 'Further reading' section.

Appendix 11.1 Mathematica Codes

The following codes give numerical solutions for the chemical kinetic equations discussed in this chapter. As in Chapter 9, NDSolve is used to obtain numerical solutions. The results can be plotted using the Plot command. Numerical output can be exported to graphing software using the Export command.

CODE A: CHIRAL SYMMETRY BREAKING

```
(* Code to show chiral symmetry breaking *)
k1f=0.5; k1r=0.01; k2f=0.5; k2r=0.2; k3=1.5; S=1.25; T=S;
R1f:=k1f*S*T; R1r:=k1r*XL[t];
R2f:=k2f*SXL[t]; R2r:=k2r*(XL[t])^2;
R3f:=k1f*S*T; R3r:=k1r*XD[t];
R4f:=k2f*SXD[t]; R4r:=k2r*(XD[t])^2;
R5f:=k3*XL[t]*XD[t];

Soln1=NDSolve[
    {XL'[t]== R1f-R1r + R2f-R2r -R5f,
     XD'[t]== R3f-R3r + R4f-R4r -R5f,
     XL[0]==0.002,XD[0]==0.0},
     {XL,XD},{t,0,100},
     MaxSteps->500]

Plot[Evaluate[{XL[t],XD[t]}/.Soln1],{t,0,50},
              AxesLabel->{"time","[XL]&[XD]"}]

{{XL -> InterpolatingFunction[{{0., 100.}}, <>],
  XD -> InterpolatingFunction[{{0., 100.}}, <>]}}
```

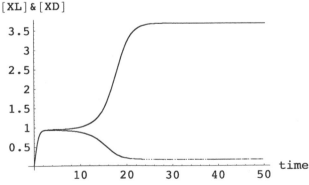

```
 - Graphics -
```

To write output files for spreadsheets use the 'Export' command and the file format List. For more detail see Mathematica help file for Export command. In the command below, the output filename is: data1.txt. This file can be read by most spreadsheets and graphing software.

The command 'X[t]/.Soln1' specifies that $X[t]$ is to be evaluated using Soln1 defined above. TableForm outputs data in a convenient form.

```
Export["data1.txt",
  Table[{t, {XL[t], XD[t]}/. Soln1}, {t, 1, 50}]//TableForm,
  "List"]

data1.txt
```

To obtain a table of t versus $X[t]$ the following command can be used:

```
Table[{t, {XL[t], XD[t]}/. Soln1}, {t, 1, 5}]//TableForm
```

```
1   0.736564   0.732883
2   0.923537   0.917889
3   0.94403    0.935624
4   0.94778    0.935304
5   0.950967   0.932455
```

CODE B: THE BRUSSELATOR

The following is the code for the Brusselator. Since no reverse reactions are involved, we shall not use the subscripts f and r for the reaction rates and rate constants.

```
(* Chemical Kinetics: The Brusselator *)

k1=1.0; k2=1.0; k3=1.0; k4=1.0; A=1.0; B=3.0;
R1:=k1*A; R2:=k2*B*X[t]; R3:=k3*(X[t]^2)*Y[t];
R4:=k4*X[t];

Soln2=NDSolve[{ X'[t]== R1-R2+R3-R4,
               Y'[t]== R2-R3,
               X[0]==1.0,Y[0]==1.0},
            {X,Y},{t,0,20},
            MaxSteps->500]

{{X → InterpolatingFunction[{{0., 20.}}, <>],
Y → InterpolatingFunction[{{0., 20.}}, <>]}} <CM>

Plot[[Evaluate[{X[t]}/.Soln2],{t,0,20}]
```

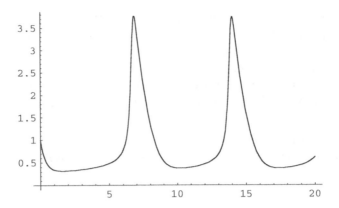

- Graphics -

```
Plot[[Evaluate[{X[t]}/.Soln2],{t,0,20}]
```

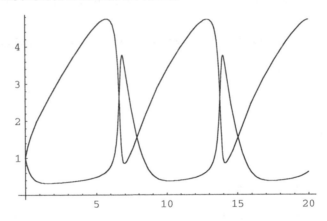

- Graphics -

```
Table[{t,Evaluate[{X[t],Y[t]}/.Soln2]},{t,0,10,1}]//TableForm
```

0	1.	1.
1	0.336806	2.13473
2	0.316948	2.83679
3	0.344197	3.48043
4	0.389964	4.0695
5	0.476017	4.55599
6	0.766371	4.68424
7	3.45347	0.851848
8	1.36828	1.64966
9	0.526015	2.63004
10	0.373263	3.36138

CODE C: THE BELUSOV–ZHABOTINSKY REACTION

The following is the FKN model of the Belousov–Zhabotinsky reaction. Since no reverse reactions are involved, we shall not use the subscripts f and r for the reaction rates and rate constants.

```
(* The Belousov-Zhabotinsky Reaction/FKN Model *)

(* X=HBrO2 Y=Br- Z=Ce4+ B=Org A=BrO3- *)

k1=1.28; k2=8.0; k3=8.0*10^5; k4=2*10^3; k5=1.0;
A=0.06; B=0.02;f=1.5;

R1:=k1*A*Y[t]; R2:=k2*A*X[t]; R3:=k3*X[t]*Y[t];
R4:=k4*X[t]^2; R5:=k5*B*Z[t];

Soln3=NDSolve[{
      X'[t]== R1+R2-R3-2*R4,
      Y'[t]== -R1-R3+(f/2)*R5,
      Z'[t]== 2*R2-R5,
      X[0]==2*10^-7,Y[0]==0.00002,Z[0]==0.0001},
    {X,Y,Z},{t,0,1000},
    MaxSteps->2000]

{{X → InterpolatingFunction[{{0., 1000.}}, <>],
  Y → InterpolatingFunction[{{0., 1000.}}, <>],
  Z → InterpolatingFunction[{{0., 1000.}}, <>]}}

Plot[Evaluate[{Z[t],10*X[t]}/.Soln3],{t,0,900},
PlotRange->{0.0,1.5*10^-3}]
```

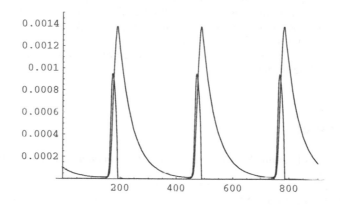

```
- Graphics -
```

References

1. Tribus, M., *Thermostatics and Thermodynamics*. 1961, New York: Van Nostrand.
2. Yourgrau, W., ven der Merwe, A. and Raw, G., *Treatise on Irreversible and Statistical Thermophysics*. 1982, New York: Dover (pp. 2–3).
3. Kleidon, A. and Lorentz, R.D., *Non-equilibrium Thermodynamics and the Production of Entropy: Life Earth and Beyond*. 2005, Berlin: Springer.
4. Prigogine, I., *Physica*, **15** (1949) 272.
5. De Groot, S.R. and Mazur, P. *Non-Equilibrium Thermodynamics*. 1969, Amsterdam: North Holland.
6. Baras, F. and Malek-Mansour, M., *Physica A*, **188** (1992) 253.

7. Jou, D., Casas-Vázquez, J., and Lebon, G. *Extended Irreversible Thermodynamics*. 1996, New York: Springer.
8. Kondepudi, D.K. and Prigogine, I. *Modern Thermodynamics: From Heat Engines to Dissipative Structures*. 1999. Chichester: Wiley.
9. Thomson, W., *Proc. Roy. Soc. (Edinburgh)*, **3** (1854) 225.
10. Onsager, L., *Phys. Rev.*, **37** (1931) 405.
11. Miller, D.G., *Chem. Rev.*, **60** (1960) 15.
12. Crick, F., *Life Itself*. 1981, New York: Simon and Schuster (p. 43).
13. Hegstrom, R. and Kondepudi, D.K., *Sci. Am.*, **262** (1990) 108.
14. Mason, S.F. and Tranter, G.E., *Chem. Phys. Lett.*, **94**(1) (1983) 34.
15. Hegstrom, R.A., Rein, D.W., and Sandars, P.G.H., *J. Chem. Phys.*, **73** (1980) 2329.
16. Frank, F.C., *Biochem. Biophys. Acta*, **11** (1953) 459.
17. Winfree, A.T., *J. Chem. Ed.*, **61** (1984) 661.
18. Prigogine, I. and Lefever, R., *J. Chem. Phys.*, **48** (1968) 1695.
19. Zhabotinsky, A.M., *Biophysika*, **9** (1964) 306.
20. Field, R.J., Körös, E., and Noyes, R.M., *J. Am. Chem. Soc.*, **94** (1972) 8649.
21. Gray, P. and Scott, K.S., *Chemical Oscillations and Instabilities*. 1990, Oxford: Clarendon Press.
22. Epstein, I. R. and Showalter, K.J., *J. Phys. Chem.*, **100** (1996) 13 132.
23. Field, R.J. and Burger, M. (eds), *Oscillations and Traveling Waves in Chemical Systems*. 1985, New York: Wiley.
24. Epstein, I., *et al.*, *Sci. Am.*, **248** (1983) 112.
25. Epstein, I.R., *J. Chem. Ed.*, **69**: (1989) 191.
26. Goldbeter, A., *Biochemical Oscillations and Cellular Rhythms: The Molecular Bases of Periodic and Chaotic Behaviour*. 1996, Cambridge: Cambridge University Press.
27. Epstein, I.R. and Pojman, J. *An Introduction to Nonlinear Chemical Dynamics*, 1998. Oxford: Oxford University Press.
28. Nicolis, G. and Prigogine, I., *Self-Organization in Nonequilibrium Systems*. 1977, New York: Wiley-Interscience.

Further Reading

Vidal, C. and Pacault, A. (eds), *Non-Linear Phenomenon in Chemical Dynamics*. 1981, Berlin: Springer.

Epstein, I., *et al.*, *Sci. Am.*, **248** (1983) 112.

Field, R.J. and Burger, M. (eds), *Oscillations and Traveling Waves in Chemical Systems*. 1985, New York: Wiley.

State-of-the-Art Symposium: Self-Organization in Chemistry. J. Chem. Ed., **66**(3) (1989). Articles by several authors.

Gray, P. and Scott, K.S., *Chemical Oscillations and Instabilities*. 1990, Oxford: Clarendon Press.

Manneville, P., *Dissipative Structures and Weak Turbulence*. 1990, San Diego: Academic Press.

Baras, F. and Walgraef, D. (eds), *Nonequilibrium Chemical Dynamics: From Experiment to Microscopic Simulation*. Physica A, 188(Special Issue) (1992).

CIBA Foundation Symposium. *Biological Asymmetry and Handedness*. 1991. London: John Wiley.

Kapral, R. and Showalter, K. (eds), *Chemical Waves and Patterns*. 1994, New York: Kluwer.

Goldbeter, A., *Biochemical Oscillations and Cellular Rhythms: The Molecular Bases of Periodic and Chaotic Behaviour*. 1996, Cambridge: Cambridge University Press.

Exercises

11.1 (a) Show the validity of (11.1.2). (b) Assume that the Gibbs relation $dU = T\,dS - p\,dV + \sum_k \mu_k dN_k$ is valid for a small volume element δV. Show that the relation $T\,ds = du - \sum_k \mu_k dn_k$ in which $s = S/\delta V$, $u = U/\delta V$ and $n_k = N_k/\delta V$ is also valid.

11.2 Calculate the relative value of fluctuations in $\tilde{N}$, $(\delta\tilde{N}/\tilde{N})$ in cell of volume $\Delta V = (1\,\mu m)^3 = 10^{-15}\,L$, filled with an ideal gas of at $T = 298\,K$ and $p = 1\,atm$.

11.3 For a positive definite 2×2 matrix, show that (11.3.3) must be valid.

11.4 Show that the equilibrium state of the set of reactions (11.4.1)–(11.4.5) must be chirally symmetric, i.e. $[X_L] = [X_D]$.

11.5 Using the definitions (11.4.8) in (11.4.6) and (11.4.7), obtain (11.4.9) and (11.4.10).

11.6 Show that the steady states of (11.5.5) and (11.5.6) are given by (11.5.7).

PART III

ADDITIONAL TOPICS

12 THERMODYNAMICS OF RADIATION

Introduction

Electromagnetic radiation that interacts with matter also reaches the state of thermal equilibrium with a definite temperature. This state of electromagnetic radiation is called **thermal radiation**; it was also called *heat radiation* in earlier literature. In fact, today we know that our universe is filled with thermal radiation at a temperature of about 2.73 K.

It has long been observed that heat transfer can take place from one body to another in the form of radiation with no material contact between the two bodies. This form of heat was called 'heat radiation'. When it was discovered that motion of charges produced electromagnetic radiation, the idea that heat radiation was a form of electromagnetic radiation was taken up, especially in the works of Gustav Kirchhoff (1824–1887), Ludwig Boltzmann (1844–1906), Josef Stefan (1835–1893) and Wilhelm Wien (1864–1928) and its thermodynamic consequences were investigated [1].

12.1 Energy Density and Intensity of Thermal Radiation

We begin by defining the basic quantities required to study the properties of thermal radiation (here, we follow the classic work of Planck on thermal radiation [1]). Radiation is associated with **energy density** u, which is the energy per unit volume, and **specific intensity** or **radiance** I, which is defined as follows (Figure 12.1a): the energy incident per unit time on a small area $d\sigma$ due to radiation from a solid angle $d\Omega$ ($=\sin\theta d\theta d\varphi$) which makes an angle θ with the surface normal equals $I\cos\theta d\Omega d\sigma$. The total amount of radiation incident on one side of the area $d\sigma$ (Figure 12.1b) is equal to $\int_{\theta=0}^{\pi/2}\int_{\varphi=0}^{2\pi} I\cos\theta \, d\Omega = \int_{\theta=0}^{\pi/2}\int_{\varphi=0}^{2\pi} I\cos\theta\sin\theta \, d\theta \, \varphi = \pi I$. The quantity πI is called the **radiation intensity**, **irradiance** or **radiant flux density**. A similar definition can be used for radiation emitted from a small surface area $d\sigma$, in which case πI is the emitted power per unit area of the surface called **radiation intensity** or **irradiance** or **radiant emittance**.

The energy density u and radiance I can also be defined as functions of frequency:

$u(v) \, dv$ is the **spectral energy density** of radiation in the frequency range v and $v + dv$ ($\mathrm{J\,m^{-3}\,Hz^{-1}}$)

Introduction to Modern Thermodynamics Dilip Kondepudi
© 2008 John Wiley & Sons, Ltd

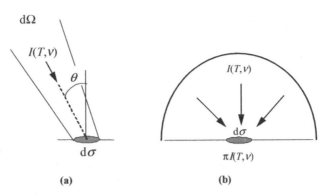

Figure 12.1 (a) Definition of spectral radiance $I(T, \nu)$. The energy flux incident on the area element $d\sigma$, from a solid angle $d\Omega = \sin\theta\,d\theta\,d\varphi$ is given by $I(T,\nu)\cos\theta\,d\Omega\,d\sigma$. Here, θ is the angle between the normal to $d\sigma$ and the incident radiation. (b) The total amount of radiation incident on $d\sigma$ from one side, the intensity of radiation, equals $\pi I(T, \nu)$. For electromagnetic radiation, the spectral intensity $\pi I(T, \nu)$ contains two independent states of polarization

$I(\nu)\,d\nu$ is the **spectral radiance** in the frequency range ν and $\nu + d\nu$ $(W\,Hz^{-1}\,sr^{-1}\,m^{-2})$

$\pi I(\nu)$ is the **spectral intensity** or **spectral irradiance** in the frequency range ν and $\nu + d\nu$ $(W\,Hz^{-1}\,m^{-2})$

There is a simple relationship between the spectral radiance $I(\nu)$ of radiation propagating at a velocity c and its energy density [1]:

$$u(\nu) = \frac{4\pi I(\nu)}{c} \qquad (12.1.1)$$

This relation is not particular to electromagnetic radiation; it is valid for any quantity that fills space homogeneously and propagates with velocity c in all directions. In addition to intensity, electromagnetic radiation has two independent states of polarization. For each independent state of polarization, (12.1.1) is valid. For unpolarized thermal radiation, the specific intensity $I(\nu)$ consists of two independent states of polarization.

As noted by Gustav Kirchhoff (1824–1887), thermal radiation that is simultaneously in equilibrium with several substances should not change with the introduction or removal of a substance.

Hence, $I(\nu)$ and $u(\nu)$ associated with thermal radiation must be functions only of the temperature T, independent of the substances with which it is in equilibrium. Therefore, we shall write thermal energy density and radiance as $u(T, \nu)$ and $I(T, \nu)$ respectively.

Gustav Kirchhoff (1824–1887) (Reproduced with permission from the Edgar Fahs Smith Collection, University of Pennsylvania Library)

A body in thermal equilibrium with radiation is continuously emitting and absorbing radiation. The **spectral absorptivity** $a_k(T, v)$ of a body k is defined as the fraction of the incident thermal radiance $I(T, v)$ that is absorbed by the body k in the frequency range v and $v + dv$ at a temperature T. The thermal radiation absorbed by the body in the solid angle $d\Omega$ equals $a_k(T, v)I(T, v)\, d\Omega$. Let $I_k(T, v)$ be the spectral radiance of the body k. Then the power emitted per unit area into a solid angle $d\Omega$ equals $I_k(T, v)\, d\Omega$. At thermal equilibrium, the radiation absorbed by the body k in the solid angle $d\Omega$ must equal the radiation it emits in that solid angle. It then follows that

$$\boxed{\frac{I_k(T, v)}{a_k(T, v)} = I(T, v)}$$

(12.1.2)

As noted above, thermal radiance $I(T, v)$ must be independent of the substances with which it is in equilibrium. Hence, the ratio of a body's radiance to its absorptivity, i.e. $I_k(T, v)/a_k(T, v)$, is independent of the substance k and is a function only of temperature T and frequency v. This fundamental observation is called **Kirchhoff's law** (Box 12.1).

For a perfectly absorbing body, $a_k(T, v) = 1$. Such a body is called a **black body**; *spectral radiance is equal to the thermal spectral radiance $I(T, v)$.* In this context, another parameter, called the **emissivity** e_k of a body k, is defined as the ratio of its spectral radiance $I_k(T, v)$ to that of a black body's, i.e. $e_k = I_k(T, v)/I(T, v)$. Thus, **Kirchhoff's law** (12.1.2) can also be stated as

emissivity e_k = absorptivity a_k

Box 12.1 Kirchhoff's Law

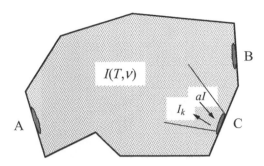

Kirchhoff's law states that, at thermal equilibrium, the ratio $I_k(T, v)/a_k(T, v)$ of emissive radiance $I_k(T, v)$ of a body k to its absorptivity $a_k(T, v)$ is independent of the body and is equal to the radiance of thermal radiation $I(T, v)$:

$$\frac{I_k(T, v)}{a_k(T, v)} = I(T, v)$$

For a perfectly absorbing body, $a_k(T, v) = 1$. Such a body is called a **black body**; its spectral radiance is equal to the thermal spectral radiance $I(T, v)$. The **emissivity** e_k of a body k is defined as the ratio of its spectral radiance to that of a black body: $e_k = I_k(T, v)/I(T, v)$. Thus, **Kirchhoff's law** can also be stated as

emissivity e_k = absorptivity a_k

at thermal equilibrium.

The **emissive power** of a body k is the power emitted per unit area into all directions in a hemisphere. It equals $\pi I_k(T, v)$. The emissivities of some materials are shown below:

Material	Emissivity
Lampblack	0.84
Polished copper	0.023
Cast iron	0.60–0.70
Polyethylene black plastic	0.92

At the end of the nineteenth century, classical thermodynamics faced the challenge of determining the exact functional form of $u(T, v)$ or $I(T, v)$. None of the deductions based on the laws of physics that were known at that time agreed with experimental measurements of $u(T, v)$. This fundamental problem remained unsolved until Max Planck (1858–1947) introduced the revolutionary quantum hypothesis. With the quantum hypothesis, according to which matter absorbed and emitted radiation in discrete bundles or 'quanta', Planck was able to derive the following expression which agreed well the observed frequency distribution $u(T, v)$:

Max Planck (1858–1947) (Reproduced courtesy of the AIP Emilio Segre Visual Archive)

$$u(T, v)\mathrm{d}v = \frac{8\pi h v^3}{c^3}\frac{\mathrm{d}v}{e^{hv/k_{\mathrm{B}}T}-1}$$
(12.1.3)

Here, $h = 6.626 \times 10^{-34}$ J s is Planck's constant and $k_{\mathrm{B}} = 1.381 \times 10^{-23}$ J K^{-1} is the Boltzmann constant. We shall not present the derivation of this formula here (which requires the principles of statistical mechanics). Finally, we note that total energy density of thermal radiation is

$$u(T) = \int_0^\infty u(T, v)\mathrm{d}v$$
(12.1.4)

When functions $u(T, v,)$ obtained using classical electromagnetic theory were used in this integral, the total energy density $u(T, v)$ turned out to be infinite. The Planck formula (12.1.3), however, gives a finite value for $u(T, v)$.

12.2   The Equation of State

It was clear even from the classical electromagnetic theory that a field which interacts with matter and imparts energy and momentum must itself carry energy and momentum. Classical expressions for the energy and momentum associated with the electromagnetic field can be found in texts on electromagnetic theory. For the purposes of understanding the thermodynamic aspects of radiation, we need an equation of state, i.e. an equation that gives the pressure exerted by thermal radiation and its relation to the temperature.

Using classical electrodynamics it can be shown [1] that the pressure exerted by electromagnetic radiation is related to the energy density u by

$$\boxed{p = \frac{u}{3}} \tag{12.2.1}$$

This relation follows from purely mechanical considerations of force exerted by electromagnetic radiation when it is reflected by the walls of a container. Though it was originally derived using classical electrodynamics, (12.2.1) can be more easily derived by treating electromagnetic radiation filling a container as a gas of photons (shown in Box 12.2). We shall presently see that when this equation of state is combined with the equations of thermodynamics, we arrive at the conclusion that the *energy density $u(T, v)$, and hence $I(T, v)$, is proportional to the fourth power of the temperature*, a result which is credited to Josef Stefan (1835–1893) and Ludwig Boltzmann (1844–1906) and called the **Stefan–Boltzmann law**. The fact that the energy density $u(T) = \int_0^\infty u(T,v)\mathrm{d}v$ of thermal radiation is only a function of temperature, independent of the volume, implies that in a volume V the total energy is

$$U = Vu(T) \tag{12.2.2}$$

Though thermal radiation is a gas of photons, it has features that are different from that of an ideal gas. At a fixed temperature T, as the volume of thermal radiation expands, the total energy increases (unlike in the case of an ideal gas, in which it remains constant). As the volume increases, the 'heat' that must be supplied to such a system to keep its temperature constant is thermal radiation entering the system. This heat keeps the energy density constant. The change in entropy due to this heat flow is given by

$$\mathrm{d}_\mathrm{e}S = \frac{\mathrm{d}Q}{T} = \frac{\mathrm{d}U + p\mathrm{d}V}{T} \tag{12.2.3}$$

Once we assign an entropy to the system in this fashion, all the thermodynamic consequences follow. Consider, for example, the Helmholtz equation (5.2.11) (which follows from the fact that entropy is a state function and, therefore, $\partial^2 S/\partial T \partial V = \partial^2 S/\partial V \partial T$):

$$\left(\frac{\partial U}{\partial V}\right)_T = T^2 \left[\frac{\partial}{\partial T}\left(\frac{p}{T}\right)\right]_V \tag{12.2.4}$$

Box 12.2 Photon Gas Pressure

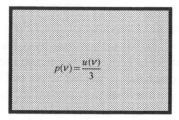

A HEURISTIC DERIVATION OF THE PRESSURE OF A PHOTON GAS

Let $n(v)$ be the number of photons per unit volume with frequency v. The momentum of each photon is hv/c. The radiation pressure on the walls is due to photon collisions. Each collision imparts a momentum $2(hv/c)$ to the wall upon reflection. Since the photons are in random motion, at any instant, the fraction of the photons that will be moving in the direction of the wall equals 1/6. Hence, the number of photons that will collide with a unit area of the wall in 1 s is $n(v)c/6$. The total momentum imparted to a unit area of the wall per second is the pressure. Hence, we have

$$p(v) = \frac{n(v)c}{6}\frac{2hv}{c} = \frac{n(v)hv}{3}$$

Now, since the energy density $u(v) = n(v)hv$, we arrive at the result

$$p(v) = \frac{u(v)}{3}$$

A more rigorous derivation, taking all the directions of the photon momentum into consideration, also gives the same result. For photons of all frequency we can integrate over the frequency v:

$$p = \int_0^\infty p(v)\,dv = \int_0^\infty \frac{u(v)}{3}\,dv = \frac{u}{3}$$

where u is the total energy due to photons of all frequencies and p is the total pressure. Note that a similar derivation for the ideal gas gives $p = 2u/3$, in which $u = n(mv_{avg}^2/2)$, where n is the number of molecules per unit volume and $mv_{avg}^2/2$ is the average kinetic energy of a molecule.

Using (12.2.2) and the equation of state $p = u/3$ in this equation we can obtain (Exercise 12.1)

$$4u(T) = T\left(\frac{\partial u(T)}{\partial T}\right) \tag{12.2.5}$$

Upon integrating this equation we arrive at the **Stefan–Boltzmann law**:

$$\boxed{u(T) = \beta T^4}$$

(12.2.6)

in which β is a constant. The value of $\beta = 7.56 \times 10^{-16}\,\mathrm{J\,m^{-3}\,K^{-4}}$ is obtained by measuring the intensity of radiation emitted by a black body at a temperature T. This law can also be written in terms of the irradiance of a black body. By integrating (12.1.1) over all frequencies ν, we arrive at $u(T) = 4\pi I(T)/c$. The irradiance $\pi I(T)$ (which is the power emitted per unit area in all directions in a hemisphere) can now be written as

$$\pi I = \frac{u(T)c}{4} = \left(\frac{\beta c}{4}\right) T^4 = \sigma T^4$$

(12.2.7)

in which the constant $\sigma = 5.67 \times 10^{-8}\,\mathrm{W\,m^{-2}\,K^{-4}}$ is called the **Stefan–Boltzmann constant**.

Using (12.2.6), we can now write the pressure $p = u/3$ as a function of temperature:

$$p(T) = \frac{\beta T^4}{3}$$

(12.2.8)

Equations (12.2.6) and (12.2.8) are the equations of state for thermal radiation. For temperatures of order $10^3\,\mathrm{K}$ or less the radiation pressure is small, but it can be quite large for stellar temperatures. In the interior of stars, where the temperatures can be $10^7\,\mathrm{K}$, we find, using (12.2.8), that the pressure due to thermal radiation is about $2.52 \times 10^{12}\,\mathrm{Pa} \approx 2 \times 10^7\,\mathrm{atm}$!

12.3   Entropy and Adiabatic Processes

For thermal radiation, the change in entropy is entirely due to heat flow:

$$dS = d_e S = \frac{dU + p\,dV}{T}$$

Considering U as a function of V and T, this equation can be written as

$$dS = \frac{1}{T}\left[\left(\frac{\partial U}{\partial V}\right)_T + p\right]dV + \frac{1}{T}\left(\frac{\partial U}{\partial T}\right)_V dT$$

(12.3.1)

Since $U = Vu = V\beta T^4$ (see (12.2.6)) and $p = \beta T^4/3$, this equation can be written as

$$dS = \left(\frac{4}{3}\beta T^3\right)dV + \left(4\beta V T^2\right)dT$$

(12.3.2)

In this equation, we can identify the derivatives of S with respect to T and V:

$$\left(\frac{\partial S}{\partial V}\right)_T = \frac{4}{3}\beta T^3 \qquad \left(\frac{\partial S}{\partial T}\right)_V = 4\beta V T^2 \qquad (12.3.4)$$

By integrating these two equations and setting $S = 0$ at $T = 0$ and $V = 0$, it is easy to see (Exercise 12.3) that

$$\boxed{S = \frac{4}{3}\beta V T^3} \qquad (12.3.5)$$

The above expression for entropy and the equations of state (12.2.6) and (12.2.8) are basic; all other thermodynamic quantities for thermal radiation can be obtained from them. Unlike the other thermodynamic systems we have studied so far, the temperature T is sufficient to specify all the thermodynamic quantities of thermal radiation: the energy density $u(T)$, the entropy density $s(T) = S(T)/V$ and all other thermodynamic quantities are entirely determined by T. There is no term involving a chemical potential in the expressions for S or U. If we consider the particle nature of thermal radiation, i.e. a gas of photons, then the *chemical potential must be assumed to equal zero*, which is a point that we will discuss in detail in Section 12.5.

In an adiabatic process the entropy remains constant. From the expression for entropy, (12.3.5), the relation between the volume and temperature in an adiabatic process immediately follows:

$$\boxed{V T^3 = \text{constant}} \qquad (12.3.6)$$

The radiation filling the universe is currently at about 2.7 K. The effect of the expansion of the universe on the radiation that fills it can be approximated as an adiabatic process. (During the evolution of the universe its total entropy is not a constant. Irreversible processes generate entropy, but the increase in entropy of radiation due to these irreversible processes is small.) Using (12.3.6) and the current value of T, one can compute the temperature when the volume, for example, is one-millionth of the present volume. Thus, thermodynamics gives us the relation between the volume of the universe and the temperature of the thermal radiation that fills it.

12.4 Wien's Theorem

At the end of nineteenth century, one of the most outstanding problems was the frequency dependence of the spectral energy density $u(T, \nu)$. Wilhelm Wien (1864–1928) made an important contribution in his attempt to obtain $u(T, \nu)$. Wien developed a method with which he could analyze what may be called the *microscopic consequences* of the laws of thermodynamics. He began by considering an adiabatic compression of thermal radiation. Such a compression keeps the system in thermal equilibrium but changes the temperature so that $VT^3 = $ constant (12.3.6). On a microscopic level, he analyzed the shift of each frequency v to a new frequency v' due to its interaction with the compressing piston. Since this frequency shift corresponds to a change in temperature such that $VT^3 = $ constant, he could obtain a relation between how $u(T, v)$ changed with v and T. This led Wien to the conclusion that $u(T, v)$ must have the following functional form (for more details see Ref. [1]):

$$\boxed{u(T, v) = v^3 f\left(\frac{v}{T}\right) = T^3 \left(\frac{v}{T}\right)^3 f\left(\frac{v}{T}\right)} \tag{12.4.1}$$

i.e. $u(T, v)$ is a function of the *ratio* v/T multiplied by T^3. This conclusion follows from the laws of thermodynamics. We shall refer to (12.4.1) as **Wien's theorem**. Note that (12.4.1) is in agreement with Planck's formula (12.1.3).

It was found experimentally that, for a given T, as a function of v, $u(T, v)$ has a maximum. Let v_{max} be the value of v at which $u(T, v)$ reaches its maximum value. Then, because $u(T, v)/T^3$ is a function of the ratio v/T, it follows that $u(v/T)$ reaches its maximum at a particular value of the ratio $v/T = C_1$. So, for a given T, as a function of v, $u(T, v)$ reaches its maximum at v_{max} when $v_{max}/T = C_1$. In other words:

$$\frac{T}{v_{max}} = C_1 \tag{12.4.2}$$

The spectral energy density $u(T, v)$ can be expressed as a function of the wavelength λ by noting that $v = c/\lambda$ and $dv = -(c/\lambda^2)\, d\lambda$. Using (12.4.1) we can write $u(T, v)\, dv = -T^3(c/\lambda T)^3 f(c/\lambda T)(c/\lambda^2)\, d\lambda = T^5 g(\lambda T)\, d\lambda$, in which $g(\lambda T)$ is an appropriately defined function of $T\lambda$ (Exercise 12.5). We can now identify $T^5 g(\lambda T)\, d\lambda = u(T, \lambda)\, d\lambda$ as the spectral energy density as a function of λ. It is a function of the product λT multiplied by T^5. The function $u(T, \lambda)/T^5$ reaches its maximum for a particular value of the product $\lambda T = C_2$. Hence, for a given T, if $u(T, \lambda)$ is plotted as a function of λ, then its maximum will occur at λ_{max} such that $\lambda_{max} T = C_2$. The values of the constants C_1 and C_2 can be obtained using Planck's formula (12.1.3). Generally, the value of C_2 is used. We thus have what is called **Wein's displacement law**, which tells us how the maximum of $u(T, \lambda)$ is displaced by changes in T:

$$\boxed{T\lambda_{max} = 2.8979 \times 10^{-3}\, \text{mK}} \tag{12.4.3}$$

As T increases, λ_{max} decreases proportionately. This conclusion is entirely a consequence of the laws of thermodynamics.

The above method of Wien is general and it can be applied, for example, to an ideal gas. Here, the objective would be to obtain the energy density u as a function of the velocity v and the temperature. It can be shown [2] that $u(T, v) = v^4 f(v^2/T)$, which shows us that thermodynamics implies that the velocity distribution is a function of v^2/T. This is consistent with the Maxwell velocity distribution (1.6.13). Wien's approach shows us how thermodynamics can be used to investigate microscopic aspects of systems, such as energy or velocity distributions.

Wien's analysis and all other classical attempts to obtain the form of $u(T, v)$ for thermal radiation not only gave results that did not agree with experiments, but also gave infinite values for $u(T, v)$ when all frequencies v (0 to ∞) were included. It is now well known that it was to solve this problem that Planck introduced his quantum hypothesis in 1901.

12.5   Chemical Potential of Thermal Radiation

The equations of state for thermal radiation are

$$p = \frac{u}{3} \quad u = \beta T^4 \tag{12.5.1}$$

where u is the energy density and p is the pressure.

If all the material particles in a volume are removed, what was classically thought to be a vacuum is not empty but is filled with thermal radiation at the temperature of the walls of the container. There is no distinction between heat and such radiation in the following sense. If we consider a volume filled with thermal radiation in contact with a heat reservoir (Figure 12.2), then, if the volume is enlarged, the temperature T and, hence, the energy density u of the system are maintained constant by the flow of heat into the system from the reservoir. The heat that flows into the system is thermal radiation.

From the particle point of view, thermal radiation consists of photons which we shall refer to as *thermal photons*. Unlike in an ideal gas, the total number of thermal photons is not conserved during isothermal changes of volume. The change in the total energy $U = uV$ due to the flow of thermal photons from or to the heat reservoir

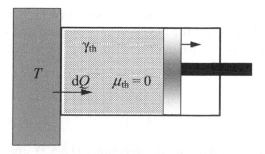

Figure 12.2   Heat radiation in contact with a heat reservoir. The energy entering or leaving such a system is thermal radiation. Though the number of photons is changing, $\mathrm{d}U = \mathrm{d}Q - p\,\mathrm{d}V$

must be interpreted as a flow of heat. Thus, for thermal radiation, in a reversible expansion at constant T we have $dS = d_e S = dQ/T$ and so we can write

$$dU = dQ - pdV = TdS - pdV \qquad (12.5.2)$$

This equation remains valid even though the number of photons in the system is changing. Comparing this equation with the equation introduced by Gibbs, $dU = TdS - pdV + \mu dN$, we conclude that the chemical potential $\mu = 0$. *The state in which $\mu = 0$ is a state in which the partial pressure or the particle density is a function only of the temperature.* Indeed, in the expression for the chemical potential, $\mu_k = \mu_k^0(T) + RT \ln(p_k/p_0)$, if we set $\mu_k = 0$ we see that the partial pressure p_k is only a function of T.

TWO-LEVEL ATOM IN EQUILIBRIUM WITH RADIATION

With the above observations that the chemical potential of thermal radiation is zero, the interaction of a two-level atom with black-body radiation (which Einstein used to obtain the ratio of the rates of spontaneous and stimulated radiation) can be analyzed in a somewhat different light. If A and A* are the two states of the atom and γ_{th} is a thermal photon, then the spontaneous and stimulated emission of radiation can be written as

$$A^* \rightleftharpoons A + \gamma_{th} \qquad (12.5.3)$$

$$A^* + \gamma_{th} \rightleftharpoons A + 2\gamma_{th} \qquad (12.5.4)$$

From the point of view of equilibrium of a chemical reaction, the above two reactions are the same. The condition for chemical equilibrium is

$$\mu_{A^*} = \mu_A + \mu_\gamma \qquad (12.5.5)$$

Since $\mu_\gamma = 0$, we have $\mu_{A^*} = \mu_A$. As we have seen in Chapter 9, if we use the expression $\mu_k = \mu_k^0(T) + RT\ln(p_k/p_0)$ for the chemical potential and note that the concentration is proportional to the partial pressure, then the law of mass action takes the form

$$\frac{[A]}{[A^*]} = K(T) \qquad (12.5.6)$$

On the other hand, looking at the reactions (12.5.3) and (12.5.4) as elementary chemical reactions, we may write

$$\frac{[A][\gamma_{th}]}{[A^*]} = K'(T) \qquad (12.5.7)$$

But because $[\gamma_{th}]$ is a function of temperature only, it can be absorbed in the definition of the equilibrium constant; hence, if we define $K(T) \equiv K'(T)/[\gamma_{th}]$, we recover equation (12.5.6), which follows from thermodynamics.

Similarly, we may consider any exothermic reaction

$$A + B \rightleftharpoons 2C + \text{Heat} \qquad (12.5.8)$$

from the viewpoint of thermal photons and write this reaction as

$$A + B \rightleftharpoons 2C + \gamma_{th} \qquad (12.5.9)$$

The condition for equilibrium can now be written as

$$\mu_A + \mu_B = 2\mu_C + \mu_\gamma \qquad (12.5.10)$$

Since $\mu_\gamma = 0$, we recover the condition for chemical equilibrium derived in the Chapter 9. For this reaction also, one can obtain $K'(T)$ similar to that defined in (12.5.7).

12.6 Matter-Antimatter in Equilibrium with Thermal Radiation: The State of Zero Chemical Potential

When we consider interconversion of matter and radiation, as in the case of particle–antiparticle pair creation and annihilation (Figure 12.3), the chemical potential of thermal photons becomes more significant. Similar thermodynamic analysis could be done for electron–hole pair production in semi-conductors by radiation. Consider thermal photons in thermal equilibrium with electron–positron pairs:

$$2\gamma \rightleftharpoons e^+ + e^- \qquad (12.6.1)$$

At thermal equilibrium we have

$$\mu_{e^+} + \mu_{e^-} = 2\mu_\gamma \qquad (12.6.2)$$

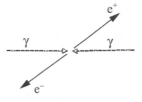

Figure 12.3 Creation of particle–antiparticle pairs by thermal photons

For reasons of symmetry between particles and antiparticles, we may assert that μ_{e^+} = μ_{e^-}. Since $\mu_\gamma = 0$, we must conclude that $\mu_{e^+} = \mu_{e^-} = 0$ for particle–antiparticle pairs that can be created by thermal photons.

It is interesting to discuss further this state of matter for which $\mu = 0$. For simplicity, let us consider the $\mu = 0$ state in an ideal monatomic gas mixture for which

$$\mu_k = \frac{U_k - TS_k + p_k V}{N_k}$$
$$= \frac{N_k[(3/2)RT + W_k] - TN_k R[(3/2)\ln T + \ln(V/N_k) + s_0] + N_k RT}{N_k} \qquad (12.6.3)$$

in which we used the internal energy $U_k = N_k [(3/2)RT + W_k]$ of component k of the ideal-gas mixture and its entropy $S_k = N_k R [(3/2)\ln T + \ln(V/N_k) + s_0]$ and the ideal-gas equation $p_k V = N_k RT$. As we have already noted in Chapter 2, the theory of relativity gives us the absolute value of energy $E^2 = p^2 c^2 + m^2 c^4$. The momentum $p = 0$ at $T = 0$, leaving the rest energy $E = mc^2$. The term W_k is the rest energy of 1 mol of the particles: $W_k = M_k c^2$, in which M_k is the molar mass of component k. Quantum theory gives us the entropy constant s_0 in the expression for entropy. Using Equation (12.6.3), we can write the molar density N_k/V as

$$\frac{N_k}{V} = z(T)e^{(\mu - M_k c^2)/RT} \qquad (12.6.4)$$

in which $z(T)$ is a function of temperature only (in Chapter 17 we can see that it is closely related to the 'partition function' of an ideal gas). When the process of pair production is in thermal equilibrium, since $\mu = 0$ the thermal particle density is given by

$$\left(\frac{N_k}{V}\right)_{\mathrm{th}} = z(T)e^{-M_k c^2/RT} \qquad (12.6.5)$$

The corresponding partial pressure is given by

$$p_{k,\mathrm{th}} = RT z(T)e^{-M_k c^2/RT} \qquad (12.6.6)$$

The physical meaning of the above equations can be understood as follows: just as photons of energy $h\nu$ are excitations of the electromagnetic field, particles of energy $E = (m^2 c^4 + p^2 c^2)^{1/2}$ are also excitations of a quantum field. In the nonrelativistic approximation, $E \approx mc^2 + p^2/2m$. According to the Boltzmann principle, in a field, the probability $P(E)$ of an excitation of energy E is given by

$$P(E) \propto \rho(E)e^{-E/kT} = \rho(E)e^{-[mc^2 + (p^2/2m)]/k_B T} \qquad (12.6.7)$$

where $\rho(E)$ is the density of states of energy E (see Chapter 17). If we approximate the statistics of theses excitations by classical Boltzmann statistics, the density of particles of mass m can be obtained by integrating (12.6.7) over all momenta p. We then obtain an expression of the form (12.6.5) in which the molar mass $M_k = N_A m_k$. Thus, Equations (12.6.5) and (12.6.6) give the density and partial pressure due to particles that appear spontaneously at temperature T as thermal excitations of a quantum field. In this state in which $\mu = 0$, there is no distinction between heat and matter; just as it is for thermal photons, the particle density is entirely determined by the temperature.

At ordinary temperatures, the thermal particle density obtained above is extremely small. Nevertheless, from the point of view of thermodynamic formalism after the advent of quantum field theory, it is important to consider this state in which the chemical potential vanishes. It is a state of thermal equilibrium that matter could reach; indeed, matter was in such a state during the early part of the universe. Had matter stayed in thermal equilibrium with radiation, at the current temperature of the universe, 2.73 K, the density of protons and electrons, given by (12.6.5) or its modifications, would be virtually zero. Indeed, the very existence of particles at the present temperatures has to be viewed as a nonequilibrium state. As a result of the particular way in which the universe has evolved, matter was not able to convert to radiation and stay in thermal equilibrium with it.

From (12.6.4) we see that assigning a nonzero value for the chemical potential is a way of fixing the particle density at a given temperature. Since we have an understanding of the **absolute zero of chemical potential**, we can write the chemical potential of ideal gas particles as

$$\mu_k = RT \ln\left(\frac{p_k}{p_{k,\text{th}}}\right) \tag{12.6.8}$$

in which $p_{k,\text{th}}$ is the thermal pressure defined above. In principle, one may adopt this scale of chemical potential for all ideal systems.

CHEMICAL POTENTIAL OF NONTHERMAL RADIATION

From the above discussion, we see how a nonzero chemical potential may be associated with nonthermal electromagnetic radiation, i.e. radiation that is not in thermal equilibrium with matter with which it is interacting. Let us consider matter at a temperature T, interacting with radiation whose spectral energy distribution is not the Planck distribution (12.1.3) at the same temperature T. For electromagnetic radiation of frequency v, whether it is in thermal equilibrium or not, for an energy density $u(v)$ the corresponding pressure is

$$p(v) = u(v)/3 \tag{12.6.9}$$

Also, as noted in Section 12.1, for any radiation that homogeneously fills space, the spectral energy density $u(v)$ is related to the intensity of radiance $I(v)$:

$$u(v) = 4\pi I(v)/c \tag{12.6.10}$$

We shall denote the Planck distribution at temperature T by $u_{th}(T, v)$ and the associated pressure and intensity by $p_{th}(T, v)$ and $I_{th}(T, v)$ respectively, with the subscript 'th' emphasizing that it is thermal radiation. Following (12.6.8), we can write the chemical potential of nonthermal radiation as

$$\mu(v) = RT \ln\left(\frac{p(v)}{p_{th}(T, v)}\right) = RT \ln\left(\frac{u(v)}{u_{th}(T, v)}\right) = RT \ln\left(\frac{I(v)}{I_{th}(T, v)}\right) \quad (12.6.11)$$

When the radiation reaches equilibrium with matter at temperature T, $u(v) = u_{th}(T, v)$ and the chemical potential will equal zero.

The same result can also be obtained from the Planck formula by introducing a chemical potential $\mu' = \mu/N_A$. When the energy density is

$$u(v, T) = \frac{8\pi h v^3}{c^3} \frac{1}{e^{(hv - \mu'(v))/k_B T} - 1} \quad (12.6.12)$$

as is done for bosons in general. Here, the chemical potential is a function of the frequency v.

If $\mu'/k_B T \ll 1$, then it is easy to see that we can approximately write

$$\frac{u(v, T)}{u_{th}(v, T)} = e^{\mu'(v)/k_B T} = e^{\mu(v)/RT} \quad (12.6.13)$$

in agreement with expression (12.6.12).

An example of nonthermal radiation is the solar radiation that reaches the Earth. The radiation has an initial Planck energy density corresponding to $T = 6000\,\text{K}$, which we write as $u_{th}(6000, v)$. As it propagates through space from the Sun's surface, the energy density decreases by a factor $(r_{Sun}/r)^2$, in which r is the distance from the center of the Sun to the Earth's surface and r_{Sun} is the radius of the Sun. When solar radiation arrives at the Earth's surface, which we shall assume is at temperature T_{earth}, its shape is that of a 6000 K Planck distribution, but the energy density is much smaller. This radiation is not in thermal equilibrium with matter on the surface of the Earth. Using (12.6.11), its chemical potential can be written as

$$\mu(v) = RT \ln\left[\frac{(r_{sun}/r)^2 u_{th}(6000, v)}{u_{th}(T_{earth}, v)}\right] \quad (12.6.14)$$

This nonzero chemical potential drives photosynthesis, a topic we will discuss in detail in Chapter 13. Ultimately, solar radiation reaches thermal equilibrium and it becomes a Planck distribution at T_{earth}; Earth emits thermal radiation at this temperature into space.

References

1. Planck, M., *Theory of Heat Radiation* (History of Modern Physics, Vol. 11). 1988, Washington, DC: American Institute of Physics.
2. Kondepudi, D.K., Microscopic aspects implied by the second law. *Found. Phys.*, **17** (1987) 713–722.

Examples

Example 12.1   Using the equation of state, calculate the energy density and pressure of thermal radiation at 6000 K (which is approximately the temperature of the radiation from the Sun). Also calculate the pressure at $T = 10^7$ K.

Solution   The energy density is given by the Stefan–Boltzmann law $u = \beta T^4$, in which $\beta = 7.56 \times 10^{-16}$ J m^{-3} K^{-4} (see (12.2.6)). Hence, the energy density is at 6000 K is

$$u(6000\,\text{K}) = 7.56 \times 10^{-16}\,\text{J m}^{-3}\,\text{K}^{-4}\,(6000\,\text{K})^4 = 0.98\,\text{J m}^{-3}$$

The pressure due to thermal radiation is given by $p = u/3 = (0.98/3)$ J m^{-3} = 0.33 Pa $\approx 3 \times 10^{-6}$ atm. At $T = 10^7$ K, the energy density and pressure are

$$u = 7.56 \times 10^{-16}\,\text{J m}^{-3}\,\text{K}^{-4}\,(10^7\,\text{K})^4 = 7.56 \times 10^{12}\,\text{J m}^{-3}$$
$$p = u/3 = 2.52 \times 10^{12}\,\text{Pa} = 2.5 \times 10^7\,\text{atm}$$

Exercises

12.1   Obtain (12.2.5) using (12.2.1) and (12.2.2) in the Helmholtz equation (12.2.4).

12.2   Using Planck's formula (12.1.3) for $u(\nu, T)$ in (12.1.4), obtain the Stefan–Boltzmann law (12.2.6) and an expression for the Stefan–Boltzmann constant β.

12.3   Show that (12.3.5) follows from (12.3.4).

12.4   At an early stage of its evolution, the universe was filled with thermal radiation at a very high temperature. As the universe expanded adiabatically, the temperature of the radiation decreased. Using the current value of $T = 2.73$ K, obtain the ratio of the present volume to the volume of the universe when $T = 10^{10}$ K

12.5   Thermal spectral radiance $I(T, \lambda)$ dλ is defined as the radiance in the wavelength range λ and $\lambda + \text{d}\lambda$ of thermal radiation at temperature T.

(a) Show that

$$I(T, \lambda)\mathrm{d}\lambda = \frac{2hc^2}{\lambda^5} \frac{\mathrm{d}\lambda}{\mathrm{e}^{hc/\lambda k_B T} - 1}$$

(b) The surface temperature of the Sun is 6000 K. Plot $I(6000, \lambda)$ as a function of λ and verify that $\lambda_{max} \approx 483$ nm for the solar thermal radiation.

(c) What will λ_{max} be if the Sun's surface temperature is 10 000 K?

12.6 The total energy of the Earth is in a steady state. This means that the flux of solar radiation absorbed by the Earth equals that emitted as thermal radiation. (a) Assuming that the average surface temperature of the Earth is about 288 K, estimate the amount of thermal radiation emitted by the Earth per second. (b)Assuming that the temperature of the solar radiation is 6000 K, estimate the total rate of entropy due to the thermal radiation flux through the Earth.

12.7 Estimate the chemical potential of solar radiation at the surface of the Earth where matter is at temperature $T = 295$ K.

13 BIOLOGICAL SYSTEMS

In seeking an understanding of the thermodynamic aspects of life, we must first recognize the inadequacy of a description of life as a *state* of matter; no description of life is complete without the inclusion of the *irreversible processes* that make life what it is. The processes bring about macroscopic features such as self-replication and adaptation that we can observe in a living cell. From the thermodynamic viewpoint, our goal is not so much to seek a precise definition of 'life' as it is to identify some characteristic features of living cells and see how we might understand them within the framework of thermodynamics. In his influential and inspiring book *What is Life* [1], Erwin Schrödinger established a thermodynamic framework for thinking about the processes in a living cell. Later, the concept of dissipative structures, pioneered by Ilya Prigogine and his coworkers [2], has shed more light on how organization could spontaneously arise in systems far from thermodynamic equilibrium. The theory of dissipative structures discussed in Chapter 11 reveals how entropy-producing irreversible processes can generate order and structure. In addition, the work of Katchalsky and Curran [3], Peacocke [4], and Caplan and Essig [5] focused on biophysical processes and elucidated how modern thermodynamics applies to biological systems. Yet fundamental questions regarding the origin of life and the evolution of complex organization from the level of individual cells to ecosystems remain and are, at best, only partially answered. In this chapter we discuss thermodynamics aspects of biological processes, in particular the flow of Gibbs energy that drives the processes associated with life. It is assumed that the reader is familiar with the basic structure of a living cell and its overall biochemical working and the molecular structure of proteins, DNA and other biomolecules.

13.1 The Nonequilibrium Nature of Life

There are many features of cells that clearly indicate the nonequilibrium nature of its state. First, they are open systems that exchange energy and matter with their exterior or environment. Plants absorb CO_2, H_2O and solar energy and expel O_2 during photosynthesis. Other organisms feed on 'food' and expel waste. Second, the complex network of chemical processes in cells is controlled by enzymes, which are protein catalysts. As we have noted in earlier chapters, catalysts have no effect on the state of equilibrium. The simple fact that enzymes can alter the state of a cell implies that the cell it is not in thermodynamic equilibrium.

There is another aspect of life that indicates, in fact, that it is a far-from-equilibrium dissipative structure, a concept introduced in Chapter 11. The entire biochemical edifice of life as we know it is founded upon a fundamental molecular asymmetry of its building blocks. Amino acids and the ribose in nucleotides are

Introduction to Modern Thermodynamics   Dilip Kondepudi
© 2008 John Wiley & Sons, Ltd

chiral molecules.* Of the two possible mirror-image structures, named L- and D-enantiomers, only one kind appears in proteins and DNA of all living cells (L stands for levo and D for dextro). With rare exceptions, the chemistry of life is dominated by L-amino acids and D-sugars. In the words of Francis Crick [6], 'The first great unifying principle of biochemistry is that the key molecules have the same hand in all organisms'. The evolutionary origin of this particular asymmetry is still an enigma, but thermodynamics gives us a framework to understand how asymmetry might arise under far-from-equilibrium conditions, through instability and symmetry-breaking transitions as described in Chapter 11. Several examples of spontaneous generation of chiral asymmetry in nonequilibrium systems are now known [7]. Chiral asymmetry, or dominance of one hand over the other, is not peculiar to biological systems: it is astonishing that chiral asymmetry pervades the whole universe, from elementary particles to the morphology of mammals.

The nonequilibrium state of an organism causes it to respond to changes in external factors (such as temperature) in complex and highly sensitive ways. In the case of alligators, for example, the sex of an offspring depends on the temperature at which the egg was incubated! In contrast, the response of equilibrium systems is all described by Le Chatelier's principle.

Having noted the nonequilibrium nature of living cells, we can now turn to their thermodynamic description. The much-discussed 'energy flow' in biological systems is a concept that is made clear only through thermodynamics; as we shell see in this chapter, it is in fact Gibbs energy flow. A proper understanding of flows of energy and matter in nature requires concepts of thermodynamic flows that are driven by thermodynamic forces. The thermodynamic forces that drive most of the 'flows' in biological systems are affinities. When affinity is the difference in chemical potential between reactants and products, the corresponding flow is a chemical reaction; when it is the difference in chemical potential from one location to another, the flow is transport of matter.

As summarized in Figure 13.1, biochemical processes are 'flows' driven by chemical affinities. At the top of the flow is the state of high chemical potential. It consists of CO_2, H_2O and solar radiation. Facilitated by enzymes, the system produces carbohydrates, which we shall indicate by (CH_2O), as it moves to states of lower chemical potential. The number of possible compounds that can be synthesized from CO_2 and H_2O is enormous; enzymes channel the synthesis towards the production of 'biomolecules'. In the following sections we shall look at some of the specific chemical steps in more detail.

Regarding the entropy of the cell, it is clear that the irreversible processes occurring in it continuously produce entropy, as indicated in Figure 13.1. This does not result in a continuous increase in the entropy of the cell, however, because there is a net outflow of entropy. The entropy flowing out of the system is larger than that

---

*Molecules that are not identical to their mirror image, i.e. molecules that have a sense of handedness are called chiral molecules.

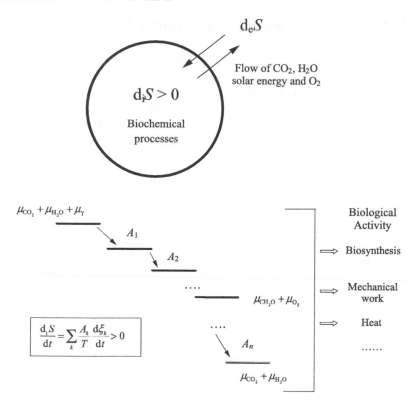

Figure 13.1 A living cell is an open system that exchanges energy and matter. The biochemical activity within the cell consists of entropy-generating irreversible processes. A complex network of biochemical reactions are driven by affinities in the cell. In each chemical reaction some Gibbs free energy is lost. In photosynthetic organisms, all the biological activity is driven by the difference between $(\mu_{CO_2} + \mu_{H_2O} + \mu_\gamma)$ and $(\mu_{CO_2} + \mu_{H_2O})$

flowing into the system. If we approximate the state of a cell to be a steady state in which its entropy remains constant or changes little, then the outflow of entropy is equal to the rate of entropy production in the organism: $d_eS/dt + d_iS/dt \approx 0$. The rate of entropy production depends on the affinities A_k and the corresponding reaction velocities $d\xi_k/dt$:

$$\frac{d_iS}{dt} = \sum_k \frac{A_k}{T} \frac{d\xi_k}{dt} > 0$$

which is positive as required by the second law. The complex and delicate state of order in biological systems is in fact maintained at the expense of high rates of entropy generation.

13.2   Gibbs Energy Change in Chemical Transformations

In Chapter 5 we saw that, at constant p and T, irreversible chemical reactions decrease the Gibbs energy or Gibbs free energy. (The usage of the term 'free energy' reminds us that it is the energy available for doing work. In this chapter we will use both terms, Gibbs energy and Gibbs free energy, though the former is recommended by the IUPAC). This is made clear in the equation $(dG/dt)_{p,T} = -T(d_iS/dt) = -\Sigma_k A_k(d\xi_k/dt)_{p,T} < 0$. To understand the meaning of this equation, let us consider the transformation of the reactants to products in the following reaction:

$$X + W \rightleftharpoons Z + Y \tag{13.2.1}$$

We shall consider three types of transformation, i.e. reversible, irreversible and what are called 'standard transformations', and calculate the corresponding changes in Gibbs free energy. Transformations from reactants to products and the associated changes in Gibbs free energy could be understood using van't Hoff reaction chambers, as shown in Figure 13.2. The reacting compounds X, W, Y and Z enter and exit the reaction chamber through a membrane that is permeable only to one of the compounds. For example, the cylinder containing X is separated from the reaction chamber by a membrane that is permeable only to X and none of the other compounds, and similarly for the other compounds. Owing to the semi-permeable

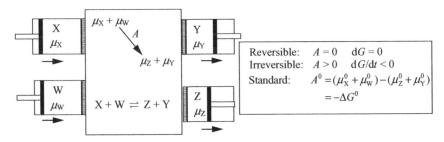

Figure 13.2   The changes in Gibbs energy during reversible and irreversible transformations can be understood using 'van't Hoff reaction chambers'. Reactants X and W are injected into the central reaction chamber using a piston and semi-permeable membrane that is permeable to only one compound; similarly, the products Z and Y are removed from the reaction chamber through a semi-permeable membrane. The affinity A can be maintained at a fixed value during such a transformation by simultaneously injecting the reactants into the reaction chamber and removing the products. In an idealized reversible transformation $A = 0$; the transformation happens at an infinitesimally slow rate. For an irreversible transformation $A \neq 0$ and the transformation takes place at a finite rate

membranes, the chemical potentials of each reactant in the reaction chamber is equal to its chemical potential in the chamber containing the pure compound (Figure 13.2). For illustrative purposes, let us assume that all the compounds are ideal gases. For an ideal gas the chemical potential $\mu(p_k, T) = \mu_k^0(T) + RT\ln(p_k/p_0)$, in which p_0 is the standard pressure. With this expression for the chemical potential it is easy to see that the partial pressure of X in the reaction chamber is equal to the pressure of X in the side chamber containing only X.

REVERSIBLE TRANSFORMATION

In an idealized reversible transformation, the reaction chamber is in equilibrium, in which case the affinity $A = 0$. The transformation of X + W → Y + Z takes place when X and W are injected into the reaction chamber by moving appropriate pistons to the right which cause an infinitesimal increase in their chemical potentials; simultaneously, Z and Y are removed from the system by moving the corresponding pistons to the right, which causes an infinitesimal decrease in their chemical potentials. Thus, the transformation X + W → Z + Y takes place with an infinitesimally small positive value of $A \approx 0$. Such a transformation takes place at a rate approaching zero. If the transformation takes place at constant p and T, then the corresponding change in the Gibbs energy is infinitesimal: $dG = -Td_iS = (A\,d\xi) \approx 0$, i.e. the products generated have the same Gibbs energy as the reactants that were consumed. In a reversible process, the reaction velocity $d\xi/dt$ is not defined.

IRREVERSIBLE TRANSFORMATION

All real chemical transformations occur at a nonzero rate with $A \neq 0$. Let us assume that $A > 0$. The corresponding transformation can occur at a constant A when X and W are injected into the reaction chamber by maintaining their partial pressures above the value of their partial pressures at equilibrium and, at the same time, keeping the partial pressures of Z and Y lower than their equilibrium partial pressures. The transformation occurs at a nonzero rate. If the process occurs at constant p and T, then the rate of change of the Gibbs energy can be written as

$$\frac{dG}{dt} = -T\frac{d_iS}{dt} = -A\frac{d\xi}{dt} < 0 \qquad (13.2.2)$$

In accordance with the second law, the Gibbs energy decreases during the transformation, i.e. the energy available to do mechanical work is reduced, hence the usage of the phrase 'decrease in "free energy"'. The change in the Gibbs energy for a given change in the extent of reaction ξ is

$$\Delta G \equiv G_f - G_i = \int_{\xi_i}^{\xi_f} -A\mathrm{d}\xi < 0 \qquad (13.2.3)$$

in which the subscripts 'i' and 'f' stand for initial and final states respectively. If the transformation proceeds at constant A, then

$$\Delta G \equiv G_f - G_i = -A(\xi_f - \xi_i) < 0 \qquad (13.2.4)$$

For the case shown in Figure 13.2, in which the reactants and products are ideal gases, for given values of partial pressures p_k we can explicitly calculate the changes in Gibbs energy. (See Example 13.3 of calculating Gibbs energy changes at the end of the chapter.) To compute $\mathrm{d}G/\mathrm{d}t$, however, we need the reaction velocity $\mathrm{d}\xi/\mathrm{d}t$, as is clear from (13.2.2). The reaction velocity depends on the mechanism of the reaction, the temperature, the pH and other such factors and it can be altered by a catalyst.

The above example clearly indicates that the change in Gibbs energy depends on the affinity that is driving the reactions. In biochemical reactions in a cell, we can estimate the rate of change in Gibbs energy in a cell only if the reaction affinities and reaction velocities *in vivo* are known.

STANDARD TRANSFORMATION AND $\Delta G^{0'}$

As we have seen in earlier chapters, the explicit from of the chemical potential is $\mu_k = \mu_k^0 + RT\ln a_k$, in which a_k is the activity; $a_k = 1$ at a defined standard state such as $p = 1.0$ bar and $T = 298.15$ K. The values of μ_k^0 for a given standard state are tabulated. Most biochemical reactions take place under physiological conditions, in which the pH is maintained at a value close to 7.0. For this reason, the biochemical standard state is defined at pH 7.0 and the Gibbs energy change between standard states is written $\Delta G^{0'}$, to distinguish it from the more generally used standard-state Gibbs energy change ΔG^0 (see Box 13.1). At this standard state, $a_k = 1$ and $\mu_k = \mu_k^0$ by definition. We shall denote the affinities corresponding to the biochemical standard states by $A^{0'}$.

It has become common practice to tabulate the Gibbs energy change per unit change of the extent of reaction ξ, i.e. $(\xi_f - \xi_i) = 1$ mol, *at the particular affinity* $A^{0'} = (\mu_X^0 + \mu_W^0) - (\mu_Y^0 + \mu_Z^0)$. We shall refer to this affinity as **standard affinity**: the corresponding change in the Gibbs energy is written as $\Delta G^{0'}$, and it is called **standard Gibbs free energy change**:

$$-\Delta G^{0'} = A^{0'} = (\mu_X^0 + \mu_W^0) - (\mu_Z^0 + \mu_Y^0)\,\mathrm{J\,mol^{-1}} \qquad (13.2.5)$$

Since changes in G in any chemical reaction depend on affinities (which may not equal standard affinities), tabulating and using $\Delta G^{0'}$ to discuss Gibbs free energy changes in biochemical reactions is somewhat arbitrary. Regarding the usage of

Box 13.1 Biochemical Standard State

In view of the conditions found in living cells, such as a pH that is maintained close to 7.0 through the use of buffers, the standard state used in a biochemical context is different from the usual standard state. The biochemical standard state is defined as follows:

$$T = 298.15\,\text{K},\ p = 1.0\,\text{bar} = 100\,\text{kPa and pH 7.0}$$

This means that the chemical potential of H$^+$ in the molarity scale, $\mu_{H^+} = \mu_{H^+}^{m0} + RT \ln(a_{H^+})$, is defined in such a way that the activity $a_{H^+} = 1.0$ when the [H$^+$] = 10^{-7} M. For dilute solutions, we may approximate the activity by molarity and $\mu_{H^+} = \mu_{H^+}^{m0} + RT\ln(c_{H^+}/10^{-7})$ in which c_{H^+} is the molar concentration.

The chemical potential of ionic species, such as ATP, and all biochemical activities are to be calculated assuming [H$^+$] = 10^{-7} M.

For biochemical reactions, the tabulated equilibrium constants are at pH 7.0. Accordingly, if a reaction includes [H$^+$], then it is assumed that [H$^+$] = 10^{-7} M.

standard Gibbs free energy change $\Delta G^{0'}$ to discuss Gibbs free energy changes in living cells, biochemist Albert Lehninger [8 (p. 33)] comments:

. . . it is clear that the actual free energy changes of metabolic reactions under the conditions existing in the cell may be quite different from the standard free energy change. Nevertheless, for consistency we must use the arbitrarily defined standard free energy changes if we are to compare the energetics of chemical reactions quantitatively.

This comment puts the use of $\Delta G^{0'}$ in perspective; it is important to bare in mind that tabulated values of $\Delta G^{0'}$ do not represent the actual changes of Gibbs free energy in a cell. For a proper understanding of the Gibbs energy changes and entropy production in a cell, reaction affinities in the conditions that exist in the cell are needed. The values of $A^{0'} = -\Delta G^{0'}$ for some biochemical reactions are shown in Table 13.1. In the same table, estimated values of the reaction affinities A_{cell} under cellular conditions for some reactions in glycolysis, a sequence of ATP-producing reactions that occurs in every cell, are also shown. All but the first reaction in Table 13.1 are reaction steps in glycolysis, which is summarized in Figure 13.6.

13.3 Gibbs Energy Flow in Biological Systems

With the thermodynamic formalism outlined in the previous section, let us look at some specific processes in the overall process of life summarized in Figure 13.3. The ultimate energy source for the biosphere is the sun. Solar energy drives photosynthesis and the cycles of life that derive from it.

Table 13.1 Standard Gibbs energy changes or standard reaction affinities $A^0 = -\Delta G^{0\prime}$ ($T = 298$ K, pH 7.0, $p = 1.0$ bar)* and estimated affinities A_{cell} under cellular conditions

Reaction	$A^0 = -\Delta G^{0\prime}$/kJ mol^{-1}	A_{cell}/kJ mol$^{-1\,†}$
$C_6H_{12}O_6(s) + 6O_2(g) \rightarrow 6CO_2(g) + 6H_2O$	2878	
$ATP^{4-} + H_2O \rightarrow ADP^{3-} + HPO_4^{2-} + H^+$	30.5	46
D-Glucose + ATP $\rightarrow$ D-Glucose-6-phosphate + ADP + H^+	16.7	33
Glucose 6-phosphate $\rightarrow$ Fructose 6-phosphate	−1.67	2.5
Fructose 6-phosphate + ATP $\rightarrow$ Fructose 1,6-bisphosphate + ADP + H^+	14.2	22
Fructose 1,6-bisphosphate $\rightarrow$ glyceraldehyde 3-phosphate + dihydroxyacetone phosphate	−23.8	1.2

* *Source*: Styer, L., *Biochemistry* (3rd edition). 1988, New York: W. H. Freeman.
† The standard affinity or Gibbs energy change used by biochemists is denoted by $A^0 = -\Delta G^{0\prime}$. It is the change in Gibbs energy at $T = 298$ K, $p = 1.0$ bar, pH 7.0, with the concentrations of the reactants and products maintained at a concentration of 1.0 M during the transformation of 1 mol of reactants to 1 mol of products.

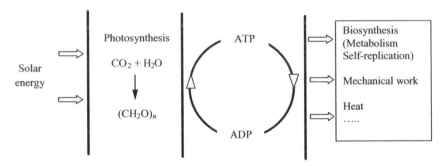

Figure 13.3 Gibbs free energy flow in a photosynthetic cell

PHOTOSYNTHESIS

Photosynthesis is a process in which solar energy drives the synthesis of large biomolecules from CO_2, H_2O and other small molecules, releasing O_2 in the process. It is a very complex process. Photosynthetic production of glucose, for example, involves over 100 steps. Among the substances produced by photosynthesis are cellulose, proteins and lipids. Cells are open systems that exchange energy and matter with their environment. Organisms that synthesize their own 'food' (molecules they need to build components of their cells) from simple inorganic substances such as CO_2 and H_2O are called *autotrophs*. If they obtain their energy from the sun, they are called *photoautotrophs*; if they obtain their energy from chemical sources they are called *chemoautotrophs*. Other organisms that obtain carbon compounds and energy from autotrophs are called *heterotrophs*.

One of the most important products of photosynthesis is glucose, $C_6H_{12}O_6$. The net reaction that generates glucose photosynthetically is

Stage I (Light-driven reactions)

$2H_2O + 2NADP^+ + \gamma \rightarrow O_2 + 2NADPH + 2H^+$

$2ADP + 2P_i \rightarrow 2ATP$

Stage II (Dark reactions)

$6(CO_2 + 2NADPH + 2ATP + 2H^+) \rightarrow (CH_2O)_6 + 6(2NADP^+ + 2ADP + 2P_i + H_2O)$

Net reaction: $6(H_2O + CO_2 + \gamma) \rightarrow (CH_2O)_6 + 6O_2$

Figure 13.4 A summary of photosynthesis. γ represents solar photons that drive the reactions. Photosynthesis produces glucose and other biomolecules through a complex series of reactions that involves over 100 steps. The above reaction scheme summarizes the overall energetics of the processes that produce glucose $(CH_2O)_6$

$$6CO_2 + 6H_2O + \gamma \rightarrow C_6H_{12}O_6 + 6O_2 \qquad (13.3.1)$$

in which we have represented solar radiation, or photons, by γ. The overall effect of this reaction is to bind H atoms to C and remove O atoms; in other words, C is 'reduced'. The H atoms in this case come from H_2O. Other sources of H atoms are also used in photosynthesis. For example, photosynthetic purple bacteria (which carry a pigment) obtain H atoms from H_2S. The reaction in this case is

$$6CO_2 + 12H_2S + \gamma \rightarrow C_2H_{12}O_6 + 12S + 6H_2O \qquad (13.3.2)$$

Several other H donors can also take part in photosynthetic reactions [8 (chapter 6)].

The complex process of photosynthesis begins with the absorption of a photon by chlorophyll. The entire process consists of two stages, as shown in Figure 13.4. Stage I occurs in the presence of light and it essentially splits H_2O and releases O_2 while capturing the H atoms; it also converts adenosine diphosphate (ADP) to adenosine triphosphate (ATP). In stage II, the H atoms are transferred to CO_2 and glucose is synthesized. The net reaction is (13.3.1).

For a thermodynamic description of photosynthesis, we need the chemical potential of solar radiation or photons. The thermodynamic properties of electromagnetic radiation or photons are described in terms of its intensity or radiance and energy density (described in detail in Chapter 12). In particular, the chemical potential can be expressed as a function of energy density or radiance. Once the chemical potential of photons is known, the affinities of photochemical reactions can be written explicitly in terms of measurable quantities and the changes in Gibbs energy can be computed using expressions such as (13.2.2). The thermodynamics of electromagnetic radiation has been discussed in Chapter 12. For the convenience of the reader, we shall summarize the main points of interest leading up to the expression for the chemical potential of radiation μ_γ.

Chemical potential of nonthermal radiation

In order to understand the affinity of reactions such as (13.3.1) quantitatively, we need the chemical potential μ_γ of the electromagnetic radiation (solar radiation) that takes part in the reaction. Towards this goal, we begin by recalling some thermo-dynamic properties of radiation that were discussed in detail in Chapter 12.

- At a given temperature T, radiation that is in equilibrium with matter is called **thermal radiation**. Its total energy density (consisting of all frequencies) depends only on T and is independent of the kind of matter with which it is interacting. At sufficiently high T, all particles, photons, electrons, positrons, protons, anti-protons, etc. can exist as thermal radiation. *The chemical potential of thermal radiation is zero.*
- The total energy density u_{th} of electromagnetic thermal radiation or thermal photons is given by the Stefan–Boltzmann law:

$$u_{th} = \beta T^4 \quad \beta = 7.56 \times 10^{-16}\,\mathrm{J\,m^{-3}\,K^{-4}} \tag{13.3.3}$$

The energy density of thermal photons $u_{th}(\nu, T)$, as function of frequency ν and temperature T, is given by the Planck formula or Planck distribution:*

$$u_{th}(T,\nu)d\nu = \frac{8\pi h\nu^3}{c^3}\frac{d\nu}{e^{h\nu/k_BT} - 1} \tag{13.3.4}$$

in which c is the velocity of light and h is Planck's constant. *As it is for all thermal radiation, the chemical potential associated with the Planck distribution is zero.*

- Any energy density $u(\nu)$ that is not a Planck distribution is **nonthermal radiation**. Only nonthermal radiation has a nonzero chemical potential. In Chapter 12 it was shown that (12.6.11)

$$\mu_\gamma(\nu) = RT\ln\left(\frac{u(\nu)}{u_{th}(T,\nu)}\right) \tag{13.3.5}$$

In practice, the spectral radiance $I(\nu)$ associated with $u(\nu)$ is the quantity that can usually be measured. The radiation power incident on a small surface $d\sigma$, from a solid angle $d\Omega = \sin\theta\,d\theta\,d\varphi$, in the frequency range ν and $\nu+d\nu$, equals $I(\nu)d\Omega\,d\sigma\,d\nu$. If it is assumed that the intensity is isotropic (same in all directions), then $u(\nu)$ and $I(\nu)$ have the following relationship:

$$u(\nu) = \frac{4\pi I(\nu)}{c} \tag{13.3.6}$$

---

*$u(T, \nu)\,d\nu$ is the energy of radiation whose frequency is in the range ν and $\nu+d\nu$, per unit volume.

in which c is the velocity of light. This relationship is valid for thermal radiation. Using the definition of radiance, it can easily be shown that the total amount of power falling on a unit area from all directions in a hemisphere, in the frequency range v and $v + dv$, is equal to $\pi I(v) \, dv$.

With the above formalism, the chemical potential for solar photons can be explicitly written in terms of the intensity of solar radiation arriving at the photosynthetic reaction site. For scattered sunlight falling on a leaf, we may, as a first approximation, assume $I(v)$ is isotropic and use (13.3.6) to obtain $u(v)$ from the measured radiance $I(v)$.

Solar radiation reaching the Earth originates as thermal radiation from the sun's surface with a corresponding intensity $I(v, T_S)$, in which T_S is the surface temperature equal to 6000 K. As solar radiation propagates through space, the shape of the curve $I(v, T_S)$ as a function of v remains essentially the same, but its magnitude decreases. (In fact, by noting the value of v at which $I(v, T)$ has its maximum, we determine the temperature of the sun's surface.) At a given point on Earth we can write the solar intensity as $\alpha I(v, T_S)$, in which α is the factor by which the intensity is reduced. Example 13.1 at the end of this chapter shows how α can be estimated; just above the Earth's atmosphere $\alpha = 2.2 \times 10^{-5}$. Given the factor α at a given point on the Earth where the temperature is T_E, we can use (13.3.6) and (13.3.5) and see that the chemical potential of the nonthermal solar radiation can be written as

$$\mu_\gamma(v) = RT \ln\left(\frac{\alpha u_{\text{th}}(v, T_S)}{u_{\text{th}}(v, T_E)}\right) \qquad (13.3.7)$$

Using this expression, the chemical potential of solar radiation reaching the Earth could be estimated as shown in Example 13.2 at the end of this chapter. Depending on the wavelength, the chemical potential is in the range 160–260 kJ mol^{-1}. If we consider the amount reaching the surface of the Earth after taking into account the reflection in the atmosphere, it is even less. The chemical potential of solar radiation at the point where photosynthesis is occurring could be even lower. Nevertheless, in a reaction such as (13.3.1), the affinity that drives the reaction is significant.

Photochemical transfer of gibbs energy

Using the chemical potential of radiation μ_γ, let us look at the first step of photosynthesis (Figure 13.5). This reaction is photo-excitation of chlorophyll followed by the transfer of its excitation energy to another molecule. We shall write this set of reactions thus:

$$Y + \gamma \rightleftharpoons X \qquad (13.3.8a)$$

$$X + W \rightleftharpoons Z_1 + Y \qquad (13.3.8b)$$

$$Z_1 \rightleftharpoons Z_2$$

$$\cdots$$

$$Z_{s-1} \rightleftharpoons Z_s \qquad (13.3.8s)$$

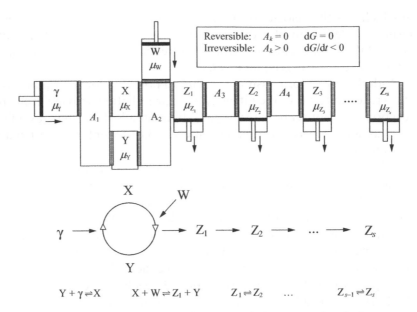

Figure 13.5 Conversion of Gibbs energy of photons to chemical energy. Driven by the high chemical potential of nonthermal radiation or photons, W is converted to compounds $Z_1, Z_2, \ldots, Z_s$. The Gibbs energy stored in these compounds is released when their chemical potentials return to their equilibrium values in the absence of nonthermal photons. The method of the van't Hoff reaction chamber described in Figure 13.2 can be extended to this series of reactions to elucidate the distinction between reversible and irreversible transformations. In each reaction chamber, a reaction is taking place with an affinity A_k. The reaction chambers are connected to chambers containing reacting compound through a semi-permeable membrane through which only that reactant can pass. Pistons attached to each reservoir can be used to maintain the affinity in each reaction chamber at a desired value. For an idealized reversible transformation, $A_k \approx 0$. All real transformations are irreversible for which $A_k \neq 0$. These are accompanied by a loss of Gibbs free energy given by expressions (13.3.11) and (13.3.12)

The exact details are quite complex and can be found in texts on biochemistry [4]. However, to understand the general thermodynamic aspects, the exact names and nature of the compounds in the reaction scheme are not essential. In (13.3.8a), X is the excited state at a higher chemical potential and Y is the ground state. If the second reaction (13.3.8b) were absent, then the absorbed photon energy would simply be reemitted and eventually the radiation will become thermal radiation and its chemical potential will become zero. When the second reaction is included, the Gibbs energy of X could result in the generation of Z_1, which may preserve or 'store' some of the Gibbs energy of the incident solar radiation. If the compound Z_1 is further converted into other compounds Z_k, then the Gibbs energy of the incident solar radiation would be 'chemically stored' in these compounds. In a cell, these energy-storing compounds correspond to glucose, ATP and other Gibbs-energy-carrying compounds.

In order to focus on the thermodynamic principles, we have represented a complex process of photosynthesis by a simple process of single-molecule conversions $Z_k \rightleftharpoons Z_j$, but this is not at the expense of thermodynamic generality. The thermodynamic analysis of (13.3.8a)–(13.3.8s) could be extended to more complex reactions that occur during photosynthesis (with many more reactants and products). During the process of respiration and biological activity, the last stage shown in Figure 13.3, the chemically stored Gibbs energy is ultimately converted to heat or thermal radiation at the surface temperature of the Earth.

Using the formalism presented in Section 13.2, we can now write the affinities and the rate at which the Gibbs free energy changes in the sequence of reactions (13.3.8a)–(13.3.8s):

$$A_1 = (\mu_Y + \mu_\gamma) - \mu_X \qquad (13.3.9a)$$

$$A_2 = (\mu_X + \mu_W) - (\mu_{Z_1} + \mu_Y) \qquad (13.3.9b)$$

$$A_3 = \mu_{Z_1} - \mu_{Z_2} \qquad (13.3.9c)$$

$$\dots$$

The change of Gibbs free energy in this series of reactions can be illustrated as was done for reaction (13.2.1). As shown in Figure 13.5, we can represent the set of reactions (13.3.8a)–(13.3.8s) as a series of van't Hoff reaction chambers each connected to chambers containing reactants on one side and products on the other. Each reacting compound is separated from the reacting chambers by a semipermeable membrane that is permeable only to that compound (as in Figure 13.2). No Gibbs free energy is lost in a reversible reaction in which $A_k \approx 0$. In this case, it is easy to see that

$$\mu_Y + \mu_W = \mu_{Z_1} = \mu_{Z_2} = \mu_{Z_3} \dots = \mu_{Z_s} \qquad (13.3.10)$$

This equation implies that the chemical potential of the compounds Z_k are higher than that of W by an amount equal to μ_γ. Thus, every mole of Z_k stores an energy equal to μ_γ, the molar Gibbs energy of the nonthermal radiation. In this case, the conversion of the Gibbs energy of the nonthermal photons to chemical energy is without any losses.

In all realistic conditions, the transformation of W to Z_k is irreversible in which the affinities $A_k > 0$. The corresponding rate of loss of Gibbs free energy is

$$dG = -T d_i S = -\sum_k A_k d\xi_k < 0 \qquad (13.3.11)$$

The rate of loss of Gibbs free energy is equal to $-\Sigma_k A_k (d\xi_k/dt)$; the remaining Gibbs energy of the initial reactants, radiation and W is converted to the Gibbs energy in the compounds Z_k. If dN_γ/dt and dN_W/dt are the rates at which photons and W are consumed, then the rate at which the Gibbs energy is channeled into the synthesis of the photosynthetic products Z_k is given by

$$\boxed{\frac{dG_{\text{synth}}}{dt} = \left(\mu_\gamma \frac{dN_\gamma}{dt} + \mu_W \frac{dN_W}{dt} \right) - \sum_k A_k \frac{d\xi_k}{dt}} \qquad (13.3.12)$$

in which G_{synth} is the Gibbs energy in the synthesized compounds Z_k.

The actual process of photosynthesis is much more complex than the simple scheme presented above, but its thermodynamic analysis could be done along the same lines, regardless of the number of reactions involved.

The products of photosynthesis are many. When a plant grows, it not only synthesizes compounds, but it also absorbs water. For this reason, to estimate the amount of substance produced in photosynthesis it is necessary to separate the 'dry mass' from the water. By measuring the Gibbs energy in the dry mass generated in the presence of a known flux of radiation, the efficiency of photosynthesis can be estimated in various conditions. When the dry mass is combusted, it is converted back to its initial reactants, CO_2, H_2O and other small compounds, and the Gibbs energy captured in photosynthesis is released. Carbohydrates release about $15.6\,\text{kJ}\,\text{g}^{-1}$ on combustion, proteins about $24\,\text{kJ}\,\text{g}^{-1}$ and fats about $39\,\text{kJ}\,\text{g}^{-1}$ [9]. Plant cells contain many other compounds that yield less energy upon combustion. On the whole, the plant dry mass yields about $17.5\,\text{kJ}\,\text{g}^{-1}$. Under optimal conditions, rapidly growing plants can pro-duce around $50\,\text{g}\,\text{m}^{-2}\,\text{day}^{-1}$, which equals $875\,\text{kJ}\,\text{m}^{-2}\,\text{day}^{-1}$. During this observed plant growth, the amount of solar energy incident on the plants averages about $29 \times 10^3\,\text{kJ}\,\text{m}^{-2}\,\text{day}^{-1}$. From these figures we can estimate that plants capture solar energy with an efficiency of about $875/(29 \times 10^3) = 0.03$, which is a rather low value. A much larger fraction, about 0.33 of solar energy that enters the Earth's atmosphere, goes into the water cycle (see Figure 2.9). A part of the reason for such a low efficiency of photosynthetic capture of solar energy is the low amounts of CO_2 in the atmosphere (about 0.04% by volume). Plants grow faster at higher levels of CO_2.

The rate of photosynthesis depends on the intensity of incident radiation I (W m^{-2}). In a leaf exposed to radiation of intensity I, the rate of photosynthesis can me measured in terms of the energy captured P (W m^{-2}). P increases with I at low intensities, but P reaches a saturation value P_{max} at high intensities (see Box 13.2). The empirical relationship between P and I can be expressed as [9]

$$P = P_{\text{max}} \frac{I}{K + I} \qquad (13.3.13)$$

in which K is a constant which is approximately $200\,\text{W}\,\text{m}^{-2}$. The similarity of this equation to the Michaelis–Menten equation that describes enzyme kinetics is obvious. A representative value of P_{max} is about $25\,\text{W}\,\text{m}^{-2}$. It is found that P_{max} depends on the temperature and the CO_2 concentration. On a bright sunny day, $I \approx 1000\,\text{W}\,\text{m}^{-2}$. Using (13.3.13) one can estimate the photosynthetic efficiency at various intensities.

Box 13.2 Facts about Photosynthesis

- Satellite measurements of **solar radiant flux** just outside the atmosphere give a value of $1370\,W\,m^{-2}$ (area perpendicular to the direction of radiation). The maximum radiation reaching the Earth's surface is about $1100\,W\,m^{-2}$. For the purposes of estimation, the maximum flux at the ground surface during a clear day is $\sim 800\,W\,m^{-2}$.
- It is estimated that 90% of photosynthesis takes place in the oceans in algae, bacteria, diatoms and other organisms. Approximately $4.7 \times 10^{15}\,mol$ of O_2 is generated per year by photosynthesis. Microorganisms in the oceans and soil consume over 90% of all the oxygen consumed by life.
- The energy captured P by photosynthesis varies with the incident solar energy intensity I according to the approximate equation shown below.

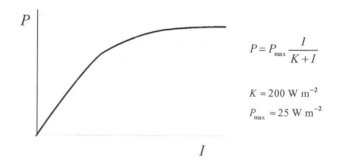

$$P = P_{max}\,\frac{I}{K+I}$$

$$K \approx 200\ W\ m^{-2}$$

$$P_{max} \approx 25\ W\ m^{-2}$$

- The rate of photosynthesis can be considered as the rate of primary production. It can be quantified as either energy captured or new biomass formed. Gross primary production is the rate at which biomass is being synthesized. The process of respiration degrades biomass into CO_2. Net primary production is the difference between the rate at which biomass is being synthesized and the rate at which it is being degraded into CO_2; it is the rate at which biomass is accumulating. For example, sugar cane growth corresponds to about $37\,g\,m^{-2}\,day^{-1}$.

Source: R. M. Alexander, *Energy for Animals*, 1999. New York: Oxford University Press.

The Gibbs energy captured by plants moves up the food chain, sustaining the process of life at the micro level and ecosystems on the macro level. Ultimately, this 'food' reacts with O_2 and turns into CO_2 and H_2O, thus completing the cycle. The cycle, however, has an awesome complexity, which is the process of life. Finally, when an organism 'dies' its complex constituents are converted to simpler molecules by bacteria.

THE ROLE OF ATP AS A CARRIER OF GIBBS ENERGY

Central to the Gibbs energy flow in biochemical systems is the energy-carrying molecule **ATP**. The role of ATP as an energy carrier came to light around 1940, by which time a significant amount of experimental data had been gathered. The ubiquitous role of ATP as a Gibbs-energy-transporting compound was noted by Fritz

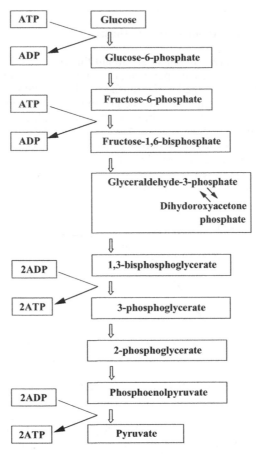

Figure 13.6 A summary of the main reaction scheme of glycolysis. Each step is catalyzed by an enzyme. Every mole of glucose that becomes pyruvate results in the conversion of 2 mol of ADP to ATP

Lipmann and Herman Kalckar in 1941. ATP is involved in an enormous number of reactions which otherwise would not occur because their affinity in the absence of ATP is negative. It provides the Gibbs energy needed for biosynthesis, muscle contraction and active transport of biomolecules. We shall now consider some typical examples of the thermodynamics of ATP-driven reactions.

ATP is synthesized in a series of reactions called **glycolysis**, shown in Figure 13.6. The production of ATP is associated with the oxidation of compounds. ATP consists of three phosphate groups linked to an [Adenine]–[D-Ribose] unit. It can react with water (hydrolysis) and lose one of the phosphate groups to form **ADP**. The different ways of writing this reaction, are shown below.

$$[\text{Adenine}]-[\text{D}-\text{Ribose}]-\text{O}-(\text{PO}_3^-)-(\text{PO}_3^-)-(\text{PO}_3^{2-})+\text{H}_2\text{O}$$
$$\rightarrow [\text{Adenine}]-[\text{D-Ribose}]-\text{O}-(\text{PO}_3^-)-(\text{PO}_3^{2-})+\text{HPO}_4^{2-}+\text{H}^+ \quad (13.3.14\text{a})$$

or

$$\text{ATP}^{4-}+\text{H}_2\text{O}\rightarrow\text{ADP}^{3-}+\text{HPO}_4^{2-}+\text{H}^+ \quad\quad (13.3.14\text{b})$$

or

$$\text{ATP}+\text{H}_2\text{O}\rightarrow\text{ADP}+\text{P}_\text{i}+\text{H}^+ \quad\quad (13.3.14\text{c})$$

$$A^{0'}=-\Delta G^{0'}=30.5\,\text{kJ}\,\text{mol}^{-1} \quad\text{and}\quad A_\text{cell}\approx 46\,\text{kJ}\,\text{mol}^{-1} \quad (13.3.14\text{d})$$

At a given pH, the phosphate group, HPO_4^{2-}, exists in different resonant forms, each corresponding to a different distribution of charge 2− within the molecule, and it is usually written as P_i in biochemical literature. (13.3.14a)–(13.3.14c) show three equivalent ways of writing the hydrolysis of ATP. For this reaction, at pH 7.0, the standard affinity $A^{0'}$ and the affinity under cellular conditions A_cell are as given in (13.3.14d). The higher affinity A_cell can be expressed in terms of the activities in the cell (a_ATP, a_ADP, etc.) as follows [10 (p. 317)]:

$$\begin{aligned}
A_\text{cell} &=A^{0'} + RT\ln(a_\text{ATP}a_{\text{H}_2\text{O}}/a_\text{ADP}a_{\text{P}_\text{i}}a_{\text{H}_+}) \\
&\approx 30.5\,\text{kJ}\,\text{mol}^{-1} + R(298)\ln(500) \\
&\approx (30.5 + 15.4)\,\text{kJ}\,\text{mol}^{-1}
\end{aligned}$$

Note that in the biochemical standard state, $a_{\text{H}^+} = 1.0$ for pH 7.0 and $a_{\text{H}_2\text{O}} \simeq 1.0$ for pure liquids. This implies $a_\text{ATP}/a_\text{ADP}a_{\text{P}_\text{i}} \simeq 500$.

An enormous number of reactions in the cell that have negative affinities in the absence of ATP find pathways with positive affinities when they couple with ATP conversion to ADP. These reactions are said to be driven by the ATP–ADP couple. The following example illustrates how ATP drives a reaction. The reaction shown below, formation of glutamine, has a negative standard affinity:

$$\text{Glutamic Acid}+\text{NH}_3\rightarrow\text{Glutamine}+\text{H}_2\text{O} \quad A^{0'}=-\Delta G^{0'}=-14.2\,\text{kJ}\,\text{mol}^{-1} \quad (13.3.15)$$

The affinity under cellular conditions is also negative. However, in the presence of ATP, glutamine is synthesized. The mechanism is as follows:

$$\text{ATP}+\text{Glutamic Acid}\rightarrow\text{Glutamyl-phosphate}+\text{ADP} \quad\quad (13.3.16)$$

$$\text{Glutamyl-phosphate}+\text{NH}_3\rightarrow\text{Glutamine}+\text{P}_\text{i} \quad\quad (13.3.17)$$

The affinities for both reactions (13.3.16) and (13.3.17) are positive. The net reaction is

$$\text{ATP}+\text{H}_2\text{O}+\text{Glutamic Acid}+\text{NH}_3\rightarrow\text{Glutamine}+\text{ADP}+\text{P}_\text{i}+\text{H}^+ \quad (13.3.18)$$

Since the affinity A_{cell} of the reaction $ATP + H_2O \rightarrow ADP + P_i + H^+$ under cellular conditions is about $46\,kJ\,mol^{-1}$ (13.3.14d), the overall reaction has a positive affinity. This example shows that ATP converts glutamic acid to glutamyl-phosphate by transferring a phosphate group. The chemical potential of glutamyl-phosphate is higher than that of glutamic acid; the higher chemical potential makes the affinity of reaction (13.3.17) positive. Increasing the chemical potential of a compound by transferring a phosphate group to it is a common theme in biochemical reactions. This increase of the chemical potential of compounds through ATP–ADP transformation drives a very large number of reactions in a cell.

The ATP–ADP cycle is continuously driving biochemical reactions. In fact, this turnover of ATP is enormous: it is estimated that $40\,kg\,day^{-1}$ of ATP is cycled in a resting human! Strenuous exercises consume about $0.5\,kg\,min^{-1}$ of ATP [10 (p. 319)]. It is estimated that ATP is present in all living cells in concentrations in the range 0.001–$0.01\,M$ [8].

TRANSPORT ACROSS CELL MEMBRANES

Lipid bilayer membranes, about 9 nm or 90 Å thick, enclose living cells; they define the boundaries of a cell. Imbedded in these lipid bilayers are many proteins, constituting about 50% of the membrane. A living cell is an open system that exchanges matter with its environment. The matter exchange takes place through the lipid–protein membrane. The membrane, however, has selective permeability. It is permeable to water, glucose, amino acids, Na^+ and K^+ ions, O_2, CO_2 and other small molecules, but it is not permeable to a large number of molecules that are within the cell, such as ATP, ADP and other molecules containing phosphate groups.

For many of the compounds that can pass through the cell membrane, their chemical potentials in the cell interior and the exterior are equal. This does not mean that the net concentrations of the compounds are equal, however. As we have noted before, *equality of chemical potentials does not imply the equality of concentrations*. For example, the membrane of seaweed cells is permeable to iodine. However, because the interior of seaweed cells contain an iodine-binding protein, the net concentration of iodine in the interior is higher than it is in seawater, but the chemical potential is the same. This is because the chemical potential of iodine is lowered when it is bound to the protein, causing more of it to accumulate in the cell until the chemical potentials of iodine in the cell's interior and exterior are equal, though the concentrations are quite different.

Membrane permeability does not always result in the establishment of equilibrium between the cell's interior and exterior for the molecules that can pass through the membrane. In most mammalian cells, as shown in the Figure 13.7, Na^+ and K^+ ions in the interior of the cell are not in thermodynamic equilibrium with those in the exterior medium: their chemical potentials in the two regions are unequal. This nonequilibrium state is maintained by ATP-driven 'ion pumps' whose mechanism is not fully understood. The concentration of K^+ is about 100 mM inside the cell and about 5 mM outside; in contrast, the concentration of Na^+ inside the cell is about 10 mM and outside it is about 140 mM. Owing to the corresponding difference in

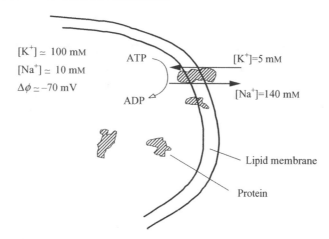

Figure 13.7 The cell membrane is permeable to some small molecules and ions, but it is not permeable to molecules such as ATP and ADP that are synthesized and used within the cell. In mammalian cells, chemical potentials of Na^+ and K^+ ions in the cell's interior are not equal to those in the exterior. This nonequilibrium state is maintained by ATP-driven 'ion pumps'. As shown in the figure, the concentration of K^+ is higher inside the cell and Na^+ is higher outside the cell. With respect to the exterior, the electrical potential of the interior of the cell is at about $-70\,\text{mV}$

chemical potential, there is a small flux of K^+ flowing out of the cell and a small flux of Na^+ flowing into the cell. These fluxes are countered by the 'active transport' of Na^+ out of the cell and K^+ into the cell through an ATP-consuming process which is thought to be a protein. Studies have indicated that the pumping of the two ions is coupled. In 1957, Jens Christian Skou discovered that transport of the two ions takes place through a protein to which they bind. He was awarded the Nobel Prize in 1997 for this discovery. This protein is called Na^+-K^+-ATPase or simply the sodium–potassium pump. As shown in Figure 13.8, the sodium–potassium pump reacts with ATP and undergoes a structural transformation in such a manner that K^+ and Na^+ are transported across the cell membrane; this transport maintains the difference in chemical potential gradients, drawing the needed Gibbs energy from ATP. As is the case with all reactions coupled to ATP to ADP conversion, the protein pump that binds the two ions must transform to an intermediate state in which the transport of Na^+ ions out of the cell and K^+ into the cell is driven by a positive affinities. We can describe the thermodynamics of this process as follows. Let us denote the intracellular and extracellular regions of the cell by superscripts 'i' and 'e' respectively. The stages summarized in Figure 13.8 can be written as the following reactions, in which [SPP] is the sodium–potassium pump:

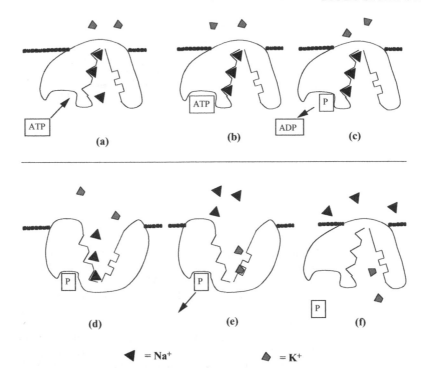

Figure 13.8 A simplified scheme showing the mechanism of the sodium–potassium pump. (a) Three intracellular Na$^+$ ions bind to the protein. (b) ATP attaches to the protein. (c) The protein is phosphorylated with the release of ADP. (d) Phosphorylation causes a structural change in the protein. The change in the structure causes a decrease in the binding affinity for Na$^+$ and an increase of binding affinity for K$^+$ ions. (e) Subsequent release of Na$^+$ into the extracellular region and the binding of two extracellular K$^+$ ions to the protein. (f) The binding of K$^+$ triggers dephosphorylation and a structural change that restores the protein to its initial structure. In this state, the two K$^+$ ions are released into the intracellular region. The overall effect is the hydrolysis of ATP, the transport of three Na$^+$ from the intracellular region to the extracellular region and transport of two K$^+$ ions in the opposite direction

$$[\text{SPP}] + 3(\text{Na}^+)^i \rightleftharpoons [\text{SPP–3Na}^+] \qquad (13.3.19a)$$

$$[\text{SPP–3Na}^+] + \text{ATP} \rightleftharpoons [\text{ATP–SPP–3Na}^+] \qquad (13.3.19b)$$

$$[\text{ATP–SPP–3Na}^+] + \text{H}_2\text{O} \rightleftharpoons [\text{P}_i\text{–SPP–3Na}^+] + \text{ADP} + \text{H}^+ \qquad (13.3.19c)$$

$$[\text{P}_i\text{–SPP} - 3\text{Na}^+] + 2(\text{K}^+)^e \rightleftharpoons [\text{P}_i\text{–SPP–2K}^+] + 3(\text{Na}^+)^e \qquad (13.3.19d)$$

$$[\text{P}_i\text{–SPP–2K}^+] \rightleftharpoons [\text{SPP}] + 2(\text{K}^+)^i + \text{P}_i \qquad (13.3.19e)$$

The net reaction is

$$2(\text{K}^+)^e + 3(\text{Na}^+)^i + \text{ATP} + \text{H}_2\text{O} \rightleftharpoons 2(\text{K}^+)^i + 3(\text{Na}^+)^e + \text{ADP} + \text{H}^+ + \text{P}_i \qquad (13.3.20)$$

The net pumping of three Na^+ ions out of the cell and two K^+ ions into the cell occurs because each of the reaction steps in (13.3.19a)–(13.3.19d) have a positive affinity. One can check whether the net reaction (13.3.20) has a positive affinity for the $[Na^+]$ and $[K^+]$ concentrations in the intra- and extra-cellular regions shown in Figure 13.8 by using the expression $\mu_k^c = \mu_k^{c0} + RT\ln(a_k)$ for the chemical potential of a solute in the concentration scale, in which μ_k^{c0} is the standard chemical potential defined for unit activity $a_k = 1$. For dilute solutions we may approximate $a_k \simeq c_k/c_0$, in which c_k is the molarity and $c_0 = 1.0\,M$. Given that the affinity of $ATP + H_2O \rightarrow ADP + P_i + H^+$ under cellular conditions is about $46\,kJ\,mol^{-1}$, it is left as an exercise (Exercise 13.2) for the reader to verify that the affinity for the net reaction (13.3.20) is positive for the $[Na^+]$ and $[K^+]$ concentrations in the intra- and extra-cellular regions shown in Figure 13.8.

The above examples illustrate the central role of ATP in the Gibbs energy flow in living cells. The ultimate source of Gibbs energy is solar radiation originating at the surface of the sun at a temperature of about $6000\,K$. This Gibbs energy is dissipated in the various biological activities, as summarized in Figure 13.3; the biological activity in turn produces heat that is ultimately radiated back into space as thermal radiation at $T = 288\,K$. The corresponding entropy production due to biological processes is small compared with the entropy generated in atmospheric processes, such as the water cycle. As described in Chapter 2, the water cycle transports enormous amount of heat from the Earth's surface to the upper atmosphere, acting like a steam engine.

13.4 Biochemical Kinetics

Biochemical reactions are catalyzed by enzymes. Catalysts have no effect on systems in thermodynamic equilibrium. In the nonequilibrium cellular conditions, enzymes control the biochemical pathways. In this section we shall look at some of the basic kinetic laws that describe enzyme-catalyzed reactions.

MICHAELIS–MENTEN RATE LAW

The early investigations of Leonor Michaelis and Maude Menten, around 1913, resulted in the formulation of a rate law based on a mechanism of enzyme catalysis they proposed. This mechanism and the resulting rate law were discussed in Chapter 9, but for convenience we shall summarize the main points here. The following notation will be used:

S = substrate   E = enzyme   ES = enzyme-substrate complex
P = product   R = rate of product formation
k_{if} = forward rate constant of reaction i k_{ir} = reverse rate constant of reaction i

The catalysis proceeds in two stages:

$$E + S \underset{k_{1r}}{\overset{k_{1f}}{\rightleftharpoons}} ES \tag{13.4.1a}$$

$$ES \underset{k_{2r}}{\overset{k_{2f}}{\rightleftharpoons}} P + E \tag{13.4.1b}$$

$$k_{1f}, k_{1r} \gg k_{2f} \quad \text{and} \quad k_{2f} \gg k_{2r} \tag{13.4.1c}$$

The enzyme–substrate complexation reaction (13.4.1a) occurs rapidly and reversibly. The second reaction, (13.4.1b), the conversion of the substrate to product, occurs essentially irreversibly, so we can ignore the reverse reaction rate. The rate equations for these reactions are:

$$\frac{d[E]}{dt} = -k_{1f}[E][S] + k_{1r}[ES] + k_{2f}[ES] \tag{13.4.2}$$

$$\frac{d[S]}{dt} = -k_{1f}[E][S] + k_{1r}[ES] \tag{13.4.3}$$

$$\frac{d[ES]}{dt} = k_{1f}[E][S] - k_{1r}[ES] - k_{2f}[ES] \tag{13.4.4}$$

$$\frac{d[P]}{dt} = k_{2f}[ES] \tag{13.4.5}$$

$$[E_0] = [E] + [ES] = \text{constant} \tag{13.4.6}$$

The Michaelis–Menten law refers to the rate at which the product is generated, $d[P]/dt$, for a given concentration of the total amount of enzyme $[E_0] = [E] + [ES]$ and the substrate concentration $[S]$. The assumption that the concentration of the enzyme substrate complex $[ES]$ is in a steady state, i.e. $d[ES]/dt = 0$, leads to the following law for the rate R at which the product is generated (Exercise 13.4):

$$R \equiv \frac{d[P]}{dt} = \frac{k_{2f}[E_0][S]}{K_m + [S]} = \frac{R_{max}[S]}{K_m + [S]} \tag{13.4.7}$$

in which

$$K_m = \frac{k_{1r} + k_{2f}}{k_{1f}} \quad R_{max} = k_{2f}[E_0] \tag{13.4.8}$$

K_m is the Michaelis–Menten constant and R_{max} is the maximum rate at which the substrate S is converted to the product P for a given $[E_0]$. These constants, K_m and R_{max}, are characteristic to an enzyme and are quite sensitive to the conditions under which the reaction takes place. A plot of $d[P]/dt$ as a function of the substrate concentration $[S]$ is shown in Figure 13.9a. At low values of $[S]$, the rate is linear, but it saturates at a high value of $[S]$ and becomes independent of $[S]$. Since the total

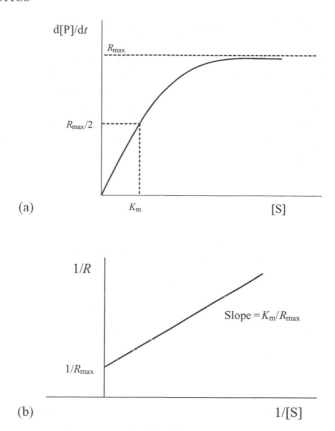

Figure 13.9 (a) The Michaelis–Menten rate of product formation as a function of the substrate concentration. (b) A plot of $1/R$ versus $1/[S]$, called a Lineweaver–Burk plot, can be used to obtain the Michaelis-Menten constant K_m and R_{max}

amount of enzyme is fixed, the rate saturates when all the available enzyme is bound to the substrate; as a result, the rate for very large $[S]$ equals $k_{2f}[E_0]$.

The values of K_m and $R_{max} = k_{2f}[E_0]$ are determined experimentally by measuring the initial rates of product formation $R = d[P]/dt$ for various values of $[S]$. From (13.4.7), it follows that

$$\frac{1}{R} = \frac{K_m}{R_{max}}\frac{1}{[S]} + \frac{1}{R_{max}} \tag{13.4.9}$$

A plot of $1/R$ versus $1/[S]$, called a **Lineweaver–Burk plot**, would be a straight line with a slope equal to K_m/R_{max} and intercept equal to $1/R_{max}$ (Figure 13.9b). From the slope and the intercept of such plots, the values of K_m and $R_{max} = k_{2f}[E_0]$ are obtained.

The Michaelis–Menten rate law (13.4.7) could be rearranged in other ways to obtain linear plots from which R_{max} and K_m could be obtained. For example, it can be shown that a plot of [S]/R versus [S] would also be straight line. We leave it as an exercise for the reader to devise alternate methods to obtain R_{max} and K_m.

The structure of an enzyme depends on factors such as pH and the temperature. Consequently, K_m and R_{max} depend on factors such as the pH and temperature. As the temperature or pH increases from a low value, the rate of product formation d[P]/dt generally reaches a maximum and then decreases. If the Arrhenius form of the rate constants k_{if} and k_{ir} for all the reactions (13.4.2)–(13.4.6) is known, then the temperature dependence of the constants K_m and R_{max} can be determined.

From the thermodynamic viewpoint, the affinity of a reaction is not changed by the catalyst, but the reaction velocity is. Consequently, the rate of entropy production increases with the introduction of a catalyst. The rate of entropy production per unit volume can be written as

$$\frac{1}{V}\frac{d_iS}{dt} = R\sum_{k=1}^{2}(R_{kf} - R_{kr})\ln(R_{kf}/R_{kr}) > 0 \tag{13.4.10}$$

in which V is the system volume, R is the gas constant and $R_{1f} = k_{1f}[E][S]$, $R_{1r} = k_{1r}[ES]$, $R_{2f} = k_{2f}[ES]$ and $R_{2r} = k_{2r}[E][P]$. In evaluating the rates of entropy production, the reverse reaction rates, however small, must be included.

MECHANISMS OF ENZYME INHIBITION

Enzymes control biochemical pathways. The control takes place not only through the production of the enzyme when needed, but also through inhibition of its action by other compounds, sometimes even by the product. We shall discuss some common mechanisms involving an inhibitor I.

Enzymes have specific sites to which the substrate binds. The specificity of the enzymes is due to the particular structure of the binding site where the recognition of the substrate molecule takes place. An inhibitor I that has a structure similar to that of the substrate can bind to the enzyme's active site and, thus, compete with the substrate for the binding site, or it could bind to another part of the enzyme and structurally alter the biding site and, thus, prevent the substrate from binding to the enzyme. Such a mechanism, called **competitive inhibition**, has the following reaction steps:

$$E + S \underset{k_{1r}}{\overset{k_{1f}}{\rightleftharpoons}} ES \quad ES \underset{k_{2r}}{\overset{k_{2f}}{\rightleftharpoons}} P + E \quad E + I \underset{k_{3r}}{\overset{k_{3f}}{\rightleftharpoons}} EI \tag{13.4.11}$$

The third reaction is the competitive binding of the inhibitor I to the enzyme. When bound to the inhibitor, forming the complex EI, the enzyme becomes inactive. The amount of the inactive complex [EI] depends on the relative values of the rate constants k_{1f}, k_{1r}, k_{3f} and k_{3r}. For the total amount of enzyme [E$_0$] we have

$$[E_0] = [E] + [ES] + [EI] \tag{13.4.12}$$

The Michaelis–Menten steady-state approximation for [ES] now takes the form

$$\frac{d[ES]}{dt} = k_{1f}([E_0] - [ES] - [EI])[S] - k_{1r}[ES] - k_{2f}[ES] = 0 \qquad (13.4.13)$$

In addition, the binding of the inhibitor is generally rapid and we may assume that the reaction $E + I \rightleftharpoons EI$ is in equilibrium:

$$k_{3f}[E][I] = k_{3r}[EI] \qquad (13.4.14)$$

As in the case of the simple Michaelis–Menten kinetics, we would like to obtain an expression for the rate of product formation as a function of the total amount of enzyme, the substrate concentration [S] and the inhibitor concentration [I]. This could be done if [ES] can be expressed as functions of [S], [I], [E_0] and the rate constants. In order to do this, first we use (13.4.12) and (13.4.14) to obtain [EI] as a function of [ES], [I], [E_0] and the rate constants:

$$[EI] = \frac{[I]([E_0] - [ES])}{(k_{3r}/k_{3f}) + [I]} \qquad (13.4.15)$$

Substituting this expression into the steady-state approximation (13.4.13), we can express [ES] as a function of [S], [I], and [E_0]. Then the rate of product formation is equal to $k_{2f}[ES]$. The result is

$$\boxed{R \equiv \frac{d[P]}{dt} = \frac{k_{2f}[E_0][S]}{[S] + K_m(1 + K_3[I])}} \qquad (13.4.16a)$$

in which

$$K_m = \frac{k_{1r} + k_{2f}}{k_{1f}} \quad \text{and} \quad K_3 = \frac{k_{3f}}{k_{3r}} \qquad (13.4.16b)$$

The expression for the rate shows that the constant K_m is modified by the factor $(1 + K_3[I])$ due to the inhibitor. For large [S], the rate R approaches its maximum $R_{max} = k_{2f}[E_0]$. A plot similar to the Lineweaver–Burk plot can be obtained by rearranging Equation (13.4.16a) in the form

$$\frac{1}{R} = \frac{K_m}{R_{max}}(1 + [I]K_3)\frac{1}{[S]} + \frac{1}{R_{max}} \qquad (13.4.17)$$

and plotting 1/[R] versus 1/[S]. Comparing this expression with the uninhibited enzyme, we see that K_m, and hence the slope, is altered by the factor $(1 + [I]K_3) = \{1 + (k_{3f}/k_{3r})[I]\}$, while the intercept remains the same.

The second type of inhibitor action is through its binding to the enzyme–substrate complex ES but not the enzyme. This is called **uncompetitive inhibition**. The enzyme bound to the inhibitor is inactive. The corresponding reaction steps for the enzyme action are

$$E + S \underset{k_{1r}}{\overset{k_{1f}}{\rightleftharpoons}} ES \quad ES \underset{k_{2r}}{\overset{k_{2f}}{\rightleftharpoons}} P + E \quad ES + I \underset{k_{3r}}{\overset{k_{3f}}{\rightleftharpoons}} ESI \qquad (13.4.18)$$

In this case the total concentration of the enzyme in its various forms is

$$[E_0] = [E] + [ES] + [ESI] \qquad (13.4.19)$$

and we may assume that the inhibitor binds reversibly and rapidly so that the reaction ES + I $\rightleftharpoons$ ESI is in equilibrium:

$$k_{3f}[ES][I] = k_{3r}[ESI] \qquad (13.4.20)$$

As before, by making the steady-state approximation for the concentration of ES and using (13.4.19) and (13.4.20), one can arrive at the following expression for the rate of product formation:

$$\boxed{R \equiv \frac{d[P]}{dt} = \frac{k_{2f}[E_0][S](1 + K_3[I])}{[S] + K_m(1 + K_3[I])}} \qquad (13.4.21)$$

in which $K_m = (k_{1r} + k_{2f})/k_{1f}$ and $K_3 = (k_{3f}/k_{3r})$.

The maximum rate and K_m are altered by the same factor $(1 + K_3[I])$. As we expect, when $[I] = 0$, rate R equals the Michaelis–Menten rate. A Lineweaver–Burk plot of $1/R$ versus $1/[S]$ will have a slope equal to $K_m/k_{2f}[E_0]$ and an intercept equal to $1/k_{2f}[E_0](1 + K_3[I])$. In this case, the inhibitor I changes the intercept but not the slope.

The third mechanism of enzyme inhibition, called **noncompetitive inhibition**, is through the binding of the inhibitor to the enzyme E as well as to the complex ES, a combination of the first two mechanisms. The mechanism is as follows:

$$E + S \underset{k_{1r}}{\overset{k_{1f}}{\rightleftharpoons}} ES \quad ES \underset{k_{2r}}{\overset{k_{2f}}{\rightleftharpoons}} P + E \quad E + I \underset{k_{3r}}{\overset{k_{3f}}{\rightleftharpoons}} EI \quad ES + I \underset{k_{4r}}{\overset{k_{4f}}{\rightleftharpoons}} ESI \quad (13.4.22)$$

The two forms of the enzyme EI and ESI are inactive, in that they cannot convert the substrate to product. The total amount of enzyme is now

$$[E_0] = [E] + [ES] + [EI] + [ESI] \qquad (13.4.23)$$

Once again, the reversible complexation of the inhibitor to the enzyme is rapid and may be assumed to be in equilibrium, i.e.

$$k_{3f}[E][I] = k_{3r}[EI] \quad k_{4f}[ES][I] = k_{4r}[ESI] \qquad (13.4.24)$$

Invoking the steady-state approximation for [ES], and using (13.4.23) and (13.4.24), one can obtain the following expression for the rate of product formation:

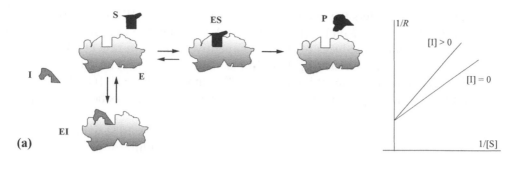

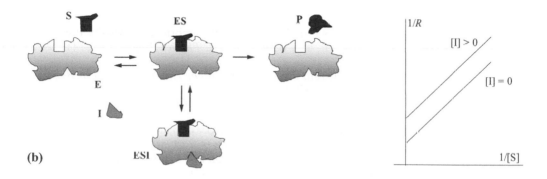

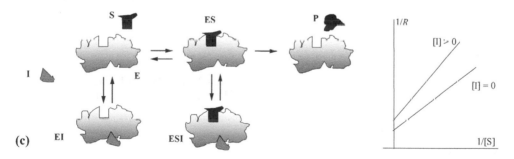

Figure 13.10   Three mechanisms of enzyme inhibition: (a) competitive; (b) noncompetitive; (c) uncompetitive. E: enzyme; S: substrate; I: inhibitor; P: product

$$R \equiv \frac{d[P]}{dt} = \frac{k_{2f}[E_0]}{1 + [I]K_4} \frac{[S]}{K_m[(1 + [I]K_3)/(1 + [I]K_4)] + [S]} \qquad (13.4.25a)$$

in which

$$K_m = \frac{k_{1r} + k_{2f}}{k_{1f}} \quad K_3 = \frac{k_{3f}}{k_{3r}} \quad K_4 = \frac{k_{4f}}{k_{4r}} \qquad (13.4.25b)$$

The inhibitor alters the maximum rate by a factor $(1 + [I]K_4)^{-1}$ and the constant K_m by a factor $(1 + [I]K_3)/(1 + [I]K_4)$. If $K_4 = 0$, i.e. when the inhibitor does not bind to ES, this equation reduces to the uncompetitive inhibition rate (13.4.21). As before, a plot of $1/R$ versus $1/[S]$ yields a straight line. The slope is equal to K'_m/R'_{max} and the intercept is equal to $1/R'_{max}$, in which $K'_m = K_m\{(1 + [1]K_3/(1 + [1]K_4)\}$ and $R'_{max} = K_{2f}[E_0]/(1 + [1]K_4)$. In this case, the inhibitor alters both the slope and the intercept.

Thus, as shown in Figure 13.10, plots of $1/R$ versus $1/[S]$ with and without the inhibitor enable us to distinguish the three types of inhibitor action.

References

1. Schrodinger, E., *What is Life*. 1945, London: Cambridge University Press.
2. Prigogine, I., Nicolis, G., and Babloyantz, A. *Physics Today*, **25**(11) (1972) 23; **25**(12) (1972) 38.
3. Katchalsky, A. and Curran, P.F., *Nonequilibrium Thermodynamics in Biophysics*. 1965, Cambridge, MA: Harvard University Press.
4. Peacocke, A.R., *An Introduction to the Physical Chemistry of Biological Organization*. 1983, Oxford: Clarendon Press.
5. Caplan, R.S. and Essig, A., *Bioenergetics and Linear Nonequilibrium Thermodynamics: The Steady State*. 1999, Cambridge, MA: Harvard University Press.
6. Crick, F., *Life Itself*. 1981, New York: Simon and Schuster (p. 43).
7. Kondepudi, D.K. and Asakura, K., *Accounts of Chemical Research*, **43** (2001) 946–954.
8. Lehninger, A.L., *Bioenergetics*. 1971, Menlo Park, CA: Benjamin Cummings.
9. Alexander, R.M., *Energy for Animals*. 1999, Oxford: Oxford University Press.
10. Stryer, L., *Biochemistry* (3rd edition). 1988. New York: W.H. Freeman.

Further Reading

R.M. Alexander, *Energy for Animals*, 1999. New York: Oxford University Press. (This book contains a wealth of data and analysis of how energy is utilized in the animal world.)

Examples

Examples 13.1   Spectral analysis of solar radiation indicates that it originates as a Planck distribution at a $T = 5780\,K$. The radius of the sun is $6.96 \times 10^5\,km$. The average distance between the Earth and the sun (orbital radius) is $1.496 \times 10^8\,km$. Using this information and the thermodynamics of thermal radiation, calculate the intensity of solar radiation just above the Earth's atmosphere (data collected by satellites give a value of $1370\,W\,m^{-2}$).

Solution   Using the Stefan–Boltzmann law we can calculate energy density of solar radiation near the surface of the sun:

$$I = cu/4 \qquad u_S = \beta T^4 \qquad T = 5780\,K \qquad \beta = 7.56 \times 10^{-16}\,J\,m^{-3}\,K^{-4}$$

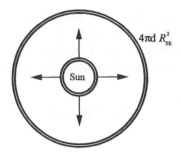

Consider the thermal energy in a thin spherical shell of thickness d at sun's surface. The energy contained in this shell is

$$U = 4\pi d R_S^2 u_S$$

where R_S is the radius of the sun.

As this radiation propagates through space and arrives at the Earth, it would have spread over a volume equal to $4\pi d R_{SE}^2$, in which R_{SE} is the average distance between the sun and the Earth. If u_E is the energy density of the radiation just above the Earth's atmosphere (before it is scattered), then, since the total energy U in the two shells must be the same, we have

$$U = 4\pi d R_S^2 u_S = 4\pi d R_{SE}^2 u_E$$

It follows that

$$u_E = \left(\frac{R_S}{R_{SE}}\right)^2 u_S = \left(\frac{R_S}{R_{SE}}\right)^2 \beta T^4$$

Substituting the numerical values $R_S = 6.96 \times 10^5\,\mathrm{km}$, $R_{SE} = 1.496 \times 10^8\,\mathrm{km}$, $T = 5780\,\mathrm{K}$ and $\beta = 7.56 \times 10^{-16}\,\mathrm{J\,m^{-3}\,K^{-4}}$ we obtain

$$I = c u_E/4 = 1370\,\mathrm{W\,m^{-2}}$$

a result in excellent agreement with the value of $1370\,\mathrm{W\,m^{-2}}$ obtained from satellite data. The factor α in Equation (13.3.7) is equal to $(R_S/R_{SE})^2 = 2.2 \times 10^{-5}$.

Example 13.2 Using the data in the previous example, estimate the chemical potential of of solar radiation as a function of the wavelength λ just above the atmosphere.

Solution The chemical potential of nonthermal radiation of frequency is given by

$$\mu(\lambda) = RT_E \ln\left(\frac{\alpha u(\lambda, T_S)}{u_{th}(\lambda, T_E)}\right)$$

in which α is the factor by which the energy density of solar radiation decreases when it reaches the point of interest; T_E is the temperature at the point of interest (just above the atmosphere) and T_S is the surface temperature of the sun.

From Example 13.1 we see that just above the atmosphere:

$$\alpha = \left(\frac{R_S}{R_{SE}} \right)^2 \qquad \text{where } R_S = 6.96 \times 10^5 \text{ km}, R_{SE} = 1.496 \times 10^8 \text{ km}$$
$$= 2.2 \times 10^{-5}$$

The Planck formula in terms of the wavelength λ is

$$u(\lambda, T)\mathrm{d}\lambda = \frac{8\pi hc}{\lambda^5} \frac{1}{\mathrm{e}^{hc/\lambda k_B T} - 1} \mathrm{d}\lambda.$$

The expression for the chemical potential then becomes:

$$\mu(\lambda) = RT \ln \left[\frac{\alpha(\mathrm{e}^{hc/\lambda k_B T_E} - 1)}{\mathrm{e}^{hc/\lambda k_B T_S} - 1} \right]$$

From the figure in Box 1.2 we see that the temperature at an altitude of 100 km is about 273 K. Hence, we take $T_E = 273$ K and $T_S = 6000$ K. The above expression for $\mu(\lambda)$ can be plotted (using Mathematica, for example). The plot below shows $\mu(\lambda)$ for λ in the range 400–700 nm.

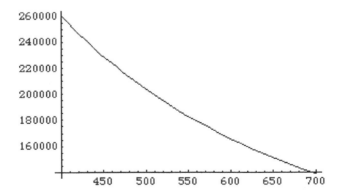

Example 13.3 Consider a reaction $X(g) + Y(g) \rightleftharpoons 2Z(g)$ taking place in a reaction chamber of the type shown in Figure 13.2 in which there is an inflow of X and Y and an outflow of Z. All the reactants may be assumed to be ideal gases. Assume that the equilibrium partial pressures of X, Y and Z in the reaction chamber are p_{Xeq}, p_{Yeq} and p_{Zeq} respectively. The inflows X and Y and the outflow of Z are such that the partial pressures in the reactions chamber are maintained at values p_X, p_Y and p_Z. What is the affinity of the reaction in the reaction chamber? What is the change in the Gibbs energy for the production of 2.6 mol of Z?

Solution   Since the reactants and products are ideal gases, their chemical potentials can be written as

$$\mu_k(p, T) = \mu_k^0 + RT\ln(p_k/p_0) \qquad k = \text{X, Y, Z}$$

The affinity is

$$A = \mu_X + \mu_Y - 2\mu_Z = \mu_X^0 + \mu_Y^0 - 2\mu_Z^0 + RT\ln(p_X p_Y/p_Z^2)$$

At equilibrium, $A = 0$. Therefore, we have

$$\mu_X^0 + \mu_Y^0 - 2\mu_Z^0 = -RT\ln(p_{X\text{eq}} p_{Y\text{eq}}/p_{Z\text{eq}}^2) = 0$$

Using this equation in the above expression for affinity, we can express the affinity as a function of equilibrium pressures and the pressures of X, Y and Z in the reaction chamber:

$$A = RT\ln(p_X/p_{X\text{eq}}) + RT\ln(p_Y/p_{Y\text{eq}}) - 2RT\ln(p_Z/p_{Z\text{eq}})$$

For the above reaction, the extent of reaction

$$d\xi = \frac{dN_X}{-1} = \frac{dN_Y}{-1} = \frac{dN_Z}{2}$$

Using Equation (13.2.4) we can write the Gibbs energy change as $\Delta G = -A(\xi_f - \xi_i)$, in which ξ_i and ξ_f are the initial and final values of the extents of reaction. For the production of 2.6 mol of Z, we have $\xi_f - \xi_i = 2.6/2 = 1.3$ mol. Hence, the Gibbs energy change is

$$\Delta G = -(1.3\,\text{mol})[RT\ln(p_X/p_{X\text{eq}}) + RT\ln(p_Y/p_{Y\text{eq}}) - 2RT\ln(p_Z/p_{Z\text{eq}})]$$

Exercises

13.1   Let $u(\nu, T_S)$ be the Planck energy distribution at temperature T_S of sun's surface. Let α be the factor by which it decreases by the time it reaches the surface of the Earth.

(a) Check whether the condition $h\nu/k_B T \gg 1$ is satisfied for wavelengths in the range 400–700 nm.

(b) When the condition $h\nu/k_B T \gg 1$, show that the chemical potential of solar radiation reaching the Earth's surface could be written approximately as

$$\mu(\nu) = RT\ln\alpha + N_A h\nu\left(1 - \frac{T}{T_S}\right)$$

in which T is the surface temperature of the Earth.

13.2 Given that the affinity of $ATP + H_2O \rightarrow ADP + P_i + H^+$ under cellular conditions is about $46 \, kJ \, mol^{-1}$, show that the net reaction $2(K^+)^e + 3(Na^+)^i + ATP + H_2O \rightleftharpoons 2(K^+)^i + 3(Na^+)^e + ADP + H^+ + P_i$ has a positive affinity under cellular conditions for the active transport of Na^+ out of the cell and K^+ into the cell for the intracellular and extracellular concentrations of Na^+ and K^+.

	Intracellular	Extracellular
$[Na^+]/mM$	10	140
$[K^+]/mM$	100	5

13.3 Write the Michaelis–Menten reaction kinetics in terms of the extents of reaction of the following two reaction steps:

$$E + S \underset{k_{1r}}{\overset{k_{1f}}{\rightleftharpoons}} ES \quad ES \underset{k_{2r}}{\overset{k_{2f}}{\rightleftharpoons}} P + E$$

13.4 For the set of rate equations (13.4.2)–(13.4.6), under the approximation $d[ES]/dt = 0$, derive the Michaelis–Menten rate law (13.4.7).

13.5 Write the complete set of rate equations for all the species in the Michaelis–Menten reaction mechanism:

$$E + S \underset{k_{1r}}{\overset{k_{1f}}{\rightleftharpoons}} ES \overset{k_{2f}}{\longrightarrow} P + E$$

(a) Write Mathematica/Maple code to solve them numerically with the following numerical values for the rate constants and initial values: $k_{1f} = 1.0 \times 10^2$, $k_{1r} = 5.0 \times 10^3$, $k_{2f} = 2.0 \times 10^3$ and at $t = 0$, $[E] = 3.0 \times 10^{-4}$, $[S] = 2 \times 10^{-2}$, $[ES] = 0$, $[P] = 0$. Using the numerical solutions, check the validity of the steady-state assumption.
(b) Plot concentration versus time graphs for each $[S]$, $[E]$, $[ES]$ and $[P]$. Use these plots to comment on the steady-state approximation $d[ES]/dt = 0$.
(c) Plot the rate of entropy production d_iS/dt as function of time.

13.6 An exercise machine's digital display has two columns. One shows 'resistance' in watts and the other shows calories/min. Depending on the speed with which the person exercises on the machine, the numbers displayed on the two columns increase and decrease in proportion. When the 'resistance' column shows $97.0 \, W$ the second column shows $7.0 \, cal \, min^{-1}$. Explain the relationship between the two.

14 THERMODYNAMICS OF SMALL SYSTEMS

Introduction

Pioneering work in formulating the thermodynamics of small systems was done by Terrell Hill [1] in the early 1960s. It could be applied to many small systems that we encounter in nature: small particles in the atmosphere called aerosols (which include small droplets of water containing dissolved compounds), crystal nuclei in supersaturated solutions, colloids, small particles in interstellar space, etc. Important as it was, thermodynamics of small systems has taken on a new significance due to the development of **nano-science**, the production and study of particles in the size range 1–100 nm. Thermodynamics applied to particles in the 'nano range' is called **nano-thermodynamics**, but, because we do not limit our discussion to this size range, we call this topic the thermodynamics of small systems.

The laws of thermodynamics are universal, valid for all systems. However, depending on the system being considered, various approximations are made. Care is necessary in applying thermodynamics to systems that are very small. First, it must be ensured that thermodynamic variables that were used for large systems have a clear physical meaning when used to describe a small system. Owing to random molecular motion, thermodynamic variable will fluctuate about their average values. We need a clear understanding of the magnitude of these fluctuations relative to the average values and whether and why the system is stable when subjected to them. Second, quantities, such as interfacial energy, that could be neglected for large systems must be taken into consideration. In Chapter 5, we have already seen how interfacial energy can be included in the thermodynamic description of a system. We shall extend this formalism to understand why some properties, such as solubility and melting point, change with size. In general, the properties of very fine powders can be significantly different from those of bulk substance, hence the current interest in nanotechnology. We shall begin by discussing the thermodynamic formalism that includes interfacial energy and then address thermodynamic stability and fluctuations.

14.1 Chemical Potential of Small Systems

Chemical potential is an important variable that enables us to understand how the properties of a system may change as its size decreases to microscopic dimensions. In this section, we will derive an expression for the chemical potential as a function of size.

Introduction to Modern Thermodynamics Dilip Kondepudi
© 2008 John Wiley & Sons, Ltd

In Section 5.6 we noted that molecules at an interface have different energy and entropy compared with molecules in the bulk. This **interfacial energy** or **surface tension** γ is generally of the order of 10^{-1}–$10^{-2}\,\mathrm{J\,m^{-2}}$. Table 14.1 lists some solid–water interfacial energies. Whether interfacial energy can be neglected or not depends on the size of the system, more precisely the area-to-volume ratio. If U_m is the molar energy, then, for a sphere of radius r, the ratio of interfacial energy to bulk energy is $4\pi r^2 \gamma/[(4\pi r^3/3 V_m) U_m] = 3\gamma V_m/r U_m$, in which V_m is the molar volume. If this quantity is very small, then the interfacial energy can be neglected, and as $r \to \infty$ it becomes zero. If this ratio is not small, then we include the interfacial energy term in the expression for dU. For a pure substance:

$$dU = T\,dS - p\,dV + \mu\,dN + \gamma d\Sigma \qquad (14.1.1)$$

in which Σ is the interfacial area. The last two terms can be combined to express the chemical potential as a function of the size of the system. For simplicity, we shall assume that the system is a sphere of radius r. Then the molar amount $N = 4\pi r^3/3 V_m$. The interfacial term $d\Sigma = d(4\pi r^2) = 8\pi r\,dr$ can be written in terms of dN by noting that $dN = 4\pi r^2\,dr/V_m = (r/2V_m)\,d\Sigma$. Thus, we can substitute $(2V_m/r)\,dN$ for $d\Sigma$ in (14.1.1) to obtain

$$dU = T\,dS - p\,dV + \left(\mu + \frac{2\gamma V_m}{r}\right)dN \qquad (14.1.2)$$

Using this equation, we see that for a pure substance we can assign an effective chemical potential that depends on the system's radius (Figure 14.1). We shall write this potential as

$$\boxed{\mu = \mu_\infty + \frac{2\gamma V_m}{r}} \qquad (14.1.3)$$

in which μ_∞ is the chemical potential as $r \to \infty$; it is the 'bulk chemical potential' that has been used in the previous chapters when interfacial energy could be ignored. The Gibbs energy of the system is

Table 14.1 Experimentally measured interfacial energies of AgCl, AgBr and AgI particles in water and their molar and molecular volumes V_m

Compound	$\gamma/\mathrm{mJ\,m^{-2}}$		$V_m/\mathrm{mL\,mol^{-1}}$	Molecular volume/10^{-23} mL
	At 10 °C	At 40 °C		
AgCl	104	100	25.9	4.27
AgBr	112	102	29.0	4.81
AgI	128	112	41.4	6.88

Source: T. Sugimoto and F. Shiba, *J. Phys. Chem. B*, **103** (1999) 3607.

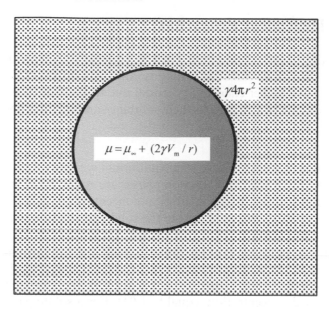

$$\gamma 4\pi r^2$$

$$\mu = \mu_\infty + (2\gamma V_m / r)$$

Figure 14.1 The chemical potential of a small spherical particle or liquid drop depends on the radius r. $\gamma\,(\mathrm{J\,m^{-2}})$ is the interfacial energy or surface tension. The energy of the interface equals $4\pi r^2\gamma$

$$G = \mu_\infty N + \gamma\Sigma \tag{14.1.4}$$

and a simple calculation shows (Exercise 14.1) that $(\partial G/\partial N)_{p,T} = \mu = \mu_\infty + (2\gamma V_m/r)$.

Equation (14.1.3) can also be understood in terms of the excess pressure in a small system. In Chapter 5 we saw that surface tension (or interfacial tension) increases the pressure in a small spherical system by an amount $\Delta p = 2\gamma/r$ (see (5.6.6)). Expression (14.1.3) is the chemical potential under this higher pressure. This can be seen by noting that

$$\mu(p + \Delta p, T) = \mu(p, T) + \int_p^{p+\Delta p} \left(\frac{\partial \mu}{\partial p}\right)_T dp = \mu(p, T) + \int_p^{p+\Delta p} V_m\, dp, \quad \Delta p = \frac{2\gamma}{r} \tag{14.1.5}$$

where we have used the relation $(\partial G_m/\partial p)_T = (\partial \mu/\partial p)_T = V_m$. For solids and liquids, the molar volume V_m does not change much with changes in pressure; hence, we can write (14.1.5) as

$$\mu(p + \Delta p, T) = \mu(p, T) + V_m \Delta p = \mu(p, T) + \frac{2\gamma V_m}{r} \tag{14.1.6}$$

which is (14.1.3). So, the increase in chemical potential of a small system by a term $2\gamma V_m/r$ is a consequence of increase in the pressure due to surface tension.

14.2   Size-Dependent Properties

Using the chemical potential (14.1.3), several size-dependent properties can be derived. We shall consider solubility and melting point. As noted above, small systems have higher chemical potential due to the fact they are under higher pressure. This causes a change in their solubility and melting point.

SOLUBILITY

We consider a solid solute Y is in equilibrium with its solution. The chemical potentials of Y in the solid and solution phases are equal. At equilibrium, the concentration of the solution is the saturation concentration, called the **solubility**; we shall denote it by $[Y]_{eq}$. We shall denote the solid and solution phases with the subscripts 's' and 'l' respectively.

As shown in (8.3.17), in the molarity scale the equilibrium chemical potential of the solute in the solution phase is $\mu_{Y,l} = \mu_Y^{c0} + RT\ln(\gamma_Y[Y]_{eq}/[Y]^0)$, in which γ_Y is the activity coefficient of Y (not to be mistaken for the interfacial energy γ) and $[Y]^0$ is the standard concentration, equal to 1.0 M. For a solute particle of radius r in equilibrium with the solution, $\mu_{Y,l} = \mu_{Y,s}$, which gives

$$\mu_{Y,l} = \mu_Y^{c0} + RT\ln\left(\frac{\gamma_Y[Y]_{eq}}{[Y]^0}\right) = \mu_{Y,s} = \mu_{Y,s\infty} + \frac{2\gamma V_m}{r} \qquad (14.2.1)$$

in which we have used (14.1.3) for the chemical potential of the solid phase. The quantity $\mu_Y^{c0} - \mu_{Y,s\infty} = \Delta G_{sol}$ is the molar Gibbs energy of solution (defined for large particles $r \to \infty$). Hence, (14.2.1) can be written as

$$RT\ln\left(\frac{\gamma_Y[Y]_{eq}}{[Y]^0}\right) = -\Delta G_{sol} + \frac{2\gamma V_m}{r}$$

i.e.

$$\frac{\gamma_Y[Y]_{eq}}{[Y]^0} = \exp\left(\frac{-\Delta G_{sol}}{RT}\right)\exp\left(\frac{2\gamma V_m}{rRT}\right) \qquad (14.2.2)$$

If we denote the equilibrium concentration for solute particles of radius r by $[Y(r)]_{eq}$ and assume that the activity coefficient γ_Y does not vary much in the concentration range of interest, (14.2.2) can be simplified to the following relation:

$$\boxed{[Y(r)]_{eq} = [Y(\infty)]_{eq} \exp\left(\frac{2\gamma V_m}{rRT}\right)} \qquad (14.2.3a)$$

or more generally as

$$\boxed{a_Y(r)_{eq} = a_Y(\infty)_{eq} \exp\left(\frac{2\gamma V_m}{rRT}\right)} \qquad (14.2.3b)$$

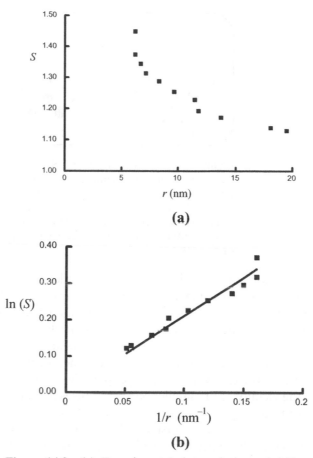

Figure 14.2 (a) Experimental data relating solubility ratio $S = [Y(r)]_{eq}/[Y(\infty)]_{eq}$ to particle size r for AgCl at 298 K. (b) Plot of $\ln(S)$ versus $1/r$ is a straight line in agreement with Equation (14.2.3a). (*Source*: T. Sugimoto and F. Shiba, *J. Phys. Chem. B*, **103** (1999) 3607)

in which a_Y is the activity of Y. These equations give solubility $[Y(r)]_{eq}$ as a function of the particle size. They tell us that the saturation concentration will be higher for smaller particles; that is, smaller particles have higher solubility. It is generally called the **Gibbs–Thompson** equation, but some also call it the **Ostwald–Freundlich** equation. The solubility of AgCl, AgBr and AgI particles whose size is in the range 2–20 nm can be satisfactorily explained using the Gibbs–Thompson equation (Figure 14.2).

The higher solubility of smaller particles has an interesting consequence. As shown in Figure 14.3, consider a supersaturated solution containing solute particles of different sizes or radii. Supersaturation means that the chemical potential of the solute in the solution phase is higher, $\mu_l > \mu_s$. So, the solute will begin to precipitate out and deposit on the solid particles. As the chemical potential in the solution phase decreases due to solute deposition on the solid phase, there will come a point at

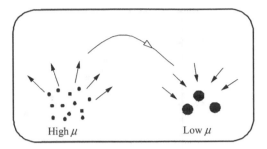

Figure 14.3 Ostwald ripening. Small particles have a higher chemical potential than
larger particles. As a consequence, in a saturated solution, small particles dissolve while
larger particles grow. The difference in chemical potential results in the effective transport
of the solute

which the solution is in equilibrium with the smaller particles, $\mu_1 \simeq \mu_s(r_{small})$, but its
chemical potential is still higher than that of the larger particles, $\mu_1 > \mu_s(r_{large})$. Hence,
solute from the solution begins to deposit on the larger particles, causing a reduction
of concentration in the vicinity of the larger particles. A concentration gradient is
thus established, with a higher concentration near smaller particles and a lower
concentration near larger particles. The solute then begins to flow from the vicinity
of the smaller particles towards the larger particles. The consequent drop in concentration in the vicinity of the smaller particles causes them to dissolve, while the larger
particles continue to grow. As the smaller particles dissolve, their solubility increases,
causing them to dissolve even faster and they ultimately disappear. Such growth of
larger particles at the expense of smaller ones is called **Ostwald ripening**. It is a very
slow process, but it can be observed.

MELTING POINT

The higher chemical potential of small particles also has the effect of reducing their
melting point. Let us consider a solid particle of radius r in equilibrium with the
melt. Let T_m be the melting point for the bulk substance; it is the temperature at
which large particles are in equilibrium with the melt. For small particles of radius
r, due to their higher chemical potential, let us assume that the melting point is T_m
$+ \Delta T$. The chemical potential of a pure substance $\mu(p, T)$ is a function of p and T.
Using (14.1.3) for the chemical potential of the solid particle, we see that solid–melt
equilibrium for large particles at T_m implies

$$\mu_{s\infty}(p, T_m) = \mu_l(p, T_m) \tag{14.2.4}$$

and the same for small particles at $T_m + \Delta T$ implies

$$\mu_s(p, T_m + \Delta T) = \mu_{s\infty}(p, T_m + \Delta T) + \frac{2\gamma V_m}{r} = \mu_l(p, T_m + \Delta T) \tag{14.2.5}$$

In this equation we can use the relation

$$\mu(p, T_m + \Delta T) = \mu(p, T_m) + \int_{T_m}^{T_m + \Delta T} \left(\frac{\partial \mu}{\partial T} \right)_p dT$$

and write it as

$$\mu_{s\infty}(p, T_m) + \int_{T_m}^{T_m + \Delta T} \left(\frac{\partial \mu_{s\infty}}{\partial T} \right)_p dT + \frac{2\gamma V_m}{r} = \mu_l(p, T_m) + \int_{T_m}^{T_m + \Delta T} \left(\frac{\partial \mu_l}{\partial T} \right)_p dT$$

Using (14.2.4) and noting that $(\partial \mu / \partial T)_p = -S_m$, the molar entropy, we can simplify this equation to

$$\int_{T_m}^{T_m + \Delta T} (S_{ml} - S_{ms}) \, dT + \frac{2\gamma V_m}{r} = 0$$

The difference in molar entropies between the liquid and the solid state $(S_{ml} - S_{ms})$ $\simeq \Delta H_{fus}/T$. The enthalpy of fusion ΔH_{fus} does not change much with T and may be assumed to be constant. With this approximation, the integral can be evaluated and we get

$$\Delta H_{fus} \ln \left(1 + \frac{\Delta T}{T_m} \right) + \frac{2\gamma V_m}{r} = 0$$

Since $\Delta T/T_m \ll 1$, we can approximate $\ln(1 + \Delta T/T_m) \simeq \Delta T/T_m$. If we write the melting point of particles of radius r as $T_m(r) = T_m(\infty) + \Delta T$, in which we have used $T_m(\infty)$ in place of T_m, the above equation can be rearranged to

$$\boxed{T_m(r) = T_m(\infty) \left(1 - \frac{2\gamma V_m}{\Delta H_{fus} r} \right)} \tag{14.2.6}$$

Sometimes this equation is written in the parametric form:

$$T_m(r) = T_m(\infty) \left(1 - \frac{\rho}{r} \right) \tag{14.2.7}$$

in which ρ is expressed in nanometers. For many inorganic materials, ρ is in the range 0.2–1.7 nm. Also for metals, the solid–melt interfacial energy can be estimated using the following formula [2]:

$$\gamma = \frac{0.59 R T_m}{a N_A}$$

in which T_m is the melting point, a is the area occupied by a single atom on the surface (approximately equal to the square of the diameter), and R and N_A are the gas and the Avogadro constants respectively.

14.3 Nucleation

The transition from a vapor to a liquid phase occurs when the corresponding affinity is positive, i.e. liquid will condense from a vapor when the chemical potential of the liquid is lower than that of the vapor; and similarly for the transition from a liquid to a solid phase. The condensation of vapor into liquid must take place through clustering of molecules that eventually grow into liquid drops. But, as we have seen, the chemical potential of a small system increases with decreasing radius. Hence, the affinity is higher for larger clusters and, indeed, can be negative for very small clusters. We can see this clearly by writing the affinity for the transformation from vapor to liquid cluster of radius r, which we write as C_r:

$$\text{Transformation} \qquad 1 \rightarrow C_r$$

$$\text{Affinity:} \qquad A = \mu_v - \left(\mu_{l\infty} + \frac{2\gamma V_m}{r} \right) = \Delta\mu - \frac{2\gamma V_m}{r} \tag{14.3.1}$$

in which the subscripts 'v' and 'l' stand for vapor and liquid respectively. Activities for nucleation of a solute from a solution or a solid from a melt will also have the same form as (14.3.1). In each case, $\Delta\mu$ is the difference between the chemical potentials of the two phases. For crystallization from solution, $\Delta\mu$ is the difference between the solution and the solid solute; in the case of solidification of a melt, it is the difference between the chemical potentials of the melt and the solid. To reflect the generality of expression (14.3.1), we can consider a phase transition from initial phase α that nucleates to phase β and write the affinity for a phase transformation as

$$A = \mu_\alpha - \left(\mu_{\beta\infty} + \frac{2\gamma V_m}{r} \right) = \Delta\mu - \frac{2\gamma V_m}{r} \tag{14.3.2}$$

We assume that initially $\Delta\mu \equiv \mu_\alpha - \mu_{\beta\infty} > 0$; that is, phase α is a supersaturated vapor or a supersaturated solution or a supercooled melt with a T below its melting point.

Equation (14.3.2) implies that the affinity A is positive only when r is larger than a critical value, r^*, i.e. $A > 0$ only when $r > r^*$ (Figure 14.4). It is easy to see that

$$r^* = \frac{2\gamma V_m}{\mu_\alpha - \mu_{\beta\infty}} = \frac{2\gamma V_m}{\Delta\mu} \tag{14.3.3}$$

r^* is called the **critical radius**. Owing to random molecular motion, the molecules in the α phase form clusters of β phase of various sizes. But most clusters of radius $r < r^*$ will evaporate or dissolve and return to the α phase. Only when a cluster's radius reaches a value $r \geq r^*$ would a β phase have 'nucleated'; since the affinity (14.3.2) is positive for such nuclei, they will grow. It is through the growth of nuclei (into liquid drops or solid particles) that phase α converts to phase β. The formation of nuclei of radius $r \geq r^*$ takes place through random energy fluctuations. It is the

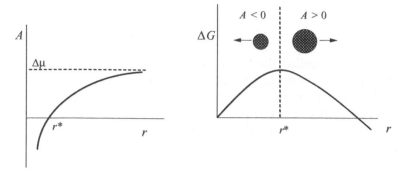

Figure 14.4 Affinity (14.3.2) and the corresponding Gibbs energy change for the process of nucleation. r^* is the critical nucleation radius. Clusters with radius $r > r^*$ will grow while clusters with $r < r^*$ will shrink

process of **nucleation**, the gateway for the transition from phase α to phase β. As is clear from (14.3.3), the critical radius r^* decreases with increasing $\Delta\mu$; that is, the critical radius decreases as the supersaturation increases.

The above understanding of affinity for the formation of clusters and the corresponding changes in Gibbs energy enable one to formulate a theory of nucleation *rate*. The theory we present here is called the **classical theory of nucleation**. In small systems, which could be subsystems of larger systems, random fluctuations in Gibbs energy occur. Since Gibbs energy reaches its minimum value at equilibrium, fluctuations in systems in equilibrium can only increase the Gibbs energy. The clustering of molecules in the α phase to form small clusters of β phase can only take place through fluctuations because the Gibbs energy change for such a transformation is positive. If the Gibbs energy of the random fluctuations is large enough, then a critical nucleus of radius r^* will form and begin to grow, thus initiating a phase transition. Therefore, we need to know the laws that govern fluctuations to understand the dynamics of nucleation. An elegant thermodynamics theory of fluctuation was formulated by Einstein. We shall discuss this theory and the theory of thermodynamic stability in the next section. According to this theory, the following formula gives the probability $P(\Delta G)$ of a fluctuation in Gibbs energy of magnitude ΔG:

$$P(\Delta G) = Z e^{-\Delta G/k_B T} \qquad (14.3.4)$$

where Z is the normalization factor such that $\int_0^\infty P(\Delta G)\mathrm{d}(\Delta G) = 1$ and k_B is the Boltzmann constant. Let $\Delta G(r^*)$ be the increase in Gibbs energy needed to form the critical nucleus. We can obtain the probability for the formation of a critical nucleus by substituting $\Delta G(r^*)$ into (14.3.4). The rate of nucleation is clearly proportional to $P[\Delta G(r^*)]$. Hence, the rate of nucleation J (number of nuclei formed per unit volume per unit time) can be written as

$$J = J_0 \exp\left[\frac{-\Delta G(r^*)}{k_B T}\right] \qquad (14.3.5)$$

in which J_0 is called the 'pre-exponential factor'; the value of J_0 depends on the particular process being considered. The Gibbs energy of a nucleus of radius r^* of the β phase containing N moles of substance is $G_\beta = \mu_{\beta\infty} N + \gamma 4\pi(r^*)^2$. The corresponding Gibbs energy in the α phase is $G_\alpha = \mu_\alpha N$. The change in Gibbs energy for this transformation from the α phase to the β phase is $\Delta G(r^*) = (G_\beta - G_\alpha)$. This can be written as

$$\Delta G(r^*) = -\frac{4\pi(r^*)^3}{3V_m} \Delta\mu + \gamma 4\pi(r^*)^2 \qquad (14.3.6)$$

where $\Delta\mu = \mu_\alpha - \mu_{\beta\infty}$. Substitution of expression (14.3.3) for the critical radius r^* into (14.3.6) gives (Exercise 14.3)

$$\Delta G(r^*) = \frac{16\pi}{3} \frac{\gamma^3 V_m^2}{\Delta\mu^2} \qquad (14.3.7)$$

Thus, the nucleation rate (14.3.5) can be written as

$$\boxed{J = J_0 \exp\left(-\frac{16\pi}{3k_B T} \frac{\gamma^3 V_m^2}{\Delta\mu^2}\right)} \qquad (14.3.8)$$

This expression shows how the nucleation rate depends on the interfacial energy γ and the supersaturation expressed through $\Delta\mu$. The pre-exponential factor J_0 depends on the details of the kinetics of nucleation and it is generally difficult to estimate its value. Reported values of J_0 are in the range $10^{25} - 10^{30} \, s^{-1} \, mL^{-1}$ for salts that are sparingly soluble. Equilibrium between the α phase and the β phase implies $\mu_{\beta\infty} = \mu_\alpha^0 + RT \ln a_{\alpha,eq}$. Since the chemical potential of the α phase $\mu_\alpha = \mu_\alpha^0 + RT \ln a_\alpha$, it follows that

$$\Delta\mu = \mu_\alpha - \mu_{\beta\infty} = RT \ln\left(\frac{a_\alpha}{a_{eq}}\right) \qquad (14.3.9)$$

Here, the equilibrium activity a_{eq} is the activity at saturation in the case of vapors and solution; for solidification of a melt it is the activity of the liquid phase (melt) at the melting point. If the vapor α is considered an ideal gas, then $\Delta\mu = RT \ln(p_\alpha / p_{sat})$, in which p_{sat} is the saturated vapor pressure. Similarly, for an ideal solution of solute Y, $\Delta\mu = RT \ln([Y]/[Y]_S)$, in which $[Y]_S$ is the saturation concentration. For solidification from a melt, the dependence of the chemical potential on T must be considered. It can be shown (Exercise 14.4) that $\Delta\mu = \Delta H_{fus}(1 - T/T_m)$.

In the above theory, the nucleation rate (number of nuclei formed per unit volume per unit time) is independent of position; it is the same everywhere in the system. This, therefore, is called **homogeneous nucleation**. According to this theory, in a supersaturated vapor or solution, we should observe nucleation in all parts of the system with some uniformity, albeit with expected statistical fluctuations. However, most of the time we do not find this to be the case. Instead, we find that nucleation occurs on small impurity particles or on the walls of the container, indicating that nucleation occurs at higher rates at particular sites. Such nucleation is called **heterogeneous nucleation**. It happens because, on impurity particles or the walls, the interfacial energy γ is smaller. The expression (14.3.8) is fundamentally correct, but the value of γ (or, more generally, the nucleation Gibbs energy G^*) depends on the site where the nucleation takes place. At these sites (called **nucleation sites**) the rate of nucleation is higher. This is the reason why, when crystals are grown from a solution, nucleation does not occur homogeneously throughout the system, but heterogeneously at certain sites.

14.4 Fluctuations and Stability

Thermodynamic variables such as pressure and temperature are clearly defined for macroscopic systems. That random molecular motion will cause these quantities to fluctuate and that the values we assign to thermodynamic variables are the average values about which they fluctuate is obvious. To be sure, our experience tells us that fluctuations in thermodynamic quantities are extremely small in macroscopic systems. When we consider very small systems, however, we must ensure that thermodynamic variables are well defined, because the size of the fluctuations relative to the average value can be significant. To look at this aspect more closely, let us consider a small system 1, of volume V_1 containing N_1 moles of substance within a larger system 2 (Figure 14.5). We shall denote the number of molecules by $\tilde{N} = NN_A$. Owing to random molecular motion, the number of molecules in this subsystem will fluctuate about its average value $\tilde{N}_1 = N_1 N_A$. As we shall show later in this section, the magnitude of these fluctuations $\delta\tilde{N}$ is of the order of the square root of the average value $\sqrt{\tilde{N}_1}$. To compare it with the average value we look at the ratio $\delta\tilde{N}_1/\tilde{N}_1 = 1/\sqrt{\tilde{N}_1}$. If this value is very small, then it means that we can meaningfully

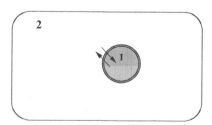

Figure 14.5 Thermodynamic quantities such as number of molecules and temperature of a subsystem 1, which is a part of a larger system 2, will fluctuate

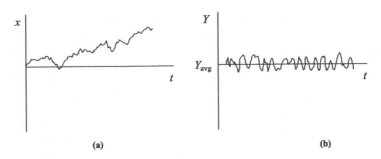

Figure 14.6 (a) Random fluctuation in position x of a free particle can drive it away from its initial position. Without a restoring force, the particle will randomly drift. (b) In contrast, a thermodynamic variable Y fluctuates about its average value Y_{avg}. When a fluctuation drives the system away from Y_{avg}, irreversible processes restore it back to Y_{avg}, thus keeping it stable

assign a value of N_1 moles to the amount of substance in system 1. Similarly, since the temperature is a measure of the average kinetic energy, its value will fluctuate and if $\delta T/T \ll 1$, the temperature of the small system is well defined.

Next, we address the question of stability in the presence of fluctuations. As shown in Figure 14.6, due to fluctuations, the position of a free particle will drift away from its initial value, which is the phenomenon of Brownian motion. In contrast, fluctuations do not cause a thermodynamic variable Y to drift randomly away from its initial value: there is a restoring force that keeps it from drifting; it is stable. We shall see that the reason for the stability of equilibrium states lies in the second law. A fundamental understanding of stability also enables us to understand instability of states that results in phase separation.

THE PROBABILITY OF A FLUCTUATION

In Chapter 3 (Box 3.2), the statistical interpretation of entropy was introduced and this was used in Sections 6.1 to calculate the entropy of mixing. Statistical interpretation of entropy is based on Ludwig Boltzmann's (1844–1906) famous formula that related entropy and probability:

$$\boxed{S = k_B \ln W} \tag{14.4.1}$$

in which $k_B = 1.38 \times 10^{-23}\,\mathrm{J\,K^{-1}}$ is the Boltzmann constant and W is the number of microscopic states corresponding to the macroscopic thermodynamic state. W is measure of the probability that a system will be in a state with entropy S. As suggested by Max Planck, W is sometimes called **thermodynamic probability** because, unlike the usual probability, it is a number larger than one; in fact it is a very, very large number! (An entropy $S = 1.0\,\mathrm{J\,K^{-1}}$ corresponds to $W = e^{S/k_B} \approx e^{7 \times 10^{22}}$.) Thus, Boltzmann's formalism relates entropy, a physical quantity, to probability. Boltzmann's idea proved to be very successful. The entire subject of statistical thermodynamics, which we will discuss in Chapter 17, is a testimony to the success of Boltzmann's approach.

Albert Einstein, in whose work thermodynamics played an important role [3], introduced a new interpretation for Boltzmann's formula. He proposed a formula for the probability of a fluctuation for thermodynamic quantities by using Boltzmann's formula in a conceptually reverse manner. Inverting (14.4.1) we can write $W = \exp(S/k_B)$. Let us assume that a fluctuation changes the entropy from its equilibrium value S_{eq} to $S = S_{eq} + \Delta S$. Then, the associated thermodynamic probability $W = \exp[(S_{eq} + \Delta S)/k_B]$. Using this relation, Einstein proposed the following formula for the probability of a fluctuation that cause a change in entropy ΔS from its equilibrium value:

$$\boxed{P(\Delta S) = Ze^{\Delta S/k_B}}$$
(14.4.2)

where Z is a normalization constant that ensures that the sum of all probabilities equals one. Though relations (14.2.1) and (14.2.2) are mathematically close, it is important to note that conceptually one is the opposite of the other. In (14.2.1), the probability of a state is the fundamental quantity and entropy is derived from it; in (14.2.2), entropy as defined in thermodynamics is the fundamental quantity and the probability of a fluctuation is derived from it. Einstein's observation was that thermodynamic entropy also gives us the probability of fluctuations. We shall obtain expressions that relate ΔS to fluctuations in temperature δT and number of molecules $\delta \tilde{N}$ after we discuss the closely related topic of stability of the equilibrium state.

STABILITY

Every fluctuation is associated with a corresponding change in entropy. For an isolated system in equilibrium, the entropy reaches its maximum value. Hence, a fluctuation can only decrease its value and drive it away from equilibrium. Also, at equilibrium, thermodynamic forces and flows are zero: $F_k = J_k = 0$. Once the system is away from equilibrium, thermodynamic forces F_k and flows J_k attain a nonzero value deviating from zero by a small amount δF_k and δJ_k. Since thermodynamic forces and flows can only increase the entropy in accord with the second law, $d_i S/dt = \Sigma_k \delta F_k \delta J_k \geq 0$, the system's entropy increases and is restored to its equilibrium value. Thus, a decrease in entropy due to fluctuations is countered by thermodynamic forces and flows keeping the equilibrium state stable. Because the change in entropy due to fluctuations and the change due to forces and flows have opposite signs, the system is stable. The stability condition for the equilibrium state can thus be expressed as:

$$\boxed{\Delta S < 0} \qquad \boxed{\frac{d\Delta S}{dt} = \sum_k \delta F_k \delta J_k > 0}$$
(14.4.3)

If this condition is satisfied, then a thermodynamic state is stable.

In classical thermodynamics, which does not contain the relation $d_iS/dt = \Sigma_k \delta F_k \delta J_k \geq 0$, the stability of a state is analyzed using a different approach called the Gibbs–Duhem theory of stability. This theory, presented in Chapter 15, is limited to constraints that extremize (maximize or minimize) one of the thermodynamic potentials, H, F or G. We recall that the Helmholtz energy F is minimized when V and T are constant. *A theory of stability that is based on positivity of entropy production does not require such constraints and is more general than the classical Gibbs–Duhem theory.* In addition, stability theory based on entropy production can also be used to obtain conditions for the stability of some nonequilibrium states.

CALCULATING ΔS

To obtain the probability of fluctuation of variables such as the temperature T or Gibbs energy G, we turn to Einstein's formula (14.4.2) and write the entropy of fluctuation $\Delta S < 0$ in terms of these variables whose fluctuations are of interest. First, we note that irreversible processes that restore the system back to equilibrium must generate entropy equal to $-\Delta S > 0$. As shown in Figure 14.7, given a fluctuation that has taken the system to the nonequilibrium state I, we can obtain the magnitude of ΔS by computing the entropy generated in restoring it back to the equilibrium state E. Thus, our computation is based on the relation

$$\Delta S = -\int_{I}^{E} d_iS \qquad (14.4.4)$$

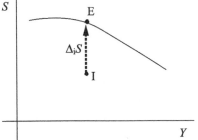

Figure 14.7 Schematic diagram illustrating the entropy change ΔS associated with a fluctuation. The figure shows equilibrium entropy S as a function of a thermodynamic variable Y. The reference equilibrium state is denoted by E. The fluctuation, which results in a decrease in entropy, drives the system to the point I. We compute the change in entropy ΔS associated with the fluctuation by computing the entropy produced Δ_iS as the system relaxes back to the equilibrium state

If we are interested in fluctuation in the Gibbs energy when T and p are constant, then we could use the relation $dG = -T \, d_iS$ in (14.4.4) and write $\Delta S = \int_I^E dG/T = -\int_E^I dG/T = -\Delta G/T < 0$. Here, ΔG is the increase in G in going from equilibrium state E to nonequilibrium state I, and $\Delta S = -\Delta G/T$. Replacing ΔS by $-\Delta G/T$ in Einstein's formula (14.4.2), we arrive at the probability for Gibbs energy fluctuation ΔG from an equilibrium state E to a nonequilibrium state I:

$$P(\Delta G) = Z \, e^{-\Delta G/k_B T} \tag{14.4.5}$$

We recall that this expression was used in Section 14.3 to obtain the nucleation rate. Similarly, the probability of a fluctuation in Helmholtz energy ΔF at constant T and V is given by $P(\Delta F) = Ze^{-\Delta F/k_B T}$.

TEMPERATURE FLUCTUATIONS

We can obtain the probability for temperature fluctuations as follows. Consider a small subsystem within a larger system at an equilibrium temperature T_{eq} (Figure 14.8a). We assume that a fluctuation has increased the temperature of the subsystem to $T_{eq} + \delta T$. To calculate the entropy associated with this change, we use Equation (14.4.4), in which d_iS is the entropy produced when heat dQ flows out of the subsystem, which is at a higher temperature, into the larger system whose temperature is T_{eq}. At any instant, if the temperature of the subsystem is $T_{eq} + \alpha$, as we have seen in Chapter 3, then the entropy change d_iS due to this heat flow dQ out of the subsystem is

$$\begin{aligned} d_iS &= \left(\frac{-1}{T_{eq} + \alpha} + \frac{1}{T_{eq}} \right) dQ \\ &= \frac{\alpha}{T_{eq}^2} dQ > 0 \end{aligned} \tag{14.4.6}$$

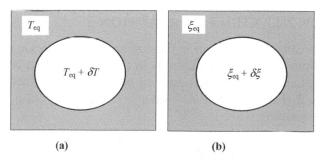

(a)   (b)

Figure 14.8 (a) A local fluctuation in temperature δT can occur in a subsystem. The probability of a fluctuation δT can be calculated using (14.4.8). (b) A local fluctuation in the extent of reaction. The entropy change associated with such a fluctuation can be calculated using the relation (14.4.14)

where we have used the approximation $(T_{eq} + \alpha)T_{eq} \simeq T_{eq}^2$ because $\alpha \ll T_{eq}$. Thus, (14.4.6) implies if $\alpha > 0$, then $dQ > 0$; that is, heat flows out of the subsystem. Since dQ is the heat flowing out of the subsystem, we must have $dQ > 0$ for $d\alpha < 0$ Hence, for small changes in temperature, we write $dQ = -C_V\, d\alpha$, in which C_V is the subsystem's heat capacity at constant volume. As the subsystem returns to equilibrium, α changes from δT to 0. Therefore:

$$\Delta S = -\int_{\delta T}^{0} d_i S = \int_{\delta T}^{0} \frac{C_V}{T_{eq}^2} \alpha d\alpha = -\frac{C_V}{T_{eq}^2}\frac{(\delta T)^2}{2} \tag{14.4.7}$$

The condition for stability (14.4.3) can now be written as

$$\Delta S = -\frac{C_V}{T_{eq}^2}\frac{(\delta T)^2}{2} < 0 \tag{14.4.8}$$

which is satisfied only if the heat capacity of the subsystem $C_V > 0$. By substituting (14.4.8) in (14.4.2) we obtain the probability of a temperature fluctuation of magnitude δT:

$$P(\delta T) = Z \exp\left(-\frac{C_V \delta T^2}{2k_B T_{eq}^2}\right) \tag{14.4.9}$$

Thus, given the heat capacity C_V and the equilibrium temperature T_{eq}, the probability of a fluctuation δT can be calculated. The normalization factor Z has to be such that the total probability of all possible fluctuations equals unity, i.e. $\int_{-\infty}^{\infty} P(\delta T)d(\delta T) = 1$. This leads to $1/Z = \int_{-\infty}^{\infty} P(\delta T)d(\delta T)$. The mean and standard deviation of δT can be calculated using (14.4.9). The probability distribution $P(\delta T)$ is a Gaussian, centered at $\delta T = 0$, so its mean equals zero, as we expect. The standard deviation can be calculated using the integrals given in Appendix 1.2. A simple calculation shows that $\sqrt{\langle \delta T^2 \rangle} = \sqrt{k_B T_{eq}^2/C_V}$.

FLUCTUATION IN EXTENT OF REACTION ξ

Diffusion causes fluctuations in the number of molecules in a subsystem, and chemical reactions can cause fluctuations in the number of reacting molecules even in a closed system. These fluctuations can be described as fluctuations in the extent of reactions ξ (Figure 14.8b). Consider a chemical reaction and the corresponding extent of reaction:

$$2X + Y \rightleftharpoons Z + W \quad d\xi = \frac{dN_X}{-2} = \frac{dN_Y}{-1} = \frac{dN_Z}{+1} = \frac{dN_W}{+1} \tag{14.4.10}$$

The fluctuations in ξ can be related to fluctuations in the corresponding values of N_X, N_Y, N_Z, and N_W. In the case of fluctuations due to diffusion of a component, say Y, $d\xi = dN_Y$ is the change in N_Y due to flow of particles into and out of the subsystem. The thermodynamic force for both of these processes is the affinity A:

for reaction (14.4.10), $A = (2\mu_X + \mu_Y - \mu_Z - \mu_W)$; for diffusion, $A = (\mu_{in} - \mu_{out})$, the difference in chemical potentials inside and outside the subsystem. The equilibrium affinity $A_{eq} = 0$.

We assume a random fluctuation has changed the value of ξ from its equilibrium value ξ_{eq} to $(\xi_{eq} + \delta\xi)$. The change in entropy associated with this fluctuation can be calculated using (14.4.4), in which $d_iS = (A/T)\,d\xi$, a fundamental relation established in Chapter 4:

$$\Delta S = -\int_{\delta\xi}^{0} d_iS = -\int_{\delta\xi}^{0} \frac{A}{T}d\xi = \int_{0}^{\delta\xi} \frac{A}{T}d\xi \qquad (14.4.11)$$

At equilibrium, the affinity $A_{eq} = 0$. For a small change $\alpha = \xi - \xi_{eq}$ of the extent of reaction from the equilibrium state we may approximate A by

$$A = A_{eq} + \left(\frac{\partial A}{\partial \xi}\right)_{eq} \alpha = \left(\frac{\partial A}{\partial \xi}\right)_{eq} \alpha \qquad (14.4.12)$$

The entropy change ΔS due to the fluctuation $\delta\xi$ can thus be expressed as

$$\Delta S = \int_{0}^{\delta\xi} d_iS = \int_{0}^{\delta\xi} \frac{A}{T}d\xi = \frac{1}{T}\int_{0}^{\delta\xi} \left(\frac{\partial A}{\partial \xi}\right)_{eq} \alpha\,d\alpha = \left(\frac{\partial A}{\partial \xi}\right)_{eq} \frac{(\delta\xi)^2}{2T} \qquad (14.4.13)$$

where we have used $d\xi = d\alpha$. The **stability condition** $\Delta S < 0$ and the probability of fluctuations take the form

$$\boxed{\Delta S = \left(\frac{\partial A}{\partial \xi}\right)_{eq} \frac{(\delta\xi)^2}{2T} < 0 \quad P(\delta\xi) = Z\exp\left[\left(\frac{\partial A}{\partial \xi}\right)_{eq} \frac{(\delta\xi)^2}{2k_BT}\right]} \qquad (14.4.14)$$

This expression can also be written in terms of the fluctuations δN_k in molar amounts N_k by rewriting (14.4.11) in terms of N_k:

$$\Delta S = -\int_{\delta N_k}^{0} d_iS = -\int_{\delta N_k}^{0} \frac{A}{T}d\xi = -\int_{0}^{\delta N_k} \frac{A}{T}\frac{dN_k}{\nu_k} \qquad (14.4.15)$$

in which ν_k are the stoichiometric coefficients shown in (14.4.10), which are negative for reactants and positive for products. In the case of a chemical reaction, all changes δN_i are related through stoichiometry and they can be expressed in terms of changes δN_k of one of the constituents k. However, the changes δN_k due to diffusion can all be independent. When molecules diffuse in out of a small volume, the chemical potential in the volume may change but in the larger surroundings its change is negligible. Taking this into consideration, it can be shown that (Exercise 14.5)

$$\Delta S = -\left(\frac{\partial \mu_k}{\partial N_k}\right)_{eq} \frac{(\delta N_k)^2}{2T} \qquad (14.4.16)$$

If the chemical potential μ_k is expressed as a function of N_k, then the above expression can be made explicit. If we consider ideal gases as an example, then the chemical potential of a component k

$$\mu_k = \mu_k^0(T) + RT \ln\left(\frac{p_k}{p_0}\right) = \mu_k^0(T) + RT \ln\left(\frac{RTN_k}{Vp_0}\right)$$

and

$$\frac{\partial \mu_k}{\partial N_k} = \frac{RT}{N_k} \tag{14.4.17}$$

Substituting (14.4.17) into (14.4.16), we see that

$$\Delta S = -\frac{R(\delta N_k)^2}{2N_{k,\mathrm{eq}}} \quad P(\delta N_k) = Z \exp\left[\frac{-R(\delta N_k)^2}{2k_\mathrm{B}N_{k,\mathrm{eq}}}\right] \tag{14.4.18}$$

This expression can also be written in terms number of molecules $\tilde{N}_k = N_\mathrm{A}N_k$; and noting $R = N_\mathrm{A}k_\mathrm{B}$, we obtain

$$\boxed{\Delta S = -\frac{k_\mathrm{B}(\delta \tilde{N}_k)^2}{2\tilde{N}_{k,\mathrm{eq}}} \quad P(\delta N_k) = Z \exp\left(\frac{-\delta \tilde{N}_k^2}{2\tilde{N}_{k,\mathrm{eq}}}\right)} \tag{14.4.19}$$

This expression shows that the probability of fluctuations in the number of molecules is independent of temperature. The mean of $\tilde{N}_k = 0$ and $\sqrt{\langle \delta \tilde{N}_k^2 \rangle} = \sqrt{\tilde{N}_{k,\mathrm{eq}}}$, as was noted at the beginning of this section.

The above results can be extended to situations in which many chemical reactions are simultaneously present. If we consider r chemical reactions, then the deviations $\delta \xi_i(\tau)$, $i = 1, 2, \ldots, r$, of the extents of reaction from their equilibrium values can be expressed as functions of a parameter τ such that $\delta \xi_i(0) = 0$ and $\delta \xi_i(\tau) \approx (\partial \xi_i/\partial \tau)_{\tau=0}\tau$ for small values of τ. Using such parameterization, it can be shown that [4]

$$\Delta S = \sum_{i,j}^{r} \frac{1}{2T}\left(\frac{\partial A_i}{\partial \xi_j}\right)_{\mathrm{eq}} \delta \xi_i \delta \xi_j \tag{14.4.20}$$

More generally, if N_k can vary independently then it can be shown that

$$\Delta S = -\sum_{i,j} \frac{1}{2T}\left(\frac{\partial \mu_i}{\partial N_j}\right)_{\mathrm{eq}} \delta N_i \delta N_j \tag{14.4.21}$$

The corresponding probability distribution can be derived following the method presented above.

GENERAL EXPRESSION FOR PROBABILITY OF FLUCTUATIONS

Following the above method, the probability of entropy associated with a fluctuation of the volume of a subsystem at a fixed pressure and molar amounts N_k of its constituents can be shown to equal $\Delta S = -(1/T_{eq}\kappa_T)(\delta V^2/2V)$, in which $\kappa_T = -(1/V)(\partial V/\partial p)_T$ is the isothermal compressibility. If we consider fluctuations in T, N_k and V, then

$$\Delta S = -\frac{C_V(\delta T)^2}{2T_{eq}^2} - \frac{1}{T_{eq}\kappa_T}\frac{(\delta V)^2}{2V_{eq}} - \sum_{i,j}\left(\frac{\partial \mu_i}{\partial N_j}\right)_{eq}\frac{\delta N_i\delta N_j}{2T_{eq}}$$

and

$$P(\delta T, \delta N_k, \delta V) = Ze^{\Delta S/k_B} \tag{14.4.22}$$

Finally, we note that this entropy term is second order in the fluctuations δT, δV and δN_k. In expression (14.4.21), the independent variables are T, V and N. A more general expression through which the entropy change for fluctuations in any other set of independent variables can be derived is the following:

$$\Delta S = \frac{-1}{2T}\left(\delta T\delta S - \delta p\delta V + \sum_i \delta\mu_i\delta N_i\right) \tag{14.4.23}$$

This relation could be derived as follows. In (14.4.21), in the first term:

$$\frac{C_V\delta T}{T} = \frac{\delta Q}{T} = \delta S$$

Similarly, in the second term:

$$\frac{\delta V}{\kappa_T V} = -\delta p$$

And in the third term:

$$\sum_j\left(\frac{\partial \mu_i}{\partial N_j}\right)dN_j = \delta\mu_i$$

At equilibrium, the entropy reaches its maximum value. Hence, its first-order variation $\delta S = 0$. So the variation ΔS due to a fluctuation is often written as $\delta^2 S/2$, especially in the classical theory of fluctuations, which is discussed in the following two chapters.

References

1. Hill, T., *Thermodynamics of Small Systems, Parts I and II* (reprinted). 2002, New York: Dover.
2. Laird, B.B., *J. Chem. Phys.*, **115** (2001) 2887.
3. Klein, M.J., *Science*, **157** (1967) 509.
4. Prigogine, I. and Defay, R. *Chemical Thermodynamics*, fourth edition. 1967, Longmans: London; 542.

Examples

Example   Consider an ideal gas at $T = 298\,K$ and $p = 1.0\,atm$. Calculate the molar amount N_1 of gas in a spherical volume of radius $1.0\,\mu m$, the average value and the magnitude of fluctuations in concentration N_1/V_1.

Solution

$$N_1 = \frac{pV}{RT} = \frac{101\,kPa \times (4\pi/3)(1.0 \times 10^{-6})^3\,m^3}{8.314\,J\,K^{-1}\,mol^{-1} \times 298\,K} = 1.7 \times 10^{-16}\,mol$$

The average concentrations

$$\left(\frac{N_1}{V_1}\right)_{avg} = 40.76\,mol\,m^{-3}\,(N_1/V_1)_{avg} = 40.76\,mol\,m^{-3}$$

The magnitude of the fluctuations

$$\delta\tilde{N} = \sqrt{N_1 N_A} = 1.02 \times 10^4$$

Fluctuation in concentrations

$$\frac{\sqrt{N_1 N_A}}{N_A V_1}\,mol\,m^{-3} = 4.02 \times 10^{-3}\,mol\,m^{-3}$$

The magnitude of

$$\frac{\delta\tilde{N}/V_1}{N_1 N_A/V_1} = \frac{1}{\sqrt{N_1 N_A}} \approx 1.0 \times 10^{-4}$$

Exercises

14.1   Using the expression $G = \mu_\infty N + \gamma\Sigma$, show that $(\partial G/\partial N)_{p,T} = \mu = \mu_\infty + (2\gamma V_m/r)$.

14.2 Using the parameters in Table 14.1, determine the size of AgBr particles whose saturation concentration $[Y(r)] = 1.3[Y(\infty)]$. Estimate the number of AgBr molecules in these particles.

14.3 N moles of the phase α form a β-phase cluster of radius r^*. For this process, assume that $G_\alpha = \mu_\alpha N$ and $G_\beta = \mu_{\beta\infty}N + \gamma 4\pi(r^*)^2$ and show that

$$G_\beta - G_\alpha = \Delta G(r^*) = \frac{16\pi}{3}\frac{\gamma^3 V_m^2}{\Delta\mu^2}.$$

14.4 For solidification from a melt, from the liquid phase α to the solid phase β, the chemical potential as a function of temperature must be analyzed. Assume $T = T_m - \Delta T$, in which $\Delta T/T \ll 1$ and show that $\Delta\mu = \mu_\alpha - \mu_\beta \approx \Delta H_{fus}(1 - T/T_m)$.

14.5 For fluctuation in N_k in a small volume due to diffusion, obtain the expression (14.4.16) for the change in entropy.

14.6 Obtain the expression

$$\Delta S = -\frac{C_V(\delta T)^2}{2T^2} - \frac{1}{T\kappa_T}\frac{(\delta V)^2}{2V} - \sum_i \frac{R(\delta N_i)^2}{2N_i}$$

for an ideal system for which $\mu_k = \mu_k^0(T) + RT\ln(x_k)$.

14.7 (a) Evaluate the normalization constant Z for (14.4.9). (b) Obtain the average values of the square of the fluctuations by evaluating $\int_{-\infty}^{\infty}(\delta T)^2 P(\delta T)d(\delta T)$. (c) In an ideal gas, estimate the value of $\sqrt{\langle\delta T^2\rangle}$ for a small cubic volume of side $1.0\,\mu m$.

14.8 Consider an ideal gas at a temperature T and $p = 1\,atm$. Assume that this ideal gas has two components X and Y in equilibrium with respect to inter-conversion, $X \rightleftharpoons Y$. In a small volume δV, calculate the number of molecules that should convert from X to Y to change the entropy by k_B in terms of the number of molecules X in the considered volume.

15 CLASSICAL STABILITY THEORY

15.1 Stability of Equilibrium States

The random motion of molecules causes all thermodynamic quantities, such as temperature, concentration and partial molar volume, to fluctuate. In addition, owing to its interaction with the exterior, the state of a system is subject to constant perturbations. The state of equilibrium must remain stable in the face of all fluctuations and perturbations. In this chapter, we shall develop a theory of stability for isolated systems in which the total energy U, volume V and mole numbers N_k are constant. The stability of the equilibrium state leads us to conclude that certain physical quantities, such as heat capacities, have a definite sign. This will be an introduction to the theory of stability as was developed by Gibbs. Chapter 16 contains some elementary applications of this stability theory.

For an isolated system, the entropy reaches its maximum value. Thus, any fluctuation can only reduce the entropy. In response to a fluctuation, entropy-producing irreversible processes spontaneously drive the system back to equilibrium. Hence, *the state of equilibrium is stable to any perturbation that results in a decrease in entropy*. The fluctuations in temperature, volume, etc. are quantified by their magnitudes, such as δT and δV. The entropy of the system is a function of these variables. In general, the entropy can be expanded as a power series in these variables, so that we have

$$S = S_{eq} + \delta S + \frac{1}{2}\delta^2 S + \dots \tag{15.1.1}$$

In such an expansion, the term δS represents the *first-order* terms containing δT, δV, etc., the term $\delta^2 S$ represents the *second-order* terms containing $(\delta T)^2$, $(\delta V)^2$ and so on. This notation will be made explicit in the examples that follow. Also, as we shall see below, since the entropy is a maximum, the first-order term δS vanishes. The change in entropy is due to the second- and higher-order terms, the leading contribution coming from the second-order term $\delta^2 S$.

We shall look at the stability conditions associated with fluctuations in different quantities such as temperature, volume and mole numbers. As stated before, we consider an isolated system in which U, V and N_k are constant.

15.2 Thermal Stability

For the fluctuations in temperature, we shall consider a simple situation without loss of generality. Let us assume that a fluctuation occurs in a small part of the

Introduction to Modern Thermodynamics   Dilip Kondepudi
© 2008 John Wiley & Sons, Ltd

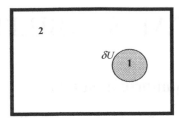

Figure 15.1 Thermal fluctuations in the equilibrium state. We consider a fluctuation that results in a flow of energy δU from one part to another causing the temperatures to change by a small amount δT

system (see Figure 15.1). Owing to the fluctuation, there is a flow of energy δU from one part to another, resulting in small temperature fluctuations δT in the smaller part. The subscripts 1 and 2 identify the two parts of the system. The total entropy of the system is

$$S = S_1 + S_2 \tag{15.2.1}$$

Here, entropy S_1 is a function of U_1, V_1, etc., and S_2 is a function of U_2, V_2, etc. If we express S as a Taylor series about its equilibrium value ΔS_{eq}, then we can express the change in entropy ΔS from its equilibrium value as

$$S - S_{eq} = \Delta S = \left(\frac{\partial S_1}{\partial U_1}\right)\delta U_1 + \left(\frac{\partial S_2}{\partial U_2}\right)\delta U_2 + \left(\frac{\partial^2 S_1}{\partial U_1^2}\right)\frac{(\delta U_1)^2}{2} + \left(\frac{\partial^2 S_2}{\partial U_2^2}\right)\frac{(\delta U_2)^2}{2} + \dots$$
$$\tag{15.2.2}$$

where all the derivatives are evaluated at the equilibrium state.

Since the total energy of the system remains constant, $\delta U_1 = -\delta U_2 = \delta U$. Also, recall that $(\partial S/\partial U)_{V,N} = 1/T$. Hence, (15.2.2) can be written as

$$\Delta S = \left(\frac{1}{T_1} - \frac{1}{T_2}\right)\delta U + \left[\frac{\partial}{\partial U_1}\frac{1}{T_1} + \frac{\partial}{\partial U_2}\frac{1}{T_2}\right]\frac{(\delta U)^2}{2} + \dots \tag{15.2.3}$$

We can now identify the first and second variations of entropy, δS and $\delta^2 S$, and write them explicitly in terms of the perturbation δU:

$$\delta S = \left(\frac{1}{T_1} - \frac{1}{T_2}\right)\delta U \tag{15.2.4}$$

$$\frac{1}{2}\delta^2 S = \left[\frac{\partial}{\partial U_1}\frac{1}{T_1} + \frac{\partial}{\partial U_2}\frac{1}{T_2}\right]\frac{(\delta U)^2}{2} \tag{15.2.5}$$

At equilibrium, since all thermodynamic forces must vanish, the entire system should be at the same temperature. Hence, $T_1 = T_2$, and the first variation of entropy $\delta S = 0$. (If it is taken as a postulate that entropy is a maximum at equilibrium, then the first variation should vanish. One then concludes that $T_1 = T_2$.) The change in entropy due to fluctuations in the equilibrium state is due to the second variation $\delta^2 S$ (the smaller higher-order terms in the Taylor series are neglected). As stated above, the equilibrium state is stable only if the fluctuation causes the entropy to decrease, i.e. $\delta^2 S < 0$; spontaneous, entropy-increasing irreversible processes then drive the system back to the state of equilibrium. Now let us write (15.2.5) explicitly in terms of the physical properties of the system and see what the condition for stability implies. First, we note that

$$\frac{\partial}{\partial U}\frac{1}{T} = -\frac{1}{T^2}\frac{\partial T}{\partial U} = -\frac{1}{T^2}\frac{1}{C_V} \tag{15.2.6}$$

in which C_V is the heat capacity. Also, if the change in the temperature of the smaller system is δT, then we have $\delta U_1 = C_{V_1}(\delta T)$, where C_{V_1} is the heat capacity of the smaller part. C_{V_2} is the heat capacity of the larger part. Using (15.2.6) for the two parts in (15.2.5) and writing $\delta U = C_{V_1}(\delta T)$, and noting that all the derivatives are evaluated at equilibrium, so that $T_1 = T_2 = T$, we obtain

$$\frac{1}{2}\delta^2 S = -\frac{C_{V_1}(\delta T)^2}{2T^2}\left(1 + \frac{C_{V_1}}{C_{V_2}}\right) \tag{15.2.7}$$

If system 1 is small compared with system 2, $C_{V_1} \ll C_{V_2}$, then the second term in the parentheses can be ignored. Thus, for stability of the equilibrium state we have

$$\frac{1}{2}\delta^2 S = -\frac{C_{V_1}(\delta T)^2}{2T^2} < 0 \tag{15.2.8}$$

This condition requires that the heat capacity $C_{V_1} > 0$. *Thus, the state of equilibrium is stable to thermal fluctuations only when the heat capacity at constant volume C_V is positive.*

15.3 Mechanical Stability

We now turn to stability of the system with respect to fluctuation in the volume of a subsystem with N remaining constant, i.e. fluctuations in the molar volume. As in the previous case, consider the system divided into two parts (see Figure 15.2), but this time we assume there is a small change in volume δV_1 of system 1 and δV_2 of

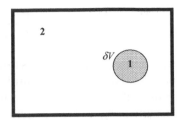

Figure 15.2 Fluctuation in the volume of a system for fixed N and U

system 2. Since the total volume of the system remains fixed, $\delta V_1 = \delta V_2 = \delta V$. For the computation of the change in entropy associated with such a fluctuation, we can write an equation similar to (15.2.3), with V taking the place of U. Since $(\partial S/\partial V)_{U,N} = p/T$, a calculation similar to the one above leads to

$$\delta S = \left(\frac{p_1}{T_1} - \frac{p_2}{T_2} \right) \delta V \tag{15.3.1}$$

$$\frac{1}{2}\delta^2 S = \left(\frac{\partial}{\partial V_1}\frac{p_1}{T_1} + \frac{\partial}{\partial V_2}\frac{p_2}{T_2} \right) \frac{(\delta V)^2}{2} \tag{15.3.2}$$

If the derivatives are evaluated at equilibrium, then $p_1/T_1 = p_2/T_2 = p/T$. Then the first variation δS vanishes (as it must if S is a maximum at equilibrium). To understand the physical meaning of the conditions for stability $\Delta S < 0$, the second variation can be written in terms of the isothermal compressibility. The isothermal compressibility κ_T is defined by $\kappa_T = -(1/V)(\partial V/\partial p)$. During the fluctuation in V we assume that T remains unchanged. With these observations it is easy to see that (15.3.2) can be written as

$$\delta^2 S = -\frac{1}{T\kappa_T}\frac{(\delta V)^2}{V}\left(1 + \frac{V_1}{V_2} \right) \tag{15.3.3}$$

As before, if one part is much larger than another, $V_2 \gg V_1$, then this expression can be simplified to

$$\delta^2 S = -\frac{1}{T\kappa_T}\frac{(\delta V)^2}{V} \tag{15.3.4}$$

The condition for stability of the equilibrium state $\delta^2 S < 0$ now means that $\kappa_T > 0$. *Thus, the state of equilibrium is stable to volume or mechanical fluctuations only when the isothermal compressibility is positive.*

15.4   Stability with Respect to Fluctuations in N

Fluctuations in the amount in moles of the various components of a system occur due to chemical reactions and due to diffusion. We discuss each of these cases separately.

CHEMICAL STABILITY

These fluctuations can be identified as the fluctuations in the extent of reaction ξ, about its equilibrium value. Considering a fluctuation $\delta\xi$, the change in entropy is

$$S - S_{eq} = \Delta S = \delta S + \frac{1}{2}\delta^2 S = \left(\frac{\partial S}{\partial \xi}\right)_{U,V}\delta\xi + \frac{1}{2}\left(\frac{\partial^2 S}{\partial \xi^2}\right)_{U,V}(\delta\xi)^2 \qquad (15.4.1)$$

We saw in Chapter 4 that $(\partial S/\partial \xi)_{U,V} = A/T$. Hence, (15.4.1) can be written as

$$\Delta S = \delta S + \frac{1}{2}\delta^2 S = \left(\frac{A}{T}\right)_{eq}\delta\xi + \frac{1}{2T}\left(\frac{\partial A}{\partial \xi}\right)_{eq}(\delta\xi)^2 \qquad (15.4.2)$$

(Here, T is constant.) In this equation, the identification of the first and second variations of entropy is obvious. At equilibrium, the affinity A vanishes, so that once again $\delta S = 0$. For the stability of the equilibrium state, we then require that the second variation $\delta^2 S$ be negative:

$$\frac{1}{2}\delta^2 S = \frac{1}{2T}\left(\frac{\partial A}{\partial \xi}\right)_{eq}(\delta\xi)^2 < 0 \qquad (15.4.3)$$

Since $T > 0$, the condition for stability of the equilibrium state is

$$\left(\frac{\partial A}{\partial \xi}\right)_{eq} < 0 \qquad (15.4.4)$$

When many chemical reactions take place simultaneously, condition (15.4.3) can be generalized to [1]

$$\frac{1}{2}\delta^2 S = \sum_{i,j}\frac{1}{2T}\left(\frac{\partial A_i}{\partial \xi_j}\right)_{eq}\delta\xi_i\delta\xi_j < 0 \qquad (15.4.5)$$

STABILITY TO FLUCTUATIONS IN N DUE TO TRANSPORT

In the above stability analysis, the fluctuations in mole numbers considered were only due to chemical reactions. The fluctuation in mole number can also occur due to exchange of matter between a part of a system and the rest (see Figure 15.3). As we did in the case of exchange of energy, we consider the total change in entropy of the two parts of the system.

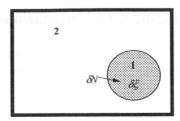

Figure 15.3 Fluctuations in mole number can occur due to chemical reactions and exchange of molecules between the two systems. The state of equilibrium is stable if the entropy change associated with such fluctuations is negative

$$S = S_1 + S_2 \tag{15.4.6}$$

$$
\begin{aligned}
S - S_{eq} = \Delta S \\
= \sum_k \frac{\partial S_1}{\partial N_{1k}} \delta N_{1k} + \frac{\partial S_2}{\partial N_{2k}} \delta N_{2k} + \sum_{i,j} \frac{\partial^2 S_1}{\partial N_{1i} \partial N_{1j}} \frac{\delta N_{1i} \delta N_{1j}}{2} \\
+ \frac{\partial^2 S_2}{\partial N_{2i} \partial N_{2j}} \frac{\delta N_{2i} \delta N_{2j}}{2} + \dots
\end{aligned}
\tag{15.4.7}
$$

Now we note that $\delta N_{1k} = -\delta N_{2k} = \delta N_k$ and $(\partial S / \partial N_k) = -\mu_k / T$. Equation (15.4.7) can then be written so that the first and second variations of the entropy can be identified:

$$
\Delta S = \delta S + \frac{\delta^2 S}{2} = \sum_k \left(\frac{\mu_{2k}}{T} - \frac{\mu_{1k}}{T} \right) \delta N_k - \sum_{i,j} \left(\frac{\partial}{\partial N_j} \frac{\mu_{1i}}{T} + \frac{\partial}{\partial N_j} \frac{\mu_{2i}}{T} \right) \frac{\delta N_i \delta N_j}{2}
\tag{15.4.8}
$$

As before, if the derivatives are evaluated at the state of equilibrium, then the chemical potentials of the two parts must be equal. Hence, the first term vanishes. Furthermore, if system 1 is small compared with system 2, then the change in the chemical potential (which depends on the concentrations) with respect to N_k of system 2 will be small compared with the corresponding change in system 1; that is:

$$
\frac{\partial}{\partial N_j} \frac{\mu_{1i}}{T} \gg \frac{\partial}{\partial N_j} \frac{\mu_{2i}}{T}
$$

if system 1 is much smaller than system 2. We then have

$$\delta^2 S = -\sum_{i,j} \left(\frac{\partial}{\partial N_j} \frac{\mu_{1i}}{T} \right) \delta N_i \delta N_j < 0 \qquad (15.4.9)$$

as the condition for the stability of an equilibrium state when fluctuations in N_k are considered.

In fact, this condition is general and it can be applied to fluctuations due to chemical reactions as well. By assuming the fluctuations $\delta N_k = \nu_k \delta \xi$, in which ν_k is the stoichiometric coefficient, we can obtain the condition (15.4.5) (Exercise 15.4). *Thus, a system that is stable to diffusion is also stable to chemical reactions.* This is called the **Duhem–Jougeut theorem** [2, 3]. A more detailed discussion of this theorem and many other aspects of stability theory can be found in Ref. [4].

In summary, the general condition for the stability of the equilibrium state to thermal, volume and mole number fluctuations can be expressed by combining (15.2.8), (15.3.4) and (15.4.9):

$$\delta^2 S = -\frac{C_V(\delta T)^2}{T^2} - \frac{1}{T\kappa_T} \frac{(\delta V)^2}{V} - \sum_{i,j} \left(\frac{\partial}{\partial N_j} \frac{\mu_i}{T} \right) \delta N_i \delta N_j < 0 \qquad (15.4.10)$$

Though we have derived the above results by assuming S to be a function of U, V and N_k, and a system in which U, V and N are constant, the results derived have a more general validity, in that they are also valid for other situations in which p or T or both p and T are maintained constant. The corresponding results are expressed in terms of the enthalpy H, Helmholtz free energy F and the Gibbs free energy G. In fact, as we saw in Chapter 14, a general theory of stability that is valid for a wide range of conditions can be formulated using the entropy production $d_i S$ as its basis.

References

1. Glansdorff, P., Prigogine, I., *Thermodynamics of Structure Stability and Fluctuations.* 1971, New York: Wiley-Interscience.
2. Jouguet, E., Notes de mécanique chimique. *Journal de l'Ecole Polytechnique (Paris), IIme Séries,* **21** (1921) 61.
3. Duhem, P., *Traité Élémentaire de Mécanique Chimique* (4 vols). 1899, Paris: Gauthier-Villars.
4. Prigogine, I. and Defay, R., *Chemical Thermodynamics.* 1954, London: Longmans.

Exercises

15.1 For an ideal gas of N_2 at equilibrium at $T = 300\,\text{K}$, calculate the change in entropy due to a fluctuation of $\delta T = 1.0 \times 10^{-3}\,\text{K}$ in a volume $V = 1.0 \times 10^{-6}\,\text{mL}$.

15.2   Obtain the expressions (15.3.1) and (15.3.2) for the first- and second-order entropy changes due to fluctuations of volume at constant N.

15.3   Explain the physical meaning of the condition (15.4.4) for stability with respect to a chemical reaction.

15.4   In expression (15.4.9), assume that the change in amount in moles N is due to a chemical reaction and obtain expression (15.4.3).

16 CRITICAL PHENOMENA AND CONFIGURATIONAL HEAT CAPACITY

Introduction

In this chapter we shall consider applications of stability theory to critical phenomena of liquid–vapor transitions and the separation of binary mixtures. When the applied pressure and temperature are altered, systems can become unstable, causing their physical state to transform into another distinct state. When the pressure on a gas is increased, for example, it may lose its stability and make a transition to liquid. Similarly, when the temperature of a two-component liquid mixture (such as hexane and nitrobenzene) changes, the mixture may become unstable to changes in its composition; the mixture then separates into two phases, each rich in one of the components. In Chapter 11 we saw that, in far-from-equilibrium systems, loss of stability can lead to a wide variety of complex nonequilibrium states. In equilibrium systems, loss of stability leads to phase separation. In this chapter, we shall also look at the response of a system that can undergo internal transformations to quick changes in temperature. This leads us to the concept of **configurational heat capacity**.

16.1 Stability and Critical Phenomena

In Chapter 7 we looked briefly at the critical behavior of a pure substance. If its temperature is above the critical temperature T_c then there is no distinction between the gas and the liquid states, regardless of the pressure. Below the critical temperature, at low pressures the substance is in the form of a gas, but liquid begins to form as the pressure is increased. We can understand this transformation in terms of stability.

As shown in Figure 16.1 by the arrows, by using an appropriate path it is possible to go from a gaseous state to a liquid state in a continuous fashion. This was noted by James Thomson, who also suggested that the isotherms below the critical point were also continuous, as shown in Figure 16.2 by the curve IAJKLBM. This suggestion was pursued by van der Waals, whose equation, as we saw in Chapter 1, indeed gives the curve shown. However, the region JKL in Figure 16.2 cannot be physically realized because it is an unstable region, i.e. it is not mechanically stable. In Section 15.3 we saw that the condition for mechanical stability is that the compressibility $\kappa_T \equiv -(1/V)(\partial V/\partial p) > 0$. In Figure 16.2, this implies that the system is stable only if

Introduction to Modern Thermodynamics   Dilip Kondepudi
© 2008 John Wiley & Sons, Ltd

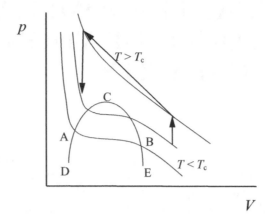

Figure 16.1 The critical behavior of a pure substance. Below the critical temperature, at a fixed temperature, a decrease in volume results in a transition to a liquid state in the region AB in which the two phases coexist. The envelope of the segments AB for the family of isotherms has the shape ECD. Above the critical temperature T_c there is no gas–liquid transition. The gas becomes more and more dense, there being no distinction between the gas and the liquid phases. By following the path shown by the arrows, it is possible to go from a gas to a liquid state without going through a transition

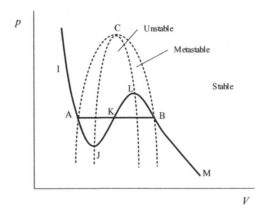

Figure 16.2 The stable, metastable and unstable regions for a liquid–vapor transition are indicated. In the region JKL, $(\partial p/\partial V)_T > 0$, which shows that the system is unstable

$$\left(\frac{\partial p}{\partial V}\right)_T < 0 \qquad (16.1.1)$$

a condition that is satisfied for the segments IA and BM and for all the isotherms above the critical temperature. These regions represent stable regions. For the segment JKL we see that $(\partial p/\partial V)_T > 0$, which means that this state is unstable. In this unstable state, if the volume of the system is kept fixed, then small fluctuations in pressure, depending on the initial state, will cause either the vapor to condense or the liquid to evaporate. The system will collapse to a point in the segment AB, where liquid and vapor coexist. As shown in Section 7.4, the amount of the substance in the two phases is given by the 'lever rule'.

In region BL of Figure 16.2, the system is a supersaturated vapor and may begin to condense if nucleation can occur. This is a **metastable state**. Similarly, in the region AJ we have a superheated liquid that will vaporize if there is nucleation of the vapor phase. The stable, metastable and unstable regions are indicated in Figure 16.2. Finally, at the critical point C, both the first and second derivatives of p with respect to V equal zero. Here, the stability is determined by the higher-order derivatives. For stable mechanical equilibrium at the critical point we have

$$\left(\frac{\partial p}{\partial V}\right)_{T_c} = 0 \quad \left(\frac{\partial^2 p}{\partial V^2}\right)_{T_c} = 0 \quad \left(\frac{\partial^3 p}{\partial V^3}\right)_{T_c} < 0 \qquad (16.1.2)$$

which is an inflection point. The inequality $(\partial^3 p/\partial V^3) < 0$ is obtained by considering terms of higher order than $\delta^2 S$.

16.2   Stability and Critical Phenomena in Binary Solutions

In solutions, depending on the temperature, the various components can segregate into separate phases. For simplicity, we shall only consider binary mixtures. This is a phenomenon similar to the critical phenomenon in a liquid–vapor transition, in that in one range of temperature the system is in one homogeneous phase (solution), but in an another range of temperature the system becomes unstable and the two components separate into two phases. The **critical temperature** that separates these two ranges depends on the composition of the mixture. This can happen in three ways, as illustrated by the following examples.

At atmospheric pressure, liquids n-hexane and nitrobenzene are miscible in all proportions when the temperature is above 19 °C. Below 19 °C, the mixture separates into two distinct phases, one rich in nitrobenzene and the other in n-hexane. The corresponding phase diagram is shown in Figure 16.3a. At about 10 °C, for example, in one phase the mole fraction of nitrobenzene is 0.18, but in the other phase the mole fraction is about 0.75. As the temperature increases, at $T = T_c$, the two liquid layers become identical in composition, indicated by the point C. Point C is called the **critical solution point** or **consolute point** and its location depends on the applied pressure. In this example, above the critical temperature the two liquids are miscible in all proportions. This is the case of an **upper critical temperature**. But this is not always the case, as shown in Figure 16.3b and c. The critical temperature can be such that below T_c the two components become miscible in all proportions. An example of such a mixture is that of diethylamine and water. Such a mixture is said to have a **lower critical solution temperature**. Binary systems can have both upper and lower critical solution temperatures, as shown in Figure 16.3c. An example of such a system is a mixture of m-toluidine and glycerol.

Let us now look at the phase separation of binary mixtures from the point of view of stability. The separation of phases occurs when the system becomes unstable with respect to diffusion of the two components; that is, if the separation of the two components results in an increase in entropy, then the fluctuations in the mole number due to diffusion in a given volume grow, resulting in the separation of the two components. As we saw in Section 15.4, the condition for stability against diffusion of the components is

$$\delta^2 S = -\sum_{i,k} \frac{\partial}{\partial N_k}\left(\frac{\mu_i}{T}\right)\delta N_k \delta N_i < 0 \tag{16.2.1}$$

At a fixed T, for binary mixtures this can be written in the explicit form

$$\mu_{11}(\delta N_1)^2 + \mu_{22}(\delta N_2)^2 + \mu_{21}(\delta N_1)(\delta N_2) + \mu_{12}(\delta N_1)(\delta N_2) > 0 \tag{16.2.2}$$

in which

$$\mu_{11} = \frac{\partial \mu_1}{\partial N_1} \quad \mu_{22} = \frac{\partial \mu_2}{\partial N_2} \quad \mu_{21} = \frac{\partial \mu_2}{\partial N_1} \quad \mu_{12} = \frac{\partial \mu_1}{\partial N_2} \tag{16.2.3}$$

Condition (16.2.2) is mathematically identical to the statement that the matrix with elements μ_{ij} is *positive definite*. Also, because

$$\mu_{21} = \frac{\partial \mu_2}{\partial N_1} = \frac{\partial}{\partial N_1}\frac{\partial G}{\partial N_2} = \frac{\partial}{\partial N_2}\frac{\partial G}{\partial N_1} = \mu_{12} \tag{16.2.4}$$

this matrix is symmetric. The stability of the system is assured if the symmetric matrix

$$\begin{bmatrix} \mu_{11} & \mu_{12} \\ \mu_{21} & \mu_{22} \end{bmatrix} \tag{16.2.5}$$

is positive definite. The necessary and sufficient conditions for the positivity of (16.2.5) are

$$\mu_{11} > 0 \quad \mu_{22} > 0 \quad \mu_{11}\mu_{22} - \mu_{21}\mu_{12} > 0 \tag{16.2.6}$$

If these are not satisfied, then condition (16.2.2) will be violated and the system becomes unstable. Note that (16.2.4) and (16.2.6) imply that $\mu_{12} = \mu_{21} < 0$ to assure stability for all positive values of μ_{11} and μ_{22}.

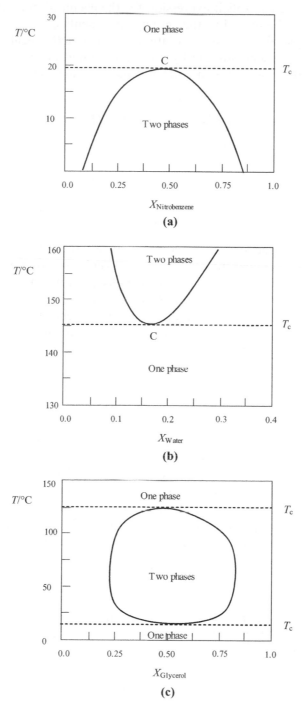

Figure 16.3 Three types of phase diagrams showing critical phenomenon in binary solutions: (a) a mixture of hexane and nitrobenzene; (b) a mixture of diethylamine and water; (c) a mixture of *m*-toluidine and glycerol

If we have an explicit expression for the chemical potential, then the conditions (16.2.6) can be related to the activity coefficients of the system. This can be done, for example, for a class of solutions called **strictly regular solutions**, which were studied by Hildebrandt and by Fowler and Guggenheim in 1939. The two components of these solutions interact strongly and their chemical potentials are of the form

$$\mu_1(T, p, x_1, x_2) = \mu_1^0(T, p) + RT \ln(x_1) + \alpha x_2^2 \qquad (16.2.7)$$

$$\mu_2(T, p, x_1, x_2) = \mu_2^0(T, p) + RT \ln(x_2) + \alpha x_1^2 \qquad (16.2.8)$$

in which

$$x_1 = \frac{N_1}{N_1 + N_2} \qquad x_2 = \frac{N_2}{N_1 + N_2} \qquad (16.2.9)$$

are the mole fractions. The factor α is related to the difference in interaction energy between two similar molecules (two molecules of component 1 or two molecules of component 2) and two dissimilar molecules (one molecule of component 1 and one of component 2). For solutions that are nearly perfect, α is nearly zero. From these expressions it follows that activity coefficients are given by $RT \ln \gamma_1 = \alpha x_2^2$ and $RT \ln \gamma_2 = \alpha x_1^2$. We can now apply the stability conditions (16.2.6) to this system. By evaluating the derivative we see that the condition $\mu_{11} = (\partial \mu_1 / \partial N_1) > 0$ becomes (Exercise 16.5)

$$\frac{RT}{2\alpha} - x_1(1 - x_1) > 0 \qquad (16.2.10)$$

For a given composition specified by x_1, if $R/2\alpha$ is positive, then for sufficiently large T this condition will be satisfied. However, it can be violated for smaller T. The maximum value of $x_1(1 - x_1)$ is 0.25. Thus, for $RT/2\alpha < 0.25$ there must be a range of x_1 in which the inequality (16.2.10) is not valid. When this happens, the system becomes unstable and separates into two phases. In this case we have an upper critical solution temperature. From (16.2.10), it follows that the relation between mole fraction and the temperature below which the system becomes unstable is

$$\frac{RT_c}{2\alpha} - x_1(1 - x_1) = 0 \qquad (16.2.11)$$

This gives us the plot of T_c as a function of x_1 shown in Figure 16.4. It is easy to see that the maximum of T_c occurs at $x_1 = 0.5$. Thus, the critical temperature and mole fractions are

$$(x_1)_c = 0.5 \quad T_c = \frac{\alpha}{2R} \qquad (16.2.12)$$

The equation $T = (2\alpha/R)x_1(1 - x_1)$ gives the boundary between the metastable region and the unstable region. The boundary between the stable region and the metastable

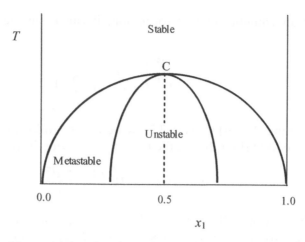

Figure 16.4 The phase diagram for strictly regular solutions

region is the coexistence curve. The coexistence curve of the two phases can be obtained by writing the chemical potentials μ_1 and μ_2 in both phases and equating them. This is left as an exercise.

16.3 Configurational Heat Capacity

The thermodynamic state of many systems may be specified, in addition to p and T and by the extent of reaction ξ. For such a system, the heat capacity must also involve changes in ξ due to the change in temperature. For example, we may consider a compound that may exist in two isomeric forms. Then the extent of reaction for the transformation is the variable ξ. The heat absorbed by such a system causes changes not only in p and T, but also in ξ, causing it to relax to a new state of equilibrium with respect to the transformation. If the system is in equilibrium with respect to the extent of reaction ξ, then the corresponding affinity $A = 0$. Now, since the heat exchanged $dQ = dU - pdV = dH - Vdp$, we can write

$$dQ = (h_{T,\xi} - V)dp + C_{p,\xi}dT + h_{T,p}d\xi \tag{16.3.1}$$

in which

$$h_{T,\xi} = \left(\frac{\partial H}{\partial p}\right)_{T,\xi} \quad C_{p,\xi} = \left(\frac{\partial H}{\partial T}\right)_{p,\xi} \quad h_{T,p} = \left(\frac{\partial H}{\partial \xi}\right)_{T,p} \tag{16.3.2}$$

At constant pressure, we can write the heat capacity $C_p = C_{pm}N$ as

$$C_p = \left(\frac{dQ}{dT}\right)_p = C_{p,\xi} + h_{T,p}\left(\frac{d\xi}{dT}\right)_p \tag{16.3.3}$$

Now for an equilibrium transformation, it can easily be shown (Exercise 16.6) that

$$\left(\frac{\partial \xi}{\partial T}\right)_{p,A=0} = -\frac{h_{T,p}}{T\left(\dfrac{\partial A}{\partial \xi}\right)_{T,p}} \tag{16.3.4}$$

By substituting (16.3.4) into (16.3.3) we obtain the following result for a system that remains in equilibrium during the time it is receiving heat:

$$C_{p,A=0} = C_{p,\xi} - T\left(\frac{\partial A}{\partial \xi}\right)_{T,p}\left(\frac{\partial \xi}{\partial T}\right)_{p,A=0}^{2} \tag{16.3.5}$$

But we have seen in Section 15.4 that the condition for the stability of a system with respect to chemical reactions is that $(\partial A/\partial \xi) < 0$. Hence, the second term on the right-hand side of (16.3.5) is positive. The term $C_{p,\xi}$ is the heat capacity at constant composition. There may be situations, however, in which the relaxation of the transformation represented by ξ is very slow. In this case, we measure the heat capacity at a constant composition. This leads us to the following general conclusion:

The heat capacity at a constant composition is always less than heat capacity of a system that remains in equilibrium with respect to ξ as it receives heat.

The term $h_{T,p}(\mathrm{d}\xi/\mathrm{d}T)$ is called the configurational heat capacity, because it refers to the heat capacity due to the relaxation of the system to the equilibrium configuration. The configurational heat capacity can be observed in systems such as glycerin near its crystalline state, where the molecules can vibrate but not rotate freely as they do in the liquid state. This restricted motion is called *libration*. As the temperature is increased, a greater fraction of the molecules begin to rotate. For this system, ξ is the extent of reaction for the libration–rotation transformation. For glycerin, there exists a state called the *vitreous state* in which the libration–rotation equilibrium is reached rather slowly. If such a system is heated rapidly, the equilibrium is not maintained and the measured heat capacity will be $C_{p,\xi}$, which will be lower than the heat capacity measured through slow heating during which the system remains in equilibrium.

Further Reading

Hildebrandt, J.M., Prausnitz, J.M., Scott, R.L., *Regular and Related Solutions.* 1970, New York: Van Nostrand-Reinhold.

Van Ness, H.C., Abbott, M.M., *Classical Thermodynamics of Nonelectrolyte Solutions.* 1982, New York: McGraw-Hill.

Prigogine, I., Defay, R., *Chemical Thermodynamics*, fourth edition. 1967, London: Longmans.

Exercises

16.1 Using the Gibbs–Duhem equation at constant p and T, and the relation $d\mu_k = \Sigma_i (\partial \mu_k / \partial N_i)_{p,T} dN_i$, show that

$$\sum_i \left(\frac{\partial \mu_k}{\partial N_i} \right)_{p,T} N_i = 0$$

This equation implies that the determinant of the matrix with elements $\mu_{ki} = (\partial \mu_k / \partial N_i)$ is equal to zero. Consequently, one of the eigenvalues of the matrix (16.2.5) is zero.

16.2 Show that, if the 2×2 matrix (16.2.5) has a negative eigenvalue, then the inequality (16.2.2) can be violated.

16.3 Show that if the matrix (16.2.5) has positive eigenvalues, then $\mu_{11} > 0$ and $\mu_{22} > 0$.

16.4 In a strictly binary solution, assuming that the two phases are symmetric, i.e. the dominant mole fraction in both phases is the same, obtain the coexistence curve by equating the chemical potentials of a component in the two phases.

16.5 Using (16.2.7) and (16.2.9), show that the condition $\mu_{11} = \partial \mu_1 / \partial N_1 > 0$ leads to Equation (16.2.10).

16.6 For an equilibrium transformation, show that

$$\left(\frac{\partial \xi}{\partial T} \right)_{p,A=0} = -\frac{h_{T,p}}{T \left(\frac{\partial A}{\partial \xi} \right)_{T,p}}$$

17 ELEMENTS OF STATISTICAL THERMODYNAMICS

Introduction

In the nineteenth century, development of kinetic theory, the ideas Daniel Bernoulli published a century earlier in his *Hydrodynamica*, came to fruition. When the atomic nature of matter became evident, James Clerk Maxwell, Ludwig Boltzmann, and others began to formulate the kinetic theory of gases. Kinetic theory demonstrated how random molecular motion gives rise to pressure in accordance with the ideal gas law, $pV = NRT$ (as discussed in Section 1.6). It gave us the precise relationship between temperature and molecular motion: the average kinetic energy of molecule is directly proportional to the temperature, $\langle mv^2/2 \rangle = (3/2)k_B T$. The concepts introduced through kinetic theory could also explain other properties of gases, such as heat conductivity, diffusion and viscosity [1], the so-called **transport properties**. Once the connection between the temperature and energy of an individual molecule was established, the relationship between energy as formulated in thermodynamics and mechanical energy of a molecule became clear. The thermodynamic energy of a system is the sum of all the energies of the molecules. Random molecular motion distributes the total energy of the system into all possible **modes of motion**, i.e. translation, rotation and vibration, and the amount of energy in each mode of motion depends on the temperature. If the average energy of a single molecule is known, then the total energy of the system can be calculated; in turn, the average energy of a molecule is related to the system's temperature. The success of these developments still left one big question unanswered: what is the microscopic explanation of entropy? What is the relationship between entropy and molecular properties? Boltzmann's answer to that question, which has already been introduced in earlier chapters, is '$S = k_B \ln W$'. This fundamental formula opened the way for the formulation of **statistical thermodynamics**, a theory that relates thermodynamic quantities to the statistical properties of molecules.

In this chapter, we introduce the reader to the basic formalism of statistical thermodynamics and illustrate how thermodynamic properties of some simple systems can be related to statistical properties of molecules. We will begin by giving the reader a brief overview of the topic.

In previous chapters the thermodynamic quantities were written as functions of moles N and gas constant R. In this chapter, it is more convenient to use molecular quantities, $\tilde{N}$ the number of particles and the Boltzmann constant k_B. Conversion to N and R may be done by noting that $\tilde{N} = NN_A$ and $R = N_A k_B$. Also, when discussing general statistical thermodynamic concepts that are valid for electrons, atoms or molecule, we shall use the term 'particles'.

Introduction to Modern Thermodynamics   Dilip Kondepudi
© 2008 John Wiley & Sons, Ltd

Ludwig Boltzmann (1844–1906) (Reproduced courtesy of the AIP Emilio Segre Visual Archive, Segre Collection)

17.1 Fundamentals and Overview

On the one hand, quantum mechanics describes the behavior of electrons, atoms and molecules with remarkable success through the concepts of quantum states, quantized energies and energy eigenstates. On the other hand, an equally successful thermodynamics describes the macroscopic behavior of matter in terms of variables such as entropy S, Helmholtz energy F, and chemical potential μ. Statistical thermodynamics relates these two theories. It enables us to calculate thermodynamic quantities such as the Helmholtz energy F, given all the energy states of constituent particles: electrons, atoms or molecules as the case might be.

In quantum theory, particles such as electrons, atoms or molecules are described by their quantum states $|\psi\rangle$. Among these states are energy eigenstates, states with definite energy. Statistical thermodynamics uses these 'energy eigenstates' $|E_k\rangle$, associated with an energy E_k, in which the subscript $k = 1, 2, 3, \ldots$ indexes the quantized energies. There could be several states that have the same energy; the energy level is then said to be 'degenerate'. A **microstate** of a system is the detailed specification of the state of every particle in the system. For a given total energy U there are a large number of different ways in which that energy can be distributed among the particles in the system. In general, there are a large number of microstates that correspond to a given thermodynamic state. Boltzmann's fundamental postulate is that entropy is related to the number of microstates W through

$$\boxed{S = k_B \ln W} \tag{17.1.1}$$

in which the constant k_B is now named after Boltzmann. W is sometimes called the **thermodynamic probability**, a term introduced by Max Planck. In Chapter 3 (Box 3.2) we considered simple examples to illustrate how W is calculated. We will discuss more such examples in the following sections. For a brief overview of statistical thermodynamics, we shall focus on two basic relations that follow when (17.1.1) is applied to systems in thermodynamic equilibrium.

- Statistical thermodynamics uses the concept of statistical ensembles, a large collection of $\tilde{N}$ identical particles or entire systems, to calculate average values. There is an alternative way of expressing (17.1.1). For an ensemble of particles or systems, if P_k is the probability that the particle or system is in state k, then in Section 17.4 we show that S can also written as

$$\boxed{S = -k_B \tilde{N} \sum_k P_k \ln} \tag{17.1.2}$$

- When a system is in thermodynamic equilibrium at a temperature T, the probability $P(E_i)$ that a particle will occupy a state with energy E_i is

$$\boxed{P(E_i) = \frac{1}{q} e^{-E_i/k_B T}} \tag{17.1.3}$$

The term

$$q = \sum_i e^{-E_i/k_B T} \tag{17.1.4}$$

is the **normalization constant**; it is introduced so that $\Sigma_i P(E_i) = 1$, as required by the very definition of probability. Expression (17.1.3) for the probability of a state k is called the **Boltzmann probability distribution**. In many situations, it is found that several distinct states have the same energy. If $g(E_i)$ is the number of states having the same energy E_i, then the probability that a particle has an energy E_i occupying any one of the $g(E_i)$ states is

$$\boxed{P(E_i) = \frac{1}{q} g(E_i) e^{-E_i/k_B T}} \quad q = \sum_i g(E_i) e^{-E_i/k_B T} \tag{17.1.5}$$

$g(E_i)$ is called the **degeneracy** of the energy level E_i.

Statistical thermodynamics of equilibrium systems is based on the fundamental expressions (17.1.2) and (17.1.3). Thus, given the quantum energy levels E_k, and their degeneracies $g(E_k)$, the average value of the energy of a single molecule, which we shall denote by $\langle E \rangle$, is calculated using (17.1.3):

$$\langle E \rangle = \sum_{i=1}^{m} E_i P(E_i) \tag{17.1.6}$$

To calculate the average energy of a system of $\tilde{N}$ particles, an ensemble of systems is used (the reason for using an ensemble of systems is explained in Section 17.4). In this case, the total energy of all the particles U_i takes the place of E_i in (17.1.6), in which $P(U_i)$ is the corresponding probability. The ensemble average $\langle U \rangle = U$, is the energy of the system. The entropy of the system can be calculated using (17.1.2). From these two quantities, the Helmholtz energy $F = U - TS$ and other thermodynamic quantities can be obtained.

In the following sections we shall see that thermodynamic quantities can be obtained from q defined in (17.1.4). Because of its importance, it is given a name, **the partition function.**[*] To be more precise with terminology, q defined above is called the 'single-particle canonical partition function'. The partition function of a system of $\tilde{N}$ particles is usually denoted by Q. For $\tilde{N}$ identical noninteracting particles, q and Q have the following relation:

$$Q = \frac{q^{\tilde{N}}}{\tilde{N}!} \qquad (17.1.7)$$

For interacting particles, Q is a more complicated function of T, V and $\tilde{N}$. Expressing Q as a function of V, T and $\tilde{N}$, and using (17.1.2), one can derive the following general relation between Q and the Helmholtz energy F:

$$F = -k_B T \ln Q(V, T, \tilde{N}) \qquad (17.1.8)$$

From the Helmholtz energy, other thermodynamic quantities can be calculated:

$$\mu(p, T) = \left(\frac{\partial F}{\partial N} \right)_{V,T} \qquad p = -\left(\frac{\partial F}{\partial V} \right)_{T,N} \qquad S = -\left(\frac{\partial F}{\partial T} \right)_{V,N} \qquad (17.1.9)$$

Statistical thermodynamics of a system usually begins with the calculation of the partition function Q. If Q can be obtained in a convenient analytic form, then all thermodynamic quantities can be calculated from it. This is the basic framework of equilibrium statistical thermodynamics. In the following sections we develop this formalism and present illustrative applications.

17.2 Partition Function Factorization

When the total energy of a particle can be written as a sum of independent energies with independent quantum numbers, the partition function can be expressed as a product of partition functions. The total energy of a molecule consists of energies in various types: energies of translation, rotation, vibration and the energies in the electronic and nuclear states. We can write the total energy E as the sum

$$E = E^{\text{trans}} + E^{\text{rot}} + E^{\text{vib}} + E^{\text{elec}} + E^{\text{nuc}} \qquad (17.2.1)$$

in which the superscripts stand for translation, rotation, etc. Each of the energies is quantized and has independent quantum numbers. (Depending on the conditions, the energies may also depend on external factors, such as gravitational and electromagnetic fields, but those terms are not included in the above expression.) The above

---

[*] The letter z is also used for the partition function, because in German the sum (17.1.4) is called Zustandsumme (which means 'state sum').

expression assumes that energy of each type of motion is independent of another. Though this may be a good approximation in many situations, it is not strictly true. For example, the rotational energy of a molecule may depend on its vibrational state; in such situations, one could deal with the combined vibrational–rotational energy levels. For simplicity, we shall assume energy levels of each type of motion have independent quantum numbers. In this case, the single molecule partition function can be factorized:

$$q = \sum_{j,k,l,m,n} g(E_j)g(E_k)g(E_l)g(E_m)g(E_n)e^{-\beta(E_j^{trans}+E_k^{rot}+E_l^{vib}+E_m^{elec}+E_n^{nuc})}$$

$$= \sum_j g(E_j)e^{-\beta E_j^{trans}} \sum_k g(E_k)e^{-\beta E_k^{rot}} \sum_l g(E_l)e^{-\beta E_l^{vib}} \sum_m g(E_m)e^{-\beta E_m^{elec}} \sum_n g(E_n)e^{-\beta E_n^{nuc}}$$

$$(17.2.2)$$

For each molecule, quantum theory gives us the energy levels of each mode of motion. As shown in Box 17.1, the spacing of energy levels increases from translation to rotation to vibration. Translational energy levels are very closely spaced compared with the average thermal energy of a molecule, which is of the order of $k_B T$. The electronic energies generally have a much larger spacing than vibrational energies do. If the ground-state energy is taken to be zero, then the electronic partition function is close to unity.

The energy of the nucleus is also quantized and the spacing is so large that transition from an excited state to the ground state is through the emission of high-energy γ rays or the ejection of α or β particles (the latter being electrons or positrons). Transitions between nuclear states do not occur as a result of thermal collisions between atoms and molecules, so we can assume that the nuclei are in their ground states (except for radioactive nuclei). However, at temperatures that are encountered in the interior of stars, transitions between nuclear states must be considered. Box 17.1 lists commonly used expressions for the energy levels in molecules. With these energy levels, the corresponding partition functions can be calculated.

17.3   The Boltzmann Probability Distribution and Average Values

To illustrate the use of the Boltzmann probability distribution (17.1.5) let us consider $\tilde{N}$ particles whose energy can be any of the m possible values $E_1, E_2, \ldots, E_m$. At equilibrium, let $\tilde{N}_1, \tilde{N}_2, \ldots, \tilde{N}_m$ be the number of particles in these energy levels, which implies $\tilde{N} = \tilde{N}_1 + \tilde{N}_2 \ldots + \tilde{N}_m$. The probability that we will find a particle in energy level E_k is proportional to $\tilde{N}_k$, the number of particles in that state. According to the Boltzmann principle:

$$P(E_k) = \frac{g(E_k)e^{-E_k/k_B T}}{q} = \frac{\tilde{N}_k}{\tilde{N}} \qquad (17.3.1)$$

$\tilde{N}_k$ is often called the **occupation number** of the state with energy E_k. From (17.3.1), it follows that the relative number of particles in energy states E_k and E_l is:

Box 17.1 Energy Levels Associated with Different Types of Motion.

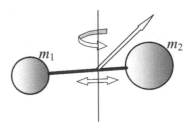

Energy levels of a molecule for various types of motion. Translational energy levels are very closely spaced compared with rotational energies, which are more closely spaced than vibrational energies. The energy level spacing shown is not to scale; these are just meant to give a qualitative idea.

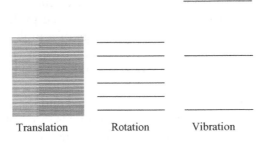

ENERGY LEVELS

- Translational energy levels of a particle of mass m in a box of sides l_x, l_y and l_z (volume $V = l_x l_y l_z$) are specified by quantum numbers n_x, n_y and n_z:

$$E_{n_x, n_y, n_z} = \frac{h^2}{8m}\left(\frac{n_x^2}{l_x^2} + \frac{n_y^2}{l_y^2} + \frac{n_z^2}{l_z^2}\right) \quad n_x, n_y, n_z = 1, 2, 3, \ldots$$

in which $h = 6.626 \times 10^{-34}$ Js is Planck's constant.

- Energy levels for rotation about an axis with moment of inertia I are specified by the quantum number L:

$$E_L = \frac{\hbar^2}{2I}L(L+1) \quad L = 0, 1, 2, 3, \ldots \quad g(E_L) = 2L+1 \quad \text{and} \quad \hbar = h/2\pi$$

- Vibrational energy levels of a diatomic molecule with reduced mass $\mu = m_1 m_2/(m_1 + m_2)$ and force constant k are specified by the quantum number v:

$$E_v = \hbar\omega\left(v + \frac{1}{2}\right) \quad \omega = \sqrt{\frac{k}{\mu}} \quad v = 0, 1, 2, 3, \ldots$$

$$\frac{\tilde{N}_k}{\tilde{N}_l} = \frac{g(E_k)}{g(E_l)} e^{-(E_k - E_l)/k_B T} \tag{17.3.2}$$

Thus, the ratio of occupation numbers is a function of the difference in the energies and the ratio of the corresponding degeneracies.

The average value of a variable or physical property can be calculated using the Boltzmann probability distribution. We shall denote the average value of a quantity X by $\langle X \rangle$. Thus, the average energy of a single particle $\langle E \rangle$ is

$$\langle E \rangle = \frac{\sum_{k=1}^{m} E_k \tilde{N}_k}{\tilde{N}} = \sum_{k=1}^{m} E_k P(E_k) \tag{17.3.3}$$

The total energy of all particles is $U = \tilde{N}\langle E \rangle$.

More generally, the average values of any physical property X can be calculated if its value in the state $|E_k\rangle$, which we denote by X_k, is known.

$$\langle X \rangle = \frac{\sum_{k=1}^{m} X_k \tilde{N}_k}{\tilde{N}} = \sum_{k=1}^{m} X_k P(E_k) \tag{17.3.4}$$

The average value of any function of X, $f(X)$, can similarly be calculated using

$$\langle f(X) \rangle = \frac{\sum_{k=1}^{m} f(X_k) \tilde{N}_k}{\tilde{N}} = \sum_{k=1}^{m} f(X_k) P(E_k) \tag{17.3.5}$$

For example, the average value of $\langle E^2 \rangle$ is

$$\langle E^2 \rangle = \sum_{k=1}^{m} E_k^2 P(E_k) \tag{17.3.6}$$

The standard deviation, ΔE, in E is defined by $(\Delta E)^2 \equiv \langle (E - \langle E \rangle)^2 \rangle$. An elementary calculation shows that

$$(\Delta E)^2 \equiv \langle (E - \langle E \rangle)^2 \rangle = \langle E^2 \rangle - \langle E \rangle^2 \tag{17.3.7}$$

In this manner, statistical quantities such as the average and standard deviation of physical variables associated with an equilibrium system can be calculated. When an ensemble of systems is considered, the energy E_k is replaced by the total energy U_i.

17.4 Microstates, Entropy and the Canonical Ensemble

A macroscopic thermodynamic state of a system corresponds to a large number of 'microstates'. For instance, if the total energy of an ensemble of $\tilde{N}$ particles

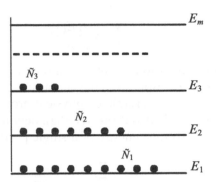

Figure 17.1 An ensemble of $\tilde{N}$ particles distributed in m energy levels. $\tilde{N}_k$ particles are in energy level E_k. $P_k = \tilde{N}_k/\tilde{N}$ is the probability that a particle will occupy a state with energy E_k. The entropy per particles

$$(S/\tilde{N}) = -k_B \Sigma_k P_k \ln P_k$$

(molecules, electrons, etc.) in a volume V is specified, then this energy can be distributed among the $\tilde{N}$ particles in a number of ways. Each distinct distribution of the energy among the $\tilde{N}$ particles corresponds to a microstate. We now show how expression (17.1.2) is derived from the fundamental formula

$$S = k_B \ln W \qquad (17.4.1)$$

in which W is the number of microstates corresponding to the given thermodynamic state (also called a macrostate). To illustrate how W is calculated, let us consider an ensemble of $\tilde{N}$ particles each of which can be in any one of the m states. These could be 'numbered particles' on a crystal lattice. A microstate specifies the energy state of each particle. As in the previous sections, we assume $\tilde{N}_k$ particles are in a state with energy E_k. The number of microstates W is the number of distinct ways in which the $\tilde{N}$ particles can be distributed in m states. W can be calculated as follows (Figure 17.1). First, we note that if a particle, say particle 26, is in energy state E_5 and another particle, say particle 14, is in energy state E_2, then an interchange of these two particles gives a different microstate; but if both particles 26 and 14 are in the same energy state, say E_5, then interchanging them does not give a new microstate. Thus, only permutations that do not correspond to interchange of particles with the same energy E_k correspond to distinct microstates. The number of all possible permutations is $\tilde{N}!$ The number of permutations of particles with the same energy E_k is $\tilde{N}_k!$ Thus, the total number of microstates W is given by

$$W = \frac{\tilde{N}!}{\tilde{N}_1! \tilde{N}_2! \dots \tilde{N}_m!} \qquad (17.4.2)$$

The entropy S is

$$S = k_\mathrm{B} \ln W = k_\mathrm{B} \ln\left(\frac{\tilde{N}!}{\tilde{N}_1! \tilde{N}_2! \dots \tilde{N}_m!} \right) \qquad (17.4.3)$$

We assume $\tilde{N}_k$ is large. Then, for the term $\ln(\tilde{N}_k!)$ we can use Stirling's approximation (see Appendix 17.1):

$$\ln(a!) \approx a \ln a - a \qquad (17.4.4)$$

Using this approximation one can show that (Exercise 17.1)

$$\ln W = -\sum_k \tilde{N}_k \ln\left(\frac{\tilde{N}_k}{\tilde{N}} \right) = -\tilde{N} \sum_k \left(\frac{\tilde{N}_k}{\tilde{N}} \right) \ln\left(\frac{\tilde{N}_k}{\tilde{N}} \right) \qquad (17.4.5)$$

Since $\tilde{N}_k/\tilde{N} = P_k$, *the probability of occupying a state with energy E_k*, we immediately see that

$$S = k_\mathrm{B} \ln W = -k_\mathrm{B} \tilde{N} \sum_k \left(\frac{\tilde{N}_k}{\tilde{N}} \right) \ln\left(\frac{\tilde{N}_k}{\tilde{N}} \right) = -k_\mathrm{B} \tilde{N} \sum_k P_k \ln P_k \qquad (17.4.6)$$

which is (17.1.2) if we replace $P(E_k)$ with P_k. We derived (17.4.6), the relationship between entropy and probability, from (17.4.1) without any assumption about the system being in equilibrium. Hence, this definition of entropy is valid for nonequilibrium systems as well. Sometimes it is considered the definition of statistical entropy and used in contexts other than thermodynamics, such as information theory.

In Chapter 5 we noted that the entropy reaches its maximum value when the energy of a system U is constant. Now, we show that the Boltzmann equilibrium distribution (17.1.3) maximizes S when the total energy is constant. In other words, we show that, with the constraint of fixed total energy, S will reach its maximum value when $P_k \propto e^{-\beta E_k}$. This result can be obtained by using Lagrange's method of finding the maximum of a function subject to constraints. Our constraints are the constancy of total energy E and the total number of particles $\tilde{N}$. They can be expressed as

$$E = \sum_k E_k \tilde{N}_k = \tilde{N} \sum_k E_k (\tilde{N}_k/\tilde{N}) = \tilde{N} \sum_k E_k P_k \qquad (17.4.7a)$$

$$\tilde{N} = \sum_k \tilde{N}_k \qquad (17.4.7b)$$

in which we have used $P_k = \tilde{N}_k/\tilde{N}$. Lagrange's method now stipulates that, to maximize $-\sum_k P_k \ln P_k$ with the constraints (17.4.7), one needs to maximize the function

$$I = -\sum_k P_k \ln P_k + \lambda\left(E - \tilde{N} \sum_k E_k P_k \right) + \xi\left(\tilde{N} - \sum_k \tilde{N}_k \right) \qquad (17.4.8)$$

in which λ and ξ are arbitrary constants whose values can be determined by additional requirements. Now it is straightforward to see that the condition $\partial I/\partial P_k = 0$ leads to the relation

$$\ln P_k = -\lambda \tilde{N} E_k + 1 - \xi$$

As a function of E_k, we can now write

$$P_k = Ce^{-\beta E_k} \qquad\qquad (17.4.9)$$

in which $C = \exp(1 - \xi)$ and $\beta = \lambda \tilde{N}$. This is essentially the Boltzmann distribution (17.1.2) once we identify $\beta = 1/k_B T$. That β must be $1/k_B T$ can deduced by calculating the average kinetic energy of a particle and equating it to $3k_B T/2$, as required by kinetic theory. Since $\Sigma_k P_k = 1$, we see that $C = 1/q$ by comparing (17.4.9) with (17.1.3). Equation (17.4.9) is valid for every state that has energy E_k. Taking into account the degeneracy $g(E_k)$, the probability that the system will occupy any one of the $g(E_k)$ states with energy E_k can be written as

$$P(E_i) = \frac{1}{q} g(E_i) e^{-E_i/k_B T} \qquad\qquad (17.4.10)$$

If each state with energy E_i is counted separately, then the degeneracy factor need not be included. *In expression (17.4.6) the P_k values are the probabilities of occupying a particular state with energy E_k.*

THE CANONICAL ENSEMBLE

In the following sections we will see that thermodynamic quantities of a system are calculated using the concept of a statistical ensemble. In deriving (17.4.10) it was assumed that the number of particles $\tilde{N}_k$ occupying a state k is large. This is a good assumption for rotational and vibrational states of molecules, but it is not valid for the occupation of translational states. Translational energies are very closely spaced. At ordinary temperatures, the average kinetic energy $3k_B T/2$ is much larger than the energy spacing of the translational quantum states. For example, if we assume $T = 298\,\text{K}$, then a simple calculation for N_2 gas in a cube of side $10\,\text{cm}$ shows (Example 17.1) that there are roughly 10^{29} states with energy less than $3k_B T/2$. At ordinary pressures, this is much larger than the number of N_2 molecules; hence, most translational states are unoccupied. Thus, we cannot assume that $\tilde{N}_k$ is large. In such cases we use the concept of an ensemble of systems. The energy U of each system in the ensemble is itself subject to fluctuations, and in that respect is similar to the energy of a single particle. The system's energy can take values U_1, U_2, ... with probabilities P_1, P_2, ... ; that is, the probability $P(U_k)$ that the total energy U of a system in the ensemble has a particular value U_k can be defined just as $P(E_k)$ was defined for a single particle. It is assumed that the thermodynamic properties of a single system are the same as the average properties of the ensemble.

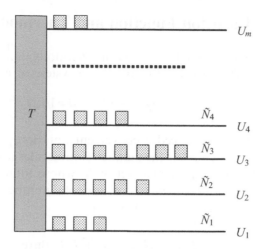

Figure 17.2 A canonical ensemble is a large set of $\tilde{N}$ identical systems in contact with a temperature reservoir. The system's total energy U can take many possible values, U_1, $U_2, \ldots, U_m$. At any instant, the ensemble of systems is distributed among the possible energy states, $\tilde{N}_k$ systems with energy U_k. $P_k = \tilde{N}_k/\tilde{N}$ is the probability that a system's energy will be U_k. The entropy of the system $S = -k_B \sum_k P_k \ln P_k$

One such ensemble is the **canonical ensemble** shown in Figure 17.2. It consists of a large number $\tilde{N}$ of identical systems in contact with a thermal reservoir at a temperature T. In this figure, $\tilde{N}_k$ is the number of systems (not particles) with total energy U_k. Each system's energy can take values U_k with probabilities $P(U_k)$. The thermodynamic energy of a system is the average energy calculated using this ensemble. With this formalism, we see that all the calculations done above for a single particle could be carried out for the canonical ensemble with the following result:

$$P(U_k) = \frac{1}{Q} e^{-U_i/k_B T} \quad Q = \sum_i e^{-U_i/k_B T} \tag{17.4.11}$$

We note here that U_i is the total energy of all the particles of the system at a given temperature T. The partition function Q of a canonical ensemble is called **canonical partition function**. The entropy of a system is

$$S = -k_B \sum_k P(U_k) \ln P(U_k) \tag{17.4.12}$$

In the following section, we shall see how thermodynamic quantities can be obtained from these two expressions.

17.5   Canonical Partition Function and Thermodynamic Quantities

There is a general scheme for calculating thermodynamic quantities from the partition functions. The partition function for a system of $\tilde{N}$ particles is

$$Q = \sum_i e^{-U_i \beta} \qquad \beta = 1/k_B T \tag{17.5.1}$$

in which we have introduced a convenient notation $\beta = 1/k_B T$. The total energy $U_i = \sum_k \tilde{N}_k^i E_k$, in which $\tilde{N}_k^i$, is the number of molecules in state k with energy E_k. The superscript i indexes a particular set of $\tilde{N}_k^i$ whose total energy adds up to U_i. The value of each $\tilde{N}_k^i$ can vary from 0 to $\tilde{N}$, the total number of molecules in the system, but $\sum_k \tilde{N}_k^i = \tilde{N}$. The entropy of the system is

$$
\begin{aligned}
S &= -k_B \sum_i P(U_i) \ln P(U_i) = -k_B \sum_i P(U_i) \ln (e^{-\beta U_i}/Q) \\
&= -k_B \sum_i P(U_i)(-\beta U_i - \ln Q) = k_B \beta \sum_i P(U_i) U_i + k_B \ln Q \sum_i P(U_i) \quad (17.5.2)\\
&= \frac{U}{T} + k_B \ln Q
\end{aligned}
$$

where we have used $U = \sum_i P(U_i) U_i$ and $\sum_i P(U_i) = 1$. From (17.5.2), it follows that $F \equiv U - TS = -k_B T \ln Q$. When we compute Q explicitly in the following sections, we will see that Q is a function of the system volume V, the temperature T and $\tilde{N}$. Making this explicit, we write

$$\boxed{F(V, T, \tilde{N}) = -k_B T \ln Q(V, T, \tilde{N})} \tag{17.5.3}$$

The total energy U can also be calculated directly from the partition function Q. It is easy to verify that

$$\boxed{U = -\frac{\partial \ln Q}{\partial \beta}} \tag{17.5.4}$$

Using (17.5.3) and (17.5.4), other thermodynamic quantities could be calculated. For example, the chemical potential $\mu = (\partial F/\partial N)_{V,T}$ and $p = -(\partial F/\partial V)_{N,T}$.

17.6   Calculating Partition Functions

For simple systems, such as an ideal gas of noninteracting particles and the vibrational and rotational states of a diatomic molecule, the partition functions can be calculated without much difficulty. In these cases, the partition function Q of the entire system can be related to the partition function of a single particle or molecule. The calculation of the translational partition function is done as follows.

TRANSLATIONAL PARTITIONS FUNCTION

For a gas of $\tilde{N}$ identical noninteracting particles the total energy $U_i = \Sigma_k \tilde{N}_k^i E_k$, in which E_k is the translational energy of state and $\tilde{N}_k^i$ are the number of particles in that state. We have already noted (Section 17.4) that translational states are sparsely occupied. Therefore, most of the $\tilde{N}_k^i$ are zero and the partition function for the translational states is a sum that looks like

$$Q_{\text{trans}} = \sum_i e^{-\beta U_i} = \sum_i e^{-\beta \Sigma_k \tilde{N}_k^i E_k}$$
$$= e^{-\beta(1 \cdot E_1 + 0 \cdot E_2 + 0 \cdot E_3 + 1 \cdot E_4 + \ldots)} + e^{-\beta(0 \cdot E_1 + 0 \cdot E_2 + 1 \cdot E_3 + 1 \cdot E_4 + \ldots)} + \ldots \qquad (17.6.1)$$

The terms in this sum can be interpreted as terms in a single-particle partition function. Each of the factors $e^{-\beta E_k}$ is a term in the single-particle partition function $q = \Sigma_k e^{-\beta E_k}$. Since the number of available translational states is much larger than the number of particles, an overwhelming number of terms correspond to only one particle in a translational state. Hence, the right-hand side of (17.6.1) can be approximated as the product of $\tilde{N}$ partition functions $q = \Sigma_k e^{-\beta E_k}$. However, as explained in Box 17.2, such a product will have permutations between particles that are not in Q_{trans}. The overcounting is corrected by dividing $q^{\tilde{N}}$ by $\tilde{N}!$. This leads to the relation

$$Q_{\text{trans}} = \frac{q_{\text{trans}}^{\tilde{N}}}{\tilde{N}!} \qquad (17.6.2)$$

Our task now is to calculate the single-particle translational partitions function q_{trans}. As shown in Box 17.1, for a gas of particles with mass m in a cubical box of sides l_x, l_y, l_z, the translational states are specified by the quantum numbers n_x, n_y, n_z with energies

$$E_{n_x,n_y,n_z} = \frac{h^2}{8m}\left[\left(\frac{n_x}{l_x}\right)^2 + \left(\frac{n_y}{l_y}\right)^2 + \left(\frac{n_z}{l_z}\right)^2\right]$$

To obtain q_{trans}, the following sum is to be evaluated:

$$q_{\text{trans}} = \sum_{n_x=1}^{\infty}\sum_{n_y=1}^{\infty}\sum_{n_z=1}^{\infty} e^{-\beta C\left[(n_x/l_x)^2 + (n_y/l_y)^2 + (n_z/l_z)^2\right]} = \sum_{n_x=1}^{\infty} e^{-\beta C(n_x/l_x)^2}\sum_{n_y=1}^{\infty} e^{-\beta C(n_y/l_y)^2}\sum_{n_z=1}^{\infty} e^{-\beta C(n_z/l_z)^2}$$

in which $C = h^2/8m$. Each of these sums can be approximated by an integral because the energy level spacing is very small. The sum over n_x can be written as the integral (which is evaluated using the table of integrals in Appendix 17.1):

$$\sum_{n_x=1}^{\infty} e^{-\beta C(n_x/l_x)^2} = \int_0^{\infty} e^{-\beta C(n_x/l_x)^2}\,\mathrm{d}n_x = \frac{1}{2}\left(\frac{\pi l_x^2}{\beta C}\right)^{1/2} = \frac{l_x}{h}(2\pi m k_B T)^{1/2}$$

Box 17.2 Relation Between q and Q

The approximation $Q_{\text{trans}} = q_{\text{trans}}^{\tilde{N}}/\tilde{N}!$ can be made clear by considering 100 translational states occupied by two identical particles. Every pair of energy states that the two particles occupy corresponds to a state of the system. In identifying distinct system states, every pair of energies should be counted only once; exchanging the two particles does not result in a different system state because the particles are identical.

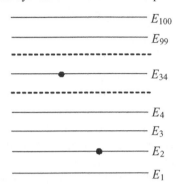

For two particles and 100 states, there are $100 \times 90/2! = 4500$ system states in which the two particles occupy different energy states, but there are only 100 in which both particles are in the same system state. The corresponding terms in Q are

$$Q = \sum_{i>k}^{100} \sum_{k=1}^{100} e^{-\beta(E_i + E_k)} + \sum_{k=1}^{100} e^{-\beta 2 E_k} \tag{A}$$

In the first term, $i > k$ assures that each pair of energy states is included only once. The single-particle partition function $q = \Sigma_{k=1}^{100} e^{-\beta E_k}$. Comparing Q with

$$q^2 = \sum_{i=1}^{100} e^{-\beta E_i} \sum_{k=1}^{100} e^{-\beta E_k} = \sum_{i=1}^{100} \sum_{k=1}^{100} e^{-\beta(E_i + E_k)}$$

we see that, when $i \neq k$, each pair of E_i and E_k occurs twice in q^2 but only once in Q. In q^2, exchange of particles is counted as a different system state. We compensate for this overcounting by dividing q^2 by 2! and get

$$\frac{q^2}{2!} = \frac{1}{2} \sum_{i=1}^{100} \sum_{k=1}^{100} e^{-\beta(E_i + E_k)} = \sum_{i>k}^{100} \sum_{k=1}^{100} e^{-\beta(E_i + E_k)} + \frac{1}{2} \sum_{k=1}^{100} e^{-\beta 2 E_k} \tag{B}$$

Comparing (A) and (B), we see that they differ only in the second term, which corresponds to two particles in the same energy state. Since such states are far fewer than those in which the two particles are in different energy states, the difference between (A) and (B) is not significant. The above argument can be extended to $\tilde{N}$ particles by replacing 2! with $\tilde{N}!$. Thus, when the number of available states far exceeds the number of particles, $Q_{\text{trans}} = q_{\text{trans}}^{\tilde{N}}/\tilde{N}!$ is a very good approximation.

When similar integrals for n_y and n_z are evaluated, the partition function can be written as

$$q_{trans} = \frac{l_x l_y l_z}{h^3}(2\pi m k_B T)^{3/2} = \frac{V}{h^3}(2\pi m k_B T)^{3/2} \tag{17.6.3}$$

in which volume of the system $V = l_x l_y l_z$. The translational partition function of the gas is thus

$$\boxed{Q_{trans} = \frac{1}{\tilde{N}!}\left[\frac{V}{h^3}(2\pi m k_B T)^{3/2}\right]^{\tilde{N}}} \tag{17.6.4}$$

This expression can be given another interpretation leading to another form in which Q_{trans} is often written. Since the average kinetic energy of a gas particle is $3k_B T/2$, the average momentum of particles at temperature T is $(3mk_B T)^{1/2}$. The de Broglie wavelength ($\lambda = h/p$) associated with this momentum equals $h/(3mk_B T)^{1/2}$. For this reason, a **thermal wavelength** $\Lambda = h/(2\pi m k_B T)^{1/2}$ is defined (replacing 3 with 2π). In terms of Λ the partition function Q_{trans} can be written in the following simple form:

$$\boxed{Q_{trans} = \frac{1}{\tilde{N}!}\left[\frac{V}{\Lambda^3}\right]^{\tilde{N}} \qquad \Lambda = \frac{h}{\sqrt{2\pi m k_B T}}} \tag{17.6.5}$$

THERMODYNAMIC QUANTITIES

For particles that have no internal structure or for particles whose internal energy at the temperature of interest can be neglected, all the energy is translational (kinetic energy). A monatomic ideal gas is an example. The Helmholtz energy of a gas of such particles is

$$F(V, T, \tilde{N}) = -k_B T \ln Q_{trans} = -k_B T \tilde{N} \ln\left[\frac{V}{h^3}(2\pi m k_B T)^{3/2}\right] + k_B T \ln \tilde{N}!$$

Using Stirling's approximation, $\ln(\tilde{N}!) \simeq \tilde{N}\ln\tilde{N} - \tilde{N}$, the above expression can be written as

$$\begin{aligned}
F(V, T, \tilde{N}) &= -k_B T\left\{\tilde{N} \ln\left[\frac{V}{h^3}(2\pi m k_B T)^{3/2}\right] - \tilde{N} \ln \tilde{N} + \tilde{N}\right\}\\
&= -k_B \tilde{N} T\left\{\ln\left[\frac{V}{\tilde{N}h^3}(2\pi m k_B T)^{3/2}\right] + 1\right\}
\end{aligned} \tag{17.6.6}$$

Since the gas constant $R = k_B N_A$ and amount in moles $N = \tilde{N}/N_A$, the above F can be expressed as

$$F(V, T, N) = -RNT\left\{\ln\left[\frac{V}{NN_A h^3}(2\pi m k_B T)^{3/2}\right] + 1\right\} \tag{17.6.7}$$

Other thermodynamic quantities can now be obtained from F. For example, since $p = -(\partial F/\partial V)_{T,N}$, it follows that

$$p = -\left(\frac{\partial F}{\partial V}\right)_{T,N} = \frac{RTN}{V} \tag{17.6.8}$$

which is the ideal gas equation. Similarly, since entropy $S = -(\partial F/\partial T)_{V,N}$, a simple calculation shows that the ideal gas entropy is

$$S = N\left\{R\ln\left[\frac{V}{NN_A h^3}(2\pi m k_B T)^{3/2}\right] + \frac{5R}{2}\right\} \tag{17.6.9}$$

This expression was obtained in 1911 by O. Sackur and H. Tetrode in the early stages of the development of quantum theory. It is called **Sackur–Tetrode equation** for the entropy of an ideal gas. It shows us that quantum theory (Planck's constant being its signature) gives the absolute value of entropy without any arbitrary constants. In Chapter 3 we derived the following expression for the entropy of an ideal gas:

$$S(V, T, N) = N\left[s_0 + R\ln\left(\frac{V}{N}\right) + C_V \ln T\right] \tag{17.6.10}$$

in which s_0 was an undetermined constant. Comparing (17.6.9) and (17.6.10), we see that $C_V = 3R/2$ and

$$s_0 = R\ln\left[\frac{(2\pi m k_B)^{3/2}}{N_A h^3}\right] + \frac{5R}{2}$$

We have noted that the energy U of the system can be obtained from Q using relation (17.5.4), $U = -(\partial \ln Q/\partial \beta)$, in which $\beta = 1/k_B T$. Because $\ln Q = -F/k_B T$, using (17.6.6), Q can be expressed in terms of β thus:

$$\ln Q = \tilde{N}\left\{\ln\left[\frac{V}{\tilde{N}h^3}(2\pi m/\beta)^{3/2}\right] + 1\right\}$$

From this, it follows that the energy of an ideal gas of particles whose energy is entirely translational is

$$U = -\frac{\partial \ln Q}{\partial \beta} = \frac{3}{2}\tilde{N}k_B T = \frac{3}{2}NRT \tag{17.6.11}$$

From the fundamental quantities U and S, all thermodynamic quantities of an ideal gas of structureless particles are obtained.

ROTATIONAL PARTITION FUNCTION

For molecules, we must consider energy and entropy associated with rotational motion. At ordinary temperatures, a large number of rotational states above the lowest energy state are occupied by molecules (this can be seen by comparing $k_B T$ with rotational energy levels). For simplicity, we consider a diatomic molecule whose atoms have masses m_1 and m_2, as shown in Box 17.1. Since the rotational energies are $E_L = (\hbar^2/2I)L(L + 1)$ with degeneracy $g(E_L) = 2L + 1$, the single-molecule partition function is

$$q_{rot} = \sum_{L=0}^{\infty}(2L+1)e^{-\beta(\hbar^2/2I)L(L+1)} \qquad (17.6.12)$$

For diatomic molecules with masses m_1 and m_2, the reduced mass μ is defined as

$$\mu = \frac{m_1 m_2}{m_1 + m_2} \qquad (17.6.13)$$

If the distance between the two nuclei (bond length) is R, then the moment of inertia I is given by

$$I = \mu R^2 \qquad (17.6.14)$$

To compare the rotational energy level spacing with $k_B T$, a characteristic temperature $\theta_{rot} \equiv \hbar^2/2Ik_B$ is defined. Then the rotational partition function q_{rot} is written as

$$q_{rot} = \sum_{L=0}^{\infty}(2L+1)e^{-L(L+1)\theta/T} \qquad (17.6.15)$$

Using bond length data, and assuming R equals bond length, the moment of inertia I and θ_{rot} can be calculated. For H_2 it is found that $\theta_{rot} = 87.5\,K$, and $\theta_{rot} = 2.1\,K$ for O_2. At very low temperature, i.e. when $T \ll \theta_{rot}$, this sum can be approximated by

$$q_{rot} = 1 + 3e^{-2\theta_{rot}/T} + 5e^{-6\theta_{rot}/T} + \ldots \qquad (17.6.16)$$

At high temperature, i.e. when $T \gg \theta_{rot}$, the sum (17.6.15) may be approximated by the following integral:

$$
\begin{aligned}
q_{rot} &= \int_0^{\infty}(2L+1)e^{-L(L+1)\theta_{rot}/T}dL \\
&= \int_0^{\infty}e^{-L(L+1)\theta/T}d[L(L+1)] = \frac{T}{\theta} = \frac{2Ik_B T}{\hbar^2}
\end{aligned} \qquad (17.6.17)
$$

For diatomic molecules with identical atoms, such as H_2 or N_2, the quantum theory of identical particles stipulates that only half the rotational states are allowed. Hence, a factor of 2 has to be introduced in the denominator of the above expression. In general, when identical atoms are present in a molecule, a **symmetry number** σ must be included in the expression for the partition function. Thus, the general expression for the partition function for a rotation around a given axis with moment of inertia I is

$$q_{rot} = \frac{2Ik_B T}{\sigma\hbar^2} \qquad (17.6.18)$$

The symmetry number σ for a larger molecule is determined by the symmetries of the molecule. It is equal to the number of proper rotations, including the identity, in the symmetry group of the molecule.

VIBRATIONAL PARTITION FUNCTION

Molecules also have vibrational motions that stretch and bend bonds. Each vibration is associated with frequency $v = \omega/2\pi$. Box 17.1 lists expressions for the energy levels for the vibrational motion:

$$E_v = \hbar\omega\left(v + \frac{1}{2}\right) \quad v = 0, 1, 2, \ldots \tag{17.6.19}$$

Using this expression, the partition function for vibrational energies can easily be calculated because the energy levels are equally spaced. We shall assume that the degeneracy of the energy levels is 1. Then, the vibrational partition function is

$$q_{\text{vib}} = \sum_{v=0}^{\infty} e^{-\beta\hbar\omega[v+(1/2)]} = e^{-\beta\hbar\omega/2}\sum_{v=0}^{\infty} x^v$$

where $x = e^{-\beta\hbar\omega}$. Since $x < 1$, the series on the right-hand side can be summed:

$$q_{\text{vib}} = e^{-\beta\hbar\omega/2}\sum_{v=0}^{\infty} x^v = e^{-\beta\hbar\omega/2}\frac{1}{1-x}$$

Thus, the single-molecule vibrational partition function is

$$q_{\text{vib}} = e^{-\beta\hbar\omega/2}\frac{1}{1-e^{-\beta\hbar\omega}} \tag{17.6.20}$$

At ordinary temperatures, the level spacing between vibrational energy states is generally larger than the thermal energy k_BT. Hence, only very few energy states higher than the ground state are occupied by molecules. As was done for rotational states, this aspect can be quantified by defining a characteristic temperature $\theta_{\text{vib}} \equiv \hbar\omega/k_B$. Then the partition function (17.6.20) can be written as

$$q_{\text{vib}} = e^{-\theta_{\text{vib}}/2T}\frac{1}{1-e^{-\theta_{\text{vib}}/T}} \tag{17.6.21}$$

The characteristic temperatures for some diatomic molecules are:[*]

$$\begin{array}{cccccccc} & H_2 & N_2 & O_2 & Cl_2 & HCl & CO & NO \\ \theta_{\text{vib}}/K & 6210 & 3340 & 2230 & 810 & 4140 & 3070 & 2690 \end{array} \tag{17.6.22}$$

Thus, at T in the range 200–400 K, only a few of the lowest vibrational states are occupied. The characteristic temperatures for electronic states are even higher, so electronic states are mostly in their lowest or ground state.

Combining all the partition functions for a diatomic molecule, we can write

$$q = q_{\text{trans}}q_{\text{rot}}q_{\text{vib}} = \frac{V}{h^3}(2\pi m k_B T)^{3/2}\frac{2Ik_BT}{\sigma\hbar^2}e^{-\theta_{\text{vib}}/2T}\frac{1}{1-e^{-\theta_{\text{vib}}/T}} \quad \text{and} \quad Q = \frac{q^{\tilde{N}}}{\tilde{N}} \tag{17.6.23}$$

[*] Source: T.L Hill, *Introduction to Statistical Thermodynamics*. 1960, Reading, MA: Addison-Wesley.

From this partition function, thermodynamic quantities U, p, μ, etc. can be calculated (see Exercises). The total energy of the system is the sum of energies in each mode of motion $U = U^{\text{trans}} + U^{\text{rot}} + U^{\text{vib}} + U^{\text{elec}}$. The heat capacity $C_V = (\partial U/\partial T)_V$. By expressing U as the sum of energies, we can know the contribution of each of the modes of motion, i.e. translation, rotation, vibration, etc., to the heat capacity C_V.

17.7 Equilibrium Constants

The formalism of statistical thermodynamics can also be used to relate equilibrium constants of chemical reactions to partitions functions. In doing so, we relate molecular energies to equilibrium constants. Let us consider the simple reaction

$$X \rightleftharpoons Y \tag{17.7.1}$$

At equilibrium, the chemical potentials of X and Y are equal. We use the subscripts X and Y to represent the quantities for the two species. The chemical potential of X is $\mu_X = (\partial F_X/\partial N_X)_{T,V}$; and since $F_X = -k_B T \ln Q_X$, in which $Q_X = q_X^{\tilde{N}_X}/\tilde{N}_X!$, we can establish a relationship between the q_X and μ_X. Here, $\tilde{N}$ is the number of molecules and N is the amount in moles. Since Q is expressed as a function of $\tilde{N}$, we note $\mu_X = (\partial F_X/\partial N_X) = (\partial F_X/\partial \tilde{N}_X)(\partial \tilde{N}_X/\partial N) = N_A(\partial F_X/\partial \tilde{N}_X)$.

When considering a system of reactants and products that interconvert, care must be taken to use the same scale of energy for all molecules when computing partition functions. In the calculations of q presented in the previous sections, generally the zero of energy was taken to be the lowest energy or ground state of that molecule. When more than one molecule is involved, their energies must be measured using a common zero. The lowest energy of a molecule can then have a nonzero value with respect to the common zero. As shown in Figure 17.3, the lowest energy states of X and Y can be different. We shall use E_X^0 and E_Y^0 to represent the lowest energies of the X and Y respectively in the common energy scale. This means that the energies

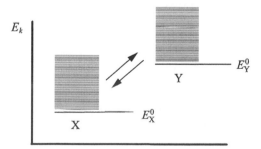

Figure 17.3 Energy levels of two molecules X and Y in a common energy scale. E_X^0 and E_Y^0 are the ground states in a common energy scale

of X will all get an additive term E_Y^0 and this in turn adds a factor $\exp(-\beta E_X^0)$ to q_X. Thus, with respect to the common zero of energy:

$$Q_X = \frac{(q_X e^{-\beta E_X^0})^{\tilde{N}_X}}{\tilde{N}_X!} = \frac{q_X^{\tilde{N}_X}}{\tilde{N}_X!} e^{-\beta \tilde{N}_X E_X^0} \tag{17.7.2}$$

The Helmholtz energy F is

$$F_X = -k_B T \ln Q_X = -k_B T (\tilde{N}_X \ln q_X - \tilde{N}_X \ln \tilde{N}_X + \tilde{N}_X - \beta \tilde{N}_X E_X^0) \tag{17.7.3}$$

and a simple calculation shows that

$$\mu_X = \left(\frac{\partial F_X}{\partial N_X}\right)_{T,V} = N_A \left(\frac{\partial F_X}{\partial \tilde{N}_X}\right)_{T,V} = -N_A k_B T \left[\ln\left(\frac{q_X}{\tilde{N}_X}\right) - \beta E_X^0\right] \tag{17.7.4}$$

This expression relates the chemical potential to the partition function and the number of molecules $\tilde{N}$. We can invert this equation and write

$$\tilde{N}_X = q_X e^{(\mu - N_A E_X^0)/RT} \tag{17.7.5}$$

or in units of moles of X:

$$N_X = \frac{q_X}{N_A} e^{(\mu - U_{0X})/RT} \tag{17.7.6}$$

in which $U_{0X} = N_A E_X^0$.

As a side remark, we note here that for a monatomic ideal gas $q_X = (V/h^3)(2\pi m k_B T)^{3/2}$. Using this expression in (17.7.6), we find

$$\frac{N_X}{V} = \frac{(2\pi m k_B T)^{3/2}}{N_A h^3} e^{(\mu - U_{0X})/RT} = \frac{1}{N_A \Lambda^3} e^{(\mu - U_{0X})/RT} \tag{17.7.7}$$

Thus, the molar density is related to the chemical potential. Equation (17.7.7) is the same as relation (12.6.4) if we identify $z(T)$ in (12.6.4) with $(1/N_A \Lambda^3)$ and U_0 with $M_X c^2$. When the chemical potential is zero, the molar density is a function of T only.

For the reaction $X \rightleftharpoons Y$, let us assume when equilibrium is reached that the moles of X and Y are $N_{X,eq}$ and $N_{Y,eq}$ respectively. Using (17.7.4) and equating the chemical potentials of the two species, we obtain

$$\mu_X = -RT\left[\ln\left(\frac{q_X}{N_{X,eq} N_A}\right) - \beta E_X^0\right] = \mu_Y = -RT\left[\ln\left(\frac{q_Y}{N_{Y,eq} N_A}\right) - \beta E_Y^0\right]$$

This expression can be rewritten as

$$\frac{N_{Y,eq}}{N_{X,eq}} = \frac{q_Y/N_A}{q_X/N_A} e^{-(U_{0Y} - U_{0X})/RT} \tag{17.7.8}$$

in which $U_{0X} = N_A E_X^0$ and $U_{0Y} = N_A E_Y^0$. Since the equilibrium concentrations $[X]_{eq} = N_{X,eq}/V$ and $[Y]_{eq} = N_{Y,eq}/V$, we can relate the equilibrium constant $K_c \equiv [Y]_{eq}/[X]_{eq}$ to the partition functions:

$$K_c = \frac{[Y]_{eq}}{[X]_{eq}} = \frac{N_{Y,eq}/V}{N_{X,eq}/V} = \frac{q_Y/N_A V}{q_X/N_A V} e^{-\Delta U_0/RT} \tag{17.7.9}$$

in which $\Delta U_0 = U_{0Y} - U_{0X}$ is the difference in the ground-state energies of the reactants and products. The above result can be generalized to the reaction

$$aX + bY \rightleftharpoons cZ + dW \tag{17.7.10}$$

$$K_c = \frac{[Z]_{eq}^c [W]_{eq}^d}{[X]_{eq}^a [Y]_{eq}^b} = \frac{(q_Z/N_A V)^c (q_W/N_A V)^d}{(q_X/N_A V)^a (q_Y/N_A V)^b} e^{-\Delta U_0/RT} \tag{17.7.11}$$

in which $\Delta U_0 = (c U_{0Z} + d U_{0W} - a U_{0X} - b U_{0Y})$. Thus, if the partition functions and the ground-state energies of the reacting molecules are known, the equilibrium constants can be calculated. The term ΔU_0 is very nearly the heat released in the reaction, i.e., it is essentially the reaction enthalpy.

Appendix 17.1 Approximations and Integrals

STIRLING'S APPROXIMATION

When N is a large number, $N!$ is a very large number. One can estimate the value of $N!$ using Stirling's approximation:

$$N! \approx N^N e^{-N} \sqrt{2\pi n} \tag{A17.1.1}$$

Using this approximation, we see that

$$\ln N! = N \ln N - N + (1/2) \ln(2\pi N) \tag{A17.1.2}$$

For large N, the last term in (A17.1.2) is small compared with the other two terms and it can be neglected. The resulting expression $\ln N! \approx N \ln N - N$ has been used in this chapter. One could also arrive at this result by using the approximation

$$\ln N! = \sum_{k=1}^{N} \ln k \approx \int_1^N \ln y \, dy = (y \ln y - y)|_1^N = N \ln N - N + 1 \tag{A17.1.3}$$

in which the sum is approximated by an integral, an approximation valid for large N.

INTEGRALS USED IN STATISTICAL THERMODYNAMICS

(a) $$\int_0^\infty e^{-ax^2} dx = \frac{1}{2} \left(\frac{\pi}{a} \right)^{1/2}$$

(b)
$$\int_0^\infty x e^{-ax^2}\,dx = \frac{1}{a}$$

(c)
$$\int_0^\infty x^2 e^{-ax^2}\,dx = \frac{1}{4a}\left(\frac{\pi}{a}\right)^{1/2}$$

(d)
$$\int_0^\infty x^3 e^{-ax^2}\,dx = \frac{1}{2a^2}$$

More generally:

(e)
$$\int_0^\infty x^{2n} e^{-ax^2}\,dx = \frac{1\times3\times5\times\cdots\times(2n-1)}{2^{n+1}a^n}\left(\frac{\pi}{a}\right)^{1/2}$$

(f)
$$\int_0^\infty x^{2n+1} e^{-ax^2}\,dx = \frac{n!}{2}\left(\frac{1}{a^{n+1}}\right)$$

Reference

1. Chapman, S., Cowling, T.G., *The Mathematical Theory of Nonuniform Gases: An Account of the Kinetic Theory of Viscosity, Thermal Conduction and Diffusion in Gases.* 1970, Cambridge, UK: Cambridge University Press.

Examples

Example 17.1   For the gas N_2 in a cube of side 10.0 cm, at $T = 298$ K, estimate the number of translational states that are below $3k_B T/2$ and compare this with the number of molecules in this cube at $p = 1.0$ bar.
Solution   The translational energies for a cube ($l_x = l_y = l_z = 1$) are given by (Box 17.1)

$$E_{n_x,n_y,n_z} = \frac{h^2}{8ml^2}(n_x^2 + n_y^2 + n_z^2)$$

The value of $n^2 = n_x^2 + n_y^2 + n_z^2$ for which the energy is $3k_B T/2$ is

$$n^2 = \frac{3k_B T}{h^2/8ml^2}$$

Substituting values $k_B = 1.38 \times 10^{-23}$ J K^{-1}, $T = 298$ K, $h = 6.626 \times 10^{-34}$ J s, $m = (28 \times 10^{-3}/N_A)$ kg and $l = 0.1$ m, we find $n^2 = 5.2 \times 10^{19}$. That means all quantum states in the sphere of radius n have energies less than $3k_B T/2$. Since only positive values of n_x, n_y, and n_z must be included:

Total number of states with energy $E < \dfrac{3k_B T}{2}$ is $\dfrac{1}{8}\dfrac{4\pi}{3}n^3 = \dfrac{1}{8}\dfrac{4\pi}{3}(5.2\times10^{19})^{3/2}$
$= 19.6 \times 10^{29}$

$$\text{Number of mulecules } \tilde{N} = \frac{N_A pV}{RT} = \frac{N_A \times 1.0 \times 10^5 \, \text{Pa} \times 10^{-3} \, \text{m}^3}{R \times 298 \, \text{K}} = 2.4 \times 10^{22}$$

This calculation shows that the number of available translational states is much higher than the number of particles in a gas at ordinary pressures and temperatures.

Exercises

17.1 Obtain (17.4.5) using Stirling's approximation.

17.2 Using an H—H bond length of 74 pm and an O=O bond length of 121 pm, calculate the characteristic rotational temperatures for H_2 and O_2.

17.3 Using $q^{vib} = e^{-\beta\hbar\omega/2}[1/(1 - e^{-\beta\hbar\omega})]$ show that

$$\left\langle E^{vib} \right\rangle = \hbar\omega\left(\frac{1}{2} + \frac{e^{-\hbar\omega/k_BT}}{1 - e^{-\hbar\omega/k_BT}}\right)$$

17.4 In earlier chapters we have expressed the chemical potential of an ideal gas $\mu = \mu^0(T) + RT\ln(p/p_0)$ (in which p_0 is the pressure of the standard state). In expression (17.7.4) the chemical potential is expressed in terms of the partition function and other molecular quantities. For a monatomic gas, rewrite (17.7.4) in the form $\mu = \mu^0(T) + RT\ln(p/p_0)$ and identify $\mu^0(T)$ as a function of T.

17.5 The bond length of H_2 is 74 pm. (a) Calculate the moment of inertia and express the rotational partition function as a function of T. (b) Obtain an expression for its molar energy as a function of T. (c) Calculate the molar heat capacity.

17.6 Calculate the equilibrium constant for the reaction $H_2 \rightleftharpoons H + H$.

LIST OF VARIABLES

Variables

a	van der Waals constant
a_k	activity of k
A	affinity
A_k	affinity of reaction k
b	van der Waals constant
$c_x, [x]$	concentration of x (mol L^{-1})
c^0	standard state concentration
C	heat capacity
C_m	molar heat capacity
C_{mV}	molar heat capacity at constant volume
C_{mp}	molar heat capacity at constant pressure
e	electron charge
E_a	activation energy
f	fugacity
F	Faraday constant
F	Helmholtz energy
F_m	molar Helmholtz energy
g	acceleration due to gravity
G	Gibbs (free) energy
G_m	molar Gibbs (free) energy
ΔG_f	Gibbs (free) energy of formation
ΔG_f^0	standard Gibbs energy of formation
ΔG_r	Gibbs (free) energy of a reaction
ΔG_{vap}	enthalpy of vaporization
ΔG_{fus}	enthalpy of fusion
$G^\dagger$	transition-state Gibbs energy
$h_{T,p}$	heat of reaction per unit ξ
h	enthalpy density
H	enthalpy
H_m	molar enthalpy
H_{mk}	partial molar enthalpy of k
ΔH_f	enthalpy of formation
ΔH_f^0	standard enthalpy of formation
ΔH_r	enthalpy of a reaction

ΔH_{vap}	enthalpy of vaporization
ΔH_{fus}	enthalpy of fusion
$H^\dagger$	transition-state enthalpy
I	electric current
J_k	thermodynamic flow
k	rate constant
k_B	Boltzmann constant
$K(T)$	equilibrium constant at temperature T
K_i	Henry's constant of i
m_k	molality, concentration (in moles of solute/kilogram of solvent)
m^0	standard state molality
M_k	molar mass of component k
n_k	concentration (mol m^{-3})
N	molar amount of substance
$\tilde{N}$	number of molecules
N_k	total molar amount of k
N_A	Avogadro constant
p^0	standard state pressure
p	total pressure
p_c	critical pressure
p_k	partial pressure of k
q	molecular partition function
Q	total partition function
R_{kf}	forward rate of reaction k
R_{kr}	reverse rate of reaction k
$s(x)$	entropy density at x
S	total entropy
S_m	molar entropy
$S^\dagger$	transition-state entropy
T	temperature
T_b	boiling point
T_c	critical temperature
T_m	melting point
u	energy density
U	total internal energy

Introduction to Modern Thermodynamics   Dilip Kondepudi
© 2008 John Wiley & Sons, Ltd

U_m	total molar energy	V_{mk}	Partial molar volume of k
v ($d\xi/dt$)	velocity of reaction or rate of conversion	V_{mc}	critical molar volume
		$[x]$, c_x	concentration of x (mol L^{-1})
V	total volume	x_k	mole fraction of k
V	voltage or potential difference	z_k	ion-number of k
V_m	molar volume	Z	compressibility factor

Greek letters

		ν_{jk}	stoichiometric coefficients of reaction k
α	coefficient of volume expansion		
β	Stefan–Boltzmann constant	π	osmotic pressure
ϕ	electrical potential	ρ	density
ϕ_k	osmotic coefficient of k	Σ	interfacial area
γ	ratio of molar heat capacities	ξ	extent of reaction
γ	surface tension	$d\xi/dt$ (v)	velocity of reaction or rate of conversion
$\gamma_\pm$	mean activity coefficient		
γ_k	activity coefficient of k	ξ_k	extent of reaction of reaction k
Γ_k	general mobility of k in a field		
κ	coefficient of heat conductivity	$d\xi_k/dt$ (v_k)	velocity of reaction k or rate of conversion k
κ_T	isothermal compressibility		
μ_k	chemical potential of k	$\tau_k\psi$	interaction energy per mole of k due to potential ψ
μ_k^0	standard chemical potential of k		
$\mu_\pm$	mean chemical potential		

STANDARD THERMODYNAMIC PROPERTIES

The following properties are listed at T = 298.15K:

$\Delta_f H^0$ Standard enthalpy of formation $\Delta_f G^0$ Standard Gibbs energy of formation
S_m^0 Standard molar entropy C_{mp} Molar heat capacity at constant pressure

The standard state pressure is 100 kPa (1 bar). An entry of 0.0 for $\Delta_f H^0$ for an element indicates the reference state of that element. Blanks indicate no data available.

Molecular formula	Name	State	$\Delta_f H^0$ kJ mol^{-1}	$\Delta_f G^0$ kJ mol^{-1}	S_m^0 J mol^{-1} K^{-1}	C_{mp} J mol^{-1} K^{-1}
Compounds not containing carbon						
Ac	Actinium	gas	406.0	366.0	188.1	20.8
Ag	Silver	cry	0.0	0.0	42.6	25.4
AgBr	Silver bromide	cry	−100.4	−96.9	107.1	52.4
AgBrO₃	Silver bromate	cry	−10.5	71.3	151.9	
AgCl	Silver chloride	cry	−127.0	−109.8	96.3	50.8
AgClO₃	Silver chlorate	cry	−30.3	64.5	142.0	
Al	Aluminum	cry	0.0	0.0	28.3	24.4
		gas	330.0	289.4	164.6	21.4
AlB₃H₁₂	Aluminium borohydride	liq	−16.3	145.0	289.1	194.6
AlBr	Aluminum bromide (AlBr)	gas	−4.0	−42.0	239.5	35.6
AlCl	Aluminum chloride (AlCl)	gas	−47.7	−74.1	228.1	35.0
AlCl₃	Aluminum trichloride	cry	−704.2	−628.8	110.7	91.8
AlF	Aluminum fluoride (AlF)	gas	−258.2	−283.7	215.0	31.9
AlF₃	Aluminum trifluoride	cry	−1510.4	−1431.1	66.5	75.1
AlI₃	Aluminum triiodide	cry	−313.8	−300.8	159.0	98.7
AlO₄P	Aluminum phosphate (AlPO₄)	cry	−1733.8	−1617.9	90.8	93.2
AlS	Aluminum sulfide (AlS)	gas	200.9	150.1	230.6	33.4
Al₂O	Aluminum oxide (Al₂O)	gas	−130.0	−159.0	259.4	45.7
Al₂O₃	Aluminum oxide (Al₂O₃)	cry	−1675.7	−1582.3	50.9	79.0
Ar	Argon	gas	0.0		154.8	20.8
As	Arsenic (gray)	cry	0.0		35.1	24.6
AsBr₃	Arsenic tribromide	gas	−130.0	−159.0	363.9	79.2
AsCl₃	Arsenic trichloride	gas	−261.5	−248.9	327.2	75.7
AsF₃	Arsenic trifluoride	lip	−821.3	−774.2	181.2	126.6
As₂	Arsenic (As₂)	gas	222.2	171.9	239.4	35.0
Au	Gold	cry	0.0	0.0	47.4	25.4
AuH	Gold hydride (AuH)	gas	295.0	265.7	211.2	29.2
B	Boron	cry (rhombic)	0.0	0.0	5.9	11.1
BCl	Chloroborane (BCl)	gas	149.5	120.9	213.2	31.7
BCl₃	Boron trichloride	liq	−427.2	−387.4	206.3	106.7
BF	Fluoroborane (BF)	gas	−122.2	−149.8	200.5	29.6
BH₃O₃	Boric acid (H₃BO₃)	cry	−1094.3	−968.9	88.8	81.4

Introduction to Modern Thermodynamics   Dilip Kondepudi
© 2008 John Wiley & Sons, Ltd

Molecular formula	Name	State	$\Delta_f H^0$ kJ mol^{-1}	$\Delta_f G^0$ kJ mol^{-1}	S_m^0 J mol^{-1} K^{-1}	C_{mp} J mol^{-1} K^{-1}
BH$_4$K	Potassium borohydride	cry	−227.4	−160.3	106.3	96.1
BH$_4$Li	Lithium borohydride	cry	−190.8	−125.0	75.9	82.6
BH$_4$Na	Sodium borohydride	cry	−188.6	−123.9	101.3	86.8
BN	Boron nitride (BN)	cry	−254.4	−228.4	14.8	19.7
B$_2$	Boron (B$_2$)	gas	830.5	774.0	201.9	30.5
Ba	Barium	cry	0.0	0.0	62.8	28.1
		gas	180.0	146.0	170.2	20.8
BaBr$_2$	Barium bromide	cry	−757.3	−736.8	146.0	
BaCl$_2$	Barium chloride	cry	−858.6	−810.4	123.7	75.1
BaF$_2$	Barium fluoride	cry	−1207.1	−1156.8	96.4	71.2
BaO	Barium oxide	cry	−553.5	−525.1	70.4	47.8
BaO$_4$S	Barium sulfate	cry	−1473.2	−1362.2	132.2	101.8
Be	Beryllium	cry	0.0	0.0	9.5	16.4
BeCl$_2$	Beryllium chloride	cry	−490.4	−445.6	82.7	64.8
BeF$_2$	Beryllium fluoride	cry	−1026.8	−979.4	53.4	51.8
BeH$_2$O$_2$	Beryllium hydroxide	cry	−902.5	−815.0	51.9	
BeO$_4$S	Beryllium sulfate	cry	−1205.2	−1093.8	77.9	85.7
Bi	Bismuth	cry	0.0	0.0	56.7	25.5
BiCl$_3$	Bismuth trichloride	cry	−379.1	−315.0	177.0	105.0
Bi$_2$O$_3$	Bismuth oxide (Bi$_2$O$_3$)	cry	−573.9	−493.7	151.5	113.5
Bi$_2$S$_3$	Bismuth sulfide (Bi$_2$S$_3$)	cry	−143.1	−140.6	200.4	122.2
Br	Bromine	gas	111.9	82.4	175.0	20.8
BrF	Bromine fluoride	gas	−93.8	−109.2	229.0	33.0
BrH	Hydrogen bromide	gas	−36.3	−53.4	198.7	29.1
BrH$_4$N	Ammonium bromide	cry	−270.8	−175.2	113.0	96.0
BrK	Potassium bromide	cry	−393.8	−380.7	95.9	52.3
BrKO$_3$	Potassium bromate	cry	−360.2	−217.2	149.2	105.2
BrLi	Lithium bromide	cry	−351.2	−342.0	74.3	
BrNa	Sodium bromide	cry	−361.1	−349.0	86.8	51.4
Br$_2$Ca	Calcium bromide	cry	−682.8	−663.6	130.0	
Br$_2$Hg	Mercury bromide (HgBr$_2$)	cry	−170.7	−153.1	172.0	
Br$_2$Mg	Magnesium bromide	cry	−524.3	−503.8	117.2	
Br$_2$Zn	Zinc bromide	cry	−328.7	−312.1	138.5	
Br$_4$Ti	Titanium bromide (TiBr$_4$)	cry	−616.7	−589.5	243.5	131.5
Ca	Calcium	cry	0.0	0.0	41.6	25.9
CaCl$_2$	Calcium chloride	cry	−795.4	−748.8	108.4	72.9
CaF$_2$	Calcium fluoride	cry	−1228.0	−1175.6	68.5	67.0
CaH$_2$	Calcium hydride (CaH$_2$)	cry	−181.5	−142.5	41.4	41.0
CaH$_2$O$_2$	Calcium hydroxide	cry	−985.2	−897.5	83.4	87.5
CaN$_2$O$_6$	Calcium nitrate	cry	−938.2	−742.8	193.2	149.4
CaO	Calcium oxide	cry	−634.9	−603.3	38.1	42.0
CaO$_4$S	Calcium sulfate	cry	−1434.5	−1322.0	106.5	99.7
CaS	Calcium sulfide	cry	−482.4	−477.4	56.5	47.4
Ca$_3$O$_8$P$_2$	Calcium phosphate	cry	−4120.8	−3884.7	236.0	227.8
Cd	Cadmium	cry	0.0	0.0	51.8	26.0
CdO	Cadmium oxide	cry	−258.4	−228.7	54.8	43.4
CdO$_4$S	Cadmium sulfate	cry	−933.3	−822.7	123.0	99.6
Cl	Chlorine	gas	121.3	105.3	165.2	21.8
ClCu	Copper chloride (CuCl)	cry	−137.2	−119.9	86.2	48.5
ClF	Chlorine fluoride	gas	−50.3	−51.8	217.9	32.1

Molecular formula	Name	State	$\Delta_f H^0$ kJ mol^{-1}	$\Delta_f G^0$ kJ mol^{-1}	S^0_m J mol^{-1} K^{-1}	C_{mp} J mol^{-1} K^{-1}
ClH	Hydrogen chloride	gas	−92.3	−95.3	186.9	29.1
ClHO	Hypochlorous acid (HOCl)	gas	−78.7	−66.1	236.7	37.2
ClH$_4$N	Ammonium chloride	cry	−314.4	−202.9	94.6	84.1
ClK	Potassium chloride (KCl)	cry	−436.5	−408.5	82.6	51.3
ClKO$_3$	Potassium chlorate (KClO$_3$)	cry	−397.7	−296.3	143.1	100.3
ClKO$_4$	Potassium perchlorate (KClO$_4$)	cry	−432.8	−303.1	151.0	112.4
ClLi	Lithium chloride (LiCl)	cry	−408.6	−384.4	59.3	48.0
ClNa	Sodium chloride (NaCl)	cry	−411.2	−384.1	72.1	50.5
ClNaO$_2$	Sodium chloride (NaClO$_2$)	cry	−307.0			
ClNaO$_3$	Sodium chlorate (NaClO$_3$)	cry	−365.8	−262.3	123.4	
Cl$_2$	Chlorine (Cl$_2$)	gas	0.0	0.0	223.1	33.9
Cl$_2$Cu	Copper chloride (CuCl$_2$)	cry	−220.1	−175.7	108.1	71.9
Cl$_2$Mn	Manganese chloride (MnCl$_2$)	cry	−481.3	−440.5	118.2	72.9
Cl$_3$U	Uraniam choride (UCl$_3$)	cry	−866.5	−799.1	159.0	102.5
Cl$_4$Si	Silicon tetrachloride	liq	−687.0	−619.8	239.7	145.3
Co	Cobalt	cry	0.0	0.0	30.0	24.8
CoH$_2$O$_2$	Cobalt hydroxide (Co(OH)$_2$)	cry	539.7	−454.3	79.0	
CoO	Cobalt oxide (CoO)	cry	−237.9	−214.2	53.0	55.2
Co$_3$O$_4$	Cobalt oxide (Co$_3$O$_4$)	cry	−891.0	−774.0	102.5	123.4
Cr	Chromium	cry	0.0	0.0	23.8	23.4
CrF$_3$	Chromium fluoride (CrF$_3$)	cry	−1159.0	−1088.0	93.9	78.7
Cr$_2$FeO$_4$	Chromium iron oxide (FeCr$_2$O$_4$)	cry	−1444.7	−1343.8	146.0	133.6
Cr$_2$O$_3$	Chromium oxide (Cr$_2$O$_3$)	cry	−1139.7	−1058.1	81.2	118.7
Cs	Cesium	cry	0.0	0.0	85.2	32.2
CsF	Cesium fluoride	cry	−553.5	−525.5	92.8	51.1
Cs$_2$O	Cesium oxide (Cs$_2$O)	cry	−345.8	−308.1	146.9	76.0
Cu	Copper	cry	0.0	0.0	33.2	24.4
CuO	Copper oxide (CuO)	cry	−157.3	−129.7	42.6	42.3
CuO$_4$S	Copper sulfate (CuSO$_4$)	cry	−771.4	−662.2	109.2	
CuS	Copper sulfide (CuS)	cry	−53.1	−53.6	66.5	47.8
Cu$_2$	Copper (Cu$_2$)	gas	484.2	431.9	241.6	36.6
Cu$_2$O	Copper oxide (Cu$_2$O)	cry	−168.6	−146.0	93.1	63.6
Cu$_2$S	Copper sulfide (Cu$_2$S)	cry	−79.5	−86.2	120.9	76.3
F$_2$	Fluorine (F$_2$)	gas	0.0	0.0	202.8	31.3
F	Fluorine	gas	79.4	62.3	158.8	22.7
FH	Hydrogen fluoride	gas	−273.3	−275.4	173.8	
FK	Potassium fluoride (KF)	cry	−567.3	−537.8	66.6	49.0
FLi	Lithium fluoride (LiF)	cry	−616.0	−587.7	35.7	61.6
FNa	Sodium fluoride (NaF)	cry	−576.6	−546.3	51.1	46.9
F$_2$HK	Potassium hydrogen fluoride (KHF$_2$)	cry	−927.7	−859.7	104.3	76.9
F$_2$HNa	Sodium hydrogen fluoride (NaHF$_2$)	cry	−920.3	−852.2	90.9	75.0
F$_2$Mg	Magnesium fluoride	cry	−1124.2	−1071.1	57.2	61.6

Molecular formula	Name	State	$\Delta_f H^0$ kJ mol^{-1}	$\Delta_f G^0$ kJ mol^{-1}	S_m^0 J mol^{-1} K^{-1}	C_{mp} J mol^{-1} K^{-1}
F_2O_2U	Uranyl fluoride	cry	−1648.1	−1551.8	135.6	103.2
F_2Si	Difluorosilylene (SiF$_2$)	gas	−619.0	−628.0	252.7	43.9
F_2Zn	Zinc fluoride	cry	−764.4	−713.3	73.7	65.7
F_3OP	Phosphoryl fluoride	gas	−1254.3	−1205.8	285.4	68.8
F_3P	Phosphorus trifluoride	gas	−958.4	−936.9	273.1	58.7
F_4S	Sulfur fluoride (SF$_4$)	gas	−763.2	−722.0	299.6	77.6
F_6S	Sulfur fluoride (SF$_6$)	gas	−1220.5	−1116.5	291.5	97.0
F_6U	Uranium fluoride (UF$_6$)	cry	−2197.0	−2068.5	227.6	166.8
Fe	Iron	cry	0.0	0.0	27.3	25.1
FeO_4S	Iron sulfate (FeSO$_4$)	cry	−928.4	−820.8	107.5	100.6
FeS	Iron sulfide (FeS)	cry	−100.0	−100.4	60.3	50.5
FeS_2	Iron sulfide (FeS$_2$)	cry	−178.2	−166.9	52.9	62.2
Fe_2O_3	Iron oxide (Fe$_2$O$_3$)	cry	−824.2	−742.2	87.4	103.9
Fe_3O_4	Iron oxide (Fe$_3$O$_4$)	cry	−1118.4	−1015.4	146.4	143.4
H_2	Hydrogen (H$_2$)	gas	0.0	0.0	130.7	28.8
H	Hydrogen	gas	218.0	203.3	114.7	20.8
HI	Hydrogen iodide	gas	26.5	1.7	206.6	29.2
HKO	Potassium hydroxide (KOH)	cry	−424.8	−379.1	78.9	64.9
HLi	Lithium hydride (LiH)	cry	−90.5	−68.3	20.0	27.9
HNO_2	Nitrous acid (HONO)	gas	−79.5	−46.0	254.1	45.6
HNO_3	Nitric acid	liq	−174.1	−80.7	155.6	109.9
HNa	Sodium hydride	cry	−56.3	−33.5	40.0	36.4
HNaO	Sodium hydroxide (NaOH)	cry	−425.6	−379.5	64.5	59.5
HO	Hydroxyl (OH)	gas	39.0	34.2	183.7	29.9
HO_2	Hydroperoxy (HOO)	gas	10.5	22.6	229.0	34.9
H_2Mg	Magnesium hydride	cry	−75.3	−35.9	31.1	35.4
H_2MgO_2	Magnesium hydroxide	cry	−924.5	−833.5	63.2	77.0
H_2O	Water	liq	−285.8	−237.1	70.0	75.3
H_2O_2	Hydrogen peroxide	liq	−187.8	−120.4	109.6	89.1
H_2O_2Sn	Tin hydroxide (Sn(OH)$_2$)	cry	−561.1	−491.6	155.0	
H_2O_2Zn	Zinc hydroxide	cry	−641.9	−553.5	81.2	
H_2O_4S	Sulfuric acid	liq	−814.0	−690.0	156.9	138.9
H_2S	Hydrogen sulfide	gas	−20.6	−33.4	205.8	34.2
H_3N	Ammonia (NH$_3$)	gas	−45.9	−16.4	192.8	35.1
H_3O_4P	Phosphoric acid	cry	−1284.4	−1124.3	110.5	106.1
		liq	−1271.7	−1123.6	150.8	145.0
H_3P	Phosphine	gas	5.4	13.4	210.2	37.1
H_4IN	Ammonium iodide	cry	−201.4	−112.5	117.0	
H_4N_2	Hydrazine	liq	50.6	149.3	121.2	98.9
$H_4N_2O_3$	Ammonium nitrate	cry	−365.6	−183.9	151.1	139.3
H_4Si	Silane	gas	34.3	56.9	204.6	42.8
$H_8N_2O_4S$	Ammonium sulfate	cry	−1180.9	−901.7	220.1	187.5
He	Helium	gas	0.0		126.2	20.8
HgI_2	Mercury iodide (HgI$_2$) (red)	cry	−105.4	−101.7	180.0	
HgO	Mercury oxide (HgO) (red)	cry	−90.8	−58.5	70.3	44.1
HgS	Mercury sulfide (HgS)	cry	−58.2	−50.6	82.4	48.4

Molecular formula	Name	State	$\Delta_f H^0$ kJ mol^{-1}	$\Delta_f G^0$ kJ mol^{-1}	S_m^0 J mol^{-1} K^{-1}	C_{mp} J mol^{-1} K^{-1}
Hg_2	Mercury (Hg_2)	gas	108.8	68.2	288.1	37.4
Hg_2O_4S	Mercury sulfate (Hg_2SO_4)	cry	−743.1	−625.8	200.7	132.0
I	Iodine	gas	106.8	70.2	180.8	20.8
IK	Potassium iodide	cry	−327.9	−324.9	106.3	52.9
IKO_3	Potassium iodate	cry	−501.4	−418.4	151.5	106.5
ILi	Lithium iodide	cry	−270.4	−270.3	86.8	51.0
INa	Sodium iodide	cry	−287.8	−286.1	98.5	52.1
$INaO_3$	Sodium iodate	cry	−481.8			92.0
K	Potassium	cry	0.0	0.0	64.7	29.6
$KMnO_4$	Potassium permanganate	cry	−837.2	−737.6	171.7	117.6
KNO_2	Potassium nitrite	cry	−369.8	−306.6	152.1	107.4
KNO_3	Potassium nitrate	cry	−494.6	−394.9	133.1	96.4
K_2O_4S	Potassium sulfate	cry	−1437.8	−1321.4	175.6	131.5
K_2S	Potassium sulfide (K_2S)	cry	−380.7	−364.0	105.0	
Li	Lithium	cry	0.0	0.0	29.1	24.8
Li_2	Lithium (Li_2)	gas	215.9	174.4	197.0	36.1
Li_2O	Lithium oxide (Li_2O)	cry	−597.9	−561.2	37.6	54.1
Li_2O_3Si	Lithium metasilicate	cry	−1648.1	−1557.2	79.8	99.1
Li_2O_4S	Lithium sulfate	cry	−1436.5	−1321.7	115.1	117.6
Mg	Magnesium	cry	0.0	0.0	32.7	24.9
MgN_2O_6	Magnesium nitrate	cry	−790.7	−589.4	164.0	141.9
MgO	Magnesium oxide	cry	−601.6	−569.3	27.0	37.2
MgO_4S	Magnesium sulfate	cry	−1284.9	−1170.6	91.6	96.5
MgS	Magnesium sulfide	cry	−346.0	−341.8	50.3	45.6
Mn	Manganese	cry	0.0	0.0	32.0	26.3
$MgNa_2O_4$	Sodium permanganate	cry	−1156.0			
MnO	Maganese oxide (MnO)	cry	−385.2	−362.9	59.7	45.4
MnS	Manganese sulfide (MnS)	cry	−214.2	−218.4	78.2	50.0
Mn_2O_3	Manganese oxide (Mn_2O_3)	cry	−959.0	−881.1	110.5	107.7
Mn_2O_4Si	Manganese silicate (Mn_2SiO_4)	cry	−1730.5	−1632.1	163.2	129.9
N_2	Nitrogen (N_2)	gas	0.0	0.0	191.6	29.1
N	Nitrogen	gas	472.7	455.5	153.3	20.8
$NNaO_2$	Sodium nitrite	cry	−358.7	−284.6	103.8	
$NNaO_3$	Sodium nitrate	cry	−467.9	−367.0	116.5	92.9
NO	Nitrogen oxide	gas	90.25	86.57	210.8	29.84
NO_2	Nitrogen dioxide	gas	33.2	51.3	240.1	37.2
N_2O	Nitrous oxide	gas	82.1	104.2	219.9	38.5
N_2O_3	Nitrogen trioxide	liq	50.3			
N_2O_4	Dinitrogen tetroxide	gas	9.16	97.89	304.29	77.28
N_2O_5	Nitrogen pentoxide	cry	−43.1	113.9	178.2	143.1
Na	Sodium	cry	0.0	0.0	51.3	28.2
NaO_2	Sodium superoxide (NaO_2)	cry	−260.2	−218.4	115.9	72.1
Na_2	Sodium (Na_2)	gas	142.1	103.9	230.2	37.6
Na_2O	Sodium oxide (Na_2O)	cry	−414.2	−375.5	75.1	69.1
Na_2O_2	Sodium peroxide (Na_2O_2)	cry	−510.9	−447.7	95.0	89.2
Na_2O_4S	Sodium sulfate	cry	−1387.1	−1270.2	149.6	128.2
Ne	Neon	gas	0.0		146.3	20.8
Ni	Nickel	cry	0.0	0.0	29.9	26.1
NiO_4S	Nickel sulfate ($NiSO_4$)	cry	−872.9	−759.7	92.0	138.0
NiS	Nickel sulfide (NiS)	cry	−82.0	−79.5	53.0	47.1

Molecular formula	Name	State	$\Delta_f H^0$ kJ mol^{-1}	$\Delta_f G^0$ kJ mol^{-1}	S_m^0 J mol^{-1} K^{-1}	C_{mp} J mol^{-1} K^{-1}
O	Oxygen	gas	249.2	231.7	161.1	21.9
OP	Phosphorus oxide (PO)	gas	−28.5	−51.9	222.8	31.8
O$_2$Pb	Lead oxide (PO$_2$)	cry	−277.4	−217.3	68.6	64.6
O$_2$S	Sulfur dioxide	gas	−296.8	−300.1	248.2	39.9
O$_2$Si	Silicon dioxide (α-quartz)	cry	−910.7	−856.3	41.5	44.4
O$_2$U	Uranium oxide (UO$_2$)	cry	−1085.0	−1031.8	77.0	63.6
O$_3$	Ozone	gas	142.7	163.2	238.9	39.2
O$_3$PbSi	Lead metasilicate (PbSiO$_3$)	cry	−1145.7	−1062.1	109.6	90.0
O$_3$S	Sulfur trioxide	gas	−395.7	−371.1	256.8	50.7
O$_4$SZn	Zinc sulfate	cry	−982.8	−871.5	110.5	99.2
P	Phosphorus (white)	cry	0.0	0.0	41.1	23.8
	Phosphorus (red)	cry	−17.6		22.8	21.2
Pb	Lead	cry	0.0	0.0	64.8	26.4
PbS	Lead sulfide (PbS)	cry	−100.4	−98.7	91.2	49.5
Pt	Platinum	cry	0.0	0.0	41.6	25.9
PtS	Platinum sulfide (PtS)	cry	−81.6	−76.1	55.1	43.4
PtS$_2$	Platinum sufide (PtS$_2$)	cry	−108.8	−99.6	74.7	65.9
S	Sulfur	cry (rhombic)	0.0	0.0	32.1	22.6
	Sulfur	cry (monoclinic)	0.3			
S$_2$	Sulfur (S$_2$)	gas	128.6	79.7	228.2	32.5
Si	Silicon	cry	0.0	0.0	18.8	20.0
Sn	Tin (white)	cry	0.0		51.2	27.0
	Tin (gray)	cry	−2.1	0.1	44.1	25.8
Zn	Zinc	cry	0.0	0.0	41.6	25.4
		gas	130.4	94.8	161.0	20.8
Compounds containing carbon						
C	Carbon (graphite)	cry	0.0	0.0	5.7	8.5
	Carbon (diamond)	cry	1.9	2.9	2.4	6.1
CAgN	Silver cyanide (AgCN)	cry	146.0	156.9	107.2	66.7
CBaO$_3$	Barium carbonate (BaCO$_3$)	cry	−1216.3	−1137.6	112.1	85.3
CBrN	Cyanogen bromide	cry	140.5			
CCaO$_3$	Calcium carbonate (calcite)	cry	−1207.6	−1129.1	91.7	83.5
	Calcium carbonate (aragonite)	cry	−1207.8	−1128.2	88.0	82.3
CCl$_2$F$_2$	Dichlorodifluoromethane	gas	−477.4	−439.4	300.8	72.3
CCl$_3$F	Trichlorofluoromethane	liq	−301.3	−236.8	225.4	121.6
CCuN	Copper cyanide (CuCN)	cry	96.2	111.3	84.5	
CFe$_3$	Iron carbide (Fe$_3$C)	cry	25.1	20.1	104.6	105.9
CFeO$_3$	Iron carbonate (FeCO$_3$)	cry	−740.6	−666.7	92.9	82.1
CKN	Potassium cyanide (KCN)	cry	−113.0	−101.9	128.5	66.3
CKNS	Potassium thiocyanate (KSCN)	cry	−200.2	−178.3	124.3	88.5
CK$_2$O$_3$	Potassium carbonate (KCO$_3$)	cry	−1151.0	−1063.5	155.5	114.4
CMgO$_3$	Magnesium carbonate (MgCO$_3$)	cry	−1095.8	−1012.1	65.7	75.5
CNNa	Sodium cyanide (NaCN)	cry	−87.5	−76.4	115.6	70.4

Molecular formula	Name	State	$\Delta_f H^0$ kJ mol^{-1}	$\Delta_f G^0$ kJ mol^{-1}	S_m^0 J mol^{-1} K^{-1}	C_{mp} J mol^{-1} K^{-1}
CNNaO	Sodium cyanate	cry	−405.4	−358.1	96.7	86.6
CNa$_2$O$_3$	Sodium carbonate (NaCO$_3$)	cry	−1130.7	−1044.2	135.0	112.3
CO	Carbon monoxide	gas	−110.5	−137.2	197.7	29.1
CO$_2$	Carbon dioxide	gas	−393.5	−394.4	213.8	37.1
CO$_3$Zn	Zinc carbonate (ZnCO$_3$)	cry	−812.8	−731.5	82.4	79.7
CS$_2$	Carbon disulfide	liq	89.0	64.6	151.3	76.4
CSi	Silicon carbide (cubic)	cry	−65.3	−62.8	16.6	26.9
CHBr$_3$	Tribromomethane	liq	−28.5	−5.0	220.9	130.7
CHClF$_2$	Chlorodifluoromethane	gas	−482.6		280.9	55.9
CHCl$_3$	Trichloromethane	liq	−134.5	−73.7	201.7	114.2
CHN	Hydrogen cyanide	liq	108.9	125.0	112.8	70.6
CH$_2$	Methylene	gas	390.4	372.9	194.9	33.8
CH$_2$I$_2$	Diiodomethane	liq	66.9	90.4	174.1	134.0
CH$_2$O	Formaldehyde	gas	−108.6	−102.5	218.8	35.4
CH$_2$O$_2$	Formic acid	liq	−424.7	−361.4	129.0	99.0
CH$_3$	Methyl	gas	145.7	147.9	194.2	38.7
CH$_3$Cl	Chloromethane	gas	−81.9		234.6	40.8
CH$_3$NO$_2$	Nitromethane	liq	−113.1	−14.4	171.8	106.6
CH$_4$	Methane	gas	−74.4	−50.3	186.3	35.3
CH$_4$N$_2$O	Urea	cry	−333.6			
CH$_4$O	Methanol	liq	−239.1	−166.6	126.8	81.1
C$_2$	Carbon (C$_2$)	gas	831.9	775.9	199.4	43.2
C$_2$Ca	Calcium carbide	cry	−59.8	−64.9	70.0	62.7
C$_2$ClF$_3$	Chlorotrifluoroethylene	gas	−555.2	−523.8	322.1	83.9
C$_2$Cl$_4$	Tetrachloroethylene	liq	−50.6	3.0	266.9	143.4
C$_2$Cl$_4$F$_2$	1,1,1,2-Tetrachloro-2, 2-difluoroethane	gas	−489.9	−407.0	382.9	123.4
C$_2$H$_2$	Acetylene	gas	228.2	210.7	200.9	43.9
C$_2$H$_2$Cl$_2$	1,1-Dichloroethylene	liq	−23.9	24.1	201.5	111.3
C$_2$H$_2$O	Ketene	gas	−47.5	−48.3	247.6	51.8
C$_2$H$_2$O$_4$	Oxalic acid	cry	−821.7		109.8	91.0
C$_2$H$_3$Cl$_3$	1,1,1-Trichlorothane	liq	−177.4		227.4	144.3
		gas	−144.6		323.1	93.3
C$_2$H$_3$N	Acetonitrile	liq	31.4	77.2	149.6	91.4
C$_2$H$_3$NaO$_2$	Sodium acetate	cry	−708.8	−607.2	123.0	79.9
C$_2$H$_4$	Ethylene	gas	52.5	68.4	219.6	43.6
C$_2$H$_4$Cl$_2$	1,1-Dichloroethane	liq	−158.4	−73.8	211.8	126.3
		gas	−127.7	−70.8	305.1	76.2
C$_2$H$_4$O$_2$	Acetic acid	liq	−484.5	−389.9	159.8	123.3
		gas	−432.8	−374.5	282.5	66.5
C$_2$H$_5$I	Iodoethane	liq	−40.2	14.7	211.7	115.1
C$_2$H$_6$	Ethane	gas	−83.8	−31.9	229.6	52.6
C$_2$H$_6$O	Dimethyl ether	gas	−184.1	−112.6	266.4	64.4
C$_2$H$_6$O	Ethanol	liq	−277.7	−174.8	160.7	112.3
C$_2$H$_6$S	Ethanethiol	liq	−73.6	−5.5	207.0	117.9
C$_2$H$_7$N	Dimethylamine	gas	−18.5	68.5	273.1	70.7
C$_3$H$_7$N	Cyclopropylamine	liq	45.8		187.7	147.1
C$_3$H$_8$	Propane	gas	−104.7			
C$_3$H$_8$O	1-Propanol	liq	−302.6		193.6	143.9

Molecular formula	Name	State	$\Delta_f H^0$ kJ mol^{-1}	$\Delta_f G^0$ kJ mol^{-1}	S_m^0 J mol^{-1} K^{-1}	C_{mp} J mol^{-1} K^{-1}
$C_3H_8O_3$	Glycerol	liq	−668.5		206.3	218.9
C_4H_4O	Furan	liq	−62.3		177.0	115.3
$C_4H_4O_4$	Fumaric acid	cry	−811.7		168.0	142.0
C_4H_6	1,3-Butadiene	liq	87.9		199.0	123.6
$C_4H_6O_2$	Methyl acrylate	liq	−362.2		239.5	158.8
C_4H_8	Isobutene	liq	−37.5			
C_4H_8	Cyclobutane	liq	3.7			
C_4H_8O	Butanal	liq	−239.2		246.6	163.7
C_4H_8O	Isobutanal	liq	−247.4			
$C_4H_8O_2$	1,4-Dioxane	liq	−353.9		270.2	152.1
$C_4H_8O_2$	Ethyl acetate	liq	−479.3		257.7	170.7
$C_4H_{10}O$	1-Butanol	liq	−327.3		225.8	177.2
$C_4H_{10}O$	2-Butanol	liq	−342.6		214.9	196.9
$C_4H_{12}Si$	Tetramethylsilane	liq	−264.0	−100.0	277.3	204.1
C_5H_8	Cyclopentene	liq	4.4		201.2	122.4
C_5H_{10}	1-Pentene	liq	−46.9		262.6	154.0
C_5H_{10}	Cyclopentane	liq	−105.1		204.5	128.8
C_5H_{12}	Isopentane	liq	−178.5		260.4	164.8
C_5H_{12}	Neopentane	gas	−168.1			
$C_5H_{12}O$	Butyl methyl ether	liq	−290.6		295.3	192.7
C_6H_6	Benzene	liq	49.0			136.3
C_6H_6O	Phenol	cry	−165.1		144.0	127.4
$C_6H_{12}O_6$	α-D-Glucose	cry	−1274.4	−910.52	212.1	
C_7H_8	Toluene	liq	12.4			157.3
C_7H_8O	Benzyl alcohol	liq	−160.7		216.7	217.9
C_7H_{14}	Cycloheptane	liq	−156.6			
C_7H_{14}	Ethylcyclopentane	liq	−163.4		279.9	
C_7H_{14}	1-Heptene	liq	−97.9		327.6	211.8
C_7H_{16}	Heptane	liq	−224.2			
C_8H_{16}	Cyclooctane	liq	−167.7			
C_8H_{18}	Octane	liq	−250.1			254.6
		gas	−208.6			
C_9H_{20}	Nonane	liq	−274.7			284.4
$C_9H_{20}O$	1-Nonanol	liq	−456.5			
$C_{10}H_8$	Naphthalene	cry	77.9		167.4	165.7
$C_{10}H_{22}$	Decane	liq	−300.9			314.4
$C_{12}H_{10}$	Biphenyl	cry	99.4		209.4	198.4
$C_{12}H_{22}O_{11}$	Sucrose	cry	−2222.1	−1544	360.2	
$C_{12}H_{26}$	Dodecane	liq	−350.9			375.8

PHYSICAL CONSTANTS AND DATA

Speed of light	$c = 2.997925 \times 10^{8}\,\mathrm{m\,s^{-1}}$
Gravitational constant	$G = 6.67 \times 10^{-11}\,\mathrm{N\,m^{2}\,kg^{-1}}$
Avogadro's number	$N_{\mathrm{A}} = 6.022 \times 10^{23}\ \text{particles/mol}$
Boltzmann's constant	$k_{\mathrm{B}} = 1.38066 \times 10^{-23}\,\mathrm{J\,K^{-1}}$
Gas constant	$R = 8.314\,\mathrm{J\,mol\,K^{-1}}$
	$= 1.9872\,\mathrm{cal\,kmol^{-1}\,K^{-1}}$
Planck's constant	$h = 6.6262 \times 10^{-34}\,\mathrm{Js}$
Electron change	$e = 1.60219 \times 10^{-19}\,\mathrm{C}$
Electron rest mass	$m_{\mathrm{e}} = 9.1095 \times 10^{-31}\,\mathrm{kg}$
	$= 5/486 \times 10^{-4}\,\mathrm{u}$
Proton rest mass	$m_{\mathrm{p}} = 1.6726 \times 10^{-21}\,\mathrm{kg}$
	$= 1.007276\,\mathrm{u}$
Neutron rest mass	$m_{\mathrm{n}} = 1.6749 \times 10^{-27}\,\mathrm{kg}$
	$= 1.008665\,\mathrm{u}$
Permittivity constant	$\varepsilon_{0} = 8.85419 \times 10^{-12}\,\mathrm{C^{2}\,N^{-1}\,m^{-2}}$
Permeability constant	$\mu_{0} = 4\pi \times 10^{-7}\,\mathrm{N\,A^{-2}}$
Standard gravitational acceleration	$g = 9.80665\,\mathrm{m\,s^{-1}} = 32.17\,\mathrm{ft\,s^{-1}}$
Mass of Earth	$5.98 \times 10^{24}\,\mathrm{kg}$
Average radius of Earth	$6.37 \times 10^{6}\,\mathrm{m}$
Average density of Earth	$5.57\,\mathrm{g\,cm^{-3}}$
Average Earth-Moon distance	$3.84 \times 10^{8}\,\mathrm{m}$
Average Earth-Sun distance	$1496 \times 10^{11}\,\mathrm{m}$
Mass of Sun	$1.99 \times 10^{30}\,\mathrm{kg}$
Radius of Sun	$7 \times 10^{8}\,\mathrm{m}$
Sun's radiation intensity at the Earth	$0.032\,\mathrm{cal\,cm^{-2}\,s^{-1}} = 0.134\,\mathrm{J\,cm^{-2}\,s^{-1}}$

Introduction to Modern Thermodynamics   Dilip Kondepudi
© 2008 John Wiley & Sons, Ltd

NAME INDEX

Note: Figures and tables are indicated by *italic* page numbers, boxes by **bold** numbers.

Andrews, Thomas 20, 23
Arrhenius, Svante 269, 285
Avogadro, Amedo 15

Bernoulli, Daniel 29, 31, 451
Berthelot, Mercellin 70, 141
Black, Joseph 11, *12*, 19, *20*, 50, 97, 128, 215
Boltzmann, Ludwig Eduard 29, 32, 124, 361, 422, 451–2
Boyle, Robert *13*, 29, 31, 50
Braun, Karl Ferdinand 277

Carnot, Hippolyte 101
Carnot, Lazare 98–9
Carnot, Sadi 51, 90, 98–9, 101–2, 136
Charles, Jacques *15*
Clapeyron, Émile 99, 102
Clausius, Rudolf 27, 108, *110*, 115–16
Crick, Francis 342, 380

Davy, Sir Humphry 11
De Donder, Théophile 119, 141–2, 144–7
de la Tour, Cagniard 20
Debye, Peter 211, 254
Duhem, Pierre 119, 173, 223

Einstein, Albert 3, 79, 423
Emanuel, Kerry **110**, 135

Faraday, Michael 49
Fermat, Pierre 163
Fourier, Jean Baptiste Joseph 51, **83**
Fowler, R.H. 446

Galileo Galilei 11
Galvani, Luigi 49
Gay-Lussac, Joseph-Louis 16–17, 107n
Gibbs, Josiah Willard 141–3, 173
Guggenheim, E.A. 306, 324, 446

Helmholtz, Hermann von 55–6, 141
Hess, Germain Henri 55, 68, *69*, 141
Hill, Terrell 411
Hückel, Erich 254

Joule, James Prescott 16, 50–1, 53, 55

Kalckar, Herman 394
Kelvin, Lord (William Thomson) 102, *106*, 333–4
Kirchhoff, Gustav 75, 361–2, *363*

Landau, Lev 231, 233
Laplace, Pierre-Simon 51, 68, 185
Lavoisier, Antoine Laurent 51, 68, 141
Le Châtelier, Henri 143, 277
Lehninger, Albert 385
Lewis, Gilbert Newton 149, 177, 206–7, 239
Lipmann, Fritz 393–4

Maxwell, James Clerk 29, 32, 179, 451
Mayer, Julius Robert von 55, 69
Menten, Maude 399
Michaelis, Leonor 399

Nernst, Walther *123*, 141

Oersted, Hans Christian 49
Onnes, Kamerlingh 200
Onsager, Lars 4, 332–3, *335*
Ostwald, Wilhelm 69, 143

Pascal, Blaise *8*
Pasteur, Louis 342
Pauli, Wolfgang 77, 80
Planck, Max 3, 55, 58, 116, 118, 125, 364–5, 422, 453
Poisson, Siméon-Denis 51
Prigogine, Ilya 4, 119–20, 329, *339*, 348, 379

Randall, Merle 149
Raoult, François-Marie 241
Reines, Frederick 80
Rumford, Count (Sir Benjamin Thompson) 52

Sackur, O. 466
Schrödinger, Erwin 379
Seebeck, Thomas 49
Shelley, Mary 49

Introduction to Modern Thermodynamics Dilip Kondepudi
© 2008 John Wiley & Sons, Ltd

Skou, Jens Christian 397
Smith, Adam 3
Stefan, Josef 361, 366

Tetrode, H. 466
Thompsen, Julius 70, 141
Thompson, Sir Benjamin (Count Rumford) 52
Thomson, James 441
Thomson, William (Lord Kelvin) 333

van der Waals, Johannes Diderik 18, 20, *22*
van Larr, J.J. 249
van't Hoff, Jacobus Henricus 245–6
Volta, Alessandro 49

Watt, James 3, 19, *97*
Wien, Wilhelm 361, 369

SUBJECT INDEX

Note: Figures and tables are indicated by *italic* page numbers, boxes by **bold** numbers, footnotes by suffix 'n'.

absolute scale of temperature 12, 102
absolute zero of chemical potential 375
achiral molecules, meaning of term 343
action, minimization of 165
activation energy 269, 285
activity 177–8
  for component of ideal solution 243
  for liquids and solids 210
  and reaction rate 272
activity coefficient(s) 240
  of electrolytes 253
adenosine diphosphate *see* ADP
adenosine triphosphate *see* ATP
adiabatic processes
  and entropy 385
  in ideal gas 68, 196
  in real gas 202–4
ADP 394
  in photosynthesis *387, 392*
affinity 143, 145–7, 149–51
  additivity of affinities 150
  in biological systems 381
  coupling between affinities 151
  and direction of reaction 150
  general properties 151–1
  liquid–vapor transformation *216*
  and osmosis 245, 247
  of real gas 208
  relation to Gibbs energy 168, 285
  relation to reaction rates 281–2
    example 281–2
  see also entropy production; Gibbs free energy
amino acids, as chiral molecules 342, 379
antiparticles 53, *54*
Arrhenius equation 269, **270**
  pre-exponential factor in 269, 286–7
Arrhenius rate constant 286, 288
Arrhenius theory 285, 288
athermal solutions 259
atmosphere *see* Earth's atmosphere
atmosphere (unit of pressure) **14**
atmospheric pressure **8**
atomic mass unit (amu) 80

ATP 394–5
  as energy carrier 393
  hydrolysis of 395
ATP–ADP cycle 396
  in photosynthesis *387*
ATP-driven 'ion pumps' 396, *397*
ATP-driven reactions 394, 396, *397*
ATP-producing reactions 385
autotrophs 386
average values 39
Avogadro number **14**, 30, **124**
Avogadro's hypothesis 30
azeotropes 224–6
  examples *226*, 260
azeotropic composition 259
azeotropic transformation 259
azeotropy 259

bar (unit of pressure) **14**
barometric formula 307, *308*
  example of application 181
Belusov–Zhabotinsky reaction **354**
  FKN model 348, 349–50
    Mathematica code for 296–7
Berthelot equation 200, 212
'big bang', events following 55, **266**
binary solutions, critical phenomena 444–6
binary systems
  liquid mixtures in equilibrium with vapor 223
  solution in equilibrium with solids 225
binding energy per nucleon **266**
binomial distribution 40
biochemical kinetics 399, 401, 403
biochemical standard state 384, **385**
biological systems 9, 380–98
  energy flows in 86
  Gibbs energy flow in 385–97
biomolecules, determination of molecular weight 248
black body 363, **364**
blackbody radiation 55
  interaction of two-level atom with 372

boiling point 220, *224*
  changes in 244–5
  listed for various substances *143, 243*
  plot vs composition of liquid mixture *224*
Boltzmann constant **14**, 31, **124**, **270**, 365, 422,
    451, 476
Boltzmann principle 33, 455
Boltzmann probability distribution 453
  average values calculated using 457
Boltzmann's relation between entropy and
    probability **125**, 422, 452–3
    application in calculation of entropy of mixing
    197
bomb calorimeter **63**
bond enthalpy 76
  listed for various bonds *77*
Boyle's law 15–16
Brownian motion 172, 422
Brusselator model 346
  Mathematica code for 350–3
bulk chemical potential 412
bulk modulus, for adiabatic process 67

caloric theory of heat 11, 51, **52**, 101
calorie **52**
calorimetry **63**
canonical ensemble 457
canonical partition function 461
  and thermodynamic quantities 462
capillary rise 185–6
carbohydrates, energy content 392
carbon dioxide concentration, effect on plant
    growth rate 482
Carnot cycle(s) *103*, 105
  combination of *111*
  hurricane as 132
Carnot's theorem (for efficiency of heat engine)
    102
  Clausius's generalizations 111–19
  hurricane wind speed estimated using 108,
    109–110, 132–7
  and second law of thermodynamics 116
catalyst, effects 269, **270**
cell diagrams, conventions used **315**
cell membranes, transport across 396
Celsius temperature scale 12, **14**
characteristic temperature
  in expression for vibrational partition function
    468
  listed for various diatomic molecules 468
Charles's law 15
chemical affinity 312
  of real gas 208
  see also affinity
chemical equilibrium 273–8
chemical kinetics 267

chemical oscillations 345–51
chemical potential 7, 112, 141, 143–4, 239
  absolute zero 375
  bulk 412
  computation of 179
  diffusion described in terms of 154
  of dilute solution 250, 253
  in fields 307–10, 319, 322
  of ideal gas 197, 374
  of ideal solution 243
  of nonideal solution 239
  of nonthermal radiation 375, 388, 390, 407
  of perfect solution 239
  of pure liquids and solids 208
  of real gas 204–7
  relation to Gibbs energy 168–9
  of small systems 411
  of solvent in osmosis 245
  see also mean chemical potential
chemical reaction rates
  Arrhenius theory 269, **270**, 285, 288
  transition state theory 269, **270**, 288
  see also reaction rates
chemical reactions
  conservation of energy in 68–71, 79
  coupled 288, 290, 294
  and entropy 144, 179
  entropy production due to 147, 149, 153, 280,
    282
  fluctuations in extent of reaction 425
  irreversibility in 144
  rate laws *332*
  rates 265–73
  reactive collisions in 330
  steady-state assumption 292, 303
  temperature range **266**
  thermodynamic forces and flows *331*
  see also extent of reaction; first-order reactions;
    second-order reactions; zero-order reactions
chemical stability 437
chemical transformations 265–6, 268, 305
  Gibbs energy change in 382, 384–5
  reaction rates 266, 269, 271–3
chemoautotrophs 386
chiral asymmetry 380
chiral molecules, meaning of term 342, 380
chirality 341
chirally autocatalytic reactions 342
chlorophyll, photo-excitation of 387, 389
Clapeyron equation 219–20
classical electrodynamics 366
classical nucleation theory 418
classical stability theory 433
  applications 411, 454
classical thermodynamics 116, 119, 122, 196, 266,
    342

Clausius–Clapeyron equation 220
  examples of application 259
Clausius inequality 118
closed systems 4, *5*
  chemical reactions in 149
  energy change in 59, **61**
  entropy changes 120
coefficient of thermal expansion 208
  listed for various liquids and solids *209*
coefficient of volume expansion 180, 202
coexistence curve (in phase diagram) *215*, 218–19,
    444
colligative properties 243, 245, 247, 249
collision cross-section 287
collision frequency 287
competitive inhibition of enzymes 402–4, *406*
compressibility factor 25–6, 206
concentration cell 317
condensation of vapor into liquid 418
condensed phase, standard state of substance
    72
condensed phases 208
  see also liquids; solids
conduction, law governing **83**
configurational heat capacity 441–2, 444, 446–8
consecutive first-order reactions 290
conservation of energy
  in chemical reactions 68–9, 71–3, 75, 77–80
  law of 55–6
    elementary applications 64–5, 67
  in nuclear reactions 79
consolute point 443
contact angle
  interfacial 183
    values listed *187*
continuous systems, entropy production in 135
convection law **83**
convection patterns 81, 132, 340–1, 344
conversion rate, in chemical reaction 75, 179
cosmic microwave background 265
cosmological principle(s) 116
coupled reactions 288
  entropy production in 151–2
critical constants 23–5, 27, 194, 200
  listed for various gases *19*
  see also molar critical volume
critical nucleation radius 419
critical parameter (in dissipative structure) *340*,
    341
critical phenomena 235, 441, 443
critical point (in phase diagram) *216*, 218, *233*,
    441
  thermodynamic behavior near 231
critical pressure 23
  listed for various gases *19*
critical solution point 443, *445*

critical temperature 23, *216*, 229, 441, *443*
  in binary solutions 443, *445*
  listed for various gases *19*
cross-effects 333–4, 338
cryoscopic constant 245
  listed for various liquids *209*
crystallization from solution 418

Dalton's law of partial pressures 16
dark energy 265
dark matter 265
Debye–Hückel theory 255
Debye's theory of molar heat capacities of solids
    210
degeneracy of energy levels 453
degrees of freedom 32, 221–3, 225, 228, 235–6, 305
density of states 33
detailed balance, principle of 278–80, 334
Dieterici equation 200, 213
diethylamine/water solutions 444, *445*
diffusion
  condition for stability 435–7, 444
  in continuous system 331
  entropy production due to 149, 153, 158
  Fick's law 310, 319, 320–3, 332
  fluctuations due to 426
  as irreversible process *117*, 158
  law governing 309, 319, *331*
  thermodynamic description 380, 387, 441
diffusion coefficient 320
  listed for various molecules *320*
diffusion equation 321–2
dilute solutions
  chemical potential 251
  thermodynamics of mixing 255
dissipative structures 9, 339, 379
  general features 340–1
drift velocity 322
Duhem–Jougeut theorem 439
Duhem's theorem 221
Dulong and Petit's law 211
Earth, physical data *485*

Earth's atmosphere 7, **8**, 44, *307*
  pressure variation **8**, *311*
  temperature variation **8**, 307, *308*
ebullioscopic constant 245
  listed for various liquids *241*
Einstein formula for probability of fluctuation that
    causes entropy change 427, 429
Einstein relation 322
electrical conduction *332*
  law governing **82**, 309, *331*
electrochemical affinity 312–14
electrochemical cell 312, **315**
  half-reactions 312

electrochemical potential 306
electrolytes 251
electrolytic cell 316
electromagnetic radiation 53
    pressure exerted by 366
electromotive force (EMF) 312
electron charge *475*
electron–positron pairs
    production of *266*, **298**, 373
    thermal photons in equilibrium with 373
electron rest mass *485*
electroneutrality 251, 307
elementary reactions **271**
elementary step (of chemical reaction) 269
    reaction rate for 282
emissive power **364**
emissivity 363, **364**
    values listed **364**
enantiomers 283, 342–3
    racemization of 283–4
energy 3–5
    as extensive function **131**
    minimization of 165, *166*
energy change, examples 56–7, 59
energy conservation
    in chemical reactions 55
    law of 55
        elementary applications 64
    in nuclear reactions 79–80
energy density, relationship to spectral radiance
        362, 388
energy flows 81, 85
    in biological systems 86, 380
    in process flows 82
    self-organized patterns and 81
    solar energy flows 85
enthalpy 71–2, 74, 170, 196
    minimization of 165–6, **169**
    variation with pressure 74
    variation with temperature 74
    see also standard molar enthalpy of formation
enthalpy of formation, of ions **252**
enthalpy of fusion 128, 417
    listed for various substances *128*
enthalpy of reaction 71
enthalpy of vaporization 128
    in hurricane system 134
    listed for various substances *128*
entropy 3, 5, 108, 112–14
    absolute value calculation 466
    as extensive function **131**, 157
    general properties 155–8
    of ideal gas 129–30, 196
    maximization of 163
    of mixing 197–9
    modern approach 119

of pure liquids and solids 208–9
of real gas 204
statistical interpretation **124**, 422
entropy changes 120
    fluctuation-associated 424–5, 433
    in irreversible processes 113–14, 119–23
        examples 125–9
    and phase transformations 128–9
entropy density 11
entropy production
    in chemical reactions 145–6, 280–5
    in continuous systems 135–6, 307–9
    in coupled reactions 151–3
    diffusion-caused 153–5
    due to electrical conduction 309–410
    effect of catalyst 402
    at equilibrium state 126, *127*, 163–4
    and heat flow 126, 368
    in irreversible chemical reactions 144–9
    in irreversible processes 121, 310
    by living cell 380–1, *381*
    local 331
    per unit length due to particle flow 309
    stability theory based on 423–4
enzyme-catalyzed reactions 289
    kinetics 399–406
enzyme inhibition mechanisms 402–6
    competitive inhibition 402
    noncompetitive inhibition 404
    uncompetitive inhibition 403, 406
equation of state 26, 365–9
    for ideal gases 195
    Mathematica codes evaluating and plotting
        pressure using 44–5
    for pure solids and liquids 208–9
    for real gases 199–201, 211
    for thermal radiation 368, 371
equilibrium constant(s) 274–5
    for electrolytes 254
    examples **276**
    relation to partition functions (in statistical
        thermodynamics) 471
    relation to rate constants 276–7
    in statistical thermodynamics 469–70
    in terms of partial pressures 275
equilibrium state(s) 10, 164
    affinity in 147
    enthalpy minimization of in 171
    entropy at 127, *127*
    entropy production in 126, *127*, 163, 164
    response to perturbation from 277–8
    restoration by irreversible processes 130, 172
    stability of 171–2, 423–4, 433
    thermal fluctuations in 433–4
equilibrium systems 7, 9
equipartition theorem 32

Euler's theorem 156, 173, 182
eutectic composition 226
   ternary systems 223, 227
eutectic point 226, *227*
eutectic temperature 226
   ternary systems 223, *227*
eutectics 225
excess enthalpy, of solutions 259
excess entropy, of solutions 258–9
excess Gibbs energy, of solutions 258
extended irreversible thermodynamics 331
extensive functions **131**, 181
extensive variables 6
extensivity **131**, 181
extent of reaction 77–80
   fluctuations in *425*, 426–8, 437
   rate equations using 269–70, 272
extremum principles
   in Nature 163
   and Second Law 163–73
   and stability of equilibrium state 171–2

Fahrenheit temperature scale 12, **14**
far-from-equilibrium nonlinear regime 327
Faraday constant 251, 305
fats, energy content 392
Fick's law of diffusion 310, 320, *332*
   applications 322
   limitations 321
Field–Körös–Noyes model *see* Belusov–
     Zhabotinsky reaction, FKN model
First Law of thermodynamics 55–71
   elementary applications 64
   as local law 331n
first-order phase transitions 232
first-order reactions **290**
   consecutive 290
   reversible 289
flow reactors 293
fluctuations
   probability of 422
   small systems 421
   thermodynamics theory 419
Fourier's law of heat conduction **83**, 126, 136,
     332, 335
free energy *see* Gibbs free energy; Helmoltz free
     energy
freezing point
   changes in 243
   listed for various liquids *245*
   see also melting point
fugacity 206
functions of state variables 5

galvanic cell 316
gas constant **14**, 16, **124**, *485*

gas laws 14–19
gas phase, standard state of substance **72**
gas thermometer 18, 107n
gases
   ideal, thermodynamics 195
   mixing of, entropy change due to 197–8
   real, thermodynamics 199–208
   solubility of 242
Gaussian distribution 40, 426
general thermodynamic relations 173
Gibbs–Duhem equation 173–7
   calculations using 173, 184–5, 206, 215, 226
Gibbs–Duhem theory of stability 424
Gibbs free energy 149, 167–9, 188
   calculation using van der Waals equation
     205
   changes during chemical transformations
     378–81
     irreversible transformation 382–3
     reversible transformation *382*, 383
     standard transformation *382*, 384
   flow in biological systems 385–97
   minimization of 165–9
   photochemical transfer of 389
   for real gas 201
   relation to affinity 164, 281
   relation to chemical potential 169
   tabulation for compounds 178
   see also affinity
Gibbs free energy of formation, of ions **252**
Gibbs free energy of reaction 270
Gibbs–Helmoltz equation 175–7, 251
Gibbs–Konovolow theorem 225
Gibbs paradox 197
Gibbs phase rule 221–3
   applications 219, 221–2
Gibbs–Thompson equation 415
glass–liquid interface
   contact angle at 186
     values listed *187*
glucose, photosynthetic production of 386
glutamine, formation of 395
glycolysis 385, 394
gravitational constant *485*

half-life **271**
half-reactions (in electrochemical cell) 312–13
handedness *see* chirality
heat
   mechanical equivalence of 55
   nature of 11, **18**, 53
heat capacity **52**
   at constant composition 448
   at constant pressure 65, 73
   at constant volume 64, 202, 435
   experimental determination of **62**, 74

solids 210–11
    see also configurational heat capacity; molar
        heat capacity
heat conduction *332*
    entropy change and 123–5
    law governing **82**, 278
heat convection, law governing **83**
heat engines 99–100
    efficiency 101–7
    hurricanes as 108, **109–10**, 132–5
heat flow
    entropy change due 120
    as irreversible process *117*
    laws governing **82**
    self-organized patterns and 81
heat radiation *see* thermal radiation
Helmholtz equation 174, 195, 201, 366
Helmholtz free energy 165–6, 187, 210
    calculation using van der Waals equation 205
    effect of liquid–glass interfacial area in capillary
        181–2
    minimization of 165–6, 184
    for real gas 201
    relation to partition function 454
Henry's law 242, *243*
    constant(s) 242
        listed for various gases *243*
    deviation from 242, *243*
    example of application 260
Hess's experiments **70**
Hess's law 70–1
heterogeneous nucleation 421
heterotrophs 386
n-hexane/nitrobenzene solutions 443, *445*
homogeneous nucleation 420
hurricane, as heat engine 108, **109–10**, 132–5
hydrogen, reaction with nitrogen 266
hydrogen bonds **266**
hydrogen–platinum electrode *318*
    standard electrode potential 318, *318*

ideal gas law **14**, 16, 195
    examples of calculations using 45
    osmotic pressure obeying 248
    van der Waals' modification 23
ideal gases
    chemical potential 197
    entropy 129, 196
    entropy of mixing 197–9
    heat capacities 196
    thermodynamic potentials 196
    total internal energy 195–6
ideal mixtures, chemical potential of each
        component 170
ideal solutions
    chemical potential 239

osmotic pressure 249
    thermodynamics of mixing 256
independent intensive variables 221
inflection point 23
integrals (listed) 39
    in statistical thermodynamics 471
intensive variables 6, 120, 173–4
interfacial energy 412, *413*
    listed for various silver halides *413*
interfacial Helmoltz energy, minimization of 184
ion numbers 251
'ion pumps' 396–7
ionic mobility 323–4
ionic solutions, solubility equilibrium 250–2
irradiance 361, 368
irreversible chemical reactions, entropy changes
        due to 144–9
irreversible expansion of gas 127
irreversible processes 6–7, 81
    entropy changes in 113, 120
        examples 125–7
    examples *121*, 125
    not considered in classical thermodynamics
        58
irreversible transformation, Gibbs energy change
        during *382*, 383
isolated systems 4, *5*
    entropy changes 122, 433
isothermal calorimeter **63**
isothermal compressibility 180, 202, 436
    listed for various liquids and solids 207, *208*
isothermal diffusion 319–24
isothermal expansion of gas
    entropy change due to 127
    work done during **61**, 89
isothermal volume change **61**

joule (SI unit of heat energy) **14**

kelvin scale of temperature **14**
kinetic theory of gases 29–37
Kirchhoff's law 75, 363, **364**

Lagrange's method 459
Landau theory of phase transitions 231, 233
    failure vs experimental data 233–4
Laplace equation 185
latent heat 11, 19–20, *20*, 128, 215
law of conservation of energy 49–50, 55–63
law of corresponding states 25–7
    limitations 27
    Mathematica code for 43
    molecular forces and 27–9
law of mass action 273–4
Le Chatelier–Braun principle 277–8, 380
least action, principle of 163

Legendre transformations 172
    listed in thermodynamics *173*
Lennard-Jones energy 27–8
lever rule 231, 443
libration 448
libration–rotation equilibrium 448
light propagation, Fermat's principle of least time
        163
linear phenomenological coefficients 310, 333
linear phenomenological laws of nonequilibrium
        thermodynamics 310, 323, 333
Lineweaver–Burk plots 401, 403, 404
lipid bilayer membranes 396
liquid drop, excess pressure in 185, 413
liquid junctions (in electrochemical cells) **315**
liquid–vapor phase equilibrium 217, 220, *221*
liquid-to-vapor transition 19, *20*, 127, *442*, 443
    enthalpies of vaporization 127, *128*, *219*
    experimental discrepancy from classical
        (Landau) theory 233–4
    transition temperatures listed for various
        substances *128*, *219*
liquids
    coefficient of thermal expansion 207, *208*
    isothermal compressibility 207, *208*
    thermodynamic quantities 208–9
living cells
    biochemical reactions in *381*
    as open systems 379, *381*, 386, 396
local conservation law 63–4
local equilibrium 119–20, 328–31
lower critical solution temperature 444, *445*

mass action law 274–5
Mathematica codes
    Belusov–Zhabotinsky reaction, FKN model
        354–5
    Brusselator 353–4
    chiral symmetry breaking 352–3
    critical constants for van der Waals equation 42
    evaluating and plotting pressure using equation
        of state 41–2
    evaluating work done in isothermal expansion
        of gas 87–8
    law of corresponding states 43
    linear kinetics 295–6
    Maxwell–Boltzmann speed distribution plot 43
    racemization kinetics 297–8
    reversible reaction kinetics 296–7
maximum entropy 164
Maxwell–Boltzmann velocity distribution 32–4,
        37, 53, 285, 286
Maxwell construction 229–31
Maxwell relations 179–81
Maxwell velocity distribution 34, *35*, 36–7, 329,
        370

mean chemical potential, ionic solutions 251–3
mean ionic activity coefficient(s), of electrolytes
        253
melting point
    listed for various substances *128*, *219*
    particle size effect 416–17
    see also freezing point
membrane permeability, living cells 396, *397*
membrane potentials 311–12
    calculation in example 324–5
metastable region
    liquid–vapor transitions *442*, 443
    strictly regular solutions 446–7
metastable state *442*, 443
Michaelis–Menten constant 400
    experimental determination of 401
Michaelis–Menten mechanism 292, 399–400
Michaelis–Menten rate law 293, 399–400
microscopic consequences of laws of
        thermodynamics, methods of analyzing
        369–70
microstates **124**, 451, 456–7
minimum energy 164, *165*
minimum Helmoltz energy 165–7
mixing of gases, entropy change due to 197–9
modern thermodynamics 4
modes of action 451
molality scale 244–5, **252**, 253, 254
molar critical volume 23, 24
    listed for various gases *19*
molar energy 16
molar enthalpy of fusion 127
molar enthalpy of vaporization 127
molar entropy of fusion 127
molar entropy of vaporization 127–8
molar Gibbs energy 169–70, 174, **178**, 188, 244
    of formation 179
    of mixing
        nonideal solutions 257
        perfect solutions 255–6
    of solution 414
molar heat capacity **52**, 62, *64*
    at constant pressure (C_{mp}) **52**, *64*
        listed for various substances 65
        real gases 202
        relation with C_{mV} 64–6, 180, 196, 202
        for solids 210
    at constant volume (C_{mV}) **52**, 62, *64*
        listed for various substances 65
        for real gases 202
        relation with C_{mp} 64–6, 180, 196, 202
        for solids 210
    experimental determination of **63**, 74
    ratio 66, 196
        listed for various gases *67*
molar volume, fluctuations in 435–6

mole numbers, fluctuations in, stability w.r.t. 437–9
molecular volume 27n

nano-science 411
nanothermodynamics 411
near-equilibrium linear regime 327, 336
Nernst equation 315
  application(s) 316
Nernst heat theorem 123–4
neutrino 80–1, 264
neutron rest mass *485*
Newton's laws of motion 50
nitrogen, reaction with hydrogen 278
noncompetitive inhibition of enzymes 404–6, *405*
nonequilibrium systems 7–9, 327–57
  entropy changes 9, *10*
  examples 9, *10*
  far-from-equilibrium nonlinear regime 327, 377
  living cells as 377–9
  near-equilibrium linear regime 327
nonideal solutions
  chemical potential 239–40
  osmotic pressure 249
  thermodynamics of mixing 257–9
nonionic solutions, solubility equilibrium 250–1
nonthermal electromagnetic radiation
  chemical potential 375–6, 388–9
  see also solar radiation
nonthermal solar radiation, chemical potential 389, 407–8
normalization constant 453
normalization factor 34
nuclear reactions, conservation of energy in 79–80
nucleation 418–21
nucleation rate equation 419–20
  pre-exponential factor 420
nucleation sites 421
nucleosynthesis 79

occupation number 455
Ohm's law **63**, 310, *332*
ohmic heat 310
Onsager coefficients 333
Onsager reciprocal relations 334
  in thermoelectric phenomena 334–8
open systems 4–5, *5*, 9
  energy change in 59
  entropy changes 120
  living cells as 379, *381*, 386
order parameter (in dissipative structure) *340*, 341
order of reaction 268–9
origins of thermodynamics 3
oscillating reactions 346–51
osmosis 245, *247*
osmotic coefficient 242

osmotic pressure 245–50
  calculations 247–8
    in example 261–2
  definition 247
  experimentally determined values compared with theory 248–9
  van't Hoff equation 248
Ostwald–Freundlich equation 415
Ostwald ripening 416
oxidation and reduction reactions (in electrochemical cell) 312–13

p–V isotherms 22, 216, *217*, *442*
  van der Waals isotherm 229, *230*
partial derivatives 37–9
partial molar enthalpy 177, 183
partial molar Gibbs energy 182–3
partial molar Helmholtz energy 183
partial molar volume 182, 208
partial pressures, Dalton's law of 16, 44
particle–antiparticle pairs, creation by thermal photons 373–4
partition function(s) 454
  calculating 462–9
  factorization of 454–5
  see also canonical partition function; rotational partition function; translational partition function
pascal (SI unit of pressure) **14**
Peltier effect 333, 337–8
Peltier heat 337
perfect binary solution, vapor pressure diagram *241*
perfect solutions
  chemical potential 239
  thermodynamics of mixing 255–7
periodic phenoena in chemical systems 346–51
permeability constant *485*
permittivity of vacuum 255, *485*
perpetual motion
  of first kind 58
  of second kind 115
phase change, thermodynamics 215–37
phase diagrams 215, *216*
  binary solutions *445*
  binary systems *216*, *218*, *227*
  strictly regular solutions *447*
  ternary systems *229*
  water *218*
phase equilibrium 216, 220, *221*
phase separation 172, 444
phase transitions 19–20, 23
  classical (Landau) theory 231, 233
    limitations 234
  entropy changes associated with 128–9
  first-order 232

general classification 232
modern theory 231, 234
second-order 232
thermodynamics 128–9, 231–4
phases 19
phosphate group, transfer of 395–6
photoautotrophs 386
photochemical transfer of Gibbs energy 389–93
photon gas pressure 366, **367**
photons 53, 55
conversion of Gibbs energy to chemical energy *390*
photosynthesis 86–7, *381*, 386–7
Gibbs free energy flow in *386*, 389–93
hydrogen donors 387
rate, factors affecting 392
physical constants and data *485*
Planck's constant 269, **270**, 365, **456**, *485*
Planck's distribution/formula 365
applications 388, 408
plasma **266**
Poisson distribution 40
positive definite matrix 334
pre-exponential factor
in chemical reaction rate equation 269, 286–8
in nucleation rate equation 420
pressure
definition **14**
in Earth's atmosphere **8**, **14**
kinetic theory 29–34
units **14**
pressure coefficient 180
principle of detailed balance 278–80, 334
principle of least action 163
probability density 32
probability distributions 40
of molecular velocity 32, *33*
probability of fluctuation 422–3
general expression for 429
probability theory, elementary concepts 39–41
process flows 82–5
proteins, energy content 392
proton rest mass *485*

quantum field theory 53
quantum theory 364, 466

racemization of enantiomers 283, 342
entropy production due to 282–5
Mathematica code fofr 297–8
radiance 361
radiant emittance 361
radiation 361–78
law governing **83**
radiation flux density 361
radiation intensity 361

radioactive elements 79
Raoult's law 241, *243*
deviation from 242, *243*
example of application 262
rate constant (for chemical reaction) 268
temperature dependence 269, **270**, 286
rate of conversion 77, 147
rate equations
extent of reaction in 269–70, 272
steady-state solutions 347
reaction enthalpy, calculation using bond enthalpies 76
reaction rate(s) **268**
and activities 272–3
Arrhenius theory 269, **270**, 285–7
determination of 269
relation to affinity 282–3
example 282–5
transition state theory 269, **270**, 288
reaction velocity 77, 141, 267
effect of catalyst 402
and reaction rate **268**
reactive collisions (in chemical reactions) 330
Redlich–Kwong equation 27, 211
reduced mass 287
reduced variables 26
reduction and oxidation reactions (in electrochemical cell) 312, 313
regular solutions 258–9
see also strictly regular solutions
relativity theory 64n, 80
renormalization group 234
residence time (in reactor) 293
reversible first-order reactions 289–90
reversible heat engines 107, 109
efficiency 102–6
reversible processes
entropy change in 113–14, 116, 118
examples 101, *117*
reversible transformation, Gibbs energy change during *381*, 382
root-mean-square (RMS) fluctuations 330
rotational energy levels **456**
rotational partition function, calculating 466–7

Sackur–Tetrode equation 466
Safir–Simpson hurricane intensity scale *133*
salt bridge (in electrochemical cell) **315**, *316*
salt effect 273
saturated vapor pressure 217, *221*
seaweed cells 396
Second Law of thermodynamics 114–16
extremum principles associated with 163–73
as local law 331n
modern statement 121
universality 130

second-order phase transitions 232
second-order reactions **271**
Seebeck effect 333, 336–7, *338*
self-organization 81–2, 130, 132, 340–1
 examples 9, *81*, 82
semi-permeable membranes 245, *247*, 382
sequential transformations 290
silver halides
 interfacial energies *412*
 solubility 415
size-dependent properties 414–17
small systems
 chemical potential 411–13
 fluctuations in 421–2
 thermodynamics 411–31
sodium–potassium pump (SPP) 397–9
 mechanism *398*
solar energy 385
 capture efficiency 392
 flows 85–6
solar radiant flux 86, 392, **393**
solar radiation 376, 389
 chemical potential of nonthermal 389, 407–8
 example calculations 406–7
solid–melt interfacial energy 417
solid-to-liquid transition 19, *20*, 128
 enthalpies of fusion 128, *129*, *219*
 transition temperatures listed for various
  substances *129*, *219*
 solidification of melt 418
solids
 coefficient of thermal expansion 208, *209*
 heat capacities 210–11
 isothermal compressibility 208, *209*
 thermodynamic quantities 209–10
solubility 250
 of AgCl (silver chloride) 255
 particle size effect 414–16
solubility equilibrium 250–5
 ionic solutions 251–3
 nonionic solutions 250–1
solubility product 254
solute in solution, standard state of **72**
solution–vapor equilibrium 240
solutions, thermodynamics 239–64
'solvent', meaning of term 241
solvent effect 288
sound intensity, unit of 67
sound propagation 67–8
sound, speed of 68
 listed for various gases *67*
specific enthalpy 85
specific heat(s) 11, **52**
 ratio 66, 89
  listed for various gases *67*
spectral absorptivity 363

spectral energy density 360
 frequency dependence 389–91
spectral intensity 362
spectral irradiance 362
spectral radiance 362
 relationship to energy density 362, 388
speed of light *485*
speed of sound 68, 89
 listed for various gases *68*
spontaneous symmetry breaking 341
stability theory 433–40
 applications 441–9
stability of thermodynamic system 423–4
standard affinity 384
standard cell potential 315
 calculation in example 325
standard deviation 39
standard electrode potentials 318–19
 listed for various electrodes *326*
standard entropy, listed for various elements and
  compounds *477–84*
standard Gibbs free energy change(s) 384–5
 listed for various bioreactions *382*
standard gravitational acceleration *481*
standard heats of formation, listed for various
  elements and compounds *477–84*
standard molar enthalpy of formation 72, **73**
 listed for various elements and
  compounds *477–84*
standard molar Gibbs energy of formation **177**,
 274
 listed for various elements and compounds
  477–84
standard state(s)
 basic definitions **72**, 384
 entropy of 116
 of ideal dilute solution **252**
standard thermodynamic properties 477–84
standard transformation, Gibbs energy change
  during *382*, 384–5
stars and galaxies 265
state of equilibrium *see* equilibrium state(s)
state functions 6, 164, 166–7, 167–8
state of system 5
state variables 5, 76–9
states of matter 19
statistical ensemble 451, *458*
 see also canonical ensemble
statistical entropy 459
statistical thermodynamics **124–5**, 422–3,
 451–73
 fundamentals 451–4
steady-state assumption (in chemical reactions)
 151, 292–3
'steam tables' 85, 91–2
Stefan–Boltzmann constant **83**, 368

Stefan–Boltzmann law 364, 368
  applications 377, 388, 406
Stirling approximation 199
Stirling's approximation 459, 469
Stokes–Einstein relation 321–4
strictly regular solutions 446–7
sugar cane, growth rate **393**
sugars, as chiral molecules 342, 379
Sun, physical data *485*
supersaturation 415
surface energy **131**, 166
surface entropy **131**
surface tension 166, 183–7
  capillary rise due to 185–7
  values listed for various liquids *187*
symmetry-breaking transition 341–5
symmetry number, in expression for rotational
      partition function 467

temperature
  absolute scale 12, 106
  in Earth's atmosphere **8**
  and heat 11
  measurement of 12, **14**, 103n
temperature fluctuations 425–6
terminal velocity 322
ternary systems 226–8
  phase diagram *229*
theorems of moderation 278
thermal equilibrium 6
thermal photons 371
  electron–positron pairs in equilibrium with 373–4
thermal radiation 52, *53*, 54, 265, 361, 388
  chemical potential of 371–3, 388
  energy density 361–5
  heat flow law governing **83**
  intensity 361–5
  pressure exerted by 368
thermal stability 433–5
thermal wavelength 465
thermochemistry, basic definitions **73**
thermodynamic equilibrium 6
thermodynamic flow(s) 120–1, 127, 273, 332
  examples *332*
  see also reaction velocity
thermodynamic force(s) 120–1, 127, 273, 332
  examples *332*
  see also affinity
thermodynamic mixing
  ideal solutions 257
  nonideal solutions 257–9
  perfect solutions 255–7
thermodynamic potentials 164, 196
  extremization of 163–73
  see also enthalpy; Gibbs energy; Helmholtz
      energy

thermodynamic probability 424, 453
  calculation of **124–5**
thermodynamic stability 172, 423–4
thermodynamic systems 4–6
thermodynamic theory of fluctuation 419, 423
thermodynamic variables 6
thermoelectric effect 333, 336–7, *338*
thermoelectric phenomena, Onsager reciprocal
      relations in 334–8
thermoelectric power 337
thermometer, early scientific use 11
theta temperature 250
'third law of thermodynamics' 124
m-toluidine/glycerol solutions 444, *445*
torr (unit of pressure) **14**
total internal energy
  ideal gas 62, 195–6
  real gas 201–2, 212
total solar radiance 85
transformation of matter 265–6
  at various temperatures **266**
  see also chemical transformations
transformation of state, energy change associated
      with 55–6
transition state theory 269, **270**, 288
  pre-exponential factor in 288
translational energies 455, **456**, 460
translational partition function, calculating 463–5
transport properties 451
tri-molecular model 346
triple point (in phase diagram) *216*, 222, *233*
two-level atom, interaction with thermal radiation
      372–3

'uncompensated heat' 119, 128, 144, 145
uncompetitive inhibition of enzymes 343–404,
      405
universal gas equation 24
upper critical solution temperature 444, *445*

van der Waals constants 18, *19*, 23–4, 201
van der Waals equation 18, 21–2, 199, 201, 441
  entropy of real gas calculated using 205
  example calculations 44
  Gibbs energy of real gas calculated using 205
  Helmholtz energy of real gas calculated using
      205
  limitations 26, 200
  p–V isotherm 229, *230*
van der Waals forces 26
van't Hoff equation 277
  deviation from 249
  example of application 299
  for osmosis 248
    example of application 261–2
van't Hoff reaction chambers 382–3, *382*, *390*, 391

vapor pressure diagrams
  perfect binary solution *241*
  real binary solution *243*
variables (listed) 475–6
velocity of reaction 77, 147
vibrational energy levels **456**
vibrational partition function, calculating 468–9
virial coefficients 200–1
virial expansion 200
  osmotic pressure of nonideal solution 249–50
vitreous state 448
volume change, mechanical work due to
     60, **61**
volume expansion, coefficient of 180, 202

volume of universe, relation to temperature of
  thermal radiation 369

water
  phase diagram *218*
  triple point 222
Wien's theorem 369–71
work, equivalence with heat **51**

zero-order reactions 289
zeroth law 7

Index compiled by Paul Nash